D0769128

NITROGEN AS AN ECOLOGICAL FACTOR

NITROGEN AS AN ECOLOGICAL FACTOR

THE 22ND SYMPOSIUM OF
THE BRITISH ECOLOGICAL SOCIETY
OXFORD 1981

EDITED BY

J. A. LEE
Department of Botany,
The University, Manchester M13 9PL

S. McNEILL
Department of Pure and Applied Biology,
Imperial College of Science and Technology,
Silwood Park, Ascot, Berkshire SL5 7PY

AND

I. H. RORISON
Unit of Comparative Plant Ecology (NERC),
Department of Botany, The University,
Sheffield S10 2TN

BLACKWELL SCIENTIFIC PUBLICATIONS
OXFORD LONDON EDINBURGH
BOSTON MELBOURNE

Editorial offices:
Osney Mead, Oxford OX2 0EL
8 John Street, London WC1N 2ES
9 Forrest Road, Edinburgh EH1 2QH
52 Beacon Street, Boston
Massachusetts 02108, USA
99 Barry Street, Carlton
Victoria 3053, Australia

First published 1983

Printed at the Alden Press, Oxford
and bound by Butler & Tanner Ltd, Frome, Somerset

DISTRIBUTORS

USA
Blackwell Mosby Book Distributors
11830 Westline Industrial Drive
St Louis, Missouri 63141

Canada
Blackwell Mosby Book Distributors
120 Melford Drive, Scarborough
Ontario, M1B 2X4

Australia
Blackwell Scientific Book Distributors
31 Advantage Road, Highett
Victoria 3190

British Library
Cataloguing in Publication Data

British Ecological Society. *Symposium (22nd: 1981 Oxford)*
Nitrogen as an ecological factor.
1. Nitrogen—Fixation—Congresses
I. Title II. Lee, J.A. III. McNeill, S.
IV. Rorison, I.H.
546′.71159 QR89.7

ISBN 0-632-01074-6

CONTENTS

PREFACE

This book is the proceedings of a symposium held in the University of Oxford from 24 to 26 March 1981. The volume comprises invited papers, together with offered papers and brief abstracts of those poster papers where the authors provided such.

In planning the 22nd Symposium of the British Ecological Society emphasis was placed on interactions between organisms and the environment rather than broad considerations of nitrogen cycling and budgets in ecosystems. A major aim has been to bring together the approaches of plant and animal ecologists to nitrogen as an ecological factor so that each may illuminate the other, particularly in the field of plant–animal interactions.

The introductory paper is concerned with nitrogen fixation and cycling within and between plants, and within a freshwater ecosystem. This is followed by three papers on symbiotic relationships. The first is concerned with angiosperm symbioses with bacteria, the second with the fate of combined nitrogen released from lichen symbioses in a subarctic ecosystem, and the third reviews the importance of mycorrhizas in the acquisition of nitrogen by host plants.

A group of papers is then concerned with nitrogen limitation of plant growth, the form in which it becomes available, and effects on ecosystem development. The first examines the evidence for nitrogen as a factor limiting plant growth. The second considers the importance of nitrogen in succession, with particular reference to china clay wastes. Nitrate utilization in some tropical communities is discussed in a paper which presents evidence from a wide range of species. The group is concluded by two papers concerned with the effects of low temperatures on nitrogen assimilation and the form of nitrogen utilized by plants.

The last paper which is devoted to plants considers the variety of nitrogen assimilation pathways and movements within plants and acts as a link with the several following papers concerned with insect–plant interactions. The first two of these look at the direct effects of nitrogen availability on insect performance in both natural vegetation and a crop, while the other two papers in this section deal with the direct and indirect effects of secondary plant substances. The two papers that then follow consider the implications of nitrogen and food quality in vertebrates with studies on large mammals and fish culture.

The final section of the book looks at the decomposer food web with a review of the nitrogen economy of termites and soil animals.

1. NITROGEN CYCLING

W. D. P. STEWART, T. PRESTON*, A. N. RAI AND P. ROWELL

Department of Biological Sciences, University of Dundee, Dundee DD1 4HN

SUMMARY

All living organisms require nitrogen. Nitrogen is limiting on Earth and there is a shuttling of nitrogenous compounds in and between cells, organisms and ecosystems. This paper considers examples of nitrogen cycling at the cellular, organismal, and ecosystem levels. The first is the case of N_2-fixing heterocystous cyanobacteria where the heterocyst is the N_2-fixing site, where nitrogen is transported out of the heterocyst as glutamine and where glutamate as well as fixed carbon is transferred from the vegetative cells to the heterocysts. The vegetative cells themselves show a cycling of nitrogen between metabolites which are used for growth and for storage. The second example taken is the N_2-fixing lichen *Peltigera aphthosa* Willd. where the N_2-fixing cyanobacterium transfers fixed nitrogen not only to vegetative cells but also to the eukaryotic partners of the symbiosis. Some of the factors underlying fixation and nitrogen transfer are considered. The third example relates to nitrogen cycling in freshwaters and results are presented on the fate of ^{15}N-labelled nitrate added to a 18 500 m^3 experimental enclosure in Blelham Tarn, a eutrophic freshwater in the English Lake District.

NITROGEN CYCLING

Nitrogen is essential for life. It is the cornerstone of the amino acids, an important component of the nucleic acid bases, and it limits productivity on Earth. Ever since Justus von Liebig (1851) advocated the use of chemical fertilizers to increase crop productivity ('The problem in agriculture, at the present time . . . is to substitute for farm-yard manure, that universal food of plants, its elements obtained from other and cheaper sources') farmers have been adding increasing amounts of nitrogenous fertilizer to agricultural crops (Gasser 1982). In many countries of the world chemical nitrogen fertilizer is

* Present address: The Scottish Universities Research and Reactor Centre, N.E.L. Estate, East Kilbride G75 0QU.

not available to meet the requirements for fixed nitrogen and this has been a contributory reason why, at present, over one-third of the world's population lives on protein-deficient diets. Nitrogen demand exceeds fixed nitrogen availability. There is thus the paradoxical situation that although 78% of the air which we breathe is N_2, only the minute fraction of living organisms (all prokaryotes) which possess the enzyme nitrogenase can tap this nitrogen source (Stewart & Gallon 1980). To multiply and survive, the others shuttle the limited amount of fixed nitrogen which is available from one biochemical process to another, from one cell to another and from one ecosystem to another with each component occupying a niche which allows self-sustenance under overall nitrogen-limited conditions. This is so on land (Gasser 1982; Hood 1982), in the sea (Fogg 1982) and in freshwaters (Stewart *et al.* 1982). Because of the multifarious interactions which occur we have entitled this paper *Nitrogen Cycling*, not *The Nitrogen Cycle*. What we shall do is emphasize, by considering three specific examples, some of the interactions which may occur.

Nitrogen cycling in and between cells—the case of N_2-fixing heterocystous cyanobacteria

Cyanobacteria (blue-green algae) are prokaryotes which photosynthesize using water as reductant source, generating ATP by photophosphorylation and fixing carbon dioxide via ribulose 1,5-bisphosphate carboxylase (RUBPCase) in a typical Calvin cycle pathway (Stanier & Cohen-Bazire 1977; Rippka *et al.* 1979). They are of worldwide distribution ranging from antarctic to tropical regions (Fogg *et al.* 1973). As tools for the study of developmental biology in photosynthetic microorganisms the unbranched heterocystous filamentous forms such as *Anabaena* and *Nostoc* are probably unsurpassed (Sutherland, Herdman & Stewart 1979). Two types of specialized cell, the heterocysts and akinetes (spores), differentiate from vegetative cells and the pattern of development of these is determined both by external and internal events. For example, heterocysts do not form in ammonium-grown cultures (Fogg 1949), and when they differentiate their position is determined by the position of preformed heterocysts (Wilcox, Mitchison & Smith 1973, 1975a, b; Carr 1979; Adams & Carr 1981).

The heterocysts of such filaments are the N_2-fixing factories which provide the oxygen-protected environment necessary for nitrogenase to function and such heterocysts provide fixed nitrogen for a suite of adjacent vegetative cells. The vegetative cells, which under aerobic conditions do not fix N_2, and which unlike heterocysts fix carbon dioxide photosynthetically, provide fixed carbon for themselves and for the heterocysts. The ways in which the heterocysts are organized to fix N_2 have recently been considered (Stewart 1980a). Briefly, the thick envelope surrounding the heterocyst reduces oxygen diffusion into the heterocysts; the heterocysts lack a capacity to evolve oxygen photosynthetically having lesions in photosystem 2 (Tel-Or & Stewart 1977; Haselkorn 1978) but generate ATP by photophosphorylation and by oxidative phosphorylation (Tel-Or & Stewart 1976, 1977; Peterson & Burris 1978). They possess a unidirectional uptake hydrogenase which recycles hydrogen evolved as a natural concomitant of nitrogenase activity, conserving reductant in this way and generating additional ATP by an oxyhydrogen reaction which also serves to scavenge oxygen (Peterson & Burris 1978; Bothe *et al.* 1980). Because heterocysts lack RUBPCase (the protein is not synthesized) the fixed carbon necessary to support nitrogenase activity is transferred from the vegetative cells to the heterocysts (Wolk 1968). Fixed carbon dissimilation is mainly via the oxidative pentose phosphate pathway, key enzymes of which (e.g. glucose-6-phosphate dehydrogenase and 6-phosphogluconate dehydrogenase) are 6–10 times more active in heterocysts than in vegetative cells (Winkenbach & Wolk 1973; Lex & Carr 1974; Apte, Rowell & Stewart 1978; Bothe *et al.* 1980). Catabolism may also occur via the incomplete tricarboxylic acid cycle according to Bothe *et al.* (1980). There is evidence that reductant may be transferred from NADPH generated in the oxidative pentose phosphate pathway and/or from the Krebs cycle (Lex & Stewart 1973; Bothe *et al.* 1980) to ferredoxin which donates electrons to nitrogenase. Several forms of such oxidoreductases are present in the vegetative cells but only one has been detected in heterocysts of *Anabaena cylindrica* (Rowell *et al.* 1981) and this may be involved in electron transfer to nitrogenase. Recent evidence from Hawkesford *et al.* (1981) suggests the involvement of a membrane potential in electron transfer to nitrogenase; this could possibly aid the thermodynamically unfavourable transfer of electrons from NADPH to ferredoxin.

Within this specialized cell, N_2 is reduced to ammonia and it is this ammonia which provides the initial pool of fixed nitrogen for the whole filament. That is, although only about 5% of the total cells fix N_2, the N_2-fixing efficiency of the heterocysts is such that growth on N_2 may be as good as when every cell is supplied with, and assimilates, nitrate.

How does such a transfer of fixed nitrogen from the heterocysts to vegetative cells occur? The ammonium must not be allowed to accumulate in the heterocysts since it directly or indirectly inhibits nitrogenase (activity, *nif* transcription and translation may all be affected, and uncoupling of phosphorylation may also occur (Ohmori & Hattori 1978)). Cyanobacteria, in general, overcome this problem because they possess the glutamine synthetase–glutamate synthase (GS-GOGAT) pathway of ammonium assimilation (Haystead, Dharmawardene & Stewart 1973; Lea & Miflin 1975; Stewart, Haystead & Dharmawardene 1975; Wolk *et al.* 1976). These two enzymes are jointly responsible for ammonium assimilation in the vegetative cells. The advantage of this pathway in cyanobacteria, as in other organisms where it is the primary route of ammonium assimilation, is that GS, in general, has a much lower K_m for ammonium than has glutamate dehydrogenase and thus it can efficiently scavenge ammonium. The disadvantage is that GS is an ATP-dependent enzyme, although this is less of a problem in a photosynthetic organism than in a heterotrophic one.

Within heterocysts the newly fixed ammonium is rapidly incorporated into glutamine which is transported via plasmodesmata which traverse the pore connecting heterocysts to vegetative cells (Lang & Fay 1971). This neutral amino acid (which is particularly suitable as a transport compound) serves as substrate for glutamate production via GOGAT in the vegetative cells (Thomas *et al.* 1977). One of the simplest ways to demonstrate ammonium production by N_2-fixation and its assimilation via GS is to inhibit the GS with analogues of glutamate such as L-methionine-DL-sulphoximine (MSX) (Stewart & Rowell 1975; Ownby 1977) or 5-hydroxylysine (Ladha, Rowell & Stewart 1978). When this happens ammonium which otherwise would have been assimilated is released extracellularly. While it is clear that glutamine is produced in heterocysts, it is less certain that glutamate is not. The initial suggestion that it was not came from the work of Thomas *et al.* (1977) who observed using ^{13}N as tracer that isolated N_2-fixing heterocysts liberated ^{13}N-labelled glutamine; they also failed to detect GOGAT in heterocysts although it was readily detectable in vegetative cells. This fitted earlier findings by Winkenbach & Wolk (1973) and Lex & Carr (1974) that certain glycolytic and Krebs cycle enzymes were undetectable in heterocysts. However, Bothe *et al.* (1980) have consistently found some enzymes of the glycolytic pathway and tricarboxylic acid cycle in heterocysts and speculate that glutamate may be produced within the heterocysts. The matter is still not resolved with any certainty. In cyanobacteria, as in other N_2-fixing organisms, it is probable that the *ntr* genes (*ntr* A, *ntr* B and *ntr* C) may be involved in *nif* regulation (Magasanik 1982). In sum, heterocyst differentiation is accompanied by the shutdown of the synthesis of various vegetative cell proteins and of various vegetative cell activities, with the concomitant synthesis of new gene products and enzymes. The end result is a heterocyst–vegetative cell nitrogen shuttle (Fig. 1.1).

In the vegetative cells the glutamate formed is subsequently used for the production of other amino acids mainly by transaminase-mediated reactions. However, when fixed nitrogen is available in excess this fixed nitrogen is stored as phycocyanin which serves primarily as a water-soluble light-harvesting

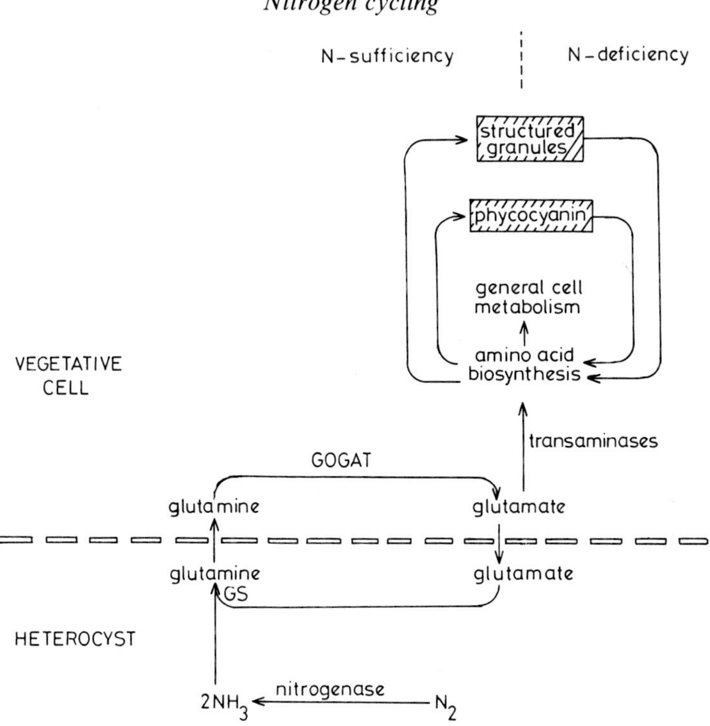

FIG. 1.1. Nitrogen cycling between heterocysts and vegetative cells of the cyanobacterium *Anabaena cylindrica* Lemm. and the storage of fixed nitrogen in phycocyanin and in structured granules (cyanophycin granules) under conditions of nitrogen sufficiency coupled with the utilization of this fixed nitrogen under conditions of nitrogen deficiency.

pigment (Stewart 1980a), and also as discrete densely osmiophilic bodies called cyanophycin granules which are copolymers of aspartic acid and arginine (Simon 1971; Stewart, Pemble & Al-Ugaily 1978) and which are degraded under conditions of nitrogen deficiency to provide an endogenous supply of fixed nitrogen for cellular metabolism. That is, not only is there nitrogen cycling between vegetative cells and heterocysts but there is also an internal cycling of fixed nitrogen within the vegetative cells. These interactions are summarized in Fig. 1.1.

Nitrogen cycling from organism to organism—the case of the N_2-fixing lichen Peltigera aphthosa

N_2-fixing lichens are symbioses of a cyanobacterium (which invariably fixes N_2) and a fungus (bipartite types), or of a cyanobacterium, a photosynthetic eukaryotic alga and a heterotrophic fungus (tripartite types) (Stewart 1980b; Stewart *et al.* 1981). Because tripartite symbioses always contain a N_2-fixing cyanobacterium, the implication is that the major role of this secondary photosynthetic organism is to fix N_2 rather than to provide additional carbon. N_2-fixing lichen symbioses extend the range of nitrogen cycling discussed above. Since the cyanobacteria are heterocystous forms, heterocyst–vegetative cell interactions again occur, with further nitrogen cycling within vegetative cells as well as shuttling of nitrogenous metabolites at the organismal level.

The system which we shall consider is the tripartite lichen *Peltigera aphthosa* Willd. which is represented diagrammatically in Fig. 1.2. *P. aphthosa*

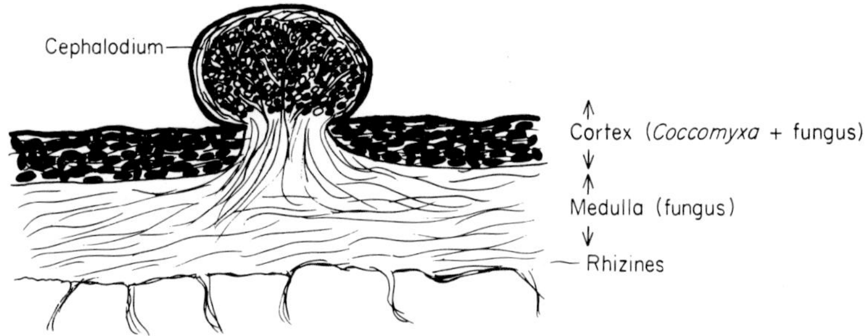

FIG. 1.2. Diagrammatic representation of a section through the cephalodia-containing region of the tripartite lichen *Peltigera aphthosa.*

is a lichen of cool temperate latitudes, where it fixes N_2 (Granhall & Selander 1973; Alexander, Billington & Schell 1974; Englund & Meyerson 1974; Crittenden 1975; Stutz & Bliss 1975; Englund 1977, 1978; Kallio & Kallio 1978; Billington & Alexander 1978) and although the rates of fixation may be only a few kg N ha^{-1} $year^{-1}$ such inputs may be highly important as sources of fixed nitrogen, bearing in mind the low rates of nitrogen turnover and growth within the ecosystems where *P. aphthosa* characteristically occurs. The major photosynthetic partner in *P. aphthosa* is a green alga (*Coccomyxa* sp.) which is distributed as a narrow discrete layer just below the dorsal surface of the thallus. The bulk of the main thallus is fungus and the N_2-fixing cyanobacterium occurs, together with interspersed fungal hyphae, in superficial packets called cephalodia on the dorsal surface of the thallus (see Fig. 1.2). The average composition of the lichen material used by us is, on a dry weight basis: *Nostoc* 2.4%, *Coccomyxa* 9.0% fungus 88.6%. The corresponding values on a total thallus protein basis are 6.0%, 14.4% and 79.6%. The cyanobacterium represents about 80% of the cephalodial protein. *P. aphthosa* has been studied extensively from a physiological and biochemical viewpoint (Millbank & Kershaw 1969; Kershaw & Millbank 1970; Millbank 1974; Feige 1976; Englund 1977, 1978; Rai, Rowell & Stewart, 1980, 1981a–c) but it is only recently that any data on nitrogen cycling within the lichen have been forthcoming.

A major input of new nitrogen to the symbiosis can be from biological N_2 fixation. The capacity of *P. aphthosa* to fix N_2 was first demonstrated by Millbank & Kershaw (1969) and the cyanobacterium is the N_2-fixing partner. The cephalodia, in fact, act like the root nodules of leguminous plants. The prokaryote within them fixes the nitrogen which is then transferred steadily and in large quantity via the surrounding eukaryotic cells of the fungus to the remainder of the thallus. Within the cephalodia, as in other N_2-fixing systems, the cyanobacterium becomes substantially modified morphologically, ultrastructurally, physiologically and biochemically (Stewart *et al.* 1981). Fig. 1.3 provides evidence, obtained using ^{15}N as tracer, of nitrogen transfer from the prokaryote to the eukaryotes.

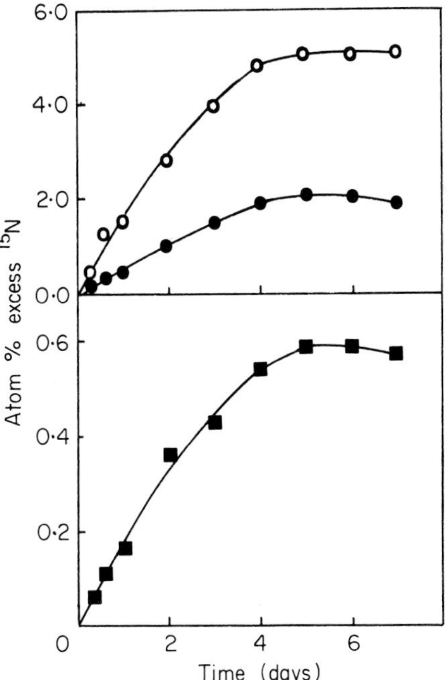

FIG. 1.3. ^{15}N-labelling of cephalodia (○—○), main thallus discs without cephalodia (●—●) and *Coccomyxa* cells of the main thallus (■—■) of *Peltigera aphthosa*, separated and analysed after exposure of cephalodia-containing thallus discs to $^{15}N_2$ for the times shown. (For further details, see Rai, Rowell & Stewart 1981b.)

The major biochemical modification of the system which brings about such a nitrogen transfer is related to the fact that in the symbiotic cyanobacterium the primary route of ammonium assimilation (the GS-GOGAT pathway) is damped down, although total nitrogenase activity remains substantial (Table 1.1). Thus fixed nitrogen which would be assimilated and used in growth in the free-living cyanobacterium is released in symbiosis and used by the eukaryotic partners; the symbiotic cyanobacterium, in effect, is trapped. We now know, in the case of the symbiotic *Nostoc* of *P. aphthosa*, that the synthesis of GS protein is substantially reduced (Stewart *et al.* 1981).

TABLE 1.1. Activities of some enzymes, involved in nitrogen metabolism, in the free-living and symbiotic *Nostoc* from *Peltigera aphthosa* Willd. n.d. = not detectable.

	Activity (nmol product formed min^{-1} mg^{-1} protein)	
Enzyme	Free-living *Nostoc*	Symbiotic *Nostoc*
Glutamine synthetase (GS)	60.0	4.5
Glutamate synthase (GOGAT)	31.29	0.72
Glutamate dehydrogenase (GDH)	<2.0	n.d.
Glutamate oxaloacetate transaminase (GOT)	25.0	12.3
Glutamate pyruvate transaminase (GPT)	2.41	n.d.
Aspartate pyruvate transaminase (APT)	2.40	2.5
Alanine dehydrogenase (ADH)	11.0	8.7
Nitrogenase	0.40	0.75

There is good evidence that much, but not necessarily all, of the nitrogen liberated by the symbiotic cyanobacterium is in the form of ammonium (Rai, Rowell & Stewart 1980; Peters, Evans & Toia 1976; Stewart & Rodgers 1977; Stewart & Rowell 1977). There are at least two possible explanations for this release of ammonium. The first is that when GS is reduced, ammonium, which otherwise would be assimilated, is released extracellularly, as happens when MSX inhibits GS in free-living cyanobacteria (Stewart & Rowell 1975). A problem with this hypothesis is that the total GS activity in the symbiotic cyanobacterium is just sufficient to assimilate all the fixed ammonium. The ammonium production hypothesis thus necessitates nitrogenase activity and

GS activity having an unequal distribution in the cells of the symbiotic cyanobacterium; this is probable, but is yet unproven (Rai, Rowell & Stewart 1981c). For example, in free-living cyanobacteria GS activity is higher in heterocysts than in vegetative cells (Dharmawardene, Haystead & Stewart 1973; Stewart, Haystead & Dharmawardene 1975) and in organisms such as *Azolla* where the age of the *Anabaena* filaments varies in relation to their position from the apex, the filaments nearest the apex show high GS activity and this activity decreases in filaments remote from the apex (Peters *et al.* 1979). The second possibility is that low GOGAT activity (Rai, Rowell & Stewart 1981c) may also contribute to the release of ammonium.

Irrespective of the metabolites involved in nitrogen transfer from the prokaryote to the cephalodial fungus, there is good evidence that within the fungus, the major route of ammonium assimilation is via NADPH-dependent glutamate dehydrogenase, the activities of which are very high in the fungus, which shows low GS activity and undetectable GOGAT activity (Stewart & Rowell 1977). ^{15}N tracer studies carried out by us (Rai, Rowell & Stewart 1981b) support the view that in the cephalodia the GS-GOGAT pathway is responsible for primary ammonium assimilation in the prokaryote whereas the GDH pathway is the primary route of ammonium assimilation in the eukaryote. We measured the sizes of the amino acid pools and the ^{15}N-label of the various amino acids after exposing the cephalodia to $^{15}N_2$. Highest initial ^{15}N-labelling was into ammonium, followed by increased label in the amide group of glutamine and glutamate; labelling of the amino group of glutamine, aspartate and alanine increased only slowly. Rather similar trends were observed when the total ^{15}N incorporated into each amino acid pool was measured.

Analogues were then used to demonstrate ammonium assimilation via cyanobacterial GS-GOGAT and via fungal GDH. On adding MSX to inhibit GS, and azaserine to inhibit GOGAT, it was found that although they completely inhibited the GS-GOGAT pathway, they had no effect on nitrogenase activity or on GDH activity. Under these conditions there was no detectable ^{15}N incorporation into glutamine but ^{15}N incorporation into glutamate continued unimpaired. Such results provide strong evidence of primary ammonium assimilation both by cyanobacterial GS-GOGAT activity and via fungal GDH. On adding amino-oxyacetate, which inhibits aminotransferase activities (Hopper & Segal 1962) the alanine and aspartate pools were substantially reduced, thus suggesting that they were formed secondarily by aminotransferase activity (Rai, Rowell & Stewart 1981b). Overall, initial fixation of N_2 was into ammonium, with subsequent incorporation of ammonium via GS-GOGAT activity in the cyanobacterium and via GDH in the fungus, and with alanine and aspartate being formed secondarily. Fixed nitrogen was then transported in organic form from the cephalodia to the main thallus.

Apart from the fungus of the main thallus, which is in direct contact with the cephalodia (i.e. no layer of *Coccomyxa* is found beneath the cephalodia), the *Coccomyxa* is the other recipient of cephalodial-produced nitrogen. In the *Coccomyxa*, the GS-GOGAT pathway is again the primary route of ammonium assimilation, although NADPH-dependent GDH is also present. (The activities of GS, GOGAT and GDH are, respectively, 37, 21 and 22 nmol product formed min^{-1} mg^{-1} protein.) We have evidence that products of ammonium assimilation by the *Coccomyxa* may, at least under laboratory conditions affect nitrogenase activity in the cephalodia. The evidence is as follows. When exogenous ammonium is added to whole thallus discs with attached cephalodia, nitrogenase activity is inhibited in the absence of MSX but not in its presence, that is, a GS-derived product inhibits nitrogenase activity. On examining the amino acid pools of the cephalodia the most noticeable change on adding ammonium was a seven-fold increase in the glutamine pool when intact discs were used; such an increase did not occur in the presence of MSX, or when ammonium was added to excised cephalodia with or without MSX. That is, main thallus GS (which was mainly in the *Coccomyxa*) seemed to produce, directly or indirectly, an inhibitor of cephalodial nitrogenase. This may be glutamine which when added to excised cephalodia is a potent inhibitor of nitrogenase activity (Rai, Rowell & Stewart 1980).

Data on *P. aphthosa* collectively show that even apparently simple symbioses exhibit nitrogen cycling between the different parts and different partners of the thallus and that genetic, physiological and morphological factors are all involved in the biochemical interactions which ensure efficient nitrogen cycling. *Nostoc* is the N_2-fixing cyanobacterium; it is never, for example, a non-heterocystous N_2-fixing form such as *Plectonema* which only fixes N_2 anaerobically—presumably because there was no selective pressure for the eukaryotes to associate with an organism which only fixes N_2 anaerobically. In the *Nostoc* there is a substantial increase in its heterocyst frequency from 3–5% to 20–30% (Hitch & Millbank 1975) thus facilitating N_2 fixation, but such an increase can occur only when a second photosynthetic partner is present; it does not occur in bipartite N_2-fixing lichens where an increase in heterocysts would decrease the photosynthetic efficiency of the cyanobiont. In the *Nostoc* which can fix N_2 aerobically (because of its heterocysts) GS-GOGAT is the sole route of ammonium assimilation of importance (cyanobacteria with high GDH activity occur, if ever, in symbiosis), while the heterotroph is a fungus, a member of the only group of plants which lack GS-GOGAT as a route of ammonium assimilation. That is, regulation of GS-GOGAT can occur without effect on eukaryotic ammonium assimilation. In the case of *P. aphthosa*, regulation of GS and possibly GOGAT is at the level of GS transcription or translation. We have shown using purified GS antibody that there is substantially reduced GS protein when compared with the free-living form. Fig. 1.4 summarizes some of the interactions which may occur.

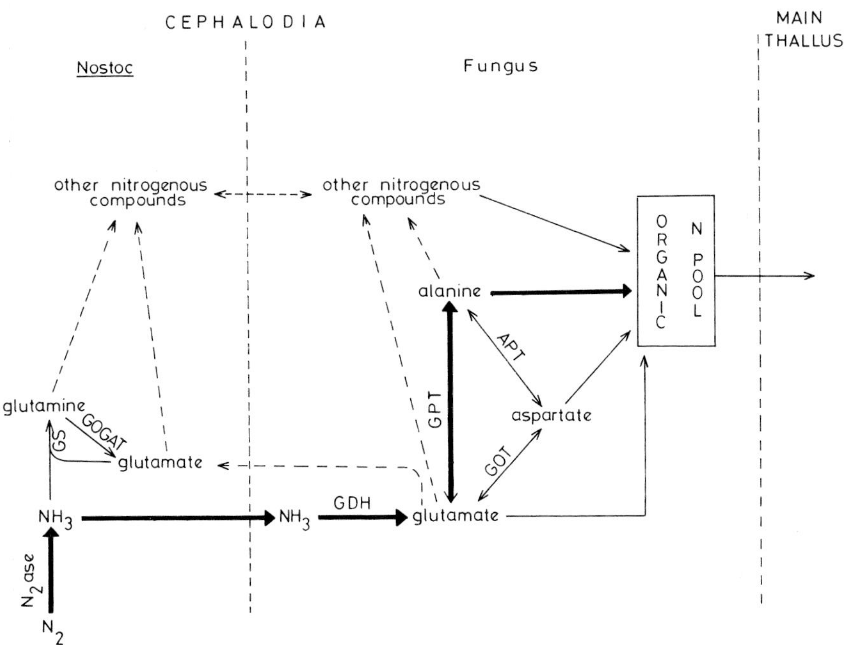

FIG. 1.4. Tentative scheme for the cycling of nitrogen compounds between the cephalodial components and main thallus of *Peltigera aphthosa*. Routes of major importance are represented by thickest arrows; dotted lines represent possibilities for which no experimental data have yet been obtained. N_2ase, nitrogenase; GS, glutamine synthetase; GOGAT, glutamate synthase; GDH, glutamate dehydrogenase, GPT, glutamate pyruvate transaminase; APT, aspartate pyruvate transaminase; GOT, glutamate-oxaloacetate transaminase.

Nitrogen cycling in eutrophic freshwaters—a ^{15}N tracer study in Blelham Tarn, English Lake District

Blelham Tarn is a eutrophic body of water in the English Lake District (Nat. Grid Ref. NY 365004) which has been extensively researched (Pearsall 1932; Lund 1972; Pennington 1974, 1978; Jones & Simon 1980). Within this lake Dr J.W.G. Lund placed an experimental enclosure of 18 500 m^3 capacity, 45 m in diameter and 12 m deep. The enclosure was open at the top but sealed into the sediment surface (Lack & Lund 1974). This provided a unique experimental system to study the fate of added nitrogenous fertilizer in an ecosystem large enough to be of relevance to what happens in nature without the drawbacks inherent in a system with inflows and outflows. The fact remains, nevertheless, that it is an artificial ecosystem. Within the body of water the main primary producers are frequently cyanobacteria (blue-green algae) and it is of interest to examine how they develop in response to the addition of exogenous fixed nitrogen, how they assimilate the fixed nitrogen and how assimilation by the algae affects nitrogen cycling within the system.

The approach used was to spray a solution containing 11 kg $NaNO_3$ (enriched with 9.63 atom % excess ^{15}N) and 789 g K_2HPO_4 (N/P atomic ratio 10:1) on to the surface waters of the enclosure on 19th August 1976 (the enclosure was thermally stratified at this time) and to monitor the subsequent fate of the added nitrogen.

The most striking immediate finding was the rapidity with which the ^{15}N-label disappeared from solution. Within seven days virtually all the nitrate-N had disappeared and there was a concomitant increase in particulate ^{15}N-labelling (Fig. 1.5). Prior to adding the fertilizer the chlorophyll *a* content

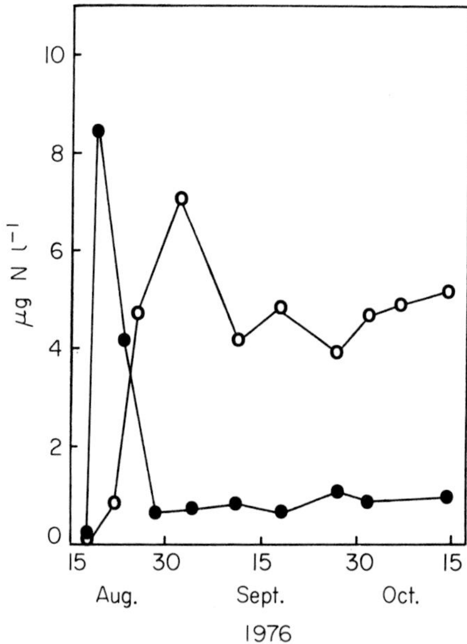

FIG. 1.5. Relationship between total dissolved ^{15}N (●—●) and total particulate ^{15}N (○—○) in the water column of the Blelham Tarn enclosure prior to and after the addition of $^{15}NO_3^-$ on 19th August 1976.

was $< 5\ \mu g$ chl *a* l^{-1}; within seven days an extensive algal bloom developed. The variation in total particulate nitrogen which occurred is shown in Fig. 1.6a and the relative abundance of the major algal groups is shown in Fig. 1.6b. The pattern of algal species development was what was predicted. When the nutrient concentrations were initially high, green algae showed their

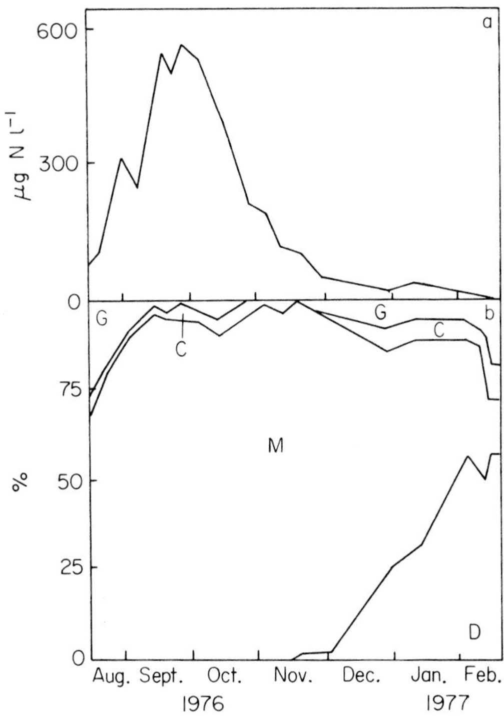

FIG. 1.6. (a) Changes in particulate nitrogen (> 90% algae) in the water column of the Blelham enclosure and (b) accompanying changes in the relative amounts of the major algal groups. G = green algae; M = *Microcystis;* C = cyanobacteria excluding *Microcystis* and D = diatoms.

maximum development but *Microcystis*, the dominant alga present, continued to increase to represent over 90% of the total algal flora by early September.

Figs 1.7a and b show the relationship which was observed between the ability of the cyanobacteria to store excess nitrogen in the form of cyanophycin granules and the availability of exogenous nitrogen. It is seen that in general there was a strong positive correlation between the availability of fixed nitrogen in the water column and the abundance of cyanophycin granules per cell (Stewart, Pemble & Al-Ugaily 1978).

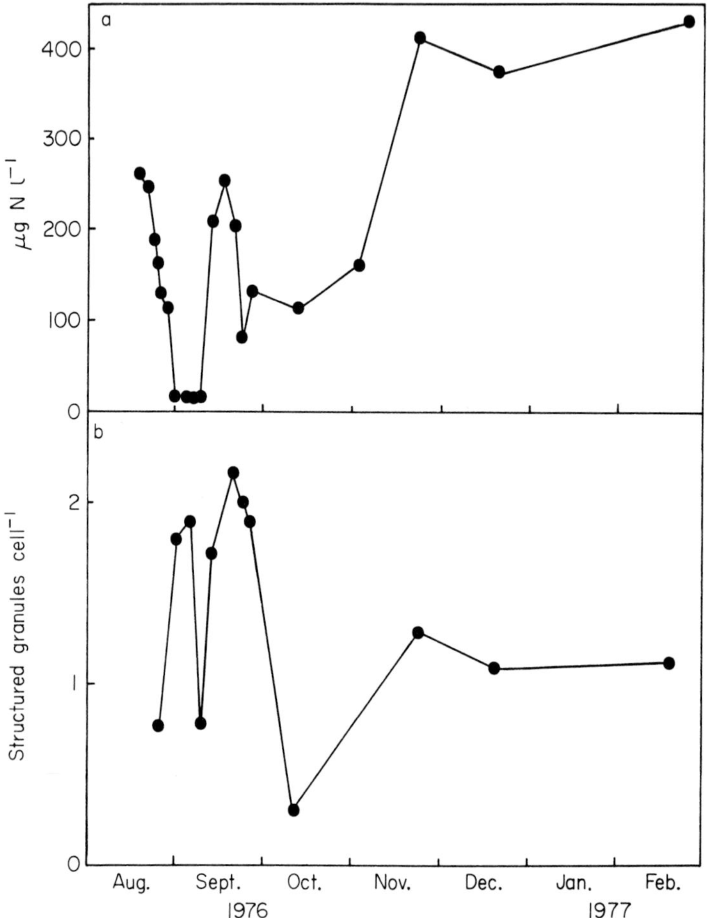

FIG. 1.7. Relationship between concentrations of NO_3^--N throughout the top 5 m of the water column of the Blelham enclosure (a) and the mean abundance of structured granules (cyanophycin granules) in the *Microcystis* cells present (b).

Evidence was obtained from the ^{15}N studies that the cyanobacteria remained remarkably resistant to predation because, despite the high ^{15}N-label in the *Microcystis*, there was little ^{15}N incorporation into the zooplankton. Throughout the experiment less than 1% of the total ^{15}N added in August (< 5% of the total particulate ^{15}N (Fig. 1.8)) was detectable at any time in the zooplankton (mainly *Daphnia hyalina*, *Eudiaptomus gracilis* and *Mesocyclops leuckarti*).

Subsequent nitrogen cycling within the total water body depended largely

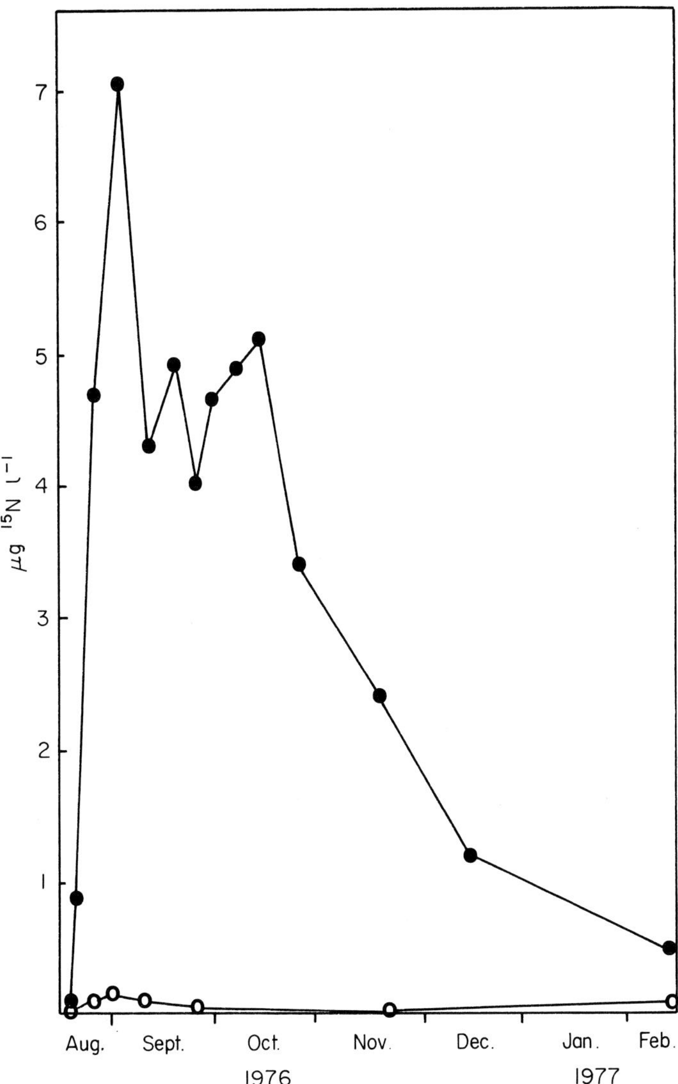

FIG. 1.8. Relative abundance of ^{15}N in the phytoplankton (●—●) and zooplankton (○—○) in the top 5 m of the water column of the Blelham enclosure during the period August 1976–February 1977.

on physical and climatic conditions and not until overturn was there a major redistribution of the fixed nitrogen. On overturn the cyanobacteria rapidly sedimented as a layer about 1 cm thick on the tarn sediment surface. The sediment surface population consisted of a mixture of cell types: dead cells, dying cells and healthy cells. Cells of the latter type accumulated various storage bodies and certain such cells served as overwintering populations, *Microcystis* having no special overwintering structures such as akinetes. Evidence that sediment-derived *Microcystis* served as inoculum for the subsequent year's growth came from the following approach. On 20th February 1977, when the *Microcystis* population was still on the sediment surface, the walls of the enclosure were lowered for 30 days in order to flush any remaining ^{15}N from the water column. After raising the enclosure walls we monitored the reappearance of particulate ^{15}N in the water column, how this correlated with the abundance of *Microcystis* in the water column, and how it correlated with the C/N ratio of the particulate material. The data obtained are summarized in Fig. 1.9.

Fig. 1.9a shows the pattern of ^{15}N-labelling in the water column from the time the enclosure was raised again until late September 1977. Particulate ^{15}N concentrations increased slowly from mid-March until early May when they increased substantially, and because soluble ^{15}N was negligible in the water column until August, it is reasonable to assume that particulate ^{15}N came from the sediment. The particulate ^{15}N increase in May corresponded to the first appearance of *Microcystis* in the water column. When stratification began in June the particulate ^{15}N-label initially decreased but during the period of late June to mid-July there was a marked movement of sediment-derived particulate ^{15}N into the water column. This ^{15}N then resedimented in September. Fig. 1.9b shows that immediately after isolating the enclosure in March the C/N ratio of the particulate material was 10.4:1. This ratio then increased to 12.2:1 in early May. This high ratio was indicative of a large detrital component in the particulate material. The ratio then declined in mid-July, particularly at depth, to less than 9:1. Such a low ratio is characteristic of living material and the C/N ratio of healthy *Microcystis* the previous year was 7–8:1. The mid-July minimum C/N ratio (8:1) occurred where the highest particulate ^{15}N occurred. This particulate material, with a low C/N ratio, spread into the surface waters in August.

Fig. 1.9. Relationship of (a) total particulate ^{15}N; (b) the C/N ratio of the particulate material and (c) *Microcystis aeruginosa* colonies throughout the water column of the Blelham enclosure during March–October 1977.
(a) ■, >80; ▩, 50–80; ▨, 35–50; ◪, 25–35; □, 0–25 (all ng l^{-1}).
(b) □ >12:1; ▨, 12:1–11:1; ▩, 11:1–9:1; ■, <9:1;
(c) ■, >7; ▩, 3–7; ▨, 1–3; ◪, 0.1–1; □ <0.1 (all colonies ml^{-1}).

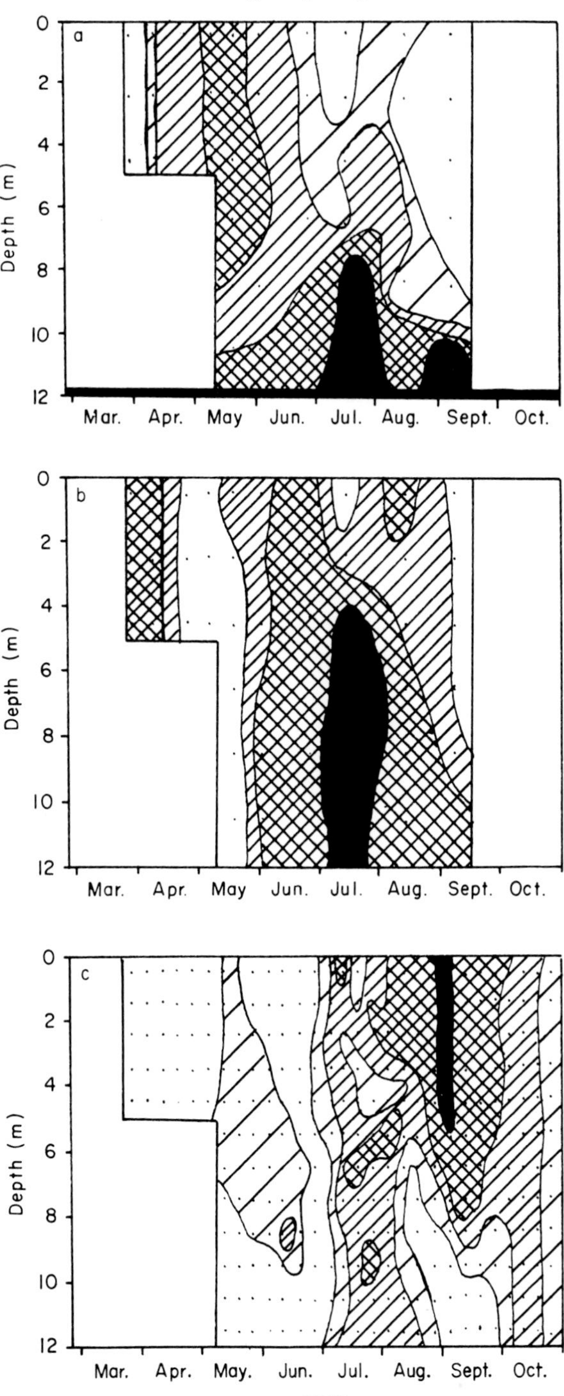
a
b
c
Depth (m)
0
2
4
6
8
10
12
Mar.
Apr.
May
Jun.
Jul.
Aug.
Sept.
Oct.
1977

On comparing the data in Figs 1.9a and b with those for the abundance of *Microcystis* (Fig. 1.9c) it is seen that the abundance of *Microcystis* correlated positively both with the high particulate ^{15}N-labelling and a low C/N ratio. Such data provide evidence that ^{15}N-labelled sediment-derived *Microcystis*, with an initial low C/N ratio, provided an inoculum for the subsequent year's bloom of *Microcystis* which developed most abundantly in surface waters during July–September. That is, a complete cycle of *Microcystis* nitrogen occurred from epilimnion→sediment→epilimnion, and no doubt the 1977 population would resediment and be subsequently recycled. Such results are discussed in more detail in Preston, Stewart & Reynolds (1980).

In such an ecosystem, cycling of particulate material from surface→sediment→surface is only part of the story. Some sedimented particulate nitrogen is lost to the deep sediment. Fig. 1.10 shows the distribution of ^{15}N-labelled particulate ^{15}N and soluble $^{15}NH_4^+$ in the sediment 8 months, 20 months and 29 months after the addition of $^{15}NO_3^-$ to the surface of the water column. It is seen that $^{15}NH_4^+$ was detected primarily in the top 2 cm after 8 months but that after 29 months, unlike particulate ^{15}N, it was almost equally distributed throughout the top 15 cm of the sediment.

Nitrogen which was not sedimented undergoes various nitrogen cycling processes in the surface sediment of which ammonification, nitrification, nitrate conversion to ammonium and denitrification are all important. Concomitant with $^{15}NH_4^+$ production in the sediment, nitrate production at

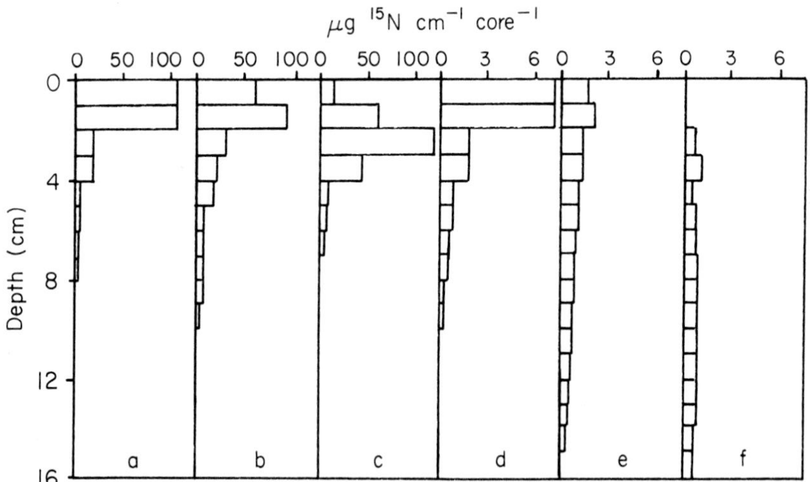

FIG. 1.10. Distribution of particulate ^{15}N (a–c) and $^{15}NH_4^+$ (d–f) in the top 16 cm of the Blelham enclosure sediment. a,d, 8 months; b,e, 20 months and c,f, 29 months after the addition of $^{15}NO_3^-$ to the surface waters.

the sediment–water interface and in the hypolimnion occurred. There is evidence this nitrate resulted from nitrification at an NH_4^+-O_2 chemocline which occurred at the sediment–water interface in spring, autumn and winter but which moved into the hypolimnion in summer when stratification occurred. ^{15}N data showed that this hypolimnetic nitrate pool was not the $^{15}NO_3^-$ added initially and which had remained in the water column but was the nitrate derived from ammonium. Such nitrification studies are detailed by Christofi, Preston & Stewart (1981) and Stewart *et al.* (1982) who consider that in terms of nitrogen sources for the growth of algae, insufficient attention has been paid in the past to endogenous nitrate production by nitrification.

The hypolimnetic and sediment nitrate pool has two main fates and we have detected both. First, nitrate may be lost from the enclosure by denitrification. Denitrification is an important route of loss of fixed nitrogen as N_2 and as nitrous oxide from both terrestrial (Dowdell 1982) and aquatic (Larsen 1977; Koike & Hattori 1978; Sorensen 1978) ecosystems to the atmosphere. There has been much controversy about the ratio of N_2 to nitrous oxide produced during denitrification (CAST 1976) and this may vary with the ecosystem, with the organisms, and with the conditions under which the organisms grow. There is also the complication that nitrous oxide may be a product of nitrification (Bremner & Blackmer 1978; Denmead, Freney & Simpson 1979) as well as of denitrification. In studies which we carried out on isolates from Scottish lochs (Stewart, May & Tuckwell 1976) we observed that N_2 accounted, in general, for over 90% of the gaseous products of nitrogen released by isolates of denitrifying bacteria. Preston & Stewart (1983) obtained similar results using sediment samples from Scottish lochs. In the Blelham Tarn enclosure, likewise, we detected negligible nitrous oxide production. Fig. 1.11 provides data on the denitrification potential (i.e. denitrification when an excess of nitrate is always available) of the enclosure sediments. The extent of nitrogen loss by denitrification at nitrate concentrations *in situ* will be substantially lower. Nitrogen lost to the atmosphere by denitrification will be compensated for by biological N_2-fixation within the enclosure. However, although various heterocystous cyanobacteria (*Anabaena* spp. and *Aphanizomenon* sp.) were detected, their numbers were low and the input by biological N_2 fixation to the water column and sediment was less than 2 kg N ha^{-1} $year^{-1}$.

The second fate of nitrate in the ecosystem is its conversion to ammonium, especially by certain fermentative bacteria. This process, first reported on by Woods (1938), has recently aroused new interest (Koike & Hattori 1978; Sorensen 1978; Caskey & Tiedje 1979; Cole & Brown 1980) particularly because the process retains nitrogen within the ecosystem while generating

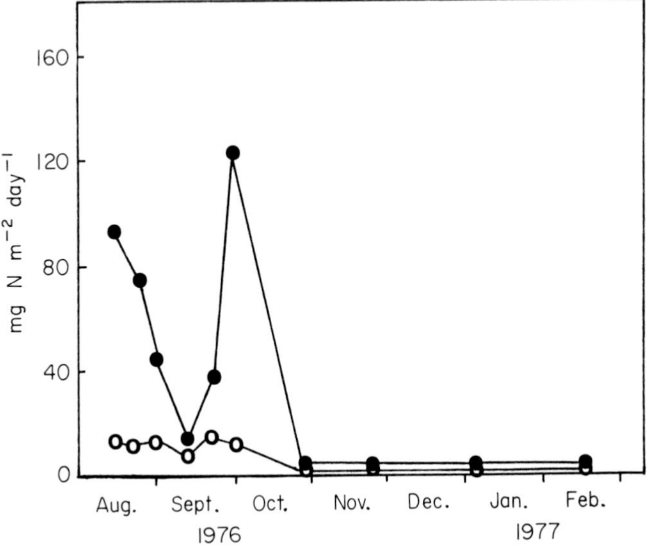

FIG. 1.11. Potential rates of denitrification (●—●) and NH_4^+ production from NO_3^- (○—○) in the top 2 cm of the sediment of the Blelham enclosure.

ATP in the process. This process was observed in the Blelham enclosure and seasonal variations in ammonium production from nitrate are shown in Fig. 1.11. Over the study period less than 25% of the nitrate present was reduced to NH_4^+, the remainder being reduced to N_2. Such nitrate reduction to ammonium which occurs particularly under acceptor-limiting conditions (Dunn, Herbert & Brown 1979) is of particular interest because, unlike denitrification, it purifies the water of nitrate but not of fixed nitrogen. Collectively, such data provide evidence of a miscellany of nitrogen-cycling processes within the Blelham enclosure. Some of the major transformations are summarized in Fig. 1.12.

PERSPECTIVE

In this paper we have attempted to demonstrate that within ecosystems, organisms and cells, there is a miscellany of nitrogen-cycling processes. Central to any nitrogen cycling process are living organisms which assimilate, harbour, transform and release nitrogen. The study of nitrogen cycling is firmly founded in the study of microbes, plants and animals. The nitrogen which provides the bulk of the new nitrogen added each year to the surface of the Earth is N_2 which is transformed to ammonium by prokaryotes possessing

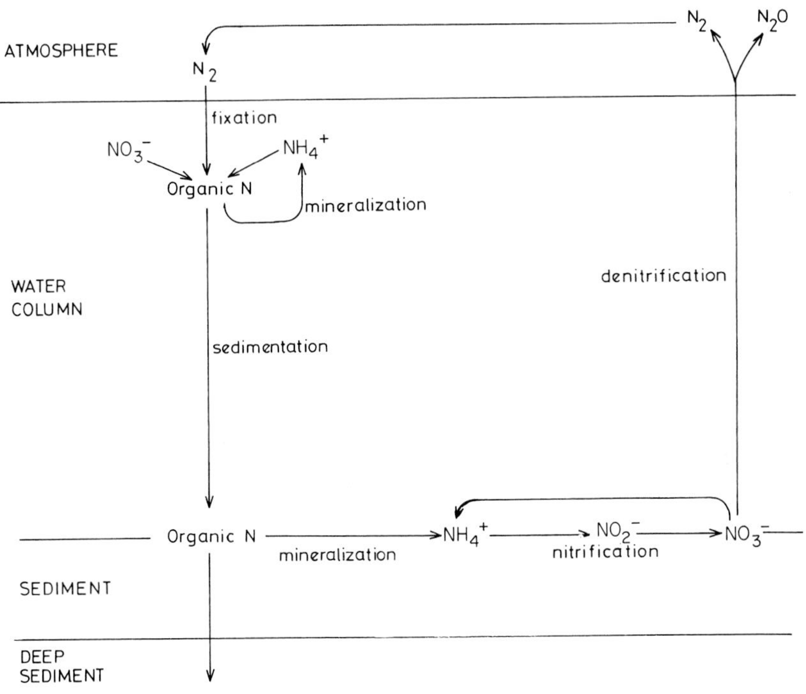

FIG. 1.12. Major nitrogen cycling processes in the Blelham enclosure.

the enzyme complex nitrogenase (certain cyanobacteria and photosynthetic bacteria; a report of an N_2-fixing green alga (Yamada & Sakuguchi 1980) is as yet unconfirmed). The heterocystous *Anabaena cylindrica* considered above is one such organism. Not only does *A. cylindrica* fix N_2 but it is a primary producer and like all microbes and plants (with the possible exception of some fungi) it assimilates ammonium taken up directly, fixed from N_2 or transformed from nitrate via the GS-GOGAT pathway. The glutamate thus produced is then transaminated to produce the amino acids essential for polypeptide and protein synthesis just as in other living cells. The cyanobacteria are archetypes in terms of nitrogen assimilation and cycling for living organisms in general.

The lichen *P. aphthosa* has a cyanobacterium, a *Nostoc*, as a symbiont which serves as an N_2-fixing factory to help sustain the nitrogen nutrition of the eukaryotes. It is only as a result of recent studies using ^{15}N as tracer that we are beginning to understand nitrogen cycling within such symbioses. From the results it is clear that such morphologically simple systems show a network of biochemical pathways and mechanisms no less complicated than those of higher plants. Now that man has emulated nature and achieved synthesis of the lichen thallus from the original components (Ahmadjian, Russell & Hildreth 1980; Culberson & Ahmadjian 1980) possibilities of examining the interactions more precisely at the biochemical and molecular levels have been opened up.

The Blelham enclosure experiment emphasizes the multiplicity of nitrogen cycling processes which occur in gross ecosystems. It emphasizes, despite the complexity of the interactions, the importance of the primary producers in providing sustenance for the secondary producers and substrate for the bacteria which dominate the nitrogen cycle: ammonifiers, nitrifiers and denitrifiers. The study emphasizes how environmental factors affect the process rates and how the gross nitrogen cycle differs only by degrees, because if one key process is damped down it is compensated for by enhanced activity of another. For example, in dissimilatory reduction more nitrate may be reduced to ammonium at low nitrate concentrations while at high nitrate concentrations more N_2 may be produced. To date the ecologist has spent much time in quantifying process rates and finding out how environmental factors affect such rates. In the future he will have to delve deeper into the regulatory mechanisms affecting ecosystems both at the biochemical and molecular genetics level. This is an opportunity which should be grasped as new techniques become available and as complex older techniques become simplified and routine. To understand ecosystems we need approaches of all kinds. In terms of nitrogen cycling, if we can, for example, provide the basic information on how to manipulate nitrate reduction to N_2 rather than to ammonium, we may be better able to explain more satisfactorily gross nitrogen cycling in lakes and find ways of improving the efficiency of sewage treatment plants; if we can understand the critical regulatory factors stimulating nitrogenase activity, we may understand better why N_2-fixing lichens persist, and how we may best exploit N_2-fixing cyanobacteria to agricultural advantage; by studying lake ecology we might hope to provide pointers to the control of algal blooms. In these and other challenges the ecologist has much to offer.

ACKNOWLEDGEMENTS

Our own work reported here was supported by the Agricultural Research Council, the Science and Engineering Research Council and the Natural Environment Research Council. We thank Gail Alexander for technical assistance.

REFERENCES

Adams D.G. & Carr N.G. (1981) Heterocyst differentiation and cell division in the cyanobacterium *Anabaena cylindrica*: effect of high light intensity. *Journal of Cell Science*, **49**, 341–52.

Ahmadjian V., Russell L.A. & Hildreth K.C. (1980) Artificial re-establishment of lichens. I. Morphological interactions between the phycobionts of different lichens and the mycobionts *Cladonia cristatella* and *Lecanora chrysoleuca*. *Mycologia*, **72**, 73–89.

Alexander V., Billington M. & Schell D.M. (1974) The influence of abiotic factors on nitrogen fixation in the Barrow, Alaska, Arctic tundra. *Report of the Kevo Subarctic Research Station*, **11**, 3–11.

Apte S.K., Rowell P. & Stewart W.D.P. (1978) Electron donation to ferredoxin in heterocysts of the N_2-fixing alga *Anabaena cylindrica*. *Proceedings of the Royal Society of London, Series B*, **200**, 1–25.

Billington M. & Alexander V. (1978) Nitrogen fixation in a black spruce (*Picea mariana* (Mill) B.S.P.) forest in Alaska. *Ecological Bulletins (Stockholm)*, **26**, 209–15.

Bothe H., Neuer G., Kalbe I. & Eisbrenner G. (1980) Electron donors and hydrogenase in nitrogen-fixing microorganisms. *Nitrogen Fixation* (eds. W.D.P. Stewart & J.R. Gallon), pp. 83–112. Academic Press, London.

Bremner J.M. & Blackmer A.M. (1978) Nitrous oxide emission from soils during nitrification of fertilizer nitrogen. *Science*, **199**, 295–6.

Carr N.G. (1979) Differentiation in filamentous cyanobacteria. In *Developmental Biology of Prokaryotes* (ed. H. Parish), pp. 167–201. Blackwell Scientific Publications, Oxford.

Caskey W.H. & Tiedje J.M. (1979) Evidence for clostridia as agents of dissimilatory reduction of nitrate to ammonium in soils. *Soil Science Society of America Journal*, **43**, 931–6.

CAST (1976) Effect of increased nitrogen fixation on stratospheric ozone. *Council for Agricultural Science and Technology, Report 53*, Iowa State University, Ames, Iowa, USA.

Christofi N., Preston T. & Stewart W.D.P. (1981) Endogenous nitrate production in an experimental enclosure during summer stratification. *Water Research*, **15**, 343–9.

Cole J.A. & Brown C.M. (1980) Nitrite reduction to ammonia by fermentative bacteria: a short circuit in the biological nitrogen cycle. *FEMS Microbiology Letters*, **7**, 65–72.

Crittenden P.D. (1975) Nitrogen fixation by lichens on glacial drift in Iceland. *New Phytologist*, **74**, 41–9.

Culberson C.F. & Ahmadjian V. (1980) Artificial re-establishment of lichens. II. Secondary products of resynthesized *Cladonia cristatella* and *Lecanora chrysoleuca*. *Mycologia*, **72**, 90–109.

Denmead O.T., Freney J.R. & Simpson J.R. (1979) Studies of nitrous oxide emission from a grass sward. *Soil Science Society of America Journal*, **43**, 726–8.

Dharmawardene M.W.N., Haystead A. & Stewart W.D.P. (1973) Glutamine synthetase of the nitrogen-fixing alga *Anabaena cylindrica*. *Archiv für Mikrobiologie*, **90**, 281–95.

Dowdell R.J. (1982) Fate of nitrogen applied to agricultural crops with particular reference to denitrification. *Philosophical Transactions of the Royal Society of London, Series B*, **296**, 363–73.

Dunn G.M., Herbert R.A. & Brown C.M. (1979) Influence of oxygen tension on nitrate reduction by a *Klebsiella* sp. growing in chemostat culture. *Journal of General Microbiology*, **112**, 379–83.

Englund B. (1977) The physiology of the lichen *Peltigera aphthosa*, with special reference to the blue-green phycobiont (*Nostoc* sp.). *Physiologia Plantarum*, **41**, 298–304.

Englund B. (1978) Effects of environmental factors on acetylene reduction by intact thallus and excised cephalodia of *Peltigera aphthosa* Willd. *Ecological Bulletins (Stockholm)*, **26**, 234–46.

Englund B. & Meyerson H. (1974) *In situ* measurement of nitrogen fixation at low temperatures. *Oikos*, **25**, 283–7.

Feige G.B. (1976) Investigations on the physiology of cephalodia from the lichen *Peltigera aphthosa* (L.) Willd. II. The photosynthetic ^{14}C-labelling pattern and the movement of carbohydrate from phycobiont to mycobiont. *Zeitschrift für Pflanzenphysiologie*, **80**, 386–94.

Fogg G.E. (1949) Growth and heterocyst production in *Anabaena cylindrica* Lemm. II. In relation to carbon and nitrogen metabolism. *Annals of Botany*, **13**, 241–59.

Fogg G.E. (1982) Nitrogen cycling in sea waters. *Philosophical Transactions of the Royal Society of London, Series B*, **296,** 511–20.

Fogg G.E., Stewart W.D.P., Fay P. & Walsby A.E. (1973) *The Blue-Green Algae*. Academic Press, London.

Gasser J.K.R. (1982) Agricultural productivity and the nitrogen cycle. *Philosophical Transactions of the Royal Society of London, Series B*, **296,** 303–14.

Granhall U. & Selander H. (1973) Nitrogen fixation in a subarctic mire. *Oikos*, **24**, 8–15.

Haselkorn R. (1978) Heterocysts. *Annual Review of Plant Physiology*, **29**, 319–44.

Hawkesford M.J., Reed R.H., Rowell P. & Stewart W.D.P. (1981) Nitrogenase activity and membrane electrogenesis in the cyanobacterium *Anabaena variabilis* Kütz. *European Journal of Biochemistry*, **115**, 519–23.

Haystead A., Dharmawardene M.W.N. & Stewart W.D.P. (1973) Ammonia assimilation in a nitrogen-fixing blue-green alga. *Plant Science Letters*, **1**, 439–45.

Hitch C.J.B. & Millbank J.W. (1975) Nitrogen metabolism in lichens. VII. Nitrogenase activity and heterocyst frequency in lichens with blue-green phycobionts. *New Phytologist*, **75**, 239–44.

Hood A.E.M. (1982) Fertilizer trends in relation to biological productivity within the UK. *Philosophical Transactions of the Royal Society of London, Series B*, **296,** 315–28.

Hopper S. & Segal H.L. (1962) Kinetic studies of rat liver glutamic alanine transaminase. *Journal of Biological Chemistry*, **237**, 3189–95.

Jones J.G. & Simon B.M. (1980) Decomposition processes in the profundal region of Blelham Tarn and the Lund Tubes. *Journal of Ecology*, **68**, 493–512.

Kallio P. & Kallio S. (1978) Adaptation of nitrogen fixation to temperature in the *Peltigera aphthosa* group. *Ecological Bulletins (Stockholm)*, **26**, 225–33.

Kershaw K.A. & Millbank J.W. (1970) Nitrogen metabolism in lichens. II. The partition of cephalodial-fixed nitrogen between the mycobiont and phycobionts of *Peltigera aphthosa. New Phytologist*, **59**, 75–9.

Koike I. & Hattori A. (1978) Denitrification and ammonia formation in anaerobic coastal sediments. *Applied and Environmental Microbology*, **35**, 278–82.

Lack T.J. & Lund J.W.G. (1974) Observations and experiments on the phytoplankton of Blelham Tarn, English Lake District. I. The experimental tubes. *Freshwater Biology*, **4**, 399–415.

Ladha J.K., Rowell P. & Stewart W.D.P. (1978) Effects of 5-hydroxylysine on acetylene reduction and NH_4^+ assimilation in the cyanobacterium *Anabaena cylindrica. Biochemical and Biophysical Research Communications*, **83**, 688–96.

Lang N.J. & Fay P. (1971) The heterocysts of blue-green algae. II. Details of ultrastructure. *Proceedings of the Royal Society of London, Series B*, **178**, 193–203.

Larsen V. (1977) Nitrogen transformation in lakes. *Progress in Water Technology*, **8**, 419–31.

Lea P.J. & Miflin B.J. (1975) Glutamine (amide): 2-oxoglutarate aminotransferase in blue-green algae. *Biochemical Society Transactions*, **3**, 381–4.

Lex M. & Carr N.G. (1974) The metabolism of glucose by heterocysts and vegetative cells of *Anabaena cylindrica. Archives of Microbiology*, **101**, 161–7.

Lex M. & Stewart W.D.P. (1973) Algal nitrogenase, reductant pools and photosystem I activity. *Biochimica et Biophysica Acta*, **292**, 436–43.

Lund J.W.G. (1972) Preliminary observations on the use of large experimental enclosures in lakes. *Verhandlungen der Internationalen Vereinigung für theoretische und angewandte Limnologie*, **18**, 71–7.

Magasanik B. (1982) Genetic control of nitrogen assimilation in bacteria. *Annual Reviews of Genetics*, **16,** 135–68.

Millbank J.W. (1974) Nitrogen metabolism in lichens. V. The forms of nitrogen released by the blue-green phycobiont in *Peltigera* spp. *New Phytologist*, **73**, 1171–81.

Millbank J.W. & Kershaw K.A. (1969) Nitrogen metabolism in lichens. I. Nitrogen fixation in the cephalodia of *Peltigera aphthosa. New Phytologist*, **68**, 721–9.

Ohmori M. & Hattori A. (1978) Transient change in the ATP pool of *Anabaena cylindrica* associated with ammonia assimilation. *Archives of Microbiology*, **117**, 17–20.

Ownby J.D. (1977) Effects of amino acids on methionine-sulphoximine-induced heterocyst formation in *Anabaena. Planta*, **136**, 277–9.

Pearsall W.H. (1932) Phytoplankton in the English lakes. II. The composition of the phytoplankton in relation to dissolved substances. *Journal of Ecology*, **20**, 241–62.

Pennington W. (1974) Seston and sedimentation formation in five Lake District lakes. *Journal of Ecology*, **62**, 215–51.

Pennington W. (1978) Response of some British lakes to past changes in land use on their catchments. *Verhandlungen der Internationalen Vereinigung für theoretische und angewandte Limnologie*, **20**, 636–41.

Peters G.A., Evans W.R. & Toia R.E. Jr. (1976) *Azolla–Anabaena azollae* relationship. IV. Photosynthetically driven nitrogenase catalyzed H_2 production. *Plant Physiology*, **58**, 119–26.

Peters G.A., Mayne B.C., Ray T.B. & Toia R.E. Jr (1979) Physiology and biochemistry of the *Azolla–Anabaena* symbiosis. In *Nitrogen and Rice*, pp. 325–44. International Rice Research Institute, Los Banos, Philippines.

Peterson R.B. & Burris R.H. (1978) Hydrogen metabolism in isolated heterocysts of *Anabaena* 7120. *Archives of Microbiology*, **116**, 125–32.

Preston T. & Stewart W.D.P. (1983) (In preparation).

Preston T., Stewart W.D.P. & Reynolds C.S. (1980) Bloom-forming cyanobacterium *Microcystis aeruginosa* overwinters on sediment surface. *Nature*, **288**, 365–7.

Rai A.N., Rowell P. & Stewart W.D.P. (1980) NH_4^+ assimilation and nitrogenase regulation in the lichen *Peltigera aphthosa* Willd. *New Phytologist*, **85**, 545–55.

Rai A.N., Rowell P. & Stewart W.D.P. (1981a) Nitrogenase activity and dark CO_2 fixation in the lichen *Peltigera aphthosa* Willd. *Planta*, **151**, 256–64.

Rai A.N., Rowell P. & Stewart W.D.P. (1981b) $^{15}N_2$ incorporation and metabolism in the lichen *Peltigera aphthosa* Willd. *Planta*, **152**, 544–52.

Rai A.N., Rowell P. & Stewart W.D.P. (1981c) Glutamate synthase activity in symbiotic cyanobacteria. *Journal of General Microbiology*, **126,** 515–18.

Rippka R., Deruelles J., Waterbury J.B., Herdman M. & Stanier R.Y. (1979) Generic assignments, strain histories and properties of pure cultures of cyanobacteria. *Journal of General Microbiology*, **111**, 1–61.

Rowell P., Diez J., Apte S.K. & Stewart W.D.P. (1981) Molecular heterogeneity of ferredoxin: $NADP^+$ oxidoreductase from the cyanobacterium *Anabaena cylindrica. Biochimica et Biophysica Acta*, **657**, 507–16.

Simon R.D. (1971) Cyanophycin granules from the blue-green alga *Anabaena cylindrica*, a reserve material consisting of copolymers of aspartic acid and arginine. *Proceedings of the National Academy of Sciences, USA*, **68**, 265–7.

Sorensen J. (1978) Capacity for denitrification and reduction of nitrate to ammonia in a coastal marine sediment. *Applied and Environmental Microbiology*, **35**, 301–5.

Stanier R.Y. & Cohen-Bazire G. (1977) Phototrophic prokaryotes: the cyanobacteria. *Annual Review of Microbiology*, **31**, 225–74.

Stewart W.D.P. (1980a) Some aspects of structure and function in nitrogen-fixing cyanobacteria. *Annual Review of Microbiology*, **34**, 497–536.

Stewart W.D.P. (1980b) Systems involving blue-green algae (cyanobacteria). In *Methods for Evaluating Biological Nitrogen Fixation* (ed. F. J. Bergersen), pp. 583–635. John Wiley & Sons, Chichester.

Stewart W.D.P. & Gallon J.R. (eds) (1980) *Nitrogen Fixation.* Academic Press, London.

Stewart W.D.P. & Rodgers G.A. (1977) The cyanophyte-hepatic symbiosis. II. Nitrogen fixation and the interchange of nitrogen and carbon. *New Phytologist*, **78**, 459–71.

Stewart W.D.P. & Rowell P. (1975) Effects of L-methionine-DL-sulphoximine on the assimilation of newly fixed NH_3, acetylene reduction and heterocyst production in *Anabaena cylindrica. Biochemical and Biophysical Research Communications*, **65,** 846–56.

Stewart W.D.P. & Rowell P. (1977) Modifications of nitrogen-fixing algae in lichen symbioses. *Nature*, **265,** 371–2.

Stewart W.D.P., Haystead A. & Dharmawardene M.W.N. (1975) Nitrogen assimilation and metabolism in blue-green algae. In *Nitrogen Fixation by Free-living Micro-organisms* (ed. W.D.P. Stewart), International Biological Programme Ser., Vol. 6, pp. 129–58. Cambridge University Press, Cambridge.

Stewart W.D.P., May E. & Tuckwell S. (1976) Nitrogen and phosphorus from agricultural land and urbanization and their fate in shallow freshwater lochs. In *Agriculture and Water Quality*, Technical Bulletin 32, pp. 276–305. Ministry of Agriculture, Fisheries & Food, London.

Stewart W.D.P., Pemble M. & Al-Ugaily L. (1978) Nitrogen and phosphorus storage and utilization in blue-green algae. *Mitteilungen der Internationalen Vereinigung für theoretische und angewandte Limnologie*, **21**, 224–47.

Stewart W.D.P., Preston T., Peterson H.G. & Christofi N. (1982) Nitrogen cycling in eutrophic freshwaters. *Philosophical Transactions of the Royal Society of London, Series B*, **296,** 491–509.

Stewart W.D.P., Rai A.N., Reed R.H., Creach E., Codd G.A. & Rowell P. (1981) Studies on the N_2-fixing lichen *Peltigera aphthosa. Current Perspectives in Nitrogen Fixation. Proceedings of the Fourth International Symposium on Nitrogen Fixation* (ed. A.H. Gibson & W.E. Newton), pp. 237–43. Australian Academy of Science, Canberra.

Stewart W.D.P., Rowell P. & Rai A.N. (1980) Symbiotic nitrogen-fixing cyanobacteria. In *Nitrogen Fixation* (eds W.D.P. Stewart & J.R. Gallon), pp. 239–77. Academic Press, London.

Stutz R.C. & Bliss L.C. (1975) Nitrogen fixation in soils of Truelove Lowland, Devon Island, Northwest Territories. *Canadian Journal of Botany*, **53**, 1387–99.

Sutherland J.M., Herdman M. & Stewart W.D.P. (1979) Akinetes of the cyanobacterium *Nostoc* PCC7524: Macromolecular composition, structure and control of differentiation. *Journal of General Microbiology*, **115**, 273–87.

Tel-Or E. & Stewart W.D.P. (1976) Photosynthetic electron transport, ATP synthesis and nitrogenase activity in isolated heterocysts of *Anabaena cylindrica. Biochimica et Biophysica Acta*, **423**, 189–95.

Tel-Or E. & Stewart W.D.P. (1977) Photosynthetic components and activities of nitrogen-fixing isolated heterocysts of *Anabaena cylindrica. Proceedings of the Royal Society of London, Series B*, **198**, 61–86.

Thomas J., Meeks J.C., Wolk C.P., Shaffer P.W., Austin S.M. & Chien W.-S. (1977) Formation of glutamine from (^{13}N) ammonia, (^{13}N) dinitrogen, and (^{14}C) glutamate by heterocysts isolated from *Anabaena cylindrica. Journal of Bacteriology*, **129**, 1545–55.

von Liebig J. (1851) *Familiar Letters on Chemistry*. Taylor, Walton & Maberly, London.

Wilcox M. Mitchison G.J. & Smith R.J. (1973) Pattern formation in the blue-green alga, *Anabaena*. I. Basic mechanisms. *Journal of Cell Science*, **12**, 707–23.

Wilcox M., Mitchison G.J. & Smith R.J. (1975a) Mutants of *Anabaena cylindrica* altered in heterocyst spacing. *Archives of Microbiology*, **103**, 219–23.

Wilcox M., Mitchison G.J. & Smith R.J. (1975b) Spatial control of differentiation in the blue-green alga *Anabaena*. In *Microbiology 1975* (ed. D. Schlessinger), pp. 453–63. American Society of Microbiology, Washington.

Winkenbach F. & Wolk C.P. (1973) Activities of the enzymes of the oxidative and reductive pentose phosphate pathway in heterocysts of a blue-green alga. *Plant Physiology*, **52**, 480–3.

Wolk C.P. (1968) Movement of carbon from vegetative cells to heterocysts in *Anabaena cylindrica*. *Journal of Bacteriology*, **96**, 2138–43.

Wolk C.P., Thomas J., Shaffer P.W., Austin S.M. & Galonsky A. (1976) Pathway of nitrogen metabolism after fixation of ^{13}N-labelled nitrogen gas by the cyanobacterium *Anabaena cylindrica*. *Journal of Biological Chemistry*, **251**, 5027–34.

Woods D.D. (1938) The reduction of nitrate to ammonia by *Clostridium welchii*. *Biochemical Journal*, **32**, 2000–12.

Yamada T. & Sakaguchi K. (1980) Nitrogen fixation associated with a hotspring green alga. *Archives of Microbiology*, **124**, 161–7.

2. ADAPTIVE VARIATION IN LEGUME NODULE PHYSIOLOGY RESULTING FROM HOST–RHIZOBIAL INTERACTIONS

JANET I. SPRENT
University of Dundee

SUMMARY

Nodulated legumes show considerable diversity in their ability to fix nitrogen and in the way in which their nodules function. Both host and rhizobial partners have evolved differently in tropical and temperate regions. The bacteroids affect the final products exported from nodules. Host cells affect bacteroid permeability and development of uptake hydrogenase. Interactions between the host and endophyte lead to flexibility in adaptation to specific environmental conditions.

INTRODUCTION

Legumes are certainly the most widely studied and arguably the most important of all the nitrogen-fixing symbioses. Because of their agricultural significance, their ecological role has generally been under studied. Research has tended to concentrate on specific grain and forage species with the result that very little is known about the physiology of nitrogen fixation in most of the world's native nodulated legumes, especially those with perennial habits. Nitrogen fixation only takes place when combined nitrogen (ammonium, nitrate) is at low levels or absent. When present, combined nitrogen reduces both nodulation and nitrogen fixation. Thus without going into the comparative energetics of the various processes, it is likely that nitrogen fixation imposes some type of burden on the plant (see also Chapter 11). To minimize this (hypothetical) burden, the metabolism of host and endosymbiont must be closely integrated. Because legumes are a large and diverse family, inhabiting all terrestrial regions of the world and with the occasional excursion into aquatic habitats (e.g. *Neptunia* in the Amazon), the strategies adopted by different species to optimize nitrogen fixation may not all be the same.

The framework of this account is a possible evolutionary tree within the subfamily Papilionoideae and the coevolution of *Rhizobium* spp. The functioning of different types of nodule, with different physiological mechanisms will be considered as a complex of interactions between the two symbionts, enabling legumes to adapt to differing ecological situations. The discussion will centre on active (mature) nodules and will be divided into three parts:

1. Taxonomic and evolutionary relationships in legume physiology.
2. Detailed consideration of host–endophyte metabolic interactions.
3. Significance of legumes in nitrogen cycling.

TAXONOMIC AND EVOLUTIONARY RELATIONSHIPS

The Leguminosae

The exact taxonomic level of this large family is still hotly disputed (Isely & Polhill 1980). For the present purpose we shall consider solely the subfamily Papilionoideae, since this is the branch with the most, and the most widely studied, nodulated species. The tribes within this group which are to be considered in particular are given in Fig. 2.1 which also shows their possible evolutionary relationships. Examples of some economically important members of these tribes are given in Table 2.1.

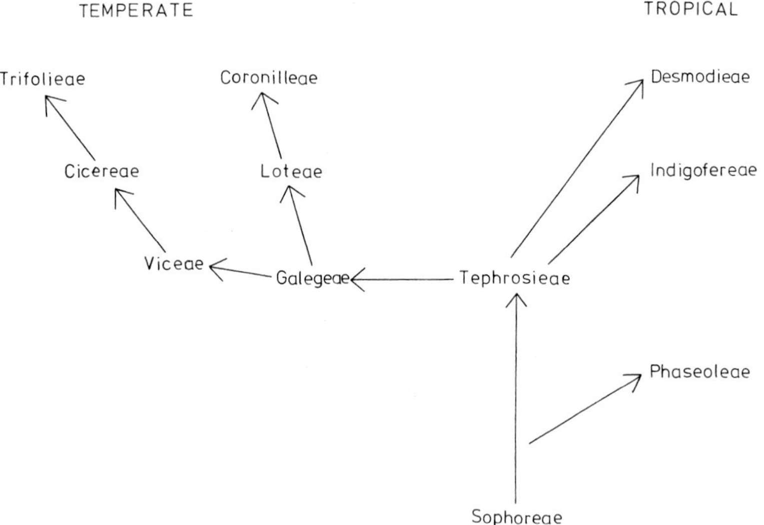

FIG. 2.1. Relationships of some tribes of the subfamily Papilionoideae. (After Polhill 1981.)

TABLE 2.1. Some examples of economically important members of tribes shown in Fig. 2.1.

Tribe	Species	Common name	Use
Trifolieae	*Trifolium* spp.	clovers	forage
Cicereae	*Cicer arietinum*	chick pea	grain
Vicieae	*Vicia faba*	broad or horse bean	grain
	Pisum sativum	garden pea	grain
	Lens esculenta	lentil	grain
Phaseoleae	*Phaseolus vulgaris*	french or navy bean	grain
	Glycine max	soybean	grain
	Vigna unguiculata	cowpea	grain
	Centrosema pubescens	centro	forage
Desmodieae	*Desmodium intortum*		forage

Rhizobium

The taxonomy of *Rhizobium* is perhaps even more in dispute than that of the Leguminosae. However, as with the latter, there are distinct tropical and temperature groupings (Table 2.2). This divergence is presenting some major barriers to the advancement of understanding how nitrogen is fixed, since those rhizobia that have been persuaded to fix nitrogen in pure culture (i.e. apart from their host) have not been amenable to genetic manipulation. Other particular attributes found in rhizobia associated with particular tribes have been summarized by Sprent (1980). The genetics of *Rhizobium* have recently been reviewed by Beringer, Brewin & Johnston (1980).

Nodule morphology

Nodules of the tropical tribes shown in Fig. 2.1, together with those of the Loteae, are more or less spherical in shape, of determinate growth and with a vascular system which is a branched loop of the main root stele, i.e. it has no open ends. They also have lenticels on the surface to assist gaseous exchange. Nodules of the other tribes are basically indeterminate, with an apical meristem. They may branch and the vascular system is branched also, differentiating as nodules grow, and thus with open ends. There are no lenticels, gaseous exchange taking place through small intercellular spaces over the whole surface. Further details of the morphology and anatomy of these types can be found in Sprent (1980). In so far as they can resume meristematic activity after periods of stress, indeterminate nodules may be at an advantage in fluctuating environments.

TABLE 2.2. Some features of tropical and temperate rhizobia.

Feature	Tropical *R. japonicum*, 'cowpea miscellany'	Temperate *R. trifolii*, *R. leguminosarum*
Growth rate on agar	slow	fast
Tolerance of low soil pH	good	poor
Success with genetic manipulations	very poor	good
Demonstration of nitrogenase in free-living cultures	frequent	not substantiated

HOST–ENDOPHYTE METABOLIC INTERACTIONS

Effects of Rhizobium *on host*

(a) *Formation of particular export products.* It is likely that all endosymbiotic rhizobia release the bulk of ammonium produced from nitrogen reduction into the host cytosol, where it is initially assimilated into glutamine and glutamate (see Sprent 1979 for a general account of this process).

Glutamine may be transported out of nodules or further reactions may take place leading to the formation of asparagine, the ureides allantoin and allantoic acid and sometimes other nitrogen compounds (Fig. 2.2). The detailed reactions are summarized in Pate, Atkins & Rainbird (1981). Within the tribes considered here we have two distinct groupings—the temperate ones export the amides glutamine and asparagine, together with smaller amounts of their amino acids, whereas the bulk of nitrogen leaving nodules of the tropical tribes is in the form of ureides. Ureides effect a considerable economy in carbon (see Chapter 11). They are, however, an order of magnitude less soluble and this may render them unsuitable for temperate regions, because at ambient temperatures insufficient of them can be dissolved in the xylem fluid leaving nodules (Sprent 1980). All the ureide-exporting species studied to date have spherical (determinate) nodules, with a closed vascular system and this may offer a lower resistance pathway for xylem transport. Transported products are not only important to the legume, but also to xylem sap-sucking insects which need to metabolize them.

One of the most intriguing features of ureide synthesis is that it occurs in the host cell cytosol, its production being triggered by the presence of nitrogen-fixing bacteroids. Neither nitrate- nor ammonium-fed plants normally produce significant quantities of ureides. In soybean (*Glycine max*) the correlation between ureide production and nitrogen fixation is so close that it

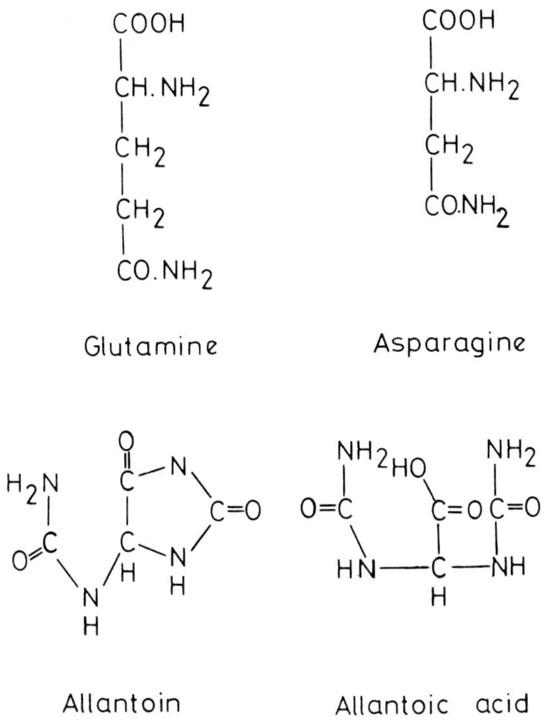

Glutamine: COOH–CH.NH$_2$–CH$_2$–CH$_2$–CO.NH$_2$

Asparagine: COOH–CH.NH$_2$–CH$_2$–CO.NH$_2$

Allantoin

Allantoic acid

FIG. 2.2. Structure of amides and ureides commonly exported by legumes.

has been suggested that ureides, which are easily assayed spectrophotometrically, could be used as a measure of nitrogen fixation (McClure, Israel & Volk 1980). The machinery for producing ureides is widespread in legumes, both tropical and temperature types (Fosse 1933). Thus the ability to stimulate ureide synthesis in nodule cells appears, in some unknown way, to be a property of the tropical rhizobia.

The correlation between ureide production and nitrogen fixation is not always as clear-cut as it can be in soybean. *Phaseolus vulgaris*, which is closely related taxonomically to soybean (see Table 2.1), shows a number of differences in its nitrogen-fixing system. First, it is infected by a different species of *Rhizobium*, *R. phaseoli*. The latter is comparatively fast growing and has been crossed with typical fast-growing species, such as *R. leguminosarum*. These are features typical of temperate rhizobia (Table 2.2). On the other hand, the nodule structure is strikingly similar to that of soybean and the morphology of the bacteroids in these two species is apparently identical and quite different to the pleomorphic forms found in *R. leguminosarum* (Sprent

1980). Although ureides are exported, significant amounts of amides are also found in the xylem exudate of nitrogen-fixing plants. Table 2.3 gives data from an experiment in which plants were grown either on nitrate or with an efficient strain of *R. phaseoli.* It can be seen that, not only has the *Rhizobium* switched on bulk synthesis of ureides, it has also changed the balance between asparagine and glutamine as has been found by Pate *et al.* (1980) for cowpea. The asparagine/glutamine balance also varied with plant age. Since glutamine is a major precursor for ureide synthesis (Sprent 1980) as well as being the first major organic product of nitrogen fixation, it is possible that under the

TABLE 2.3. Levels of some compounds in xylem exudate from root systems of *Phaseolus vulgaris* cv Glamis, grown either on 10 mM nitrate or inoculated with *R. phaseoli* strain 3622. (Unpublished data of Miller & Sprent.)

N source	Ureides	Asparagine	Glutamine
Nitrate	low	high	low
N_2 fixation	high	low	high

particular host–rhizobial combination and/or growing conditions used (greenhouse with supplementary light in a Scottish winter) the ureide synthesizing system could not process all the glutamine produced and therefore some was exported. However, it indicates that there is considerable flexibility in this case and this may be important in view of the wide geographical range in which *P. vulgaris* is grown. Also, the fact that *R. phaseoli* has features in common with temperate rhizobia and can be crossed with, for example, *R. leguminosarum* opens up the possibility that hybrids can be produced which will help elucidate the control of ureide synthesis in nodules. There is the further possibility that genetic manipulation may lead to *P. vulgaris–Rhizobium* combinations which are principally amide exporters and these may be more adapted to temperate conditions. Nature may well have already produced such variants in other species.

Preliminary evidence also indicates that bacterial strain can influence the balance between allantoin and allantoic acid produced in the cytosol. Both of these components are an order of magnitude less soluble than amides and under temperate conditions the water flow through the nodule may be insufficient to remove all the fixed nitrogen (Sprent 1980). By using both molecular species, the problem is halved. However, allantoic acid requires a counter ion, probably potassium. The solubility of potassium allantoate is much greater than that of the free acid, although stability may be a problem at normal xylem pH (Sprent, unpublished data). Table 2.4 shows the ureide and potassium contents and sap pH of various *Cajanus cajan* (pigeon pea, a member of the tribe Phaseoleae)–*Rhizobium* strain pairs. Although further combinations need to be studied, the data strongly suggest that the *Rhizobium* has an important effect on the ureide species.

These results have further implications. It has recently been pointed out that nitrogen-fixing plants must take up cations in excess of anions, when compared with nitrate-fed plants, and thus must produce an organic anion such as malate to maintain charge balance: this may be exported in xylem sap (Israel & Jackson 1981). This is an added burden on the carbon supplies of a nodulated plant, since up to half the carbon (by weight) in sap may be organate. Cation uptake and organic anion synthesis occur in roots. What happens to the charge balance when a shot of allantoic acid is injected to the xylem mainstream from the nodule remains to be elucidated.

TABLE 2.4. Composition of xylem sap from nodulated *Cajanus cajan* (mmol).

Host cv	*Rhizobium* strain	Sap pH	Allantoin	Allantoic acid	K^+
8103	3806	5.88	1.64	28.79	10.6
8101	3871	5.51	0.76	16.41	9.9
8101	3871	5.70	1.33	15.78	9.2
8101	3862	5.84	5.18	6.06	7.8
8101	3862	—	5.81	4.67	2.1
8103	3862	6.11	3.72	5.18	8.6
8103	3862	—	1.26	3.79	8.3

(b) *Growth factors*. Most nitrogen-fixing bacteria produce growth factors, which in the case of associative systems (e.g. grass/*Azospirillum*) have been claimed to have effects as important as the supply of fixed nitrogen (see van Berkum & Bohlool (1980) for discussion). *Rhizobium* is no exception, and recent work indicates that different bacterial strains may vary in their production of growth factors within active nodules. Triplett *et al.* (1981) showed that the effects of bacterial strain on internode length of *Phaseolus lunatus* L. could be mimicked by various applications of gibberellic acid.

Effects of host on Rhizobium

(a) *Uptake hydrogenase activity*. One of the apparently inevitable consequences of nitrogen reduction is the parallel reduction of protons (H^+) to hydrogen gas. This wastes ATP energy and reductant. Some of this energy may be retrieved by an uptake hydrogenase which recycles the hydrogen, oxidizing it to water and probably replacing about 43% of the energy wasted in hydrogen evolution (Pate, Atkins & Rainbird 1981). Net hydrogen evolution by nodules is thus a measure of their inefficiency (Schubert & Evans 1976). In a wide survey, Schubert & Evans showed that many of the natural combinations of host–endophyte (particularly the non-leguminous-actinomycete nodules) showed a high efficiency whereas agricultural crops (normally selected on well-fertilized fields) were much less efficient. More recently, workers in Evans' laboratory (Zablotowicz, Russell & Evans 1980) have shown that legumes inoculated with rhizobia having low hydrogen evolution can out-yield those with high net evolution. These findings imply that natural selection may have acted to minimize hydrogen evolution in plants wholly dependent on fixed nitrogen. However, this may be too facile an explanation, for hydrogen evolution is yet another example of a very complex interaction between host and endophyte. A close examination of the data of Schubert & Evans (1976) shows that two host species, cowpea (referred to as *Vigna sinensis* in this paper, but more usually *V. unguiculata*) and mung bean (*Phaseolus mungo*, now usually classified as *Vigna radiata*) inoculated with the same rhizobial strain, 32HI, gave nodules with different levels of net hydrogen evolution. This difference could have resulted from various factors, one of which is an effect of the host on the expression of uptake hydrogenase in the bacteroid. Table 2.5 shows data of Gibson & Sprent (unpublished) in which four species of *Vigna* were inoculated with three different strains of *Rhizobium* sp., all of which were known to possess the genes for uptake hydrogenase (Hup^+). With two of these strains, large quantities of hydrogen were evolved from nodules formed on *V. radiata*. Further work showed that all those nodules with low rates of net hydrogen evolution could take up exogenous hydrogen. *V. radiata* nodules formed with strains CB756 and 32HI continued to evolve hydrogen at a high rate in the presence of 1% exogenous hydrogen.

TABLE 2.5. Hydrogen evolution by *Vigna* nodules inoculated with different rhizobia, μmol H_2 g fresh wt. nodule^{-1} h^{-1}. (Unpublished data of Gibson & Sprent.)

Rhizobial strain	Host species *angularis*	*radiata*	*mungo*	*umbellata*
61B9	1.07	0.10	0.50	1.44
CB756	1.04	5.78	0.58	1.59
32H1	0.87	8.63	1.47	2.31

Clearly, we have here an effect of the host on the expression of the bacterial genotype.

(b) *Viability and detergent sensitivity*. There have been conflicting reports in the literature about whether or not rhizobia retain their viability after they have differentiated into active nitrogen-fixing bacteroids. Sutton & Peterson (1980) tested a wide range of legume/*Rhizobium* pairs to test whether bacteroids could be cultured after treatment with mild detergents. They concluded that the host genotype controlled detergent sensitivity. Some of their data, regrouped according to the tribal classification of Polhill (1981) are given in Table 2.6. There is a clear correlation between detergent sensitivity and tribe. In addition, those species which bear nodules of determinate

TABLE 2.6. Effect of plant host on detergent sensitivity of bacteroids.* (After Sutton & Petersen 1980.)

Subtribe of Papilionoideae	Species	Detergent sensitivity
Trifolieae	*Medicago sativa*	high
	M. scutella	high
	M. trunculata	high
	Trifolium repens	high
	T. resupinatum	high
	T. subterraneum	high
Vicieae	*Lathyrus japonicus*	moderate
	Pisum sativum	high
	Vicia dasycarpa	high
Galegeae	*Clianthus puniceus*	moderate
Loteae	*Lotus tenuis*	moderate
	L. schoelleri	moderate
	L. corniculatus	slight
	L. pedunculatus	slight
	L. angustissimus	low
Coronilleae	*Onobrychis vicii folia*†	moderate
	Ornithopus sativus	low
Phaseoleae	*Glycine max*	low
	Macroptilium atrupurpureum	low
	Phaseolus vulgaris	slight
	Phaseolus aureus‡	low

* Only those species included by Polhill (1981) in the tribes shown in Fig. 2.1 are included.

† Possibly in a separate tribe, Hedysareae (Polhill, pers. comm.) indeterminate nodules.

‡ Often called *Vigna radiata*, ssp. *radiata* or *V. radiata* var. *aureus*.

growth, with the exception of two *Lotus* spp. show low or slight detergent sensitivity. Sutton & Peterson (1980) interpret their data as showing that the host affects the bacterial cell wall–cell membrane complex. When this is modified during maturation, cells become sensitive to detergent as well as less viable.

A similar inference has been drawn by Sen & Weaver (1980). These workers noted that nitrogen-fixing activity of strain 32HI was much greater in peanuts (*Arachis hypogea*) than cowpea (*Vigna unguiculata*). Since bacteroids in peanuts are essentially reduced to spheroplasts (i.e. lacking cell walls), whereas in cowpea they are rod-shaped with a defined wall, it was suggested that peanut bacteroids are more permeable and hence could exchange metabolites more freely with the host cells. Sutton & Peterson (1980) found peanut bacteroids to have high detergent sensitivity, in agreement with the interpretation of Sen & Weaver (1980). At present, it is not possible to relate these observations to exact metabolic differences, but again they suggest the evolution of different mechanisms in different legume tribes.

Interactions between host and bacteroids

It has long been known that some host–rhizobial cultivars fix more nitrogen than others and that the amount of nitrogen fixed is related to environmental factors such as light, temperature and water supply (Sprent 1979). However, how these factors act at the host–bacterial interface is by no means clear. In order to fix nitrogen optimally a bacteroid needs the following:

1. Energy in the form of ATP, derived from oxidative phosphorylation using a high flux/low concentration of oxygen supplied by leghaemoglobin. Carbon substrates from the host cytosol are used.
2. An energized membrane (Laane *et al.* 1980). This is probably maintained by a membrane potential and a pH gradient, with the former being regulatory.
3. A supply of low redox potential reductant to reduce nitrogen to ammonia. The membrane potential discussed in (2) probably regulates the reductant supply. Once ammonia is produced it is handed on to the host for assimilation as we have seen.

The substrates generally found to most actively support nitrogen fixation by isolated bacteroids are organic acids, particularly succinic acid; this ultimately derives from host photosynthesis. Thus it is not surprising that legumes are uncompetitive under shade conditions (Sprent & Silvester 1973).

At the cellular level, factors limiting nitrogen fixation may vary with host–rhizobial combination. For example, in soybeans nitrogen fixation is very sensitive to ATP/ADP ratio. Switching on and off the oxygen supply affects nitrogenase activity in this way (Ching 1976). It appears that reductant supply is not limiting, as large quantities of the highly reduced polymer poly-β-hydroxybutyrate (PHB) accumulate in bacteroids (Wong & Evans 1971). *Vigna* spp., which are closely related to soybean do not accumulate PHB in their bacteroids (Gibson, Turner & Sprent, unpublished). Perhaps here the metabolisms of host and *Rhizobium* are better balanced and it may not be a coincidence that specific nitrogenase activities in *Vigna* are generally much higher than in soybean.

Bacteroid respiration is clearly crucial to nitrogen-fixing activity and as mentioned above, a low level/high flux oxygen system is used. The fact that host cells have low oxygen as well may switch their metabolism to a more anaerobic type leading to the production of the organic acids which the bacteroids prefer (de Vries, Veld & Kijne 1980). Both bacterial and host respiration produce carbon dioxide. Emerich *et al.* (1980) have shown that carbon dioxide and also organic acids such as succinate can inhibit hydrogen uptake by soybean bacteroids. Thus when respiration is low, hydrogen uptake may be stimulated. This could lead to production of ATP to offset any lack of energy from respiratory sources. Whether or not high concentrations of soil carbon dioxide affect this process is not known. The whole question is complicated by the fact that nodule host cells may incorporate carbon dioxide into organic acids used as acceptor molecules for fixed nitrogen (see Chapter 11). Some of these relationships are summarized in Fig. 2.3.

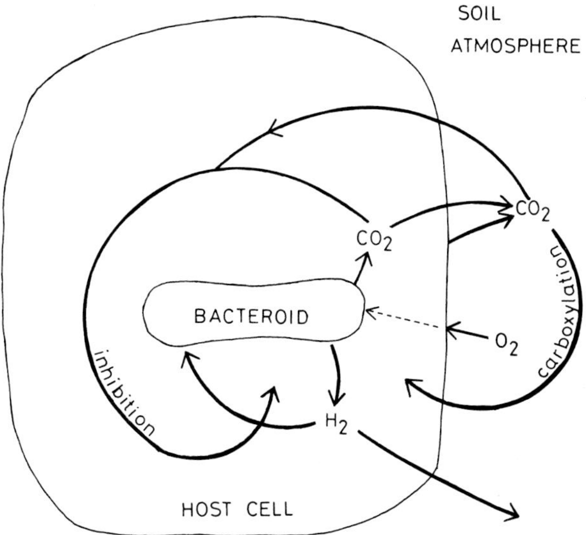

FIG. 2.3. Summary of some of the interrelationships between O_2, CO_2 and H_2 in nodule cells and soil atmosphere.

Carbon dioxide inhibition of uptake hydrogenase could be involved in the *V. radiata* nodules discussed above (see Table 2.5). Gibson, Sprent & Turner (unpublished) were unable to detect any differences in nitrogen fixation (^{15}N incorporation) between *Vigna* nodules with and without an uptake hydrogenase, in contrast to the results of Zablotowicz *et al.* (1980) for soybean. It may be that respiratory supplies of ATP were adequate in *Vigna* but not in soybean.

SIGNIFICANCE OF LEGUMES IN ECOSYSTEMS

Legumes fix nitrogen to give them a competitive advantage in nutrient-poor conditions. Obviously, the more efficient they are at doing this under a particular set of conditions the more competitive they will be. The actual strategies employed may vary considerably as we have seen. We are accustomed to considering the plants' photosynthetic system in ecological terms, C_3 *v* C_4 *v* CAM. The information is gradually becoming available to enable us to think in parallel terms of nitrogen fixation—for example, amide or ureide production. The coevolution of legume host and *Rhizobium* has adapted to different environmental conditions and this may even extend to tolerance of specific soil minerals such as nickel and magnesium (Pegtel 1980).

Nitrogen-poor soils in nature normally occur in areas being newly colonized, such as sand dunes, volcanic lava or following man-made depletion such as in agriculture and after mining activity. Legumes play a positive role in developing ecosystems and in grasslands where their specific location may be closely related to that of complementary species, especially grasses, as Turkington and co-workers have elegantly shown (Turkington 1979). They are comparatively rare as dominants in climax systems, especially forests. It is noteworthy that most of the tree legumes belong to the sub-family Caesalpinioideae which is nodulated only sporadically. Climax systems are usually considered stable, and little nitrogen is thought to be lost from them. Little nitrogen is needed to replace this and the input from sources such as rainfall may be sufficient. Most is recycled both within plants (see Sprent, Scott & Perry (1978) for non-legume *Myrica gale*) and within communities.

Nitrogen fixed by legumes only becomes available to other plants when some or all of the plant dies. The evidence for nitrogen excretion from nodules is scanty and largely circumstantial. There is a continuous turnover of root and nodule material in soil, and combined nitrogen from this source may become available to associated species. This turnover may be aggravated by adverse conditions such as drought (Wilson 1931) or as a result of grazing or defoliation by insects, man or other agents. Fallen leaves are an obvious source. If a system develops fully to mature woodland, shading will generally kill the nitrogen-fixing legumes. Nitrogen released in this way can be appreciable and may be exploited in forestry (Sprent & Silvester 1973). As soil nitrogen builds up, legumes lose their competitive advantage and even if they persist, they fix little nitrogen.

ACKNOWLEDGEMENTS

I should like to thank Miss Shona McInroy for technical assistance, Professor J. A. Raven for many useful discussions, Dr R. D. Summerfield (Reading) for *Cajanus cajan* seed, Mr H. Taylor (SCRI) for *Phaseolus vulgaris* seed and the curator of the Rothamsted Culture Collection for rhizobial strains.

REFERENCES

Beringer J.E., Brewin N.J. & Johnston A.W.B. (1980) The genetic analysis of *Rhizobium* in relation to symbiotic nitrogen fixation. *Heredity*, **45**, 161–86.

Ching T.M. (1976) Regulation of nitrogenase activity in soybean nodules by ATP and energy charge. *Life Science*, **18**, 1071–6.

de Vries G.E., Veld P. in 'T & Kijne J.W. (1980) Production of organic acids in *Pisum sativum* root nodules as a result of oxygen stress. *Plant Science Letters*, **20,** 115–23.

Emerich D.W., Ruiz-Argüeso T., Russell S.A. & Evans H.J. (1980) Investigation of the H_2 oxidation system in *Rhizobium japonicum* 122 DES bacteroids. *Plant Physiology*, **66**, 1061–6.

Fosse R., de Grave P. & Thomas P.E. (1933) Role de l'acide allantoique chez les végétaux supérieurs. *Comptes Rendus Hebdomadaires des Séances de l'Académie des Sciences, Paris*, **196**, 883–6.

Isely D. & Polhill R. (1980) Leguminosae sub-family Papilionoideae. *Taxon*, **29**, 105–20.

Israel D.W. & Jackson W.A. (1981) Ion balance, uptake, and transport processes in N_2-fixing and nitrate- and urea-dependent soybean plants. *Plant Physiology*, **69,** 171–8.

Laane C., Krone W., Konings W.N., Haaker H. & Veeger C. (1980) The involvement of the membrane potential in nitrogen fixation by bacteroids of *Rhizobium leguminosarum*. *FEBS Letters*, **103**, 328–31.

McClure P.R., Israel D.W. & Volk R.J. (1980) Evaluation of the relative ureide content of xylem sap as an indicator of N_2 fixation in soybeans. *Plant Physiology*, **66**, 720–5.

Pate J.S., Atkins C.A. & Rainbird R.M. (1981) Theoretical and experimental costing of nitrogen fixation and related processes in nodules of legumes. *Proceedings of 4th International Congress on N_2 Fixation*, Canberra, Australia.

Pate J.S., Atkins C.A., White S.T., Rainbird R.M. & Woo K.C. (1980) Nitrogen nutrition and xylem transport of nitrogen in ureide-producing legumes. *Plant Physiology*, **65**, 961–5.

Pegtel D.M. (1980) Evidence for ecotypic differentiation in *Lupinus*-associated *Rhizobium*. *Acta Botanika Neerlanda*, **29**, 429–41.

Polhill R. (1981) Papilionoideae. In *Recent Advances in Legume Systematics* (eds R. M. Polhill & P. H. Raven). pp. 191–208. Royal Botanic Gardens, Kew, Surrey.

Schubert K.R. & Evans H.J. (1976) Hydrogen evolution: a major factor affecting the efficiency of nitrogen fixation in nodulated symbionts. *Proceedings of the National Academy of Sciences, USA*, **73**, 1207–11.

Sen P. & Weaver R.W. (1980) Nitrogen-fixing activity of rhizobial strain 32HI in peanut and cowpea nodules. *Plant Science Letters*, **18**, 315–18.

Sprent J.I. (1979) *The Biology of Nitrogen Fixing Organisms*. McGraw-Hill, London.

Sprent J.I. (1980) Root nodule anatomy, type of export product and evolutionary origin in some Leguminosae. *Plant Cell and Environment*, **3**, 35–43.

Sprent J.I. & Silvester W.B. (1973) Nitrogen fixation by *Lupinus arboreus* grown in the open and under different aged stands of *Pinus radiata*. *New Phytologist*, **72**, 991–1003.

Sprent J.I., Scott R. & Perry K.M. (1978) The nitrogen economy of *Myrica gale* in the field. *Journal of Ecology*, **66**, 409–20.

Sutton W.D. & Peterson A.D. (1980) Effects of plant host on the detergent sensitivity and viability of *Rhizobium* bacteroids. *Planta*, **148**, 287–92.

Triplett E.W., Heitholt J.J., Evenson K.B. & Blevins D.E. (1981) Increase in internode length of *Phaseolus lunatus* L. caused by inoculation with a nitrate reductase-deficient strain of *Rhizobium* sp. *Plant Physiology*, **67**, 1–4.

Turkington R. (1979) Neighbour relationships in grass–legume communities. IV. Fine scale biotic differentiation. *Canadian Journal of Botany*, **57**, 2711–16.

van Berkum P. & Bohlool B.S. (1980) Evaluation of nitrogen fixation by bacteria in association with tropical grasses. *Microbiological Reviews*, **44**, 491–517.

Wilson J.K. (1931) Shedding of nodules by beans. *Journal of the American Society for Agronomy*, **23**, 670–4.

Wong P.P. & Evans H.J. (1971) Poly-β-hydroxybutyrate utilization of soybean (*Glycine max* Merr.) nodules and assessment of its role in maintenance of nitrogenase activity. *Plant Physiology*, **47**, 750–5.

Zablotowicz R.M., Russell S.A. & Evans H.J. (1980) Effect of hydrogenase system in *Rhizobium japonicum* on the nitrogen fixation and growth of soybeans at different stages of development. *Agronomy Journal*, **72**, 555–9.

3. THE ROLE OF LICHENS IN THE NITROGEN ECONOMY OF SUBARCTIC WOODLANDS: NITROGEN LOSS FROM THE NITROGEN-FIXING LICHEN *STEREOCAULON PASCHALE* DURING RAINFALL

P. D. CRITTENDEN

Department of Botany

*University of Sheffield, Sheffield S10 2TN**

SUMMARY

Concentrations of ammonium-N, organic-N and potassium in rainfall were monitored before and after its passage through monospecific mats of the nitrogen-fixing lichen *Stereocaulon paschale* (L.) Hoffm. and the non-nitrogen-fixing lichen *Cladonia stellaris* (Opiz) Pouzar & Vězda. The sampling was conducted during natural rainfall events in lichen-rich birch woodland near Kevo in Finnish Lapland. During rainfall, both species absorbed ammonium-N from rain and released organic-N. On a unit area basis losses of organic-N from mats of *S. paschale* were up to 6.5 times greater than those from mats of *C. stellaris*: this factor was 2.1 on a dry weight basis and *c.* 0.7 when organic-N released was related to unit weight of thallus nitrogen. Losses of organic-N from *S. paschale* were maximal following the resumption of rainfall after rain-free intervals during which the lichen mats remained moist, and they were not associated with concurrent losses of potassium. The total cumulative ammonium-N and organic-N contents of the rain collected during the seven consecutive rainfall events studied were 5.8 ± 0.2 and 2.8 ± 0.2 (total 8.6 ± 0.8) mg m^{-2} respectively; the corresponding values for lichen throughfall were 0.87 ± 0.04 and 17.2 ± 0.6 (total 18.0 ± 0.6) mg m^{-2} for *S. paschale*, and 1.1 ± 0.1 and 5.2 ± 0.1 (total 6.4 ± 0.2) mg m^{-2} for *C. stellaris*. The possible form of the organic-N lost from *S. paschale*, the possible mechanism by which it is released and the ecological significance of the process are discussed.

* Present address: Department of Botany, University of Nottingham, Nottingham NG7 2RD.

INTRODUCTION

Three lines of evidence suggest that nitrogen-fixing lichens make a significant contribution to the nitrogen economy of boreal-arctic ecosystems. First, lichen species containing blue-green algae are ubiquitous and locally abundant both in the ground vegetation of northern boreal forests that develop on well-drained sites and in tundra communities on drier terrain with at least a moderate winter snow cover. Secondly, under suitable environmental conditions rates of nitrogenase activity that can be developed by cyanophilic lichens are high in comparison with rates of activity associated with other potential sites of nitrogen fixation such as soils and the phyllosphere. Thirdly, the availability of soil nitrogen is frequently a key factor ultimately limiting plant production in cold regions (Tamm 1964; Warren Wilson 1966; Haag 1974; Tieszen 1978).

The nitrogen-fixing lichen *Stereocaulon paschale* (L.) Hoffm. occurs in abundance in lichen woodlands throughout the continental regions of the northern boreal zone. Lichen woodlands (Dry Series of Hustich 1949) are seral communities in the post-fire recovery of vegetation on well-drained sites and are typically composed of open tree stands with a ground cover dominated by lichens, usually species of *Cladonia*, *Cetraria* and *Stereocaulon*. *Stereocaulon*-rich woodlands are found throughout north central Canada (*Forest-Tundra* and *Northwestern Transition* zones of Rowe 1972; Kershaw 1977), western and central Siberia and the forest-tundra ecotone of the USSR (Lavrenko & Sochava 1956; Karev 1961), and the continental subalpine regions of northern Fennoscandia (Hämet-Ahti 1963) (Plate 3.1a & b). Maikawa & Kershaw (1976) estimated that in the black spruce (*Picea mariana*) woodlands on drier terrain in the Abitau-Dunvegan Lakes region of the Canadian Northwest Terrorities *S. paschale* achieved a mean cover of 22% and frequently occurred as uninterrupted ground cover (Plate 3.1c). In more mature close-stand forests on well-drained land lichens form only a minor component of the ground vegetation which is typically dominated by feather mosses (Moist Series of Hustich 1949). Nevertheless, the cyanophilic lichens *Peltigera* spp. (e.g. *P. aphthosa*, *P. canina*, *P. polydactyla*, *P. malacea*, *P. scabrosa*) and *Nephroma arcticum* are found consistently in these communities, usually with a mean cover of less than 10% (Moss 1953; Lavrenko & Sochava 1956; Ritchie 1956; Hämet-Ahti 1963; Damman 1964; Payette & Filion 1975; Payette, Samson & Lagarec 1976; Billington & Alexander 1978; Dyrness & Grigal 1979; Orlóci & Stanek 1979).

The environmental control of nitrogenase activity in lichens has received considerable attention and is comparatively well understood. In contrast there are few experimental data concerning the movement of nitrogen subsequent to fixation in the lichen thallus; little is known, for example, about decomposition rates of lichens (Franklin 1974). Hitch & Stewart (1973) suggested that, in addition to the mobilization of nitrogen during decomposition of thalli, viable lichens might liberate extracellular nitrogenous compounds similar to those produced by free-living blue-green algae. This opinion was clearly shared by Pike *et al.* (1972) and Denison (1973) who made the presumption that the nitrogen content of rainfall is enriched during flow over epiphytic nitrogen-fixing lichens. Evidence in support of these views has come from Farrar & Smith (1976) who showed that when air-dry *Hypogymnia physodes* is submerged in water appreciable quantities of soluble metabolites are liberated from the thallus into the bathing solution within several minutes, the rate of subsequent release declining rapidly. Similar results have since been obtained with a range of lichen species (Lang, Reiners & Heier 1976; Buck & Brown 1979). The results of recent studies with cyanophilic lichens have demonstrated that, indeed, nitrogenous compounds can be leached from apparently healthy lichen thalli by a variety of washing procedures, the release being greatest when dry thalli are rewetted (Millbank 1978; Pike 1978; Buck & Brown 1979). Simon (1974) has interpreted this rewetting phenomenon in terms of differences in cell membrane conformation in dry and wet tissues. He suggested that, on rapid reintroduction to excess water, the porous conformation of dehydrated cell membranes will permit soluble metabolites to leak out of dry cells, the leakage being arrested when the impermeable bilayer structure of hydrated membranes is re-established.

A corollary to these observations is that in nature soluble forms of nitrogen might be leached from nitrogen-fixing lichens when they are rewetted by rainfall that follows a period of dry weather. However, a shortcoming of rewetting experiments conducted in the laboratory is that the pretreatments to which the experimental material is exposed and the washing procedures adopted fail to simulate adequately many characteristics of field conditions during the onset of rainfall. Factors such as increases in relative humidity that often precede rain and the variable intensity of rainfall may influence the extent of nutrient release by lichens following rewetting.

The objective of the present work was to quantify the leakage of nitrogen from *S. paschale* under field conditions during natural rainfall events. Fluxes of ammonium-N, total Kjeldahl-N (and organic-N by difference) and potassium from *S. paschale* mats were monitored during rainfall and compared with those from mats of the non-nitrogen-fixing lichen *Cladonia stellaris* (Opiz) Pouzar & Vězda (= *C. alpestris* (L.) Nyl.). The fieldwork was conducted during the summer of 1980 in lichen-rich birch (*Betula pubescens*) woodland 1 km south of the University of Turku's Subarctic Research Station at Kevo in Finnish Lapland (69° 45′ N, 27° 00′ E).

MATERIALS AND METHODS

Collection of water samples

Stereocaulon paschale was gathered from the ground cover around the study site, either in the moist or the air-dry state, and the decaying or non-photosynthetic basal parts of individual pseudopodetia were broken off and discarded. The pruned samples, which usually consisted of the upper 20–35 mm of pseudopodetia, were arranged on circular grilles of stainless steel mesh (260 mm diameter, 3 × 3 mm mesh) to form a single-layered, closed lichen canopy on each. The thalli in these reconstructed lichen mats (*c*. 860 individuals) were supported in upright positions comparable with the growth habit of *S. paschale in situ*. The grilles were then superimposed on specially constructed funnels which were supported over trenches in the ground with the mouth of each funnel (250 mm i.d.) at ground level (Fig. 3.1). The collecting surfaces of the funnels were constructed from polyester film (0.15 mm gauge), a material

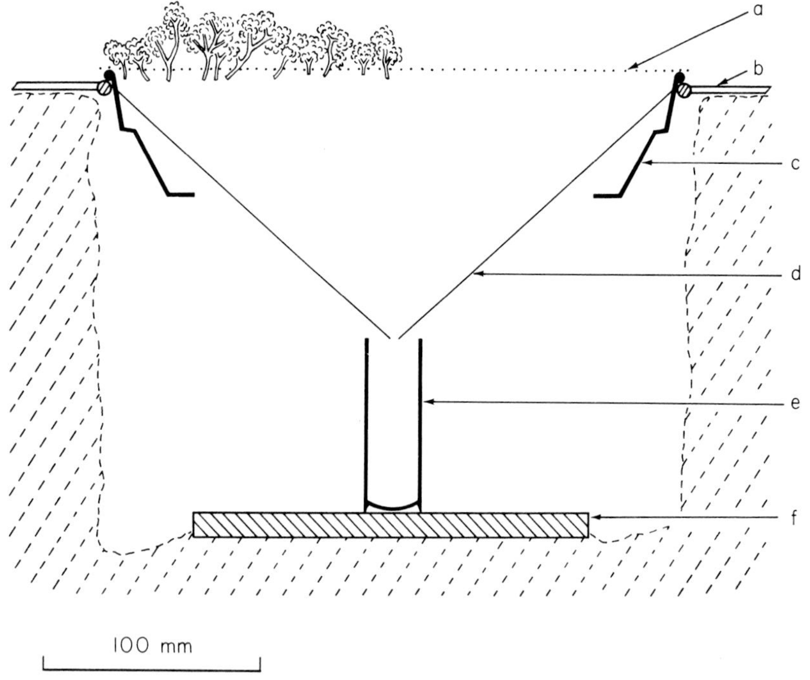

FIG. 3.1. Arrangement of collecting funnels supported above trenches in the ground. (a) Stainless steel mesh (partially covered by *S. paschale* in diagram), (b) support rods, (c) plastic supporting frame, (d) polyester film, (e) sample tube, 30 ml capacity, (f) wooden platform.

(a)

(b)

(c)

PLATE 3.1. *Stereocaulon paschale*-rich subarctic plant communities.

(a) *S. paschale*-rich dwarf birch (*B. nana*) heath, Finnmarksvidda, northern Norway (69° 30′ N, 24° 30′ E).

(b) Birch (*Betula pubescens*)-lichen woodland, Finnmarksvidda, northern Norway (69° 30′ N, 24° 30′ E) with *S. paschale* as a principal component of the ground cover.

(c) Black spruce (*Picea mariana*)-*Stereocaulon* woodland in the Abitau-Dunvegan Lakes region of the Northwest Territories, Canada (60° 21′ N, 106° 54′ W); *S. paschale* forms an almost uninterrupted ground cover.

found not to adsorb or release the elements under investigation. *Cladonia stellaris* was always collected in the wet state since the podetia are particularly brittle when dry, but otherwise this species (*c.* 225 individuals per funnel) was treated in a similar manner to *S. paschale*. Control funnels were covered with stainless steel grilles only. At the end of the study period the mean* dry weight of lichen per funnel was 37.7 ± 0.2 and 12.1 ± 0.2 g ($n = 5$) for *S. paschale* and *C. stellaris* respectively.

During natural rainfall events effluent from each of the funnels was collected in a series of 30 ml capacity polystyrene sample tubes. The times at which tubes were changed were noted and charged tubes were sealed with polyethylene screw caps. The major rainfall events between 19th June and 6th August were sampled quantitatively (Fig. 3.2). During dry periods the collecting surfaces of the funnels were washed with distilled water at 3–4 day intervals to remove dry deposits such as pollen grains.

Growth measurements

Growth of the lichen supported over the funnels was estimated by placing tagged and preweighed thalli in the canopies. The marked thalli were placed in position on 26th June and reweighed on 3rd August at the end of the experiments, thallus water content being estimated on both dates by measuring the oven-dry weight of 'dummy' lichen samples.

Pretreatment and storage of water samples

At the end of each rainfall event the sample tubes were immediately returned to the laboratory and weighed. Each water sample was filtered through a cellulose acetate membrane filter (0·2 μm pore size) that had been pretreated by washing in three changes of deionized water (1 hour each wash) to remove readily leachable nitrogen compounds (see Appendix, p. 64). The filtrate was collected in a second polystyrene tube containing *c.* 0.6 mg pentachlorophenol coated on the internal walls and the pH of the water lowered to *c.* 3.5 by the addition of 0.1 N H_2SO_4. The sample tubes were stored in the dark at 5–8°C until analysis with the exception of a four day shipment period. Storage times prior to analysis ranged between 40 and 100 days. Pentachlorophenol was included for its antibiotic properties (Bevenue & Beckman 1967): it did not interfere with subsequent chemical analyses. Dilute aqueous solutions, both natural waters and laboratory preparations, if pretreated in the manner outlined above, can be stored for at least 100 days without significant changes in concentration of total Kjeldahl-N, ammonium-N or potassium (Crittenden, unpublished data).

* Unless stated otherwise, mean values are cited with ±1 standard error.

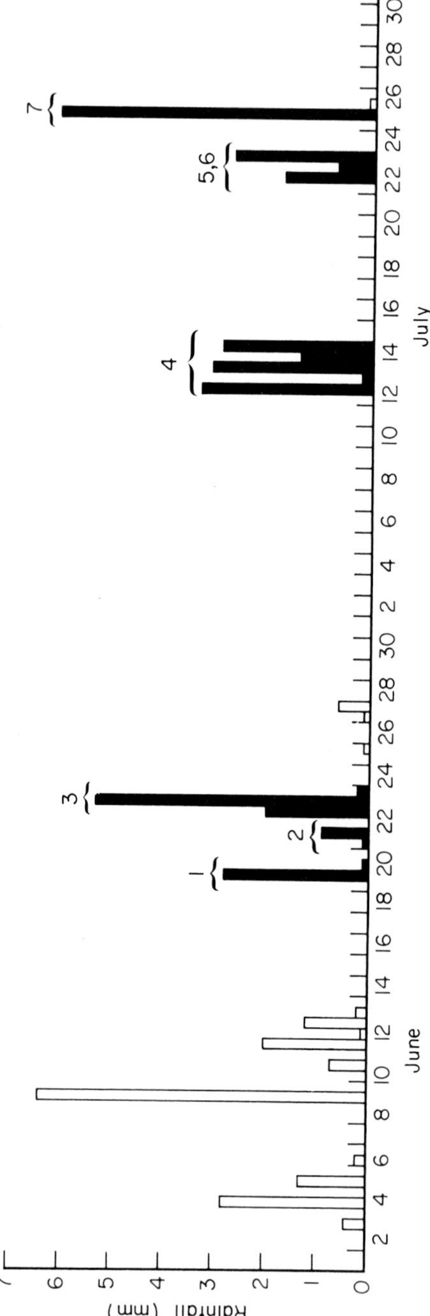

FIG. 3.2. Rainfall recorded at the Kevo Meteorological Station during 12-hour periods between 1st June and 6th August 1980. Filled columns with numbers indicate rainfall events sampled quantitatively at the research site.

Chemical analyses

Ammonium-N and Kjeldahl-N. Ammonium-N in water samples was determined, both before and after digestion, using a modification of the indo-phenol blue method described by Allen (1974). Analyses were performed in glass test-tubes with screw caps and silicon rubber liners. Micro-Kjeldahl digestions were performed to convert organic-N to ammonium-N as follows. Rainfall or throughfall (1–5 ml) was placed in a test-tube followed by 1 ml 4 N H_2SO_4 containing 2.66 g l^{-1} sodium selenate (modified after Johnson 1941). The tube was transferred to an aluminium heating block and the solution warmed under a stream of warm ammonia-free air until all the water had evaporated. The air stream was then removed, the tube covered with a loose glass cap and the temperature raised to 280°C for *c*. 6 h. The subsequent colour development was performed in the digestion tube. By use of this procedure the recovery of nitrogen from standard solutions of tryptophan and bovine serum albumin was found to be quantitative.

Total Kjeldahl-N in lichens was determined by the method of Bremner (1965).

Potassium. Potassium was determined by atomic absorption spectrophotometry allowing for interference from sodium.

Analysis of data

The computer programs of Hunt & Parsons (1974) and Parsons & Hunt (1981) were used to analyse the data. These programs respectively fit simple curves or splined series of curves to the natural logarithms of sequential determinations of variates, 95% confidence limits being calculated for each of the resultant fitted values. In the present work fitted values and 95% confidence intervals have been back-transformed from logarithms.

RESULTS

Seven rainfall events were studied with a total delivery of 36 mm and on all occasions the lichen canopies were air-dry before precipitation commenced. The designation of a period of rainfall as a single event was arbitrary since Events 3 and 4 were periods of intermittent rainfall with rain-free intervals of up to 14 hours. For each individual rainfall event there was a good agreement between cumulative rainfall collected in fractions at the study site and quantities recorded in a standard rain gauge at the Kevo Meteorological Station 1 km to the north (Fig. 3.3).

Figure 3.4 records data collected during the course of Rainfall Event 7.

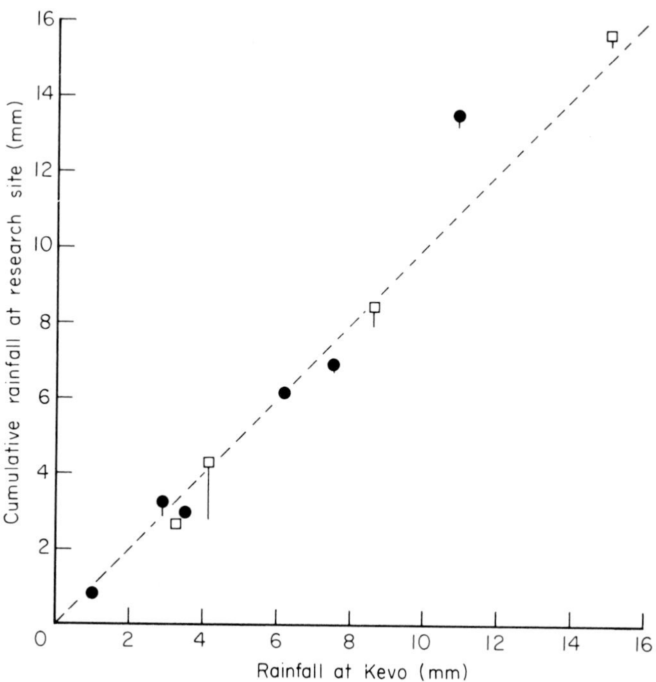

FIG. 3.3. Relationship between rainfall recorded at the Kevo Meteorological Station in a standard rain gauge and rainfall collected in fractions in five funnels at the research site during the present work, 1980 (●), and during a preliminary investigation, 1978 (□). Maximum values are plotted together with the range of values.

Throughfall from *C. stellaris* and *S. paschale* did not occur until about four and five hours respectively after the onset of rainfall (i.e. after *c.* 0.8 and *c.* 3.2 mm rainfall). Concentrations of nitrogen and potassium in both rainfall and lichen throughfall decreased with time. Concentrations of total Kjeldahl-N were consistently higher in the throughfall from *S. paschale* than in the other waters collected, and levels in the throughfall from *C. stellaris* were initially higher than in rainfall but the latter difference diminished rapidly and eventually disappeared (Fig. 3.4b). These trends were reflected in the progress of cumulative Kjeldahl-N collected in each of the funnels, there being a net loss of 111 ± 4 μg N from the *S. paschale* canopies compared with 9 ± 1 μg N per funnel from *C. stellaris* (Fig. 3.4c). Concentrations of ammonium-N were at first lower in the throughfall from both species than in the rainfall (Fig. 3.4d) and, although the differences later disappeared, this resulted in appreciable net gains in ammonium-N by both species (Fig. 3.4e). The nitrogen leached from the lichens was apparently in organic form (Fig. 3.4f & g)

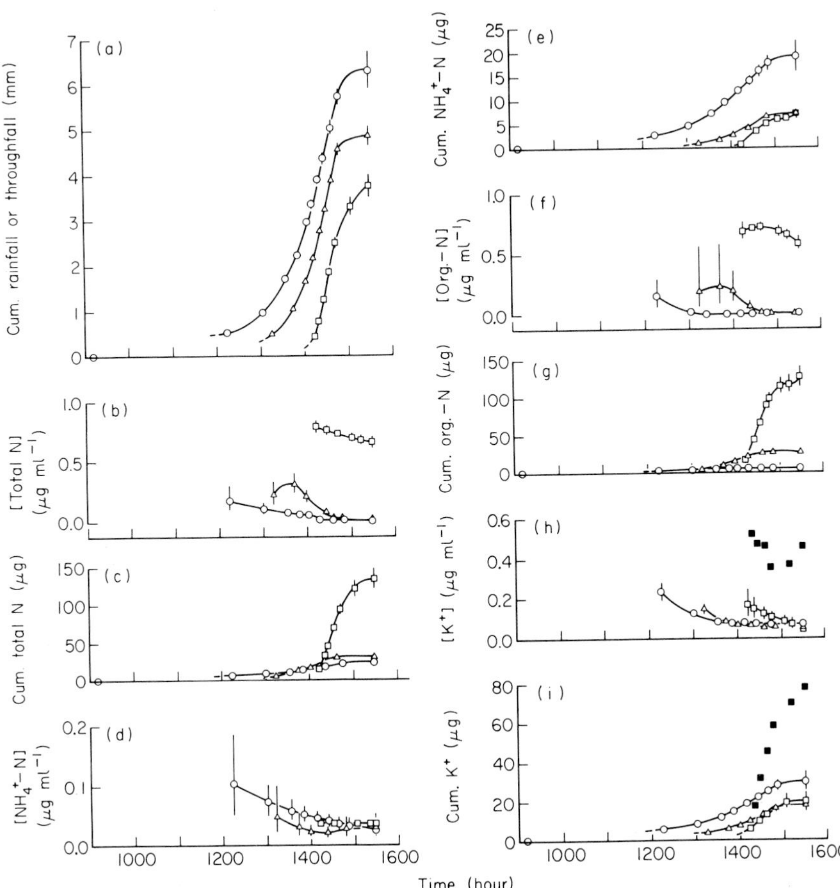

FIG. 3.4. Progress curves in Rainfall Event 7 (25th July) of (a) cumulative rainfall and throughfall, (b) concentration of total Kjeldahl-N, (c) cumulative total Kjeldahl-N, (d) ammonium-N concentration, (e) cumulative ammonium-N. (f) organic-N concentration, (g) cumulative organic-N, (h) potassium concentration, and (i) cumulative potassium. (○) Rainfall (control) (n = 4); (Δ) *Cladonia* throughfall (n = 4); (□) *Stereocaulon* throughfall (n = 5); (■) anomalous data for throughfall from one of the *Stereocaulon* canopies. For the sake of clarity 95% confidence limits are plotted for selected fitted values only. Cum. = cumulative, [ ] = concentration, and org. = organic. Air temperature at ground level ranged between 14–18°C during the collection period.

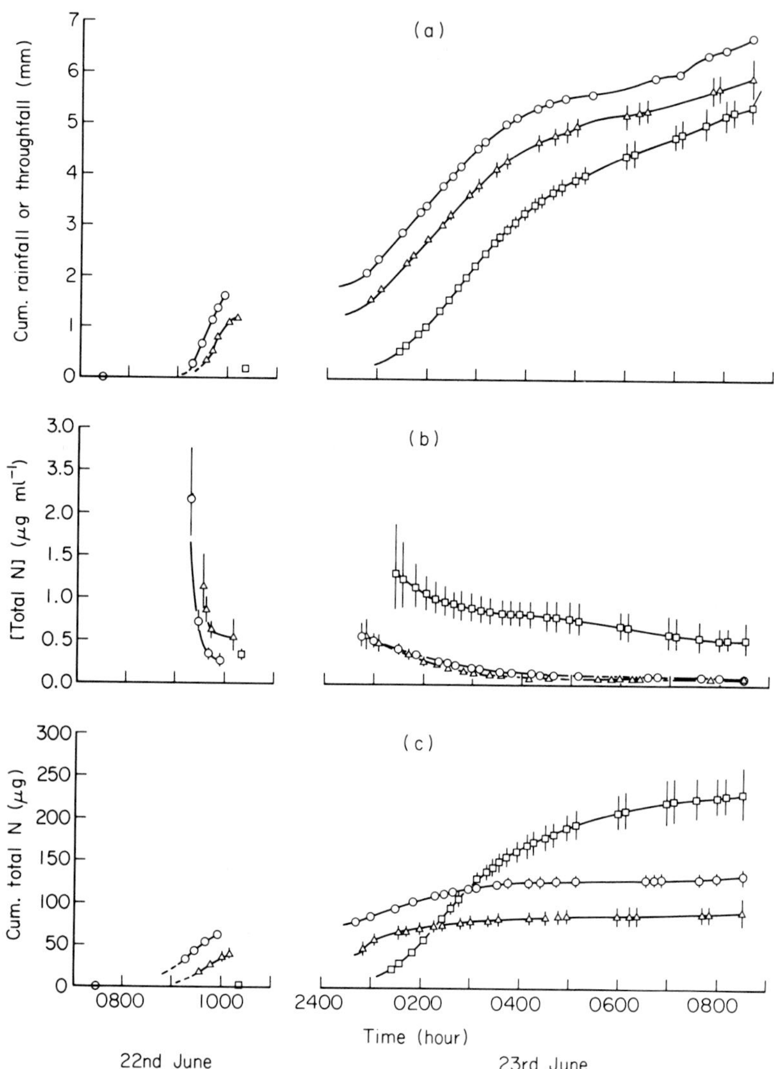

FIG. 3.5. Selected data from Rainfall Event 3. (a)–(g) as in Fig. 3.4. (○) Rainfall (control) ($n = 5$); (△) *Cladonia* throughfall ($n = 3$); (□) *Stereocaulon* throughfall ($n = 5$). For the sake of clarity 95% confidence limits are plotted for selected fitted values only. Cum. = cumulative, [ ] = concentration, and org. = organic. Air temperature at ground level ranged between 6–10°C during the collection period.

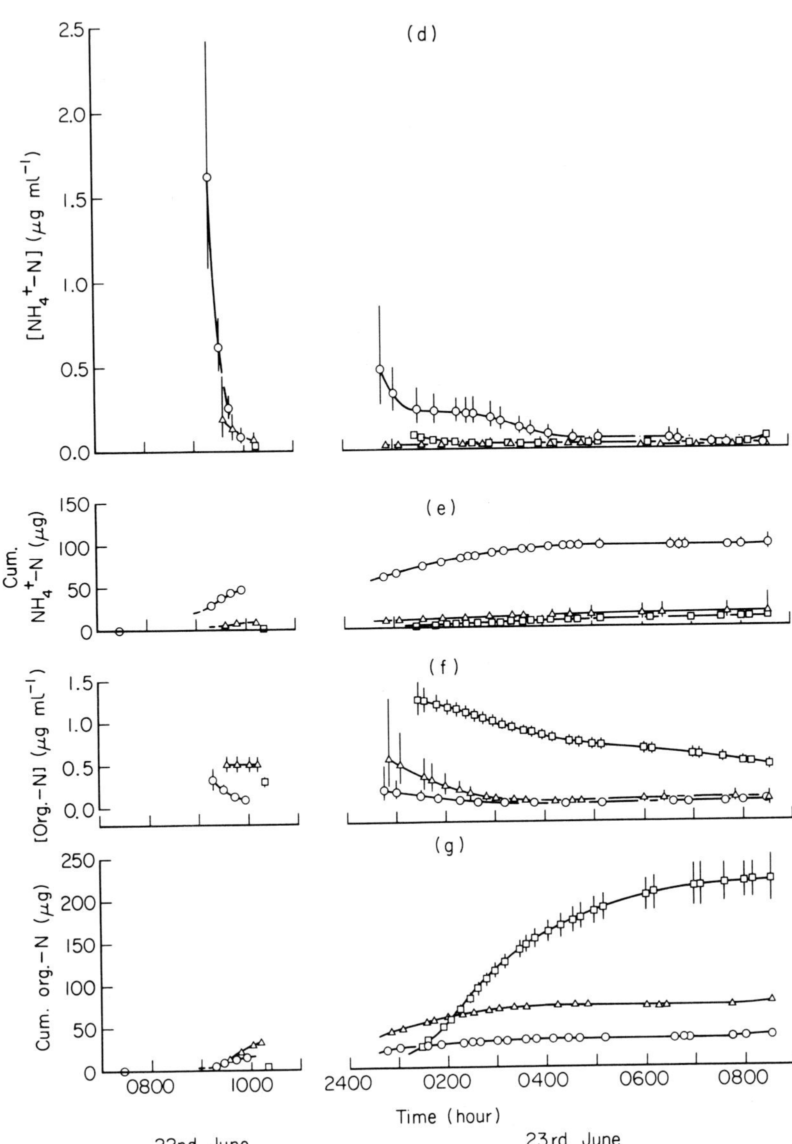
(d)
2.5
2.0
1.5
1.0
0.5
0.0
[NH4+-N] (μg ml−1)
(e)
150
100
50
0
Cum. NH4+-N (μg)
(f)
1.5
1.0
0.5
0.0
[Org.-N] (μg ml−1)
(g)
250
200
150
100
50
0
Cum. org.-N (μg)
0800
1000
2400
0200
0400
0600
0800
Time (hour)
22nd June
23rd June

FIG. 3.5. (*cont.*)

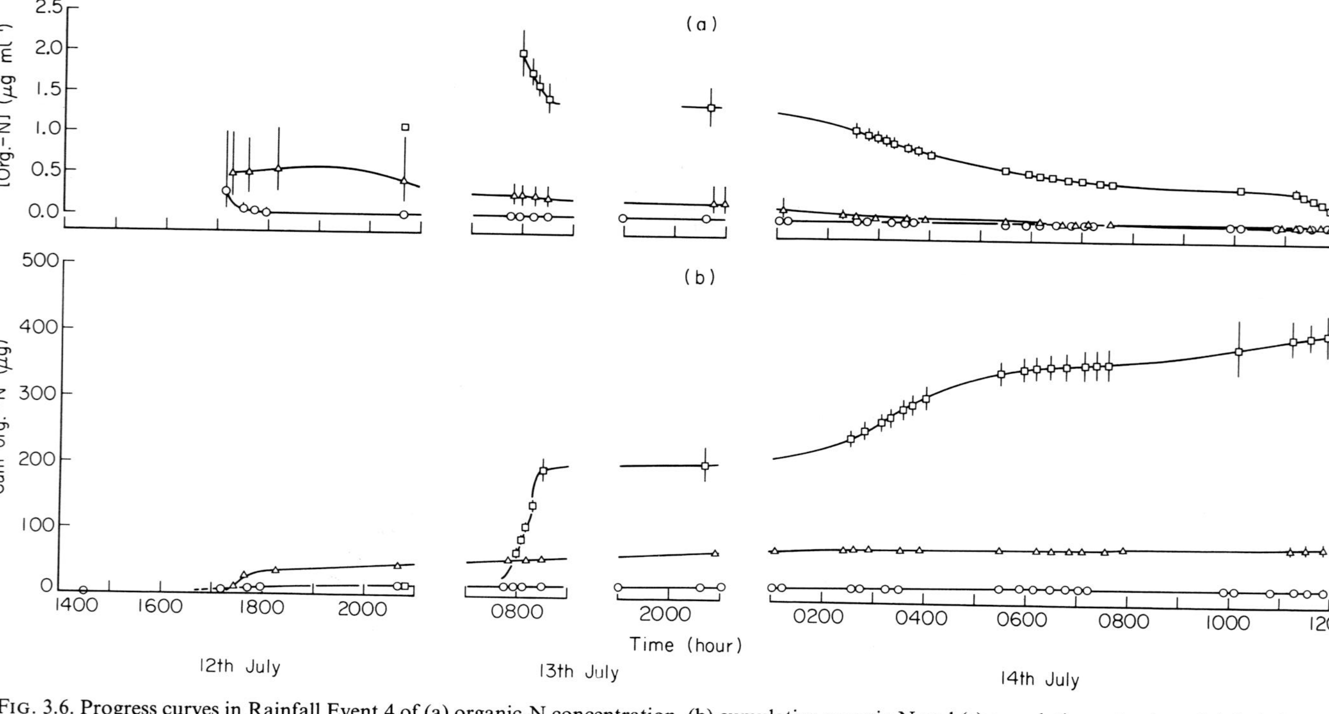

FIG. 3.6. Progress curves in Rainfall Event 4 of (a) organic-N concentration, (b) cumulative organic-N and (c) cumulative potassium. (○) Rainfall (control) (n = 5); (△) *Cladonia* throughfall (n = 5) and (□) *Stereocaulon* throughfall (n = 5); (■) anomalous data for throughfall from one of the *Stereocaulon* canopies. For the sake of clarity 95% confidence limits are plotted for selected fitted values only. Cum. = cumulative, [ ] = concentration, and org. = organic. Air temperature at ground level during the collection period ranged between 7–16°C.

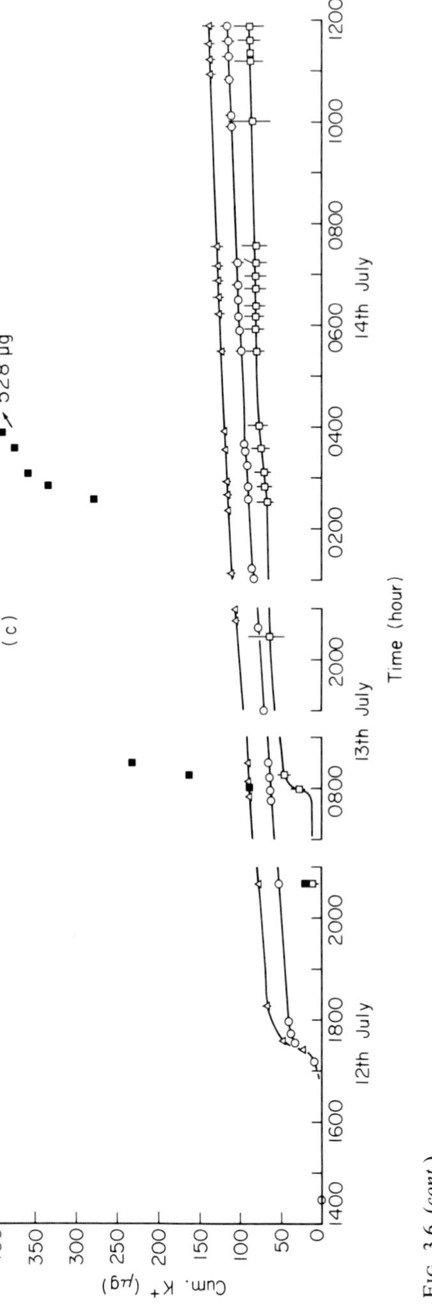
(c)
528 µg
Cum. K⁺ (µg)
400
350
300
250
200
150
100
50
0
1400
1600
1800
2000
0800
2000
0200
0400
0600
0800
1000
1200
12th July
13th July
14th July
Time (hour)

FIG. 3.6. *(cont.)*

whilst losses from the *C. stellaris* mats were less than 0.16 times those from the mats of *S. paschale* and these losses were largely offset by absorption of ammonium-N.

Differences in potassium concentration between lichen throughfall and rainfall were small with the exception of throughfall from one of the *Stereocaulon* canopies which consistently contained more potassium than other water fractions during all rainfall events after Event 2 (Fig. 3.4h). The anomalous result from this funnel was probably associated with the development of a bleached area *c.* 30 mm diameter in the lichen canopy shortly after the start of the study. Discoloured patches can be seen frequently in *S. paschale* mats *in situ*: these perhaps indicate damage of unknown cause. The potassium results from this collecting funnel have been ignored in subsequent calculations. The progress of cumulative potassium in the three treatments shows that, in general, the lichens achieved small gains in potassium during rainfall (Fig. 3.4i).

Selected data from Rainfall Events 3 and 4 appear in Figs 3.5 and 3.6. On these occasions rain fell for comparatively long periods but the trends in losses and gains of nitrogen and potassium are similar to those seen in Fig. 3.4. Concentrations of organic-N were consistently higher in the *Stereocaulon* throughfall than in the other waters collected but decreased during the course of the events. Evaporative water loss from the lichens occurred in the rain-free intervals but the lichens became air-dry only on one occasion (between 0900 h and 1800 h on 13th July). It is of interest that in both these rainfall events the concentrations of organic-N in *Stereocaulon* throughfall were higher after a rain-free interval in which the lichens remained moist than immediately after rewetting. In contrast, levels of organic-N in the *Cladonia* throughfall were always maximal in the first samples collected. The losses of organic-N from *S. paschale* and *C. stellaris* canopies were 176 ± 20 and 32 ± 7 μg per funnel respectively for Event 2 and 398 ± 22 and 61 ± 5 μg per funnel for Event 3.

The net cumulative gains and losses of nitrogen and potassium by the two species (mg m^{-2}) for each rainfall event are recorded in Fig. 3.7 and reveal the following:

1. Both species effectively scavenged ammonium-N from rain,
2. Both species lost organic-N but only *S. paschale* suffered substantial net losses of nitrogen.
3. *S. paschale* consistently made gains in potassium but *C. stellaris* incurred losses in potassium during the larger deliveries of rainfall (although these losses were not significant at $P < 0{\cdot}05$).

For both species there was a strong relationship between the quantity of rainfall in any one event and the quantity of organic-N leached (Fig. 3.8).

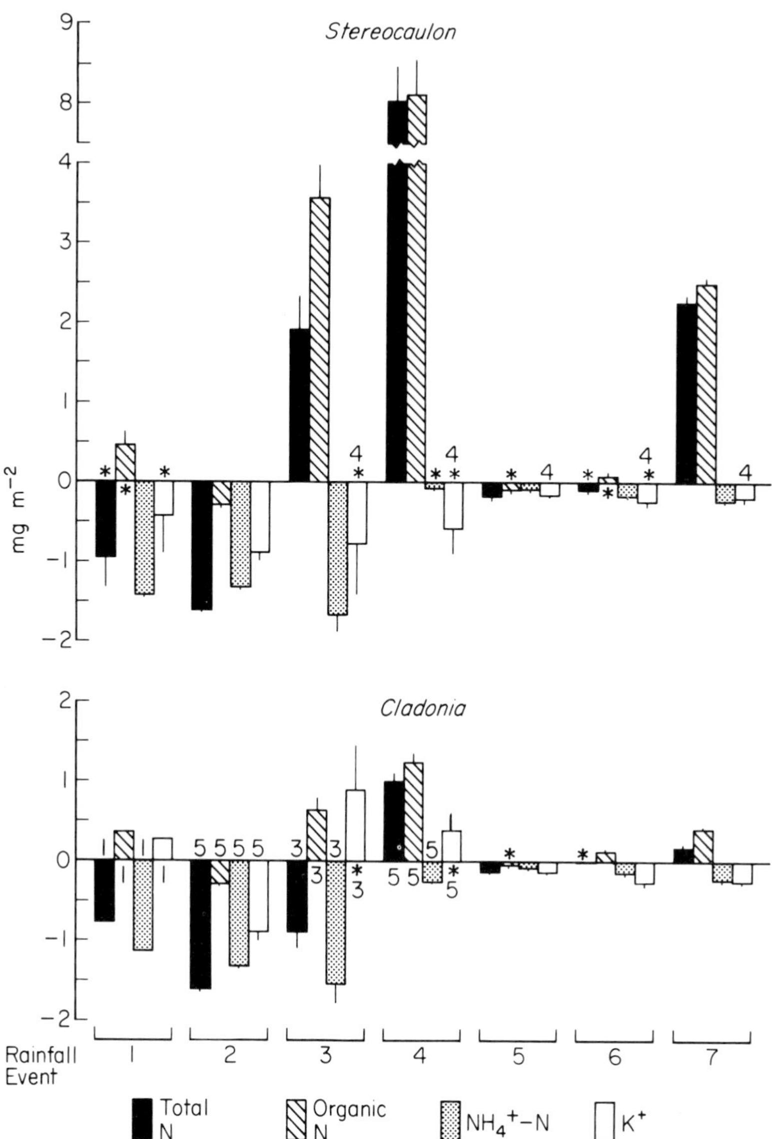

FIG. 3.7. Cumulative net release (+) or uptake (−) of nitrogen and potassium by pure mats of *Stereocaulon paschale* and *Cladonia stellaris* for all rainfall events studied. Mean values are plotted together with one standard error (assuming no error to be associated with replicate cumulative values). Replication, unless indicated otherwise: *S. paschale*, n = 5, *C. stellaris*, *n* = 4 (data from one funnel only collected for *C. stellaris* during Rainfall Event 1); * = not significantly different from zero at $P < 0.05$.

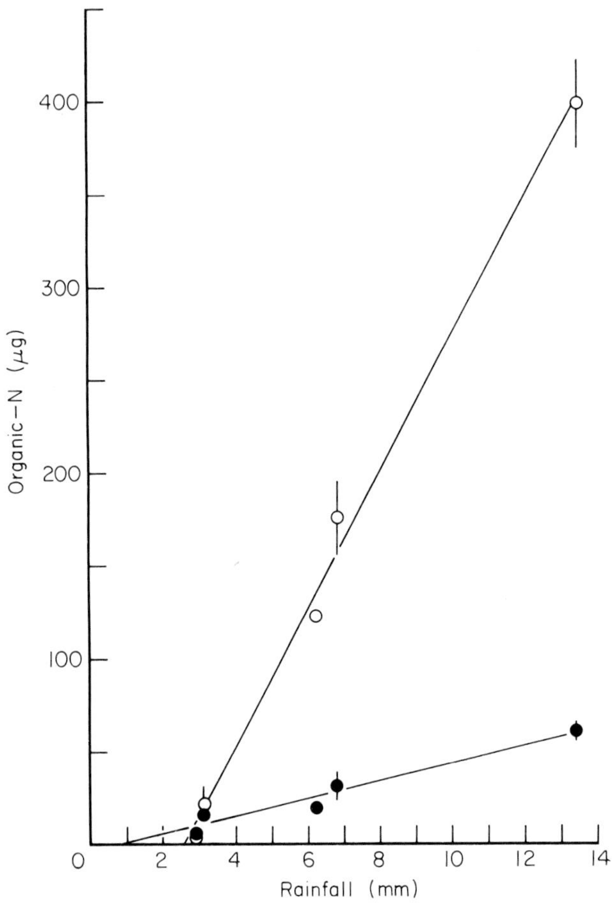

FIG. 3.8. Relationship between quantity of rain intercepted by lichen canopies during a single rainfall event and the quantity of organic-N lost by leaching. (○) *Stereocaulon paschale*; (●) *Cladonia stellaris*. Mean values are plotted together with standard errors. Regression equation of y on x for upper line is $y = -95 + 37.1\ x$ and for lower line is $y = -3 + 4.7\ x$.

The total ammonium-N and organic-N contents of the rainfall collected at the experimental site over the 49 day study period were 5.8 ± 0.2 and 2.8 ± 0.2 (total 8.6 ± 0.8) mg m^{-2} respectively; the corresponding values for lichen throughfall were 0.87 ± 0.04 and 17.2 ± 0.6 (total 18.0 ± 0.6) mg m^{-2} for *S. paschale* and 1.1 ± 0.1 and 5.2 ± 0.1 (total 6.4 ± 0.2) mg m^{-2} for *C. stellaris*. Release of organic-N from pure mats of *S. paschale* (mean density 768 ± 4 g m^{-2}) was up to 6.5 times greater than that from pure mats of *C. stellaris* (mean density 245 ± 4 g m^{-2}). On a dry weight basis this factor was 2.1 while, per unit weight of nitrogen, losses of organic-N from *C. stellaris* were *c.* 1.4 times greater than those from *S. paschale*.

The tagged lichen thalli interposed in the lichen mats over the funnels made readily measurable gains in dry weight during the 38 day 'growth period': the mean percentage dry weight increases for *C. stellaris* and *S. paschale* were 9.0 ± 4.2 ($n = 5$) and 5.3 ± 1.5 ($n = 10$) respectively. These increments correspond to mean relative growth rates of 0.0023 ± 0.0004 day^{-1} and 0.0014 ± 0.0002 day^{-1} which are in close agreement with rates recorded for the same species at Kevo by Kärenlampi (1971).

DISCUSSION

The results of these field experiments show that the organic-N content of rainfall increases as it percolates through mats of *S. paschale* and *C. stellaris* while the ammonium-N content decreases. The nature of the organic-N and the mechanism by which it is released are not known.

Free amino acids were present in lichen leachates in trace quantities only and free amino sugars were not detected. However, hydrolysis of bulked and concentrated samples of leachates yielded a wide range of amino acids accounting for *c.* 15 and *c.* 25% of the organic-N fractions in throughfall from *S. paschale* and *C. stellaris* respectively. At least a part of these amino acid complements probably originated from contamination because water collected in control funnels also contained combined amino acids (*c.* 13% of organic-N fraction). Dry pollen grains and fungal spores are known to release proteins and free amino acids into bathing solutions (Simon 1974) and may have been a source of interference while glycine, alanine and histidine were present in hydrolysed washings from membrane filters through which all samples were passed (see Appendix, p. 64). Analyses also proved negative for mono-, di- and trimethylamine, compounds found in cyanophilic lichens of the genera *Sticta* and *Lobaria* by Bernard & Goas (1968), and an amino hexitol (ninhydrin positive) isolated from *S. paschale* by Solberg (1979) was not revealed during amino acid analysis. Further studies will be necessary in order to discover the nature of the organic-N released from *S. paschale* during rainfall. Quaternary ammonium compounds such as betaines and choline are candidates for further analytical tests. Betaines occur widely in fungi (Guggenheim 1958; Towers 1976) and Bernard *et al.* (1980) have reported that sticticin, the methylated ester of the betaine of dihydroxyphenylalanine, comprises as much as 6.5% of the dry weight of *Lobaria laetevirens*. The total choline content of lichens varies between 0.03 and 0.63% of dry weight (DaSilva & Jensen 1971). Since the concentrations of organic-N in the lichen throughfall were not markedly lower after rain-free periods during which the lichen canopies remained moist (See Figs 3.5 & 3.6) it can be concluded that the nitrogen released from the lichens was not in a form that could be subsequently reabsorbed.

The progressive decrease in the nitrogen concentration in the *Stereocaulon* throughfall is compatible with the hypothesis that leakage of solutes from the lichen occurred during the initial rewetting phase only and that the solutes were subsequently eluted from the apparent free space as rainfall continued. However, the absence of a concomitant leakage of potassium is quite inconsistent with this view. Potassium is largely located in solution within protoplasts and readily leaks out of both lichens and bryophytes when membrane damage is induced (Puckett *et al.* 1977; Tomassini *et al.* 1977) or when air-dried thalli are rewetted rapidly by submersion in water (Farrar & Smith 1976; Lang, Reiners & Heier 1976; Buck & Brown 1979; Brown & Buck 1979). Indeed, rewetting freshly collected air-dry *S. paschale* in this manner was found to result in a potassium efflux of 32 ± 2 μg g^{-1} ($n = 10$) after 5 min, although the contribution by surface contamination to this quantity is not known. However, there is evidence that leakage of solutes from dry lichens upon abrupt reintroduction to excess water is a stress phenomenon that only occurs on a much smaller scale, if at all, when tissues are rehydrated gradually (Simon 1974; Buck & Brown 1979; Brown & Buck 1979). Under field conditions the rewetting and resaturation of lichen and bryophyte communities can be a protracted process; rainfall intensity is usually low at the onset of rainfall (Norbury & White 1975) and is associated with increases in relative humidity. This is especially true of subarctic regions where much of the summer rainfall occurs as drizzle (Dolgin 1970); even the rainfall intensities recorded for the thunderstorms sampled in the present study (Rainfall Events 1, 5 & 6) were less than 0.05 times the maximum rates recorded for storms of similar duration in the UK (Rodda 1970). Loss of potassium from *S. paschale* may be negligible when lichen mats are rewetted slowly under field conditions. Alternatively, small quantities of potassium intially released into the apparent free space may be subsequently reabsorbed before the lichen mats become fully saturated and leaching occurs, a period that can extend to several hours (see Figs 3.4 & 3.6). Furthermore, there is also evidence that the magnitude of rewetting losses from lichens and bryophytes is related to water availability in the habitats from which the plants are collected, leakage being minimal in species from dry situations (Buck & Brown 1979; Brown & Buck 1979). Communities rich in *S. paschale* and *C. stellaris* develop in regions and on terrain subject to frequent and prolonged summer drought, and both species occur in open situations where they intercept all incident rainfall (cf. Plate 3.1). Consequently both species may be resistant to ‘rehydration stress’, and the potential for solute loss upon rewetting may be small. In contrast, the rewetting of cyanophilic lichens such as *Peltigera* spp. and *Nephroma arcticum* which in boreal forests typically occur in the ground vegetation beneath tree canopies will be retarded due to interception of rain by the foliage above. *Peltigera* species have been found to be particularly leaky when rewetted rapidly (Millbank 1978; Buck & Brown 1979), a stress that they are unlikely to encounter in woodland habitats.

Also at variance with the rewetting hypothesis is the observation that concentrations of organic-N in the first samples of *Stereocaulon* throughfall collected were sometimes very much lower than those in later samples. In fact, the greatest efflux of organic-N occurred with the resumption of rainfall after comparatively long rain-free periods in which the lichen canopies remained moist (see Figs 3.5 & 3.6). The pool of leachable nitrogen appears to have increased during periods suitable for metabolic activity, possibly as a result of nitrogen fixation. Furthermore, if leakage was a short-lived rehydration phenomenon, rapid rewetting by thunderstorms might have resulted in high initial solute losses. Because of the uncontrolled nature of the experiments it is not possible to make valid comparisons of the quantities of organic-N lost during rainfall events of differing intensities. However, there was no relationship between the rate at which *S. paschale* canopies were rewetted and the *concentration* of organic-N in the resultant leachate. For example, during the thunderstorm on 19th June (Rainfall Event 1) 2.3 mm of rain fell in *c.* 14 min on an initially dry lichen mat yet the organic-N concentrations in the *Stereocaulon* throughfall (*c.* 1.0 $\mu g\ ml^{-1}$) were comparable with those recorded on other occasions when lichen mats became saturated gradually over periods of several hours (e.g. 12th and 25th July).

The apparent retention of potassium by *S. paschale* during rainfall attests the assumption that the release of organic-N was a natural process and not a result of structural or metabolic damage caused by the treatment of the lichen during the experiments. With the larger deliveries of rainfall net losses of potassium occurred from *C. stellaris* concomitant with the losses in organic-N and may indicate that this lichen did not survive transplantation to the steel grilles as successfully as *S. paschale*. However, both species made appreciable gains in dry weight during the study period suggesting that in general they remained healthy.

Rainfall during the 49 day investigation resulted in a net nitrogen loss from the *S. paschale* canopies of 9.4 ± 0.7 mg m^{-2} of pure lichen cover. The cumulative net nitrogen loss for the summer of 1980 can be derived from 12-hourly rainfall records from the Kevo Metereological Station using the relationship between the quantity of rainfall delivered and the quantity of organic-N leached in a single event (see Fig. 3.8) and assuming that the mean ratio of net nitrogen loss to gross loss of organic-N observed during the study period (0.66) applies generally for seasonal net nitrogen loss by leaching. The value so derived is 46 mg N m^{-2}. However, this figure is likely to be low because the summer of 1980 was unusually dry in interior Lapland; rainfall recorded in June and July (28 and 21 mm respectively) was only 70% and 31% of the average values for these months during the period 1962–1980. Application of the same predictive procedure to rainfall data for 1964, the wettest year of the 1962–1980 period (44 and 126 mm received in June and July respectively), yields a value of 151 mg m^{-2}. Although pure carpets of *S. paschale* are known to develop in lichen woodlands (cf. Plate 3.1c) it more frequently occurs in association with mosses, dwarf shrubs and other lichens (usually spceies of *Cladonia*, *Cetraria* and *Alectoria*). In the pine and birch forests in the Kevo district *S. paschale* achieves a biomass of *c.* 40 g m^{-2} (Kjelvik & Kärenlampi 1975), about 5% of the biomass represented by the reconstructed lichen mats used in this study. The predicted nitrogen losses from *S. paschale*, which relate to monospecific lichen mats, must therefore be adjusted accordingly to produce appropriate mean values for the ground cover at Kevo.

These estimates of nitrogen release from *S. paschale* can be compared with other nitrogen fluxes in the ecosystem. They undoubtedly represent only a small fraction of the annual nitrogen demand of systems in which *S. paschale* is abundant. For example, the following estimates of annual nitrogen uptake by plant communities have been produced for subarctic or subalpine systems: subalpine birch (*Betula pubescens*) forest and alpine lichen heathland, 8100 and 2100 mg m^{-2} (Wielgolaski, Kjelvik & Kallio 1975); subarctic mire, 670 mg m^{-2} (Rosswall & Granhall 1980); spruce-feathermoss forest 3500 mg m^{-2} (Havas 1979); subalpine Pacific silver fir (*Abies amabilis*) forest, 1190 mg m^{-2} (Turner & Singer 1976); *Cladonia*-jack pine (*Pinus banksiana*) forest, 1540 mg m^{-2} (Weetman & Webber 1972). Crittenden & Kershaw (1979) have tentatively calculated quantities of nitrogen fixed by pure mats of *S. paschale* ($>95\%$ cover, 591 ± 36 g m^{-2}) in spruce-lichen woodland (Northwest Territories, Canada) over periods of 19–27 h following rainfall. From these data it can be roughly estimated that the measured wetting losses of nitrogen from *S. paschale* during single rainfall events could represent between 2 and 12% of the total nitrogen fixed in the period of suitable moisture conditions initiated by the onset of rainfall. Clearly then, leaching by rainfall is a comparatively minor pathway by which fixed nitrogen is transferred from *S. paschale* into the soil. On the other hand, the measured wetting losses were roughly equivalent to the input of Kjeldahl-N by rainfall.

The nitrate contents of rainfall and throughfall were not measured in the present study. Concentrations of nitrate and ammonia in rainfall at Kevo are broadly comparable (Kallio & Veum 1975) but the available evidence suggests that lichens, particularly cyanophilic species, absorb nitrate substantially less efficiently than they absorb ammonia (Smith 1960; Lang, Reiners & Heier 1976). Consequently net losses of total nitrogen from *S. paschale* are unlikely to be very much less than the net losses of Kjeldahl-N reported here.

Organic forms of nitrogen produced extracellularly by surface soil encrusting blue-green algae can be utilized by vascular plants with roots near to the soil surface and this nitrogen transfer is thought to contribute significantly to primary production in habitats in which blue-green algal crusts are well developed (Mayland & McIntosh 1966; Stewart 1967). The extent to which organic-N leached from nitrogen-fixing lichens can be utilized by neighbouring plants is of particular interest in view of the oligotrophic nature of podzolic soils in lichen woodlands and soils of boreal-arctic regions in general (Moore 1980). The results of two recent studies suggest that nitrogen fixed by lichenized blue-green algae can subsequently become available to adjacent mosses. J. W. Millbank (pers. comm.) has demonstrated that during continuously moist conditions with episodes of simulated rainfall a part of the total ^{15}N previously incorporated into *Lobaria pulmonaria* was transferred to moss thalli intimately associated with the lichen, and P. Kallio & B. Saxén (pers. comm.) have found that the nitrogen content of moss thalli growing in contact with *Peltigera aphthosa* and *Nephroma arcticum* in subarctic birch woodland was as much as 2.3 times greater than that of mosses sampled at 100 mm from the lichens. It has not been established, however, whether the absorption of lichen-derived nitrogen results in increased growth rates of the mosses. Depending on the nature of the organic compounds involved, it is possible that organic-N leached from *S. paschale* may be available to vascular plants with absorbing roots near to the soil surface (Persson 1980), uptake being mediated by mycorrhizas. Dwarf shrubs such as *Empetrum nigrum*, *Vaccinium myrtillus*, *V. uliginosum*, *V. vitis idaea* and *Betula nana* which are often spatially closely associated with *S. paschale* typically have well-developed mycorrhizal systems as do northern boreal forest trees. Mycorrhizal infections are generally best developed under nutrient-poor conditions (Harley 1969) and, in the Ericaceae at least, they are known to facilitate the utilization of simple organic-N sources by the host plants (Stribley & Read 1980). Brown & Mikola (1974) and Fisher (1979) found evidence of an inhibitive effect of *Cladonia stellaris* and *C. rangiferina* ground cover on the formation of ectomycorrhizas in pine and spruce and the former suggest that other lichen species may have a similar effect. However, we have found that ericaceous plants rooted beneath almost pure mats of *S. paschale* in lichen woodland are heavily infected with mycorrhizal endophyte (Crittenden & Read, unpublished data). In addition to organic-N leached from viable lichens, absorbing root systems beneath *S. paschale*-rich ground cover may receive nitrogenous decomposition products from basal parts of lichen thalli. The extent to which *S. paschale* mats influence rooting patterns of associate vascular plants may be worthy of investigation.

ACKNOWLEDGEMENTS

I am particularly indebted to Professor P. Kallio (Kevo Subarctic Research Institute, University of Turku) for the generous provision of facilities at the Kevo Subarctic Research Station and to the station staff for their friendly and invaluable cooperation during the field work. Thanks are also due to Dr R. Hunt, Dr D.H. Lewis, Dr D.J. Read, Professor A.J. Willis (Department of Botany, University of Sheffield) and Dr J. W. Millbank (Department of Botany, Imperial College, London) for useful discussion, Mr R.J.B. Williams (Department of Soils and Plant Nutrition, Rothamsted Experimental Station) for suggesting the use of pentachlorophenol, and Nick Rhodes (Department of Biochemistry, University of Sheffield) for conducting the amino acid analyses. The work was supported by grants from the Royal Society and the Sheffield University Research Fund.

APPENDIX

Correction for interference from membrane filters

Comparatively large quantities of nitrogen can be leached from membrane filters by shaking with water (*c.* 7 μg Kjeldahl-N per filter used in the present investigation) and for this reason Allen (1974) and Golterman, Clymo & Ohnstad (1978) have cautioned against the use of membrane filters prior to nitrogen analyses. Accordingly, laboratory tests were conducted to examine the effect of prewashing cellulose acetate membrane filters on their capacity for modifying nitrogen and potassium concentrations in water samples during filtration. Of the prewashing and prefiltering treatments investigated, shaking in several changes of excess deionized water (after which the filters were dried) proved the most expedient; the use of filters *immediately* after this pretreatment resulted in insignificant changes in nitrogen and potassium levels in small volumes (3 ml) of test solutions filtered. Consequently all membrane filters used in the present work were prewashed in this manner and then repacked and stored until required. However, despite this precaution unusually high concentrations of organic-N were recorded in rainfall samples, particularly in the smaller volumes collected, and subsequent tests revealed that these could be attributed to nitrogen release from the filters; the soluble nitrogen content of the prewashed filters had apparently increased during storage.

A correction curve (Fig. 3.9) was therefore used to rectify all measurements of total nitrogen for this interference. The membrane filters used in the construction of the correction curve were taken from the same batch as those used in the field experiments but whereas the latter had postwash ages of between 4 and 10 weeks the former had a postwash age of 24 weeks. The details of the relationship, if real, between storage time and soluble nitrogen content of washed filters are not known.

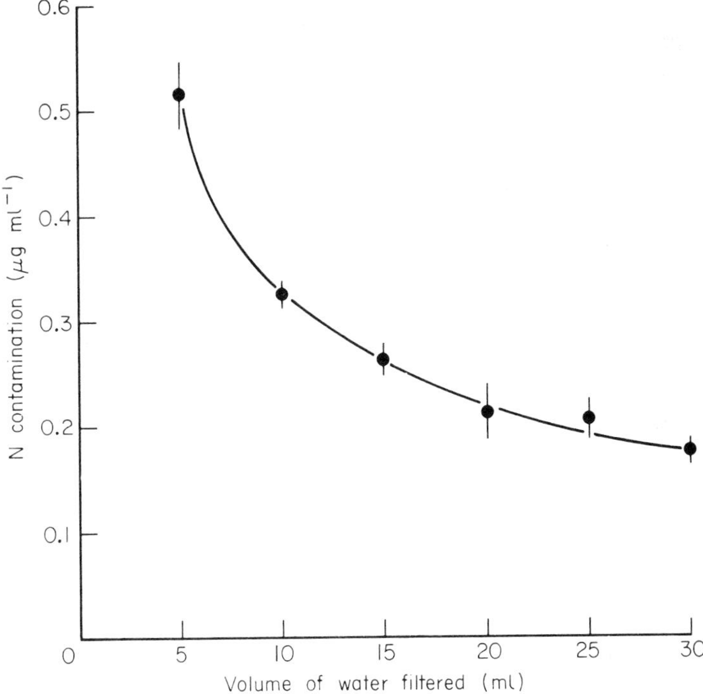

FIG. 3.9. Relationship between volume of water[1] passed through cellulose acetate membrane filters[2] and nitrogen[3] enrichment of sample due to contamination[4]. Mean values ($n = 10$) are plotted together with 95% confidence intervals.

[1] Oligotrophic stream water containing only trace quantities of Kjeldahl-N.

[2] Filters were pretreated by shaking in three changes of excess deionized water, then dried, repacked and stored for 24 weeks prior to tests.

[3] Kjeldahl-N.

[4] Nitrogen leached from filters was entirely organic in nature.

REFERENCES

Allen S.E. (ed.) (1974) *Chemical Analysis of Ecological Materials.* Blackwell Scientific Publications, Oxford.

Bernard T. & Goas G. (1968) Contribution à l'étude du métabolisme azoté des Lichens: caractérisation et dosages des méthylamines de quelques espèces de la famille des Stictacées. *Comptes Rendus Hebdomadaires des Séances de l'Académie des Sciences, Paris, Série D*, **267**, 622–4.

Bernard T., Joucla M., Goas G. & Hamelin J. (1980) Caractérisation de la sticticine chez le lichen *Lobaria laetevirens. Phytochemistry*, **19**, 1967–9.

Bevenue A. & Beckman H. (1967) Pentachlorophenol: a discussion of its properties and its occurrence as a residue in human and animal tissues. *Residue Reviews*, **19**, 83–134.

Billington M. & Alexander V. (1978) Nitrogen fixation in a black spruce (*Picea mariana* [Mill] B.S.P.) forest in Alaska. In *Environmental Role of Nitrogen-fixing Blue-green Algae and Asymbiotic Bacteria (Ecological Bulletins, 26)* (ed. U. Granhall), pp. 209–15. Swedish Natural Science Research Council, Stockholm.

Bremner J.M. (1965) Total nitrogen. In *Methods of Soil Analysis* (ed. C.A. Black), pp. 1149–78. American Society of Agronomy, Madison.

Brown D.H. & Buck G.W. (1979) Desiccation effects and cation distribution in bryophytes. *New Phytologist*, **82**, 115–25.

Brown R.T. & Mikola P. (1974) The influence of fruticose soil lichens upon the mycorrhizae and seedling growth of forest trees. *Acta Forestalia Fennica*, **141**, 1–23.

Buck G.W. & Brown D.H. (1979) The effect of desiccation on cation location in lichens. *Annals of Botany*, **44**, 265–77.

Crittenden P.D. & Kershaw K.A. (1979) Studies on lichen-dominated systems. XXII. The environmental control of nitrogenase activity in *Stereocaulon paschale* in spruce-lichen woodland. *Canadian Journal of Botany*, **57**, 236–54.

Damman A.W.H. (1964) Some forest types of central Newfoundland and their relation to environmental factors. *Society of American Foresters, Forest Science Monograph No. 8.* pp. 1–62.

DaSilva E.J. & Jensen A. (1971) Choline, ergosterol and tocopherol content of some Norwegian lichens. *Journal of the Science of Food and Agriculture*, **22**, 308–11.

Denison W.C. (1973) Life in tall trees. *Scientific American*, **228**, 74–80.

Dolgin I.M. (1970) Subarctic meteorology. In *Ecology of the Subarctic Regions*, pp. 41–61. UNESCO, Paris.

Dyrness C.T. & Grigal D.F. (1979) Vegetation–soil relationships along a spruce forest transect in interior Alaska. *Canadian Journal of Botany*, **57**, 2644–56.

Farrar J.F. & Smith D.C. (1976) Ecological physiology of the lichen *Hypogymnia physodes*. III. The importance of the rewetting phase. *New Phytologist*, **77**, 115–25.

Fisher R.F. (1979) Possible allelopathic effects of reindeer-moss (*Cladonia*) on jack pine and white spruce. *Forest Science*, **25**, 256–60.

Franklin J. (1974) Decomposition of lower plants. In *Biology of Plant Litter Decomposition, Vol. 1* (eds C.H. Dickinson & G.J.F. Pugh), pp. 3–36. Academic Press, London.

Golterman H.L., Clymo R.S. & Ohnstad M.A.M. (1978) *Methods for Physical and Chemical Analysis of Fresh Waters (IBP Handbook, 8)*, 2nd edn. Blackwell Scientific Publications, Oxford.

Guggenheim M. (1958) Die biogenen amine in der Pflanzenwelt. In *Handbuch der Pflanzenphysiologie, Band VIII* (ed. W. Ruhland), pp. 889–988. Springer-Verlag, Berlin.

Haag R.W. (1974) Nutrient limitations to plant production in two tundra communities. *Canadian Journal of Botany*, **52**, 103–16.

Hämet-Ahti L. (1963) Zonation of the mountain birch forests in northernmost Fennoscandia.

Annales Botanici Societatis Zoologicæ Botanicæ Fennicæ 'Vanamo', **34**, (No. 4), 1–127.

Harley J.L. (1969) *The Biology of Mycorrhiza.* Leonard Hill, London.

Havas P. (1979) Paksusammaltyypin kuusimetsän typpitaloudesta. *Acta Universitatis Ouluensis, Series A, Scientiae Rerum Naturalium, 68 (Biologica, 4)*, 135–40.

Hitch C.J.B. & Stewart W.D.P. (1973) Nitrogen fixation by lichens in Scotland. *New Phytologist*, **72**, 509–24.

Hunt R. & Parsons I.T. (1974) A computer program for deriving growth-functions in plant growth-analysis. *Journal of Applied Ecology*, **11**, 297–307.

Hustich I. (1949) On the forest geography of the Labrador Peninsula. A preliminary synthesis. *Acta Geographica*, **10** (No. 2), 1–63.

Johnson M.J. (1941) Isolation and properties of a pure yeast polypeptidase. *Journal of Biological Chemistry*, **137**, 575–86.

Kallio P. & Veum A.K. (1975) Analysis of precipitation at Fennoscandian tundra sites. In *Fennoscandian Tundra Ecosystems. Part 1. Plants and Micro-organisms* (*Ecological Studies, 16*) (ed. F.E. Wielgolaski), pp. 333–8. Springer-Verlag, Berlin.

Kärenlampi L. (1971) Studies on the relative growth rate of some fruticose lichens. *Reports from the Kevo Subarctic Research Station*, **7**, 33–9.

Karev G. I. (1961) Kormovaya baza severnogo olenevodstva. In *Severnoe Olenevodstvo*, 2nd edn (ed. P. S. Zhigunov), pp. 188–254. Izdatel'stvo Sel'skokhozystvennoy Literatury, Zhurnalov i Plakatov, Moskva. (Translated into English by Israel Program for Scientific Translations, 1968.)

Kershaw K.A. (1977) Studies on lichen-dominated systems. XX. An examination of some aspects of the norther boreal lichen woodlands in Canada. *Canadian Journal of Botany*, **55**, 393–410.

Kjelvik S. & Kärenlampi L. (1975) Plant biomass and primary production of Fennoscandian subarctic and subalpine forests and of alpine willow and heath ecosystems. In *Fennoscandian Tundra Ecosystems. Part 1. Plants and Micro-organisms (Ecological Studies, 16)* (ed. F.E. Wielgolaski), pp. 111–20. Springer-Verlag, Berlin.

Lang G.E., Reiners W.A. & Heier R.K. (1976) Potential alteration of precipitation chemistry by epiphytic lichens. *Oecologia*, **25**, 229–41.

Lavrenko E.M. & Sochava V.B. (eds) (1956) *Rastitel' nyy pokrove SSSR (Descriptio Vegetationis URSS).* Akademia nauk SSSR, Moskva-Leningrad.

Maikawa E. & Kershaw K.A. (1976) Studies on lichen-dominated systems. XIX. The postfire recovery sequence of black spruce–lichen woodland in the Abitau Lake Region, N.W.T. *Canadian Journal of Botany*, **54**, 2679–87.

Mayland H.F. & McIntosh T.H. (1966) Availability of biologically fixed atmospheric nitrogen-15 to higher plants. *Nature*, **209**, 421–2.

Millbank J.W. (1978) The contribution of nitrogen fixing lichens to the nitrogen status of their environment. In *Environmental Role of Nitrogen-fixing Blue-green Algae and Asymbiotic Bacteria (Ecological Bulletins, 26)* (ed. U. Granhall), pp. 260–5. Swedish Natural Science Research Council, Stockholm.

Moore T.R. (1980) The nutrient status of subarctic woodland soils. *Arctic and Alpine Research*, **12**, 147–60.

Moss E.H. (1953) Forest communities in northwestern Alberta. *Canadian Journal of Botany*, **31**, 212–52.

Norbury J.R. & White W.J.K. (1975) Intensity-time profiles of high-intensity rainfall. *Meteorological Magazine*, **104**, 221–7.

Orlóci L. & Stanek W. (1979) Vegetation survey of the Alaska Highway, Yukon Territory: types and gradients. *Vegetatio*, **41**, 1–56.

Parsons I.T. & Hunt R. (1981) Plant growth analysis: a program for the fitting of lengthy series of data by the method of *B*-splines. *Annals of Botany*, **48**, 341–52.

Payette S. & Filion L. (1975) Écologie de la limite septentrionale des forêts maritimes, baie d'Hudson, Nouveau-Québec. *Naturaliste canadien*, **102**, 783–802.

Payette S., Samson H. & Lagarec D. (1976) The evolution of permafrost in the taiga and in the forest-tundra, western Quebec-Labrador Peninsula. *Canadian Journal of Forest Research*, **6**, 203–20.

Persson H. (1980) Spatial distribution of fine-root growth, mortality and decomposition in a young Scots pine stand in Central Sweden. *Oikos*, **34**, 77–87.

Pike L.H. (1978) The importance of epiphytic lichens in mineral cycling. *Bryologist*, **81**, 247–57.

Pike L.H., Tracy D.M., Sherwood M.A. & Nielsen D. (1972) Estimates of biomass and fixed nitrogen of epiphytes from old-growth Douglas fir. In *Research on Coniferous Forest Ecosystems: First Year Progress in the Coniferous Forest Biome, US/IBP* (eds J.F. Franklin, L.J. Dempster & R.H. Waring), pp. 177–87. Pacific Northwest Forest and Range Experimental Station, Forest Service, US Department of Agriculture, Portland.

Puckett K.J., Tomassini F.D., Nieboer E. & Richardson D.H.S. (1977) Potassium efflux by lichen thalli following exposure to aqueous sulphur dioxide. *New Phytologist*, **79**, 135–45.

Ritchie J.C. (1956) The vegetation of northern Manitoba. I. Studies in the southern spruce forest zone. *Canadian Journal of Botany*, **34**, 523–61.

Rodda J.C. (1970) Rainfall excesses in the United Kingdom. *Transactions of the Institute of British Geographers*, **49**, 49–60.

Rosswall T. & Granhall U. (1980) Nitrogen cycling in a subarctic ombrotrophic mire. In *Ecology of a Subarctic Mire (Ecological Bulletins, 30)* (ed. M. Sonesson), pp. 209–34. Swedish Natural Science Research Council, Stockholm.

Rowe J.S. (1972) *Forest Regions of Canada.* Information Canada, Ottawa.

Simon E.W. (1974) Phospholipids and plant membrane permeability. *New Phytologist*, **73**, 377–420.

Smith D.C. (1960) Studies in the physiology of lichens. I. The effects of starvation and of ammonia absorption upon the nitrogen content of *Peltigera polydactyla. Annals of Botany*, **24**, 52–62.

Solberg Y. (1979) Studies on the chemistry of lichens, XIX. New amino compounds from *Anaptychia fusca* and several *Stereocaulon* species. *Zeitschrift für Naturforschung*, **34C**, 493–7.

Stewart W.D.P. (1967) Transfer of biologically fixed nitrogen in a sand dune slack region. *Nature*, **214**, 603–4.

Stribley D.P. & Read D.J. (1980) The biology of mycorrhiza in the Ericaceae. VII. The relationship between mycorrhizal infection and the capacity to utilize simple and complex organic nitrogen sources. *New Phytologist*, **86**, 365–71.

Tamm C. O. (1964) Determination of nutrient requirements of forest stands. *International Review of Forest Research*, **1**, 115–70.

Tieszen L.L. (ed.) (1978) *Vegetation and Production Ecology of an Alaskan Arctic Tundra (Ecological Studies, 29)*. Springer-Verlag, New York.

Tomassini F.D., Lavoie P., Puckett K.J., Nieboer E. & Richardson D.H.S. (1977) The effect of time of exposure to sulphur dioxide on potassium loss from and photosynthesis in the lichen, *Cladina rangiferina* (L.) Harm. *New Phytologist*, **79**, 147–55.

Towers G.H.N. (1976) Secondary metabolites derived through the shikimate-chorismate pathway. In *The Filamentous Fungi.* Vol. 2. *Biosynthesis and Metabolism* (eds J.E. Smith & D.R. Berry), pp. 460–74. Edward Arnold, London.

Turner J. & Singer M.J. (1976) Nutrient distribution and cycling in a subalpine coniferous forest ecosystem. *Journal of Applied Ecology*, **13**, 295–301.

Warren Wilson J. (1966) An analysis of plant growth and its control in arctic environments. *Annals of Botany*, **30**, 383–402.

Weetman G.F. & Webber B. (1972) The influence of wood harvesting on the nutrient status of two spruce stands. *Canadian Journal of Forest Research*, **2**, 351–69.

Wielgolaski F.E., Kjelvik S. & Kallio P. (1975) Mineral content of tundra and forest tundra plants in Fennoscandia. In *Fennoscandian Tundra Ecosystems. Part 1. Plants and Micro-organisms (Ecological Studies, 16)* (ed. F.E. Wielgolaski), pp. 316–32. Springer-Verlag, Berlin.

4. THE SIGNIFICANCE OF ECTOMYCORRHIZAS IN THE NITROGEN CYCLE

I. J. ALEXANDER

Department of Botany, University of Aberdeen, Aberdeen AB9 2UD

SUMMARY

Ectomycorrhizas are relatively specialized with a distinctive morphology and physiology. Although largely confined to long-lived, woody, perennial hosts in cool temperate, boreal or montane regions they also occur in a restricted number of important tropical families. Thus many dominant forest trees are ectomycorrhizal as are shrubs in some artic-alpine and mediterranean ecosystems. The fungus is known to exploit and store non-mobile soil nutrients such as phosphorus. However, much ectomycorrhizal vegetation is nitrogen-stressed rather than phosphorus-stressed and both the intensity of mycorrhizal infection and their absolute numbers are highest in nitrogen-rich organic layers at the soil surface.

Cultural studies indicate that ectomycorrhizal fungi and intact ectomycorrhizas are unlikely to utilize recalcitrant organic nitrogen and the balance between net mineralization and nitrogen uptake in some examples of ectomycorrhizal vegetation supports this conclusion. Simpler organic nitrogen, such as peptides or amino acids, available in humus layers are probably absorbed thus reducing the residence time of nitrogen in the soil and the possibility of its loss through leaching or immobilization.

Ammonium nitrogen is readily absorbed and uptake may be influenced by the identity of the fungal symbiont. Nitrate is less readily absorbed and may indeed inhibit mycorrhizal formation. Profuse ectomycorrhizal development is often associated with low rates of nitrification. Under field conditions infection may make the plant a more efficient competitor for nitrogen and increase the exploitation of the available pool. In forests the nitrogen concentration of the soil solution drops by about 70% passing through the intensely mycorrhizal surface horizons.

Ectomycorrhizas are also a quantitatively important component of the nitrogen cycle. Nitrogen concentration in living mycorrhizal tissue is high and turnover results in a nitrogen input to the decomposition process from half to six times that in litter fall.

The ready supply of carbohydrate to ectomycorrhizal fungi probably contributes to their ability to compete for nitrogen. Where they are active, available ammonium will be low and autotrophic nitrification in turn reduced. Lowered rates of litter decomposition may also result from mycorrhizal absorption of available nitrogen. Thus ectomycorrhizas help to create and maintain conditions of which they are the most efficient exploiters. Controlled and virtually closed nitrogen cycles may operate, dominated by ectomycorrhizal fungi. Only plants capable of forming an association can break into the cycle and become established.

INTRODUCTION

The overwhelming majority of higher plants are mycorrhizal. The organs through which they absorb nutrients and water are composed of both fungal and plant tissue and these 'fungus roots', or mycorrhizas, are the most metabolically active part of the root system. The diverse structures arising from such fungal–plant symbioses can be placed in natural groupings on the basis of their biology, morphology and physiology (Lewis 1975). Of these by far the most common and widely distributed in the plant kingdom are vesicular–arbuscular (VA) mycorrhizas where the fungal symbiont, one of a number of species of the Endogonaceae (Phycomycetes), penetrates the cortical cells of the host root forming characteristic coils, vesicles or arbuscules which are in time digested. Although there may be much mycelium external to the root no specialized fungal tissue develops and there is no marked change in host root morphology. Such VA associations are very ancient and are known from the fossil record of early land plants in the Devonian (Harley 1969; Baylis 1974).

At a much later stage in the development of the land flora, possibly in the Cretaceous, the apparently more specialized ectomycorrhizal association evolved (Baylis 1974; Malloch, Pirozynski & Raven 1980). Ectomycorrhizas are formed by Basidiomycetes or Ascomycetes and have a distinctive morphology and physiology. Infection gives rise to marked changes in the morphology and branching pattern of the host root system, partly due to the release of fungal hormones into host tissue (Slankis 1973). The fungus does not normally penetrate active host cortical cells but passes between them. Moreover, in addition to varying amounts of external mycelium, strands and rhizomorphs, it produces around the outside of the host root a compact fungal tissue, the sheath, which commonly accounts for up to 30–40% of the dry weight of the mycorrhiza (Harley 1975).

Ectomycorrhizas also have a much more restricted distribution than VA mycorrhizas (Meyer 1973). The host plants are almost invariably long-lived woody perennials and are largely, but not exclusively, restricted to cool temperate, boreal or montane regions (Moser 1967). Ectomycorrhizas are characteristic of Pinaceae and some Cupressaceae and, among others, of Fagaceae, Betulaceae, Salicaceae and many Myrtaceae. Thus many of the forest dominants of the world are ectomycorrhizal. Assemblages of ectomycorrhizal genera are also found in artic-alpine dwarf shrub communities (e.g. *Salix*, *Dryas*, *Polygonum*) and Mediterranean/Chaparral vegetation (e.g. *Cistus*, *Quercus*) often in association with, and probably sharing fungal symbionts with, genera which form arbutoid (Lewis 1975) mycorrhizas (e.g. *Arbutus*, *Arctostaphylos*) (Dominik, Nespiak & Pachlewski 1954; Zak 1973; Largent, Sugihara & Wishner 1980). Malloch, Pirozynski & Raven (1980) have recently compiled a list of reported associations which greatly extends the range of host plants among lowland tropical taxa. Some may refer to casual infections but many are undoubtedly constant. Of particular interest is the widespread occurrence of ectomycorrhizas in the Dipterocarpaceae (Singh 1966; de Alwis & Abeynayake 1980) which attain family dominance in many South East Asian forests and in the Leguminosae (Caesalpinioideae) (Fassi & Fontana 1962; Redhead 1968; Thomazini 1974) which characteristically form monospecific stands on infertile soils.

The distinctive morphology of ectomycorrhizas and their specialized distribution in both the taxonomic and ecological sense has given rise to much speculation about the selection pressures leading to their evolution and the role they play in the ecosystems in which they habitually occur. A considerable body of evidence (Harley 1969, 1978; Bowen 1973) has demonstrated their significance in the mineral nutrition of the host plants and emphasized the part played by the fungus in exploiting sources of non-mobile nutrients in the soil. Absorbed elements are first accumulated in the fungal sheath which seems to act as a storage organ not only for soil-derived nutrients but also for the plant-derived carbohydrate which fuels the vegetative and reproductive activity of the fungus. These features have given rise to the theory that ectomycorrhizas are an adaptation to sharply seasonal climates, short growing seasons and nutrient stress (Harley 1975; Malloch, Pirozynski & Raven 1980). Although attractive, this theory does not fully explain the observed pattern of ectomycorrhizal distribution.

There are other known or postulated benefits of ectomycorrhizas such as disease resistance (Marx 1973; Bowen 1978) or water uptake (Dudderidge, Malibari & Read 1980) but it is the uptake of minerals, particularly phosphorus, which has received most attention. To some extent this may reflect the encouragement to experimentation provided by the existence of suitable isotopes of phosphorus and the analogies which can be drawn with VA mycorrhizas where efficient exploitation of the soil labile pool of phosphorus can be shown to be of paramount importance (Tinker 1975). These analogies are useful and there are undoubtedly parallels to be made but the distinctive nature and distribution of ectomycorrhizas demand some further explanation. In fact many ectomycorrhizal vegetation systems are probably nitrogen-stressed rather than phosphorus-stressed (Tamm 1979; Kramer & Kozlowski 1979). This is open to a number of interpretations. Miller *et al.* (1979), for example, have interpreted mineral cycles in pine on poor soils as demonstrating the efficiency of ectomycorrhizas in phosphorus absorption and their inability to prevent nitrogen stress. On the other hand, one could suggest that they might be an adaptation to nitrogen stress or even that they may promote or maintain it. In these circumstances it seems appropriate to examine their role in the nitrogen cycle more closely.

DISTRIBUTION AND INTENSITY OF INFECTION IN NATURAL SOILS

The intensity of ectomycorrhizal infection and its distribution in natural soils depends on prevailing conditions (Harley 1969; Meyer 1973; Slankis 1974; Fogel 1980). In general, low or imbalanced levels of available nutrients in the soil, particularly nitrogen, phosphorus or potassium, affect the internal nutrient status of the host and lead to increased intensity of infection. In soils not subject to prolonged drought both the absolute numbers of mycorrhizas and the intensity of infection of the absorbing roots are greatest in the surface horizons particularly where these are organic. When a tree species is examined on a range of soil types the number of mycorrhizas is greatest, and their tendency to surface accumulation most pronounced, on podsolic soils or soils with surface humus horizons. Meyer (1967) and Meyer & Göttsche (1971) have given particularly convincing demonstrations of this for *Picea abies* (L.) Karsten and *Fagus sylvatica* L. Numerical estimates of the change of intensity of infection in different soils are more difficult to obtain (Harley 1969) and generalizations more difficult to make. Thus Melin (1917), Harley (1940) and others have shown that in raw humus soils roots are almost entirely mycorrhizal but in mull soils less so. This conforms to the 'mineral nutrition' theory of formation as outlined above, and is supported by many pot experiments. On the other hand, Meyer (1962, 1973, 1974) has claimed that while the absolute number of mycorrhizas of *F. sylvatica* is indeed inversely related to the nutritional state of the soil, *i.e.* high in mor, low in mull, the mycorrhizal frequency (the percentage of root tips converted to mycorrhizas) is in fact lower in mor, where a proportion of the root tips are instead infected by weak parasitic fungi. There is some support for this in the work of Melin (1927) and Hatch (1937) who showed that infection is depressed in raw humus where nitrogen mobilization is low. Fertilizer experiments in natural soils also give conflicting results (Harley 1969; Slankis 1974) and obviously depend on the original nutrient status of the soil and the availability of the added nutrient to the mycorrhizal plants. However, it is possible to apply levels of nitrogen which depress mycorrhizal formation. In summary, intense ectomycorrhizal development can be expected in organic horizons where nitrogen availability is relatively low but where nitrogen mineralization can still be demonstrated. In this context it is interesting to note that auxin production by mycorrhizal fungi in pure culture is reduced by high levels of nitrogen (Moser 1959). If the morphogenetic effects of mycorrhizal infection are, as has been suggested (Slankis 1973), auxin-based, then the extremely high numbers of root tips in some mor humus horizons may in part be induced by the fungus. The pattern of root production in these soils may therefore be in response to a combination of factors; the maintenance of a functional balance between root and shoot (Kern, Moll & Braun 1961; Thornley 1977) which leads to relatively greater absorbing root production on nitrogen-poor sites; proliferation of roots in zones of local enrichment, *i.e.* the humus layers (Drew 1975); and an exaggeration of these processes as a result of mycorrhizal infection because of hormonal effects and the loss of gravitational response.

In view of the prevalence of ectomycorrhizas in humus layers it is not surprising that early in their investigation it should be proposed that their primary role lay in the absorption of organic nitrogen (Frank 1894). This 'nitrogen theory' had considerable support among early workers until Hatch's (1937) studies directed attention more towards phosphorus and the internal nutrient status of the plant. Since then there has been considerable speculation and assertion about ectomycorrhizas and nitrogen cycling and rather less information.

UTILIZATION OF ORGANIC NITROGEN

For mycorrhizas to exploit organic sources of nitrogen the fungi concerned must have a suitable enzyme potential and suitable substrates must be available. Enzymes for the decomposition of humus-bound nitrogen are apparently absent (Slankis 1974) and Lundeberg (1970) showed that 24 isolates of ectomycorrhizal fungi were unable to utilise ^{15}N from gamma-sterilized raw humus. Thus in order to envisage an important role for ectomycorrhizal fungi in the release of complex organic nitrogen from humus we have to postulate that they behave very differently in symbiosis than in pure culture. What little direct evidence there is suggests that this is not the case (Lundeberg 1970; Todd 1979). Nevertheless the assumption lingers, often supported by reference to the 'direct mineral cycling' theory developed by Went & Stark

(1968a, b) for South American rainforest. Two points have to be borne in mind when assessing the relevance of this theory. The type of mycorrhiza involved is not defined but likely to be vesicular–arbuscular, and although Herrera *et al.* (1978) present an electron micrograph showing the passage of labelled phosphate through a septate hyphae from litter to a mycorrhizal root, there is no evidence that enzymes from the mycorrhizal fungus participate directly in the decomposition process (Swift, Heal & Anderson 1979).

Ectomycorrhizal fungi in pure culture do utilize a number of simpler forms of organic nitrogen. There is considerable variation in response between and within species but in general peptone, casein hydrolysate and, in particular, the amino acids provide a nitrogen source for many isolates equal to or better than ammonium (Lundeberg 1970). Occasionally other substances, urea (Melin 1925; Laiho 1970), nucleic acids (Melin 1925; Mikola 1948; Laiho 1970), legumin (protein) (Melin 1925; Mikola 1948) and purines (Melin 1925, 1959) have been found to support good growth. Amino acids may serve as a sole source or give rise to increased growth in the presence of ammonium. Mixtures are normally more effective than single amino acids but their relative proportions are important. The interaction between ammonium and amino acids is complex and probably involves regulation of both transport and assimilation (Harley 1969; Whitaker 1976). Proteinase activity has also been found in several ectomycorrhizal isolates (Lyr 1963; Lundeberg 1970; Pachlewski & Chrusciak 1979).

It is important to know whether the obvious fungal potential to utilize simple organic nitrogen extends to the symbiotic state and whether under natural conditions substrates are available. In laboratory studies Melin & Nilsson (1953) have shown that ^{15}N from ^{15}N-glutamic acid may be transferred to the tissue of *Pinus sylvestris* L. seedlings through the hyphae of the fungal symbiont and Carrodus (1966) and Marčenko (1967) have observed uptake of glutamic and aspartic acid and their amides into excised *Fagus sylvatica* and *Picea abies* mycorrhizas. Miller (1967) in Bowen (1973) reported that mycorrhizal *Pseudotsuga menziesii* (Mirb.) Franco and *Pinus radiata* D. Don absorbed organic nitrogen from soil and that differences occurred between mycorrhizal fungi in use of amino acids. Stribley & Read (1980) have demonstrated efficient utilization of amino acids as a nitrogen source by *Vaccinium macrocarpon* Ait. infected with ericoid endomycorrhizas.

Fig. 4.1 shows the results of an experiment where aspartic acid and serine were supplied to mycorrhizal and non-mycorrhizal *Picea sitchensis* (Bong.) Carr. seedlings growing in vermiculite under sterile conditions with an ammonium source of nitrogen. The isolate of *Lactarius rufus* Fries used for inoculation was known to grow on serine or aspartic acid as sole sources of nitrogen. Aspartic acid had depressed its growth when supplied in addition to

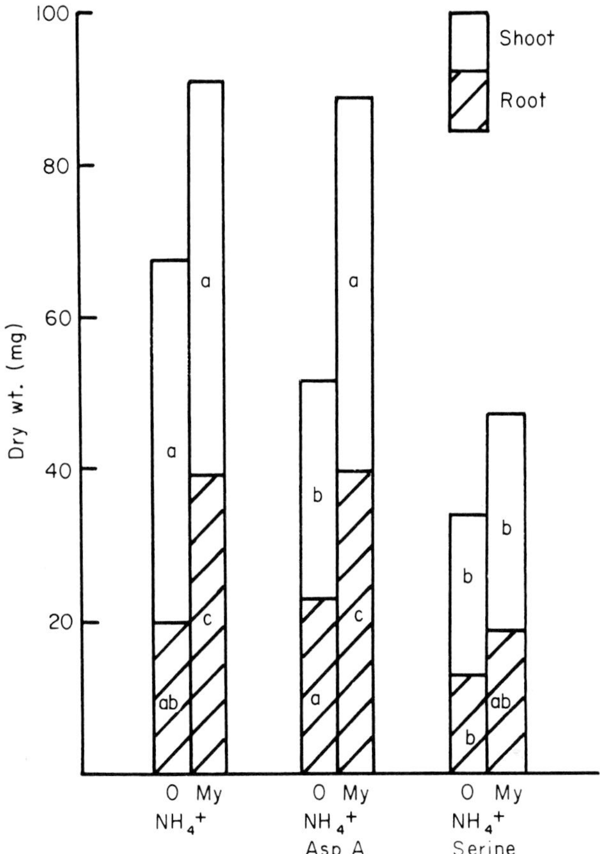

FIG. 4.1. Effect of 0.5 mM L-aspartic acid and L-serine on the growth of 12-week-old *Picea sitchensis* (Bong.) Carr. seedlings with (My) and without (0) ectomycorrhizas formed in aseptic culture by *Lactarius rufus* Fries. Values designated with the same letter are not significantly different at the 5% level. There was no significant difference in inoculated treatments in the percentage of root tips converted to mycorrhizae. (After Alexander 1973.)

ammonium. Both acids depressed seedling growth. The inhibitory effect of aspartic acid was removed by mycorrhizal infection; that of serine was reduced. The effects of other amino acids, mixtures and differing concentrations require investigation; nevertheless, it is clear that the pattern of nitrogen assimilation by ectomycorrhizas cannot easily be deduced from that of the constituent organisms and that the fungus should not only be considered as having a nutritional role but one of detoxification also.

Natural leachates of forest humus may contain substantial amounts of soluble organic nitrogen (Sollins *et al.* 1980; H. Miller pers. comm.) but detailed information on the forms of nitrogen and their rates of release are not available. Free amino acids are known to occur but quantitative and qualitative evaluation is hampered by problems of extraction and the difficulties of knowing whether extracts represent amino acids which are in soil solution, or are absorbed to colloids etc., or are constituents of living microbial cells (Kowalenko 1978). Results from a number of extracts of organic horizons are shown in Table 4.1. Amounts are greatly in excess of the value of 2 $\mu g\ g^{-1}$ commonly quoted for mineral soil (Bremner 1967; Wainwright & Pugh 1975) and in some cases may be comparable with extractable mineral (ammonium + nitrate) nitrogen (Alexander 1973; Labroue & Carles 1977). A substantial soluble protein or peptide fraction is also present in the extracts and increases in amino nitrogen of 50–80% (Alexander 1973; H_2O extract) and 200–300% (Sowden & Ivarson 1966; CCl_4 extract) follow hydrolysis. Van der Linden (1971) reported that from 57–82% of the total nitrogen in decomposing hazel leaves could be mineralized by a proteolytic enzyme.

A significant pool of organic nitrogen is therefore likely to be available to ectomycorrhizas. The speed with which added amino acids disappear (Schmidt, Putnam & Paul 1960; Dedeken & Voets 1965) indicates that the pool is highly labile. Distribution in the profile supports the conclusion of Nykvist (1963) that soluble nitrogen in fresh litter is an important input. The

TABLE 4.1. Free amino acid content of some organic soil horizons (n.d. = not determined).

Horizon		Amino acid extracted in μg amino-N g^{-1} dry soil			Reference
		H_2O	$H_2O + Ba(OH)_2$	Organic solvent	
Podsol	AoL	365	375	1169*	Alexander (1973)
(Spruce)	AoF	77	324	292	
	AoH	2	283	123	
Podsol	AoF	n.d.	n.d.	506†	Grov (1963 a, b)
(Pine)	AoH	n.d.	n.d.	683	
	A	n.d.	n.d.	58	
Brown forest	Ah	<6	n.d.	154‡	Sowden & Ivarson (1966)
Podsol	H	<4	n.d.	106‡	
Alpine soils		100§	n.d.	n.d.	Labroue & Carles (1977)

* 20% Ethanol, † Ether, ‡ CCl_4, § 10 min at 80°C

size of the pool at any given time presumably represents the balance between inputs from plant and microbial sources and losses through mineralization and immobilization or resynthesis. The mycorrhizal fungus will compete with free-living immobilizers and mineralizers and uptake will reduce both the residence time of that nitrogen in the soil and the possibility of its loss through leaching or its incorporation into recalcitrant forms.

The utilization of organic nitrogen can also be assessed by comparing the uptake of nitrogen by ectomycorrhizal vegetation with independently measured rates of nitrogen mineralization for the same vegetation. If uptake, plus leaching losses, are equal to, or less than, release in mineralization, plus whatever is added in precipitation, then it is unlikely that any recalcitrant organic nitrogen is being used. This approach (Table 4.2) indicates that mineralization and precipitation inputs do satisfy all, or the great majority of, the uptake requirement of the vegetation for which data are available. The comparison is crude and depends not only on the acceptance of 'potential mineralization' data but also on a number of assumptions, particularly that mineralization and precipitation are additive and that mineralization rates are unaffected by the presence of living roots. Seasonal effects are also averaged out. Nevertheless, the evidence favours the conclusion that ectomycorrhizal plants absorb nitrogen primarily from mineral sources or from sources which, in the absence of roots, would be easily mineralized.

UTILIZATION OF NITRATE AND AMMONIUM NITROGEN

In many of the situations characterized by profuse ectomycorrhizal development the ammonium ion is the predominant form of mineral nitrogen and nitrification is slow. Lee & Stewart (1978) have discussed the factors thought to lower nitrification rates. These include soil acidity, competition for ammonium or phosphorus between nitrifiers and other microorganisms, and biotic influences, particularly allelochemicals associated with late successional or climax vegetation. Rice & Pancholy (1972) have argued that this latter phenomenon promotes ecosystem efficiency as leaching losses of the mobile nitrate ion are minimized and energy is conserved by the avoidance of the nitrate reduction step in nitrogen assimilation (see also Chapters 5 and 8). However, it would be an oversimplification to assume that nitrate is never available to ectomycorrhizal vegetation and that its utilization is only of academic interest. Runge (1974 a, b), for example, has shown considerable nitrification in acid beechwood soil. Significantly, measured rates were high only in the H horizon and mineral soil and not in the L and F horizons where ectomycorrhizal activity was highest. Substantial amounts of nitrate have also been found in soil solution and drainage water from a number of forest

TABLE 4.2. A comparison between nitrogen uptake by ectomycorrhizal vegetation and net nitrogen mineralization measured by field incubation (n.d. = not determined).

Vegetation	Annual flux of nitrogen in $kg\ ha^{-1}$					References
	Uptake[1]	Leaching loss	Net mineralization	Precipitation input	(C+D)−(A+B)	
	A	B	C	D		
Fagus sylvatica 125 years Solling, Germany	58[2]	6	93	29[3]	+58	Runge (1974 a, b) Prenzel (1979)
Fagus sylvatica 130 years Mirwart, Belgium	73	n.d.	107	10	+44	Denaeyer de Smet & Duvigneaud (1972), Van Praag *et al.* (1973), Duvigneaud & Denaeyer de Smet (1970)
Picea abies 55 years Mirwart, Belgium	52	n.d.	34[4]	10	−8	Denaeyer de Smet & Duvigneaud (1972), Froment & Remacle (1975)
Quercus ilex Montpellier, France	46	n.d.	26	15	−5	Rapp (1971), Loissant (1973)
Quercus coccifera Montpellier, France	29	n.d.	< 22	15	< +8	Rapp (1971), Loissant (1973)
Salix herbacea Mt Patscherkofel, Austria	2[5]	n.d.	> 1[6]	n.d.	> −1	Rehder & Schäfer (1978)

[1] Measured as retention in biomass + return in litter and leachate
[2] Above-ground parts only
[3] Includes canopy leachate
[4] From incubations *in vitro*
[5] Measured as differences between maximum and minimum content of above-ground parts during growing season
[6] Growing season only

systems (Vitousek 1977; Vitousek & Reiners 1975; Vitousek *et al.* 1979). Further investigation of nitrate production in a greater range of ectomycorrhizal communities is required, particularly in the tropics where nitrification rates in non-ectomycorrhizal lowland forest are high (de Rham 1970; Lamb 1981).

In culture ectomycorrhizal fungi, in common with many higher fungi, make better growth on ammonium salts than on nitrates. Lundeberg (1970) found that only 60% of the isolates he tested could utilize nitrate–N and of these only 45% made comparable growth on nitrate and ammonium sources. Trappe (1967) and Ho & Trappe (1980) found nitrate reductase activity to be low or absent in the isolates they tested. In view of the apparently limited ability of ectomycorrhizal fungi to utilize nitrate, the effect of nitrate on mycorrhizal formation and the capacity of intact mycorrhizas or mycorrhizal plants to absorb nitrate have been questioned.

There are reports (Lindquist 1932; Richards 1961; Richards & Wilson 1963; Theodorou & Bowen 1969) that nitrate inhibits mycorrhizal infection in natural soils. In some respects their distribution (see above) lends support to this theory. However, attempts to differentiate between effects of nitrate *per se* and correlated factors, notably pH, have not been very satisfactory. In addition it is not clear whether the effect is on the fungi alone (as a result of their reduced ability to utilize nitrate) or on the infection process. Bigg (1981) has recently clarified the situation. He grew conifer seedlings in aseptic culture conditions permitting periodic renewal of nutrients (Bigg & Alexander 1981). Shifts in pH associated with nitrogen source were therefore minimized. The fungi used for inoculation responded differently to nitrate in pure culture: *Paxillus involutus* (Fries) Karsten grew almost as well on nitrate as on ammonium; *Lactarius rufus* Fries was unable to utilize nitrate at all. The uninfected tree hosts also showed nitrogen source preferences: at pH 5.0 *Pseudotsuga menziesii* grew equally well on either source while *Picea sitchensis* made better growth with ammonium. Under the experimental conditions nitrate greatly reduces mycorrhizal infection (Table 4.3) even when both host and fungus can utilize that nitrogen source. Bigg (1981) also showed that inhibition of infection is not due to bound or diffusible inhibitors produced by the host under nitrate nutrition and that mycorrhizas formed under ammonium nutrition are disrupted by a switch to nitrate.

If nitrate does inhibit mycorrhizal formation and nitrate levels are low in soils where mycorrhizal infection is high, then naturally occurring mycorrhizas might be expected not to absorb nitrate readily. This is indeed the case with excised beech mycorrhizas (Carrodus 1966, 1967; Smith 1972). Data of Haines (1977) indicating a low uptake of nitrate applied *in situ* (in contrast to the ready uptake of ammonium) in *Pinus*/*Quercus* woodland also support this view. On the other hand, France & Reid (1978) have demonstrated short-term uptake of nitrate by excised pine mycorrhizas. However, these mycorrhizas were synthesized under aseptic conditions on an ammonium source of nitrogen. Nitrate was generally absorbed at a lower rate than ammonium, and the relative rates differed between fungal symbionts. Further investigations are obviously required.

Ammonium uptake by ectomycorrhizas has been studied using excised beech mycorrhizas (Carrodus 1966, 1967). Ammonium salts are readily utilized, the rate of absorption being dependent on the incorporation of absorbed ammonium into organic nitrogen compounds in the tissue. Thus exogenous carbohydrate, bicarbonate or tricarboxylic acids can stimulate ammonium uptake (Harley 1964). From 65–75% of the absorbed ammonium is at first accumulated as glutamine in the fungal sheath (Harley 1978) and a pattern of uptake, accumulation, and transfer similar to that already worked out for phosphorus may also apply to nitrogen. Reid & Lewis (see Lewis 1976) have shown that nitrogen moves to the host as glutamine.

The excised pine mycorrhizas used by France & Reid (1978) also absorbed ammonium rapidly and to a greater extent than uninfected roots on a dry weight basis. This was attributed to the increased surface area of mycorrhizas and/or to increased rates of uptake. As with nitrate, the identity of the fungal symbiont influenced uptake. This is not surprising in view of the range of behaviour of ectomycorrhizal fungi in pure culture. Bigg (1981) has demonstrated this by comparing the growth and nitrogen uptake of *P. involutus* and

TABLE 4.3. Effect of ammonium and nitrate sources of nitrogen on the growth and mycorrhizal infection of 100-day-old *Picea* and *Pseudotsuga* seedlings grown under aseptic conditions and inoculated with *Paxillus involutus* (Fr.) Karsten or *Lactarius rufus* Fr. Means ($n = 10$) for each tree species denoted by the same letter are not significantly different at the 5% level. (After Bigg 1981.)

Tree species	Mycorrhizal fungus	N source	Total dry weight in mg	No. of root tips	% tips converted to mycorrhizas
Pseudotsuga	*Paxillus*	NH_4^+	261a	162a	69a
menziesii	*involutus*	NO_3^-	268ab	165a	20b
(Mirb.) Franco	*Lactarius*	NH_4^+	316b	149a	69a
	rufus	NO_3^-	241a	111a	0c
Picea	*Paxillus*	NH_4^+	261a	371a	63a
sitchensis	*involutus*	NO_3^-	154b	284a	8b
(Bong.) Carr.	*Lactarius*	NH_4^+	248a	382a	80c
	rufus	NO_3^-	169b	314a	0d

L. rufus in liquid culture supplied with 25 mg l^{-1} nitrogen as ammonium and either 0.5, 1.0 or 2.0 g l^{-1} glucose (Fig. 4.2). *L. rufus* attained the greater dry weight per unit glucose supplied, particularly at the lower levels of supply. It also showed more rapid growth during the rapid growth phase. *P. involutus*, on the other hand, accumulated more nitrogen per unit dry weight produced. Again this was most pronounced at low levels of glucose supply. When glucose supply was exhausted *P. involutus* rapidly goes into autolysis and ammonium is released into the culture medium. This was never the case with *L. rufus* over the duration of these experiments. The ability of fungal symbionts to absorb and retain nitrogen relative to the carbohydrate drain imposed on the host is of obvious importance and could be a factor influencing the distribution of mycorrhizal associations in the field. *P. involutus*, for example, is reported to fruit most abundantly on good sites and 'the amount of available nitrogen must be regarded as the most important edaphic factor regulating (its) growth' (Laiho 1970).

THE EFFICIENCY OF NITROGEN ABSORPTION UNDER FIELD CONDITIONS

From the preceding sections ectomycorrhizas can be seen as efficient organs for the absorption of ammonium and simple organic nitrogen from solution. The question now arises as to their efficiency under field conditions. In the introduction the importance of increased surface area, radiating hyphae and strands in the uptake of immobile elements such as phosphorus was stressed. Although the ammonium ion is much more mobile than phosphate the work of Prenzel (1979) in the Solling beech forests suggests that the demand for nitrogen by the trees is not satisfied by mass flow in the transpiration stream. In these circumstances increased exploitation of the soil by mycorrhizas will increase uptake. This work is likely to have general applicability to ectomycorrhizal vegetation. Where the more mobile nitrate ion is the major available form of nitrogen the mycorrhizal effect would be much less important (Bowen 1973).

Increased uptake as a result of extension into unexploited zones of the soil is only one aspect of nitrogen absorption by plants growing in closed communities. The other is competition for available nitrogen both between and within plant species and between plants and microorganisms. In this respect both the rate of depletion of the available supply of nitrogen and the extent of that depletion become important. There is theoretical and experimental evidence that rooting density influences both the extent and rate of depletion of all but the least mobile ions and that quite low root densities compete for nitrogen (Barley 1970; Andrews & Newman 1970; Nye & Tinker

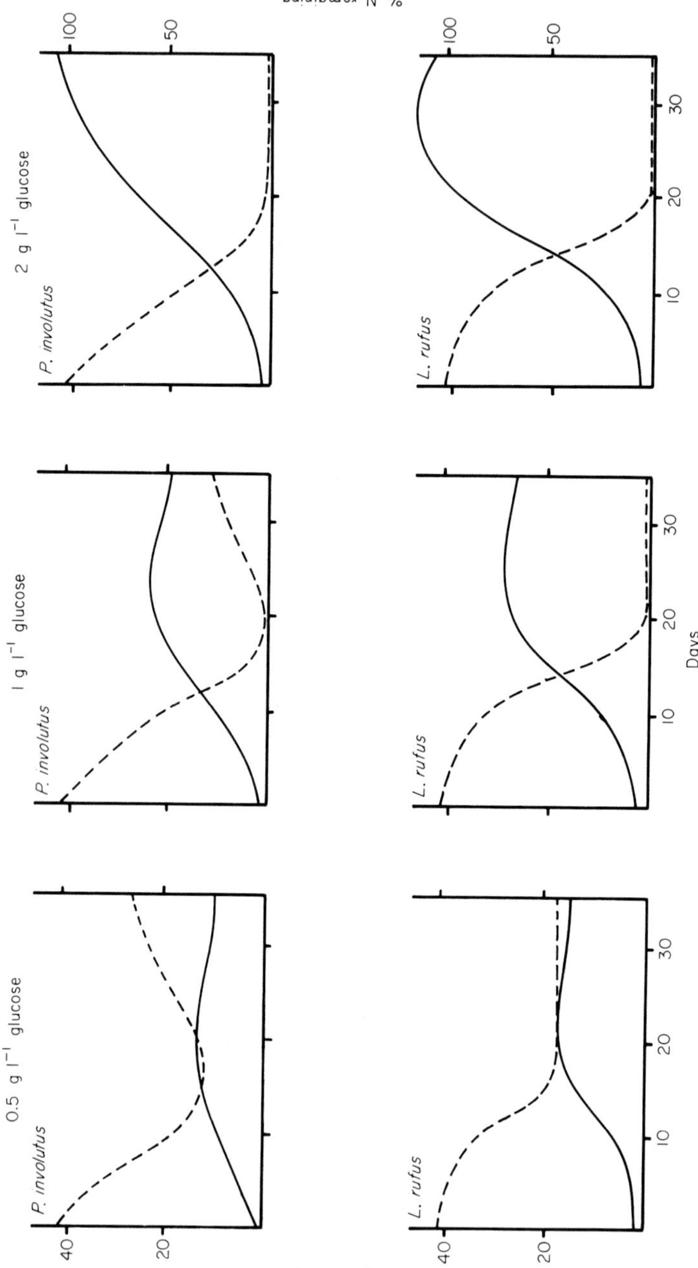

FIG. 4.2. Growth and nitrogen uptake of *Paxillus involutus* (Fr.) Karsten and *Lactarius rufus* Fr. at pH 4.5 and 20°C in shaking liquid culture with 0.5, 1 and 2 g l^{-1} glucose and 25 mg l^{-1} nitrogen as $(NH_4)_2SO_4$. Smoothed curves for dry weight (——) and percentage of original nitrogen remaining in the culture solution (- - - -) are fitted to means of three replicate samples taken every five days. (After Bigg 1981.)

1977). Bowen (1973) has commented on the sparsity of the roots of woody species relative to herbaceous roots and echoed the views of Went & Stark (1968a) and Harley (1969) that ectomycorrhizal infection converts such root systems into highly efficient and competitive exploiters of available nutrient particularly in the surface layers of soil. Harley (1978) has added that in many situations release of nutrients into the available pool is likely to be periodic and that interception of flushes of released nutrients may be particularly important.

Demonstrating the extent and efficiency of exploitation under natural conditions is more difficult. Bowen (1973, 1980) has doubted if interception approaches 100%. A rough estimate can be obtained from lysimeter data extracted from nutrient budget studies. Prenzel's (1979) data for beech show a 65% drop in nitrogen concentration in the soil solution from 7.22 mg l^{-1} in unrooted litter to 2.5 mg l^{-1} at 10 cm depth in the mineral soil. Vitousek (1977) gives data on the ammonium and nitrate concentration in leachate from below the A horizon of trenched and untrenched lysimeter plots in spruce-fir and fir forest. These show a 60–70% drop in concentration below live roots. Sollins *et al.* (1980) estimate that 77% of the flux of nitrogen out of the 'litter solution' compartment in their model of nitrogen cycling in old growth Douglas fir is plant uptake. These examples only serve to indicate possible levels of exploitation and this is an area of ectomycorrhizal research which urgently requires further study.

Overall the evidence suggests that ectomycorrhizas are efficient exploiters of, and competitors for, nitrogen.

ACCUMULATION AND TURNOVER OF NITROGEN IN MYCORRHIZAS

Consideration has so far been given to the formation of ectomycorrhizas and the process of nitrogen uptake but quantitative data on the accumulation and turnover of nitrogen in ectomycorrhizal tissue is also required. The importance of fine root dynamics to nutrient cycles has been stressed in a number of recent articles (Coleman 1976; Cox *et al.* 1977; Santantonio, Hermann & Overton 1977) and studies on forest systems have shown that most of the organic matter input to the decomposition process results from fine root production and that this is a major pathway for returning nitrogen from living vegetation to soil (Henderson & Harris 1975; Wells & Jorgensen 1975). Fogel (1980) has recently reviewed the contribution of mycorrhizas to fine root dynamics. He points out the formidable problems in investigating and quantifying this component of the ecosystem and the difficulties in making comparison between different studies. Unfortunately data are only available for temperate and boreal forests and not for other ectomycorrhizal communities. Some general points emerge from Fogel's review. The annual production of fine roots and mycorrhizas may be up to twice the annual litter fall. Annual throughput or 'die-back' values are from 40–90% and suggest that mycorrhizas last on average from 1–3 years. Such throughputs may result in a nitrogen input to dead organic matter from one to six times that from litterfall.

These points are illustrated by R. Fairley's recent work on mycorrhizal numbers and fine root biomass in the L, F, and H horizons of a 36-year-old Sitka spruce stand (Fig. 4.3). The numbers of living mycorrhizal roots show seasonal fluctuations with maxima in spring and autumn and minima in summer and winter. The biomass of living fine roots (< 1 mm diameter) also follows this pattern and indeed the great majority is mycorrhizal tissue. By assessing the biomass of living small roots (1–5 mm diameter) and the biomass (necromass) of dead roots (< 5 mm diameter) as well, and then balancing the monthly transfers (Santantonio 1978) between these three compartments a picture of the root dynamics in the stand emerges. The greatest activity takes place in the fine (< 1 mm) root compartment with a mean annual standing crop of around 130 kg ha^{-1}, mean annual production of 615 kg ha^{-1} and mortality of 595 kg ha^{-1}. The turnover of 97% indicates that on average the mycorrhizas do not last for more than one year. Corresponding figures for the small (1–5 mm) root compartment are 880 kg ha^{-1}, 300 kg ha^{-1} year^{-1} and 225 kg ha year^{-1}. The standing crop of dead roots (< 5 mm) is 1800 kg ha^{-1}, about 2.2 times the annual input.

The nitrogen content of the fine roots also shows some seasonal fluctuations. This may represent redistribution of nitrogen from roots before they are shed or the accumulation of nitrogen storage compounds during the winter months (Taylor 1957; Van den Driessche & Weber 1977). Mean nitrogen content of living fine roots is 1·8% and of dead roots 1.3%. Their standing crops are 2.3 kg ha^{-1} and 23 kg ha^{-1} nitrogen respectively, with an annual flux of nitrogen into the dead root compartment (assuming some withdrawal before shedding) of about 11 kg ha^{-1} year^{-1}. This is between 30–50% of the likely input of nitrogen in litter fall for a stand of this age (Owen 1954; Adams, Dickson & Quinn 1980). These estimates show that standing crop and turnover of nitrogen in mycorrhizas on this study site are considerable. Even so they are small compared to those quoted by Fogel (1980).

It would be useful to know the extent to which ectomycorrhizas and ectomycorrhizal fungi in the field accumulate and store nitrogen and whether nitrogen in dead mycorrhizal tissue is preferentially reabsorbed by living mycorrhizas. Unfortunately, these are areas in which little information is available. Fairley's data (quoted above) shows seasonal increases in nitrogen

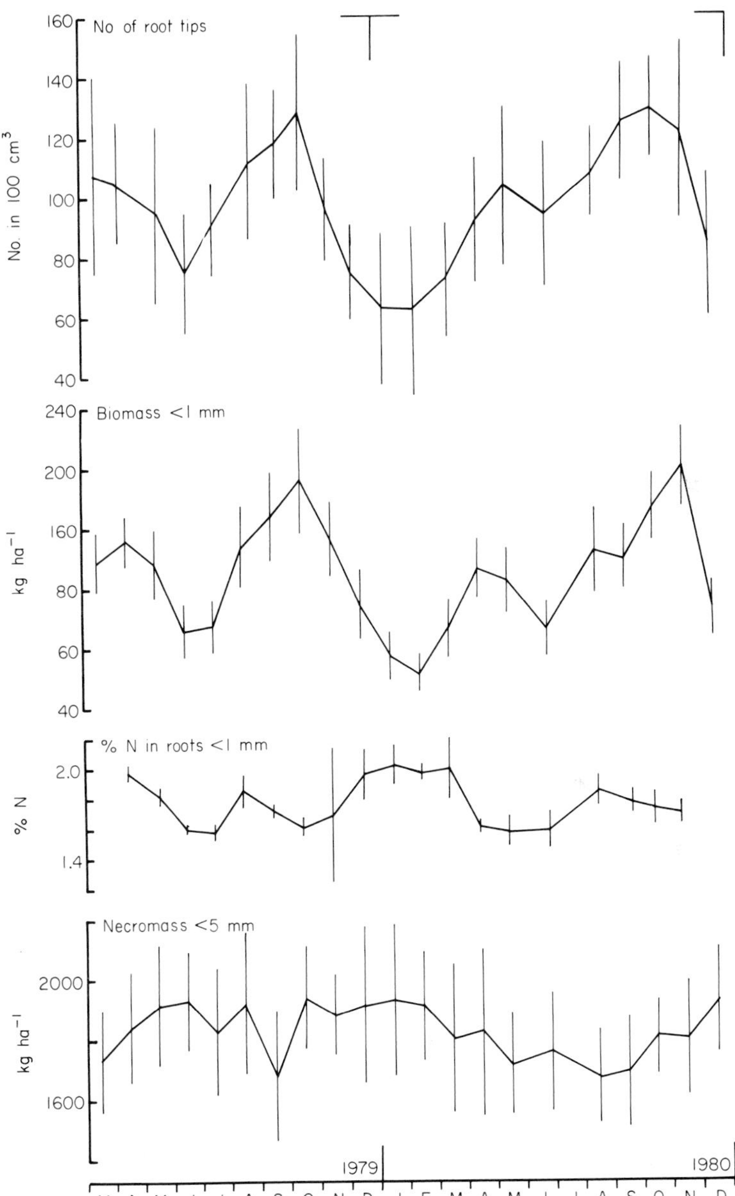

FIG. 4.3. Seasonal fluctuations (with standard errors) in number of living root tips, biomass and percentage nitrogen content of living roots and mycorrhizas < 1 mm diameter and biomass (necromass) of dead roots < 5 mm diameter. Unpublished data of R.Fairley from the L, F, H and Ae horizons of a 36-year-old *Picea sitchensis* (Bong.) Carr. stand at Kirkhill, Aberdeenshire.

content of mycorrhizas and indicates the possibility of resorption. Harley (1940) reported that the nitrogen content of beech mycorrhizas (2.0–2.5%) was greater than that of uninfected fine roots (1.7%). Uptake into the sheath of excised mycorrhizas (Harley 1978) and measured concentrations (2.5–4.0%) in sporocarps (Vogt & Edmonds 1980) indicate that the fungal component in particular has a high potential to accumulate nitrogen. Transfer from dead or dying ectomycorrhizas has not been demonstrated but the colonization of senescing cortical cells by the mycorrhizal fungus in, for example, beech mycorrhizas (W. Bigg & I. Alexander, unpublished data) and the demonstration of phosphorus transfer in VA systems (Heap & Newman 1980) indicates that such movement is likely.

CONCLUSIONS

Although research on nitrogen uptake by ectomycorrhizas has been fragmented and almost exclusively restricted to temperate forest trees, some general conclusions can be drawn. They appear to be efficient competitors for ammonium and probably soluble organic nitrogen. Their success is related to the spatial exploitation of the substrate by the fungal partner and the proximity of hyphae to the sites of release of available nitrogen. Their comparative longevity allows exploitation over a period of time (Grime 1979) and the accumulation of nutrients outside the growing season (Waring & Franklin 1979). Success must also be due to the fact that, in contrast to saprotrophic soil organisms which derive their energy from resistant carbon in detritus, the acquisition of nitrogen by ectomycorrhizal fungi is fuelled by simple carbohydrates obtained from the autotrophic host. A striking demonstration of this is given by Turner (1977). An application of sucrose and sawdust to the floor of a 45-year-old, nitrogen-limited, Douglas fir stand reduced uptake of nitrogen into the trees by 80%. Nitrogen level in the forest floor solution did not appear to be significantly affected and remained at a low level ($< 1\ \mu g\ ml^{-1}$). Apparently the soil microflora were now able to compete effectively with ectomycorrhizal fungi and available nitrogen was not entering the trees but was instead immobilized in microbial tissue.

Competitive withdrawal of nitrogen from soil solution by ectomycorrhizas may have other consequences. Autotrophic nitrifiers are poor competitors for ammonium ions compared with heterotrophic microorganisms, particularly when energy is readily available (Jones & Richards 1977; Johnson & Edwards 1979; Lamb 1980). Under conditions of nitrogen stress ammonium availability to nitrifiers will be low in those soil horizons where ectomycorrhizas are active and, as a consequence, initial populations of nitrifiers may be low and nitrification difficult to demonstrate. Temporal variation may be similarly explained. Where ammonium availability is higher as a result of greater fertility and/or reduced ectomycorrhizal activity, nitrification will increase. The observed patterns of mycorrhizal infection, mycorrhizal density and nitrification in beechwood soils conform to this hypothesis (see above). These conclusions are in agreement with those of Lamb (1981) who has suggested that the pattern of nitrification is dependent on the availability of ammonium nitrogen which in turn is a function of nutrient stress.

A reduction in the rate of litter decomposition in the presence of ectomycorrhizal roots has been conclusively demonstrated for *Pinus radiata* plantations by Gadgil & Gadgil (1971, 1975) and possibly in Swedish forests (B. Berg quoted by Tamm 1979). This too can be explained by competition for available nitrogen, in this case between mycorrhizas and decomposers, a hypothesis first put forward by Romell in 1935! Under conditions of nitrogen stress the mycorrhizal fungi with their ready source of energy are in an advantageous position; however, when nitrogen stress is relieved by addition of fertilizer the rate of litter decomposition rises (Turner 1977). Delayed decomposition is viewed as a stabilizing mechanism in nutrient cycles by many ecologists (Witkamp & Ausmus 1976) as nutrient release is controlled, leaching losses minimized and a slowly recycling pool of nutrient created. On the other hand, in extreme cases, an ever-growing proportion of the limited nitrogen capital of the site may become immobilized in the organic layers, the vegetation becomes progressively more dependent on rainfall input and nitrogen deficiency and declining productivity may result. From an evolutionary and ecological (rather than forest management) viewpoint this need not be disadvantageous. As long as the vegetation can regenerate, self-induced nitrogen stress may in fact be of selective advantage in that no other species can cope better with these conditions. Ectomycorrhizas may then be seen as contributing to the creation and maintenance of conditions of which they are the most efficient exploiters. Moreover, periodic disturbance by wildfire or climatic agents occurs in many ecosystems in which ectomycorrhizal plants are important components and serves to reduce surface organic accumulation and expose mineral soil. In this respect it is interesting that the establishment and mycorrhizal infection of the seedlings of some ectomycorrhizal trees is most successful in mineral soil (Bakshi *et al.* 1972; Alvarez, Rowney & Cobb 1979).

Evidence presented above has shown that the uptake of nitrogen into mycorrhizas is rapid and the quantity of nitrogen free in the soil solution usually very small. In a number of vegetation types with a strong ectomycorrhizal component (Mooney & Rundell 1979; Waring & Franklin 1979) redistribution of nitrogen within plant tissue is an important feature of their nitrogen economy. Taken together these features suggest that a controlled and virtually closed nitrogen cycle may be in operation, dominated to a considerable extent by the ectomycorrhizal fungi. Only plants capable of forming an association with these fungi can break into the cycle and become established. The degree of host/fungus specificity known to exist at least at the generic or family level may heighten this effect and in part explain the occurrence of ectomycorrhizal plants in monocultures and situations of family dominance.

ACKNOWLEDGEMENTS

I wish to thank W.L. Bigg and R.I. Fairley for allowing me to use their unpublished data.

REFERENCES

Adams S.N., Dickson E.L. & Quinn C. (1980) The amount and nutrient content of litterfall under Sitka spruce on poorly drained soils. *Forestry*, **53**, 65–70.

Alexander I.J. (1973) Unpublished Ph.D. thesis. University of Edinburgh.

Alvarez I.F., Rowney D.L. & Cobb F.W. Jr (1979) Mycorrhizae and growth of white fir seedlings in mineral soil with and without organic layers in a California forest. *Canadian Journal of Forest Research*, **9**, 311–15.

Andrews R.E. & Newman E.I. (1970) Root density and competition for nutrients. *Oecologia Plantarum*, **5**, 319–34.

Bakshi B.K., Ram Reddy M.A., Thapar H.A. & Khan S.N. (1972) Studies of silver fir regeneration. *Indian Forester*, **98**, 135–44.

Barley K.P. (1970) The configuration of the root system in relation to nutrient uptake. *Advances in Agronomy*, **22**, 159–201.

Baylis G.T.S. (1974) The evolutionary significance of phycomycetous mycorrhizas. *Mechanisms of Regulation of Plant Growth* (eds R.L. Bieleski, M. Cresswell & A.R. Ferguson). *Royal Society of New Zealand Bulletin*, **12**, 191–3.

Bigg W. L. (1981) Unpublished Ph.D. thesis. University of Aberdeen.

Bigg W.L. & Alexander I.J. (1981) A culture unit for the study of nutrient uptake by intact mycorrhizal plants under aseptic conditions. *Soil Biology and Biochemistry*, **13**, 77–8.

Bowen G.D. (1973) Mineral nutrition of ectomycorrhizae. *Ectomycorrhizae: Their Ecology and Physiology* (eds G.C. Marks & T.T. Kozlowski), pp. 151–205. Academic Press, New York & London.

Bowen G.D. (1978) Dysfunction and shortfalls in symbiotic response. *Plant Disease—An Advanced Treatise* (eds J.G. Horsfall & E.B. Cowling) Vol. 3, pp. 231–56. Academic Press, London & New York.

Bowen G.D. (1980) Mycorrhizal roles in tropical plants and ecosystems. In *Tropical Mycorrhiza Research* (ed. P. Mikola), pp. 165–90. Clarendon Press, Oxford.

Bremner J.M. (1967) Nitrogenous compounds. *Soil Biochemistry* (eds A.D. McLaren & G.H. Peterson), pp. 19–66. Edward Arnold, London.

Carrodus B.B. (1966) Absorption of nitrogen by mycorrhizal roots of beech. I. Factors affecting the assimilation of nitrogen. *New Phytologist*, **65**, 358–71.

Carrodus B.B. (1967) Absorption of nitrogen by mycorrhizal roots of beech. II. Ammonium and nitrate as sources of nitrogen. *New Phytologist*, **66**, 1–4.

Coleman D.C. (1976) A review of root production processes and their influence on soil biota in terrestrial ecosystems. In *The Role of Terrestrial and Aquatic Organisms in Decomposition Processes* (eds J.M. Anderson & A. MacFayden), pp. 417–34. Blackwell Scientific Publications, Oxford.

Cox T.L., Harris W.F., Ausmus B.S. & Edwards N.T. (1977) The role of roots in biogeochemical cycles in eastern deciduous forests. In *The Belowground Ecosystem: A Synthesis of Plant-associated Processes* (ed. J.K. Marshall), pp. 231–6. Range Science Department Science Series No. 26. Colorado State University, Fort Collins.

de Alwis D.P. & Abeynayake K. (1980) A survey of mycorrhizae in some forest trees of Sri Lanka. *Tropical Mycorrhiza Research* (ed. P. Mikola), pp. 146–3. Clarendon Press, Oxford.

Dedeken M. & Voets J.P. (1965) Studies on the metabolism of amino acids in soil. I. Metabolism of glycine, alanine, aspartic acid and glutamic acid. *Supplément, Annales Institut Pasteur, Paris*, **109**, 103–11.

Denaeyer de Smet S. & Duvigneaud, P. (1972) Comparison du cycle des polyéléments biogènes dans une hêtraie (Fagetum) et une pessière (Piceetum) établies sur même roche-mère, à Mirwart. *Bulletin de la Société Royale de Botanique de la Belgique*, **105**, 197–205.

Dominik T., Nespiak A. & Pachlewski R. (1954) Investigations on the mycotrophy of vegetal associations on calcareous rocks in the Tatra Mts. *Acta Societatis Botanicorum Poloniae*, **23**, 471–85.

Drew M.C. (1975) Comparison of the effects of a localized supply of phosphate, nitrate, ammonium and potassium on the growth of the seminal root system, and the shoot, in barley. *New Phytologist*, **75**, 479–90.

Dudderidge, J.A., Malibari A. & Read D.J. (1980) Structure and function of mycorrhizal rhizomorphs with special reference to their role in water transport. *Nature*, **287**, 834–6.

du Rham P. (1970) L'azote dans quelques forêts, savanes et terrains de culture d'Afrique tropicale humide (Côte d'Ivoire). *Veroffentlichungen Geobotanischen Institutes Rübel, Zurich*, **45**, 1–124.

Duvigneaud P. & Denaeyer de Smet S. (1970) Biological cycling of minerals in temperate deciduous forests. In *Analysis of Temperate Forest Ecosystems* (ed. D.E. Reichle), Ecological Studies Vol. I, pp. 199–225. Chapman & Hall, London.

Fassi B. & Fontana A. (1962) Micorrhize ectotrofiche di *Brachystegia laurentii* e di alcune altre Caesalpiniaceae minori del Congo. *Allionia*, **8**, 121–31.

Fogel R. (1980) Mycorrhizae and nutrient cycling in natural forest ecosystems. *New Phytologist*, **86**, 199–212.

France R.C. & Reid C.P.P. (1978) Absorption of ammonium and nitrate by mycorrhizal and non-mycorrhizal roots of pine. In *Root Physiology and Symbiosis, Proceedings of IUFRO Symposium, Nancy* (ed. A. Reidaker), pp. 410–24.

Frank A.B. (1894) Die bedeutung der Mykorrhizapilze für die gemeine Kiefer. *Forstwissenschaftliches Zentralblatt*, **16**, 185–90.

Froment A. & Remacle J. (1975) Evolution de l'azote minéral et de la microflore dans le sol d'une pessière (Piceetum) à Mirwart. *Bulletin de la Société Royale de Botanique de la Belgique*, **108**, 53–64.

Gadgil R.L. & Gadgil P.D. (1971) Mycorrhiza and litter decomposition. *Nature*, **233**, 133.

Gadgil R.L. & Gadgil P.D. (1975) Suppression of litter decomposition by mycorrhizal roots of *Pinus radiata*. *New Zealand Journal of Forest Science*, **5**, 33–41.

Grime J.P. (1979) *Plant Strategies and Vegetation Processes*. John Wiley, Chichester.

Grov A. (1963a) Amino acids in soil. II. Distribution of water soluble amino acids in a pine forest soil profile. *Acta Chemica Scandinavica*, **17**, 2316–18.

Grov A. (1963b) Amino acids in soil. III. Acids in hydrolyzates of water extracted soil and their distribution in a pine forest soil profile. *Acta Chemica Scandinavica*, **17**, 2319–24.

Haines B.L. (1977) Nitrogen uptake. Apparent pattern during old field succession in southeastern US. *Oecologia (Berlin)*, **26**, 295–303.

Harley J.L. (1940) A study of the root system of the beech in woodland soils, with especial reference to mycorrhizal infection. *Journal of Ecology*, **28**, 107–17.

Harley J.L. (1964) Incorporation of carbon dioxide into excised beech mycorrhizas in the presence and absence of ammonia. *New Phytologist*, **63**, 203–8.

Harley J.L. (1969) *The Biology of Mycorrhiza.* 2nd edn. Leonard Hill, London.

Harley J.L. (1975) Problems of mycotrophy. In *Endomycorrhizas* (eds F.E. Sanders, B. Mosse & P.B. Tinker), pp. 1–24. Academic Press, London & New York.

Harley J.L. (1978) Nutrient absorption by ectomycorrhizas. *Physiologie Végétale*, **16**, 533–45.

Hatch A.B. (1937) The physical basis of mycotrophy in the genus *Pinus. Black Rock Forest Bulletin*, **6**, 1–168.

Heap A.J. & Newman E.I. (1980) The influence of vesicular–arbuscular mycorrhizas on phosphorus transfer between plants. *New Phytologist*, **85**, 173–9.

Henderson G.S. & Harris W.F. (1975) An ecosystem approach to characterisation of the nitrogen cycle in a deciduous forest watershed. In *Forest Soils and Forest Land Management* (eds B. Bernier & C.H. Wright), pp. 179–93. Les Presses de l'Université Laval, Québec.

Herrera R., Merida T., Stark M. & Jordan G.F. (1978) Direct phosphorus transfer from leaf litter to roots. *Naturwissenschaften*, **65**, 208–9.

Ho I. & Trappe J.M. (1980) Nitrate reductase activity of non-mycorrhizal Douglas fir rootlets and some associated mycorrhizal fungi. *Plant & Soil*, **54**, 395–9.

Johnson D.W. & Edwards N.T. (1979) The effects of stem girdling on biogeochemical cycles within a mixed deciduous forest in eastern Tennessee. II. Soil nitrogen mineralisation and nitrification rates. *Oecologia (Berlin)*, **40**, 259–71.

Jones J.M. & Richards B.N. (1977) Effect of reforestation on turnover of ^{15}N-labelled nitrate and ammonium in relation to changes in soil microflora. *Soil Biology & Biochemistry*, **9**, 383–92.

Kowlaenko C.G. (1978) Organic nitrogen, phosphorus and sulphur in soils. In *Soil Organic Matter* (eds M. Schnitzer & S.V. Khan), pp. 95–136. Elsevier Scientific Publishing Co., Amsterdam.

Kramer P.J. & Kozlowski T.T. (1979) *Physiology of Woody Plants.* Academic Press, London & New York.

Labroue L. and Carles J. (1977) Le cycle de l'azote dans les sols alpins du Pic du Midi de Bigorre (Hautes Pyrénées). *Oecologia Plantarum*, **12**, 55–77.

Laiho O. (1970) *Paxillus involutus* as a mycorrhizal symbiont of forest trees. *Acta Forestalia Fennica*, **106**, 1–72.

Lamb D. (1980) Soil nitrogen mineralisation in a secondary rainforest succession. *Oecologia (Berlin)*, **47**, 257–63.

Largent D.L., Sugihara N. & Wishner C. (1980) Occurrence of mycorrhizae on ericaceous and pyrolaceous plants in northern California. *Canadian Journal of Botany*, **58**, 2274–9.

Lee J.A. & Stewart G.R. (1978) Ecological aspects of nitrogen assimilation. *Advances in Botanical Research* Vol. 6, pp. 1–43. Academic Press, London & New York.

Lewis D.H. (1975) Comparative aspects of the carbon nutrition of mycorrhizas. In *Endomycorrhizas* (eds F.E. Sanders, B. Mosse & P.B. Tinker), pp. 119–49. Academic Press, London & New York.

Lewis D.H. (1976) Interchange of metabolites in bitrophic symbioses between angiosperms and fungi. *Perspectives in Experimental Biology*, **2**, 207–19.

Lindquist B. (1932) Den sydskandinaviska kulturgromomskogens reproduktionsförhällanden. *Svenska Skogsvardsforeningens Tidskrift*, **30**, 17–38.

Loissant P. (1973) Soil–vegetation relationships in Mediterranean ecosystems of Southern France. In *Mediterranean Type Ecosystems. Origin and Structure* (eds F. Di Castri & H.A. Mooney), Ecological Studies, Vol. 7, pp. 199–210.

Lundeberg G. (1970) Utilisation of various nitrogen sources in particular bound soil nitrogen, by mycorrhizal fungi. *Studia Forestalia Suecica*, **79**, 1–95.

Lyr H. (1963) Zur Frage des Streuabbaus durch ectotrophe Mykorrhizapilze. In *Mykorrhiza*, International Mykorrhiza Symposium, Welmar 1960 (eds W. Rawald & H. Lyr), pp. 123–45.

Malloch D.W., Pirozynski K.A. & Raven P.H. (1980) Ecological and evolutionary significance of mycorrhizal symbioses in vascular plants (a review). *Proceedings of the National Academy of Sciences of the USA*, **77**, 2113–18.

Marčenko L.A. (1967) [The utilization of free amino-acids by the roots of adult spruce in natural conditions.] Materialy k naučno-techničeskoj konferencii (Maj 1967 goda), pp. 82–85. Lesotehniceskaja Academija, Leningrad. (*cit. Forestry Abstracts*, **29**, (1968) No. 1967).

Marx D.H. (1973) Mycorrhizae and feeder root diseases. In *Ectomycorrhizae: Their Ecology and Physiology* (eds G.C. Marks & T.T. Kozlowski), pp. 351–82. Academic Press, London & New York.

Melin E. (1917) Studier över de norrländska myrmarkernas vegetation. In *Nörrlandskt Handbibliolet* Vol. 7, pp. 1–426. Almqvist & Wiksells, Uppsala.

Melin E. (1925) Untersuchungen uber die Bedeutung der Baummykorrhiza. Gustav Fischer Verlag, Jena.

Melin E. (1927) Studies över Barrträdsplantans utreckling i råhumus. *Meddelanden Statens Skogsförsöksanstalt*, **23**, 433–94.

Melin, E. (1959) Mycorrhiza. In *Handbuch der Pflanzenphysiologie* (ed. W. Ruhland), pp. 605–38. Springer-Verlag, Berlin.

Melin E. & Nilsson H. (1953) Transfer of labelled nitrogen from glutamic acid to pine seedlings through the mycelium of *Boletus variegatus* (Sw.) Fr. *Svensk Botanisk Tidskrift*, **48**, 555–8.

Meyer F.H. (1962) Die Buchen und Fichten Mykorrhiza in verschiedenen Bodentypen, ihre Beeinflussung durch Mineraldüngung sowie für die Mykorrhizabildung wichtige Faktoren. *Mitteilungen die Bundesforschungsanstalt für Forst und Holzwirtschaft*, **54**, 73.

Meyer F.H. (1967) Feinwurzelverteilung bei Waldbäumen in Abhängigkeit vom Substrat, *Forstarchiv*, **38**, 286–90.

Meyer F.H. (1973) Distribution of ectomycorrhizae in native and man-made forests. In *Ectomycorrhizae: Their Ecology and Physiology* (eds G.C. Marks & T.T. Kozlowski), pp. 79–105. Academic Press, London & New York.

Meyer F.H. (1974) Physiology of mycorrhiza. *Annual Review of Plant Physiology*, **25**, 567–86.

Meyer F.H. & Göttsche D. (1971) Distribution of root tips and tender roots of beech. *Integrated Experimental Ecology* (ed. H. Ellenberg), pp. 48–52. Chapman & Hall, London.

Mikola P. (1948) On the physiology and ecology of *Cenococcum graniforme* especially as a mycorrhizal fungus of birch. *Communicationes Instituti Forestalis Fennicae*, **36**, 1–104.

Miller H.G., Cooper J.M., Miller J.D. & Pauline O.J.L. (1979) Nutrient cycles in pine and their adaptation to poor soils. *Canadian Journal of Forest Research*, **9**, 19–26.

Mooney H.A. & Rundell P.W. (1979) Nutrient relations of the evergreen shrub *Adenostoma fasciculatum*, in the California chaparral. *Botanical Gazette*, **140**, 109–13.

Moser M. (1959) Beiträge zur Kenntnis der Wuchsstoffbeziehungen im Bereich ectotrophe Mycorrhizen. *Archiv für Mikrobiologie*, **34**, 251–69.

Moser M. (1967) Die ectotrophe Ernährungsweise an der Waldgrenze. *Mitteilungen Forstlichen Bundesversuchsanstalt, Wien*, **75**, 357–80.

Nye P.H. & Tinker P.B. (1977) *Solute Movement in the Soil–Root System*. Blackwell Scientific Publications, Oxford.

Nykvist N. (1963) Leaching and decomposition of water-soluble organic substances from different types of leaf and needle litter. *Studia Forestalia Suecica*, **3**, 1–31.

Owen T.H. (1954) Observations on the monthly litter fall and nutrient content of Sitka spruce litter. *Forestry*, **27**, 7–15.

Pachlewski R. & Chrusciak E. (1979) Aktywnosc enzymatyczna grzybow mikoryzowych. *Acta Mycologica*, **15**, 3–9.

Prenzel J. (1979) Mass flow to the root system and mineral uptake from a beech stand calculated from 3 yr field data. *Plant and Soil*, **51**, 39–49.

Rapp M. (1971) Cycle de la matière organique et des éléments mineraux dans quelques écosystèmes méditerranéens. *IBP Recherche Cooperative Programme CNRS, 40, Ecologie du Sol, 184S.*

Redhead J.F. (1968) Mycorrhizal associations in some Nigerian forest trees. *Transactions of the British Mycological Society*, **51**, 377–7.

Rehder H. & Schäfer A. (1978) Nutrient turnover studies in alpine ecosystems. IV. Communities of the Central Alps and comparative survey. *Oecologia (Berlin)*, **34**, 309–27.

Rice E.L. & Pancholy S.K. (1972) Inhibition of nitrification by climax vegetation. *American Journal of Botany*, **59**, 1033–40.

Richards B.N. (1961) Soil pH and mycorrhiza development in *Pinus*. *Nature*, **190**, 105–6.

Richards B.N. & Wilson G.L. (1963) Nutrient supply and mycorrhiza development in Caribbean pine. *Forest Science*, **9**, 405–12.

Romell, L.G. (1935) Ecological problems of the humus layer in the forest. In *Cornell University Agriculture Experiment Station Memoir*, p. 170.

Runge M. (1974a) Die Stickstoff-Mineralisation im Boden eines Sauerhumus—Buchenwaldes. I. Mineral stickstoff-Gehalt und Netto-Mineralisation. *Oecologia Plantarum*, **9**, 201–18.

Runge M. (1974b) Die Stickstoff-Mineralisation im Boden eines Sauerhumus—Buchenwaldes. II. Die Nitratproduktion. *Oecologia Plantarum*, **9**, 219–30.

Santantonio D. (1978) Seasonal dynamics of fine roots in mature stands of Douglas fir of different water regimes—a preliminary report. In *Root Physiology and Symbiosis. Proceedings of IUFRO Symposium, Nancy* (ed. A. Reidaker), pp. 190–203.

Santantonio D., Hermann R.K. & Overton W.S. (1977) Root biomass studies in forest ecosystems. *Pedobiologia*, **17**, 1–31.

Schmidt E.L., Putnam H.D. & Paul E.A. (1960) Behaviour of free amino acids in soil. *Proceedings of the Soil Science Society of America*, **24**, 107–9.

Singh K.G. (1966) Ectotrophic mycorrhiza in equatorial rain forests. *Malayan Forestry*, **29**, 13–18.

Slankis V. (1974) Soil factors influencing formation of mycorrhizae. *Annual Review of Phytopathology*, **12**, 437–57.

Slankis V. (1973) Hormonal relationships in mycorrhizal development. In *Ectomycorrhizae: Their Ecology and Physiology* (eds G.C. Marks & T.T. Kozlowski), pp. 231–98. Academic Press, London & New York.

Smith F.A. (1972) A comparison of the uptake of nitrate, chloride and phosphate by excised beech mycorrhizas. *New Phytologist*, **71**, 875–82.

Sollins P., Grier C.C., McGorison F.M., Cromack K. Jr & Fogel R. (1980) The internal element cycles of an old-growth Douglas fir ecosystem in Western Oregon. *Ecological Monographs*, **50**, 261–85.

Sowden F.J. & Ivarson K.C. (1966) The 'free' amino acids of soil. *Canadian Journal of Soil Science*, **46**, 109–14.

Stribley D.P. & Read D.J. (1980) The biology of mycorrhiza in the Ericaceae VII. The relationship between mycorrhizal infection and the capacity to utilize simple and complex organic nitrogen sources. *New Phytologist*, **86**, 365–71.

Swift M.J., Heal O.W. & Anderson J.M. (1979) *Decomposition in Terrestrial Ecosystems.* Blackwell Scientific Publications, Oxford.

Tamm C.O. (1979) Productivity of Scandinavian forests in relation to changes in management and environment. *Irish Forestry*, **36**, 111–20.

Taylor B.K. (1957) Storage and mobilization of nitrogen in fruit trees: a review. *Journal of the Australian Institute of Agricultural Science*, **33**, 23–9.

Theodorou C. & Bowen G.C. (1969) The influence of pH and nitrate on mycorrhizal associations of *Pinus radiata* D. Don. *Australian Journal of Botany*, **17**, 59–67.

Thomazini K.I. (1974) Mycorrhiza in plants of the 'Cerrado'. *Plant and Soil*, **41**, 707–11.

Thornley J.H.M. (1977) Root: shoot interactions. In *Integration of Activity in the Higher Plant*, SEB Symposium 31 (ed. D.H. Jennings), pp. 367–89. Cambridge University Press, Cambridge.

Tinker P.B. (1975) Soil chemistry of phosphorus and mycorrhizal effects on plant growth. In *Endomycorrhizas* (eds F.E. Sanders, B. Mosse, & P.B. Tinker), pp. 353–71. Academic Press, London & New York.

Todd A.W. (1979) Decomposition of selected soil organic matter components by Douglas fir ectomycorrhizal associations. *Abstracts. Fourth North American Conference on Mycorrhiza.* Colorado State University.

Trappe J.M. (1967) Principles of classifying ectotrophic mycorrhizae for identification of fungal symbionts. *Proceedings of the 14th IUFRO Congress, Munich*, **5**, 46–59.

Turner J. (1977) Effect of nitrogen availability on nitrogen cycling in a Douglas fir stand. *Forest Science*, **23**, 307–16.

Van den Driessche R. & Weber J.E. (1977) Seasonal variations in a Douglas fir stand in total and soluble nitrogen in inner bark and root and in total and mineralizable nitrogen in soil. *Canadian Journal of Forest Research*, **7**, 641–7.

Van der Linden M.J.H.A. (1971) Availability of protein in leaf litter—an enzymological approach. *Organismes du Sol et Production Primaire*, Proceedings of the 4th Colloquium on Soil Zoology, pp. 337–48. INRA, Paris.

Van Praag J., Weissen F., Brigode N. & Dufour J. (1973) Évaluation de la quantité d'azote minéralisé par an dans un sol de hêtraie ardennaise. *Bulletin de la Société Royale de Botanique de la Belgique*, **106**, 137–46.

Vitousek P.M. (1977) The regulation of element concentrations in mountain streams in the Northeastern United States. *Ecological Monographs*, **47**, 65–87.

Vitousek P.M. & Reiners W.A. (1975) Ecosystem succession and nutrient retention—a hypothesis. *Bioscience*, **25**, 276–81.

Vitousek P.M., Gosz J.R., Grier C.C., Melillo J.M., Reiners W.A. & Todd R.L. (1979) Nitrate losses from disturbed ecosystems. *Science*, **204**, 469–74.

Vogt K.A. & Edmonds R.L. (1980) Patterns of nutrient concentration in basidiocarps in western Washington. *Canadian Journal of Botany*, **58**, 694–8.

von Kern K.G., Moll W. & Braun H.J. (1961) Wurzeluntersuchungen in Rein und Mischbeständen des Hochschwarzwaldes. *Allgemeine Forst- und Jagdzeitung*, **32**, 241–59.

Wainwright M. & Pugh G.J.F. (1975) Changes in the free amino acid content of soil following treatment with fungicides. *Soil Biology and Biochemistry*, **7**, 1–4.

Waring R.H. & Franklin J.F. (1979) Evergreen coniferous forests of the Pacific Northwest. *Science*, **204**, 1380–6.

Wells C.G. & Jorgensen J.R. (1975) Nutrient cycling in loblolly pine plantations. In *Forest Soils and Forest Land Management* (eds B. Bernier & C.H. Winget), pp. 137–58. Les Presses de l'Université Laval, Québec.

Went F.W. & Stark N. (1968a) Mycorrhiza. *Bioscience*, **18**, 1035–9.

Went F.W. & Stark N. (1968b) The biological and mechanical role of soil fungi. *Proceedings of the National Academy of Sciences of the USA*, **60**, 497–504.

Whitaker A. (1976) Amino acid transport into fungi: an essay. *Transactions of the British Mycological Society*, **67**, 365–76.

Witkamp M. & Ausmus B.S. (1976) Processes in decomposition and nutrient transfer in forest systems. In *The Role of Terrestrial and Aquatic Organisms in Decomposition Processes* (eds J.M. Anderson & A. MacFadyen), pp. 375–96. Blackwell Scientific Publications, Oxford.

Zak B. (1973) Classification of ectomycorrhizae. In *Ectomycorrhizae: Their Ecology and Physiology* (eds G.C. Marks & T.T. Kozlowski), pp. 43–78. Academic Press, London & New York.

5. NITROGEN AS A LIMITING FACTOR IN PLANT COMMUNITIES

J. A. LEE, R. HARMER AND R. IGNACIUK
Department of Botany, The University,
Manchester M13 9PL

SUMMARY

The importance of nitrogen availability as a factor limiting plant growth in semi-natural habitats is reviewed and illustrated for saltmarsh and sand dune communities. Factors influencing nitrogen availability in soils from a wide range of environments are discussed with reference to the potential for differences in the supply of nitrate and ammonium. The efficacy of these ions as nitrogen sources for plant growth is reviewed and related to their probable availability in natural situations. The possible interaction of the available nitrogen forms, the rate of supply and its effects on plant growth are discussed in relation to plant succession. The evidence suggests that generalizations concerning the importance of the form of nitrogen available in climax communities cannot be made.

INTRODUCTION

Nitrogen is the element obtained by plants from the soil in greatest quantities, and many plants commonly contain between 1 and 4% of their leaf dry weight as nitrogen. It can be expected, therefore, that the supply of nitrogen often limits plant growth in natural ecosystems. This expectation is not always realized, and in British soils much emphasis has been placed on phosphorus as the primary growth-limiting element, even in nitrophilous communities (see Pigott & Taylor 1964). This in part explains the relative lack of attention given to nitrogen as an edaphic factor by British plant ecologists, but a contributory cause is the difficulty of adequately investigating the nitrogen economy of plants in the field. This is compounded by the fact that for most plants a supply of combined nitrogen is dependent on the activities of the soil microflora mineralizing nitrogen from the soil organic matter. There is no easy way of identifying what portion of the soil organic nitrogen is available to plants, although estimates commonly assume that *c.* 2% per annum of soil organic nitrogen becomes available. This is, however, likely to vary widely between different soils. It is not only the absolute rate of supply but its timing which may be critical to the survival and growth of plants. The key role of the soil microflora in determining the supply of combined nitrogen imposes the limitation that the factors which affect the activities of these organisms may indirectly affect plant growth. A close coupling of the activities of the soil microflora and the growth and nutrition of plants in natural ecosystems is therefore likely. This coupling can only be adequately investigated by time-consuming seasonal sampling of plants and soils, and even then the necessary disturbance of soils in the sampling process may affect the activities of the microorganisms and pose difficulties in the interpretation of the results. It is perhaps no surprise that many plant ecologists have preferred to investigate less intractable problems.

NITROGEN AS A LIMITING FACTOR

Fertilizer addition to plant communities represents the simplest method of demonstrating the importance of nitrogen as an ecological factor. Four results can follow such addition:

1. No change in total yield or species composition.
2. An increase in yield with little change in species composition or relative abundance.
3. A change in species composition or relative abundance with little or no change in overall yield.
4. A change in species composition and a marked increase in overall yield.

It can be argued that only result (2) demonstrates nitrogen limitation of a particular community since results (3) and (4) effectively produce new communities. Thus Willis (1963) demonstrated that nitrogen addition to dune slacks resulted in a marked increase in yield, but *Agrostis stolonifera* L. dominated and the other species composition of treated plots was much reduced. Where a yield response to fertilizer addition is observed, a change in species composition is usual. Thus nitrogen limitation in the strictest sense can only be demonstrated in monocultures or in communities of few species with similar growth strategies.

A natural community which can be used to demonstrate the potential importance of nitrogen limitation is the vegetation of sandy strand-lines. In a sense this represents amongst the simplest of all systems to investigate because the 'soil' is a freshly deposited mixture of sand and detached thalli of sublittoral algae. Organic matter input occurs principally twice a year (around the equinoxes), but if the beach is showing marked accretion the organic detritus does not accumulate in the same position. Thus plants colonizing the drift-lines are presented with a potential source of nitrogen (and other nutrients) at one instance in time, and this organic matter is rapidly broken down (Ignaciuk, unpublished). Drift-lines in Britain are colonized by a small number of annual nitrophilous species and the perennials *Elymus farctus* (Viv.) Runemark ex Melderis and *Honkenya peploides* (L.) Ehrh. The contribution of the annuals *Atriplex glabriuscula* Edmondston, *A. laciniata* L., *Cakile maritima* Scop. and *Salsola kali* L. to the total biomass varies considerably from place to place and from year to year. At Morfa Harlech, North Wales (SH 557 340), where there is a particularly well-developed strand-line community, the annuals predominate and in some years, for example, *Salsola kali* and *Cakile maritima* form almost pure stands of seedlings over large areas. Germination of these annual species is delayed until late April, a month after the major spring tides, when the salinity of the sand has fallen to low levels (Ignaciuk & Lee 1980). However, seedlings are not confined to the drift, but may occur on sand which has had little or no organic matter addition. Fig. 5.1. shows the biomass of *S. kali* plants on a transect through the drift-line. The performance of the plants is markedly dependent on the position of establishment, with those plants on the drift having the

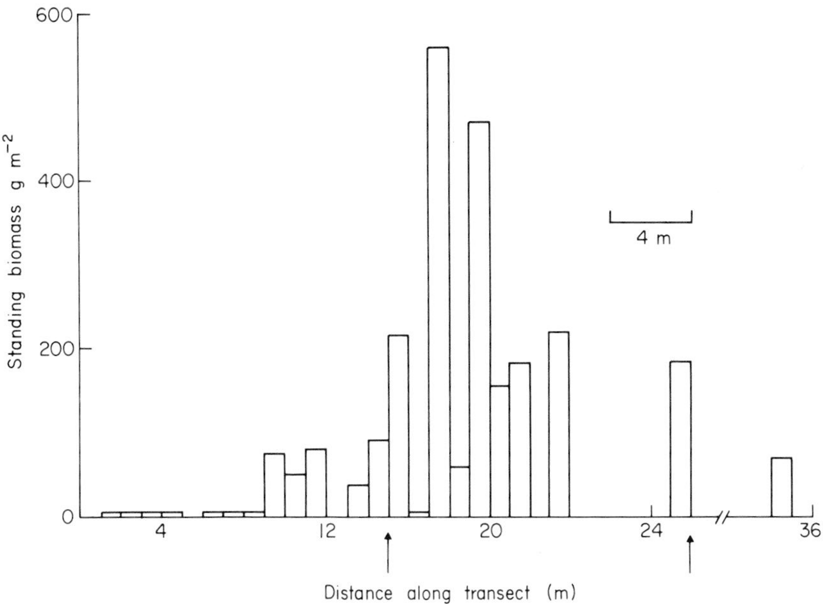

FIG. 5.1. Standing biomass g m^{-2} of *Salsola kali* L. along a transect through the drift zone at Morfa Harlech. The origin indicates the seaward part of the beach; the extent of the drift zone is the region between the arrows.

greatest biomass. Fig. 5.2 compares the growth of seedlings on the drift-line and on sand above the drift throughout a season. This demonstrates the much poorer growth of *S. kali* seedlings away from the drift-line, and these remain small and stunted, setting little seed. The cause of this stunted growth can be shown to be primarily nitrogen limitation. The result of a factorial fertilizer addition to sand from a position where no organic matter deposition had occurred is shown in Fig. 5.3. All four strand-line annual species show a primary response to nitrogen addition in both root and shoot growth.

Another habitat in which annual species may form almost pure stands is the saltmarsh. Plants of *Salicornia europaea* L. colonize extensive areas of bare mud on the submergence (lower) marsh of many European salt marshes, but are also found in areas of emergence (upper) marsh as well, where they usually show poorer growth. Fig. 5.4 demonstrates the effect of nitrate fertilizer addition on shoot length of *S. europaea* seedlings growing in the emergence marsh. Nitrate caused a marked stimulation of shoot growth, and after six weeks plants were of similar size to those of the lower marsh. Salt marshes may

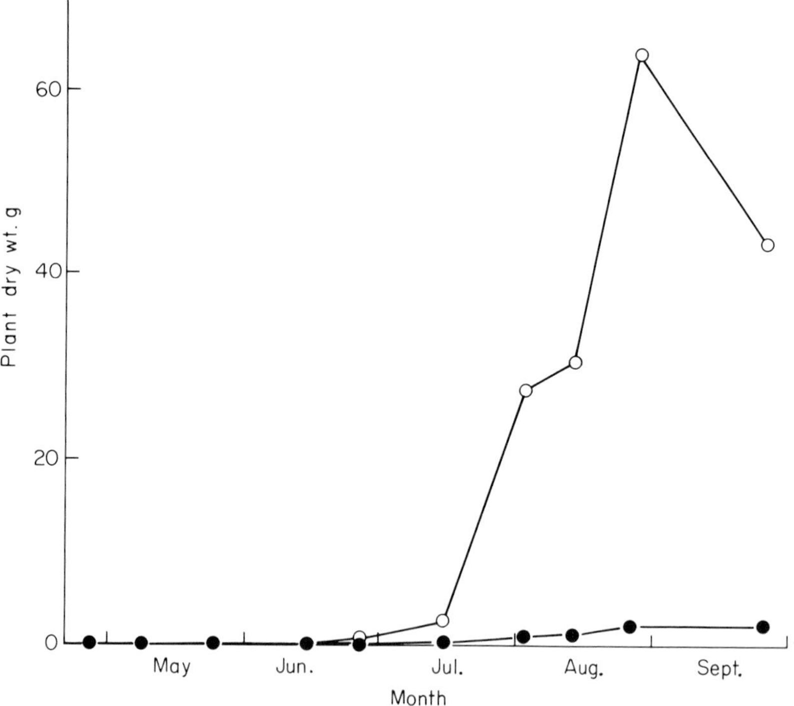

FIG. 5.2. Mean dry weight (g) of *Salsola kali* L. plants throughout a season from sites above (—●—) and within (—○—) the drift zone.

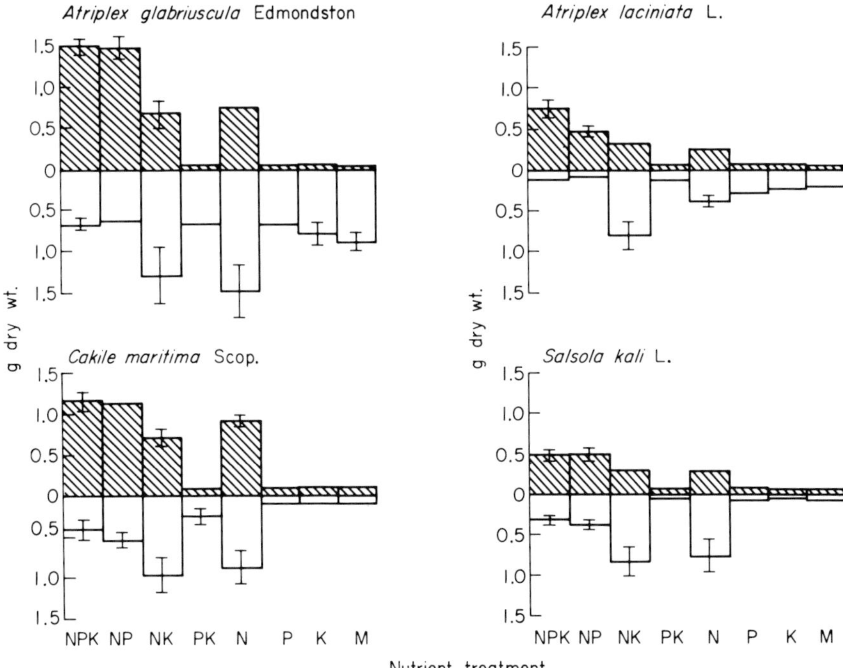

FIG. 5.3. Dry weights (g) of 4 strand-line annuals grown in sand and treated factorially with NPK. Hatched areas indicate shoot weight, plain areas root weight. Standard errors are indicated where they are large enough to be drawn. M is the control treatment receiving no nutrient addition.

also contain almost pure stands of perennials, notably in the *Spartina alterniflora* Loisel. marshes of the eastern coast of North America. Addition of inorganic nitrogen to stands of *S. alterniflora* stimulated growth whereas addition of phosphorus had no effect (Patrick & Delaune 1976; Sullivan & Daiber 1974). Many workers have demonstrated the importance of nitrogen limitation in salt marshes, e.g. Tyler (1967), Pigott (1969), Stewart, Lee & Orebamjo (1972, 1973) and Valiela & Teal (1974), and this may in part reflect the importance of soluble nitrogenous compounds in the salt tolerance mechanism of many halophytes (Stewart *et al.* 1979). In contrast Jefferies & Perkins (1977) were unable to demonstrate any marked response of salt marsh plants to nitrogen.

In general, fertilizer additions which result in a growth response cause a change in species composition, because fast-growing species are encouraged at the expense of slow-growing ones. Indirect evidence for this comes, for example, from the vegetation of bird cliffs in the Arctic. The fertilizing effect of

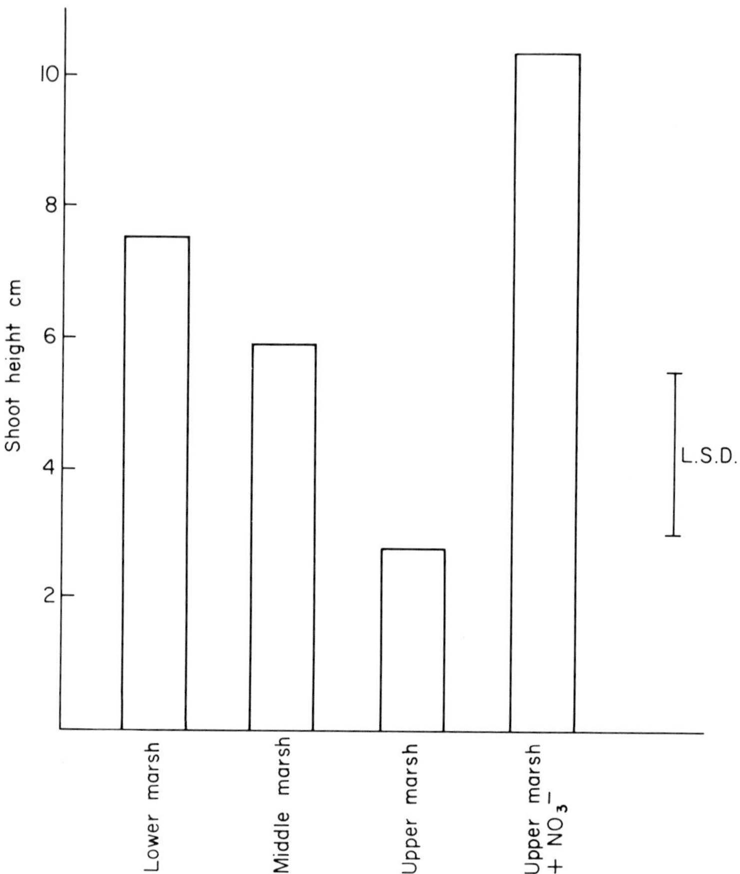

FIG. 5.4. Height (cm) of *Salicornia europaea* L. shoots at various sites on a salt marsh. Sampling undertaken six weeks after treatment of the upper marsh with nitrate fertilizer.

the birds results in a luxuriant vegetation with a different species composition compared to that of unoccupied cliffs (Summerhayes & Elton 1928). Soils from these cliffs have a high nitrogen availability (Russell 1940), and nitrogen has been shown to be an important growth-limiting factor in arctic soils. For example Haag (1974) found that in both a lowland sedge meadow and a birch–willow upland heath growth was stimulated by nitrogen but not by phosphorus. Warren-Wilson (1966) has suggested that nitrogen deficiency is a major cause of infertility of arctic soils, but few field experiments have been attempted to substantiate this generalization.

Although fertilizer addition experiments have demonstrated a nitrogen response in a number of other semi-natural or man-made ecosystems, the majority of observations have been of very restricted duration designed to investigate gross nutritional deficiencies and to identify ways of increasing yield or plant cover. Relatively few attempts have been made to observe long-term effects. Wight & Black (1979) found that over a ten-year period nitrogen had only a small effect on the species composition of rangeland in Montana. The species did show different responses to applied nitrogen but there was, in general, as much variation between years as between fertilizer treatments. The oldest and best documented fertilizer experiment is the Park Grass experiment at Rothamsted (Thurston 1969). This experiment clearly demonstrates one of the difficulties of long-term fertilizer experiments. Unfertilized plots contain *c*. 60 species of plants and yield about 1020 kg dry matter per acre. Continued application of ammonium sulphate alone has markedly reduced yield and species number as the result of soil acidification. However, sodium nitrate addition alone has increased average herbage yield by approximately 50% with only a small effect on species composition. The increased acidity of ammonium-treated plots means that no direct comparison between nitrate and ammonium as nitrogen sources can be made, and indeed no unequivocal comparison exists from the field. Tyler (1967) demonstrated a response to ammonium but not to nitrate in a saltmarsh, but this could relate to the rapid removal of nitrate from the root environment rather than to any preference of the species for ammonium. The lack of unequivocal experiments on semi-natural vegetation demonstrating a response to different nitrogen sources is a pity, because there is a wealth of laboratory evidence which suggests species may differ in their growth on nitrate and ammonium and also that soils may differ markedly in their ability to supply these ions.

NITROGEN SUPPLY

The vast majority of higher plants are dependent on the microbial breakdown of soil organic matter for their nitrogen supply. In some soils this results in an appreciable accumulation of quantities of amino acids (Saebo 1970), and there is laboratory evidence that some amino acids may be taken up and utilized by plants (Wright 1962; Watson & Fowden 1975), but there are few data that can be used to assess the importance of amino acid sources in the nitrogen economy of plants in the field. It is usual to assume that there are potentially only two major sources of nitrogen, ammonium and nitrate ions. These result from ammonification, a process carried out by a wide variety of heterotrophic microorganisms, and the subsequent oxidation of ammonium to nitrate by the nitrifying bacteria or by a number of heterotrophs. Thus microbial activity determines the release of nitrate and ammonium to higher plants, and any environmental factor which affects the microorganisms will indirectly determine the supply of these ions for plant growth. Thus in many soils there is a marked seasonal variation in the rate of supply. In temperate zones there is usually a peak of availability in late winter or early spring (Davy & Taylor 1974). Spring peaks of nitrogen mineralization have been attributed to the partial sterilization of soil during the winter, mineralization being stimulated by frost action followed by short periods of favourable weather (Harmsen & van Schreven 1955; Williams 1969). Similar periodicity in mineralization has been found in tropical systems (Pfadenhauer 1979) and is often associated with the start of the rains following the dry season. Greenland (1958) found this pattern of nitrogen availability and mineralization in a majority of the 19 different cropped and uncropped forest and savannah sites he studied in Ghana; this was attributed to a partial sterilization of the soil. During the dry season nitrifiable nitrogen was converted only slowly to nitrate and thus accumulated in the soil. When the rains began, and conditions became favourable, a rapid production of large amounts of nitrate occurred. Drought may also have an effect on nitrogen mineralization in temperate soils. Grimme (1975) studying seasonal trends of availability in a rendzina under a limestone beechwood found that nitrification was stopped when soil water contents fell below the permanent wilt point, but concentrations of nitrate and ammonium increased after rain.

The biological processes of ammonification and nitrification are differentially affected by a number of environmental features. Thus low temperatures, high acidity and anaerobic conditions depress nitrification to a greater extent than ammonification. The adverse effects of low pH on nitrification and the low levels of nitrate which can be measured in acid soils has led to the conclusion that nitrate may not be an important nitrogen source in these soils. However, there is considerable evidence that this may be untrue. Kriebitzch (1976) found that many acid soils in northern Germany had the potential to produce nitrate. In a six-week laboratory incubation study of 121 soils with pH values in the range 2.7–4.4 from a variety of habitats, only 12% showed no ability to produce nitrate. The remaining soils produced nitrate in amounts varying from less than 10% to greater than 90% of the mineral nitrogen present. Runge (1974) calculated that on an annual basis 50–60% of the mineral nitrogen in an acid soil supporting *Fagus sylvatica* L. forest was in the form of nitrate. Taylor (1979) has observed changes in available nitrogen and mineralization in an acidic and calcareous soil in the south of England. Both soils showed similar changes; maximum levels of soluble nitrogen were found in spring, the period of highest mineralization activity. Both soils contained nitrate, but greater quantities were found in the acid soil. The nitrate reductase activity of *Deschampsia flexuosa* (L.) Trin, *Zerna erecta* (Hudson) S.F. Gray and *Poterium sanguisorba* L. growing in these soils was also estimated and found to be related to the quantities of nitrate present, indicating that in both acid and calcareous soils plants were utilizing the nitrate which became available. Whilst there is evidence that ericaceous species have impaired ability to utilize nitrate (Lee & Stewart 1978), many species growing in acid soils show nitrate reductase activity above constitutive levels indicating the presence of nitrate which plants are able to use (Havill, Lee & Stewart 1974).

Low temperatures also have a deleterious effect on the nitrification process, and observations suggest that ammonium may be the major source of nitrogen in cold soils. Although Flint & Gersper (1974) found very low concentrations of nitrate compared to ammonium in wet meadow tundra, it is not necessarily true for all arctic soils, and in some situations quantities of nitrate may exceed ammonium (Russell 1940). Field incubation studies in Spitzbergen have indicated that soils in a variety of habitats are capable of producing nitrate, greatest quantities accumulating in soils from bird cliffs, but even scree soils of generally low nutrient status were able to form nitrate (Ignaciuk, unpublished). Many Spitzbergen plants were found to have high nitrate reductase activities which could be enhanced by feeding (Ignaciuk unpublished), a result similar to that found in Norwegian tundra (Lee & Stewart 1978).

In tropical habitats nitrate is often the predominant form of nitrogen found, and in some soils ammonium is undetectable (Greenland 1958). However, in savannah grasslands concentrations of nitrate are very low compared to ammonium. As with temperate and arctic situations the varying availability of nitrate in tropical soils is reflected in the nitrate reductase activities of native plants (see Chapter 8).

There is therefore marked seasonality in the supply of nitrate and ammonium in many soils, and this may to some extent influence the growth strategies of plants, particularly those of the annuals. The absolute nitrogen-supplying power of a soil may also be critical—thus the very poor nitrogen-supplying power of most arctic soils may be an important determinant of the paucity of annuals in arctic floras. But perhaps the most intriguing aspect of nitrogen supply is the consequence of different nitrogen sources. The different availabilities of nitrate and ammonium in soils raises the prospect that plants may be principally adapted to utilize only one nitrogen source, and this may result in the form of available nitrogen being an important factor limiting species distributions.

NITRATE AND AMMONIUM AS NITROGEN SOURCES

The effectiveness of different nitrogen sources has attracted the attention of plant physiologists and agronomists for many years, and their studies have been reviewed periodically (Pardo 1935; Street & Sheat 1958; Hewitt 1966), but there is still no adequate theory explaining the relative merits of ammonium and nitrate. It is generally assumed that the energy conserved by the utilization of ammonium in preference to nitrate should result in some form of growth stimulation, but this effect has not been demonstrated conclusively in higher plants (Cox & Reisenauer 1973). Carbohydrates are required in the assimilation of both nitrate and ammonium ions, not only as a source of carbon skeletons but also to supply respiratory energy for reductive amination. When nitrate is assimilated there is a further energy requirement for the reduction of ammonia. However, a recent thermodynamic study of the nitrogen nutrition of *Lolium perenne* L. has indicated that the assimilation of nitrate requires only 8% more energy than the utilization of ammonium, the latter requiring more oxygen and producing more water (Middleton & Smith 1979). A similar situation prevails in the alga *Chlorella* where phototrophic growth on ammonium requires 4.4 units of reductant per atom of carbon assimilated whereas growth on nitrate requires 5.6 units (Raven 1976).

Plants growing on nitrate or ammonium show differences in their chemical constitution which may be related to the form of nitrogen present. Plants grown on ammonium usually accumulate amino acids and amides and contain much lower levels of organic acids than nitrate-grown plants (Kirkby 1968) and this may have important consequences for insect herbivores. Vladimov (1945) has suggested that nitrate nutrition should result in the accumulation of oxidized compounds, such as organic acids whereas ammonium nutrition should favour accumulation or reduced compounds. It has been proposed that the carboxylate to amide ratio of root tissues may give a measure of the level of ammonium being assimilated (Reisenaur 1978) but Hiatt (1978) suggests this may not be useful as dramatic changes in amides only occur at yield-limiting levels of ammonium, and the addition of yield-stimulating levels of ammonium to plants growing on optimum levels of nitrate has little effect on the quantities of amide present. Similarly, Van Egmond (1978) indicates that this proposition is only feasible for plants with an active recycling of carboxylates. No proven non-experimental method of distinguishing the relative contributions of nitrate and ammonium to the nutrition of plants in the field exists.

Many studies have attempted to distinguish the preferences of plants for nitrate or ammonium ions (Pardo 1935; Hewitt 1966; Bogner 1968; Gigon & Rorison 1972), and the responses of plants to these ions have often been correlated with the presumed availability of nitrogen in their natural habitats. Bogner (1968) was able to demonstrate that a group of calcifuge species grew better on ammonium in sand culture than on nitrate. Generally fewer plants have been shown to have a preference for ammonium than nitrate (Pardo 1935), but not all of these are identifiable as coming from habitats in which ammonium ions predominate. A further complication is that in many cases the preferred nitrogen form is dependent on the acidity of the medium (Pardo 1935; Gigon & Rorison 1972) and may vary with the age of the plant (Pardo 1935; Hewitt 1966). Two factors have complicated the interpretation of many of the nitrate versus ammonium growth experiments. The first is that many experiments have utilized high concentrations of these ions, and the plants often show higher total nitrogen contents than is normal in the field as the result of luxury consumption. The ammonium ion is more toxic than nitrate at high concentrations, and in some cases ammonium toxicity may have masked growth responses apparent at lower concentrations of this ion. The second is that the assimilation of nitrate and ammonium leads to a pH drift in the medium. This is particularly serious in the case of ammonium where the medium can quickly become acid, thus introducing a further variable into the experiment. More recent experimentation using continuous flow cultures has attempted to get round these problems (see Chapter 9), but there is still a dearth of good experiments which demonstrate clearly the ecological preference of species for nitrate or ammonium.

A further factor which must be taken into account in determining a species response to nitrate and ammonium ions is seasonal variation in utilization and uptake of nitrogen, and this demands careful field experimentation. A good example of this is a study by Grasmanis & Nicholas (1970) on apple trees. They found that for much of the year uptake of $^{15}NH_4^+$ and $^{15}NO_3^-$ was nearly equivalent, but during winter ammonium was observed in much greater amounts than nitrate. This was ascribed to a low temperature inhibition of nitrate reduction, but it may also be due to a differential effect of temperature on the absorption processes. Lycklama (1963) found similar rates of uptake of nitrate and ammonium by *Lolium perenne* seedlings at 30°C, but much lower rates of nitrate uptake than ammonium at 5°C. A similar response was obtained by Clarkson & Warner (1979) for *Lolium multiflorum* Lam. An extension of these observations might be that cold climates favour growth supported by ammonium ions, and McCown (1978) has suggested that arctic grasses are adapted to utilize ammonium. Fig 5.5 shows that this may not necessarily be so. Populations of *Poa annua* L. collected from a lowland site in the south of England and an upland site in Sweden showed no consistent preference for either nitrate or ammonium.

Despite the difficulties of establishing species preferences for nitrate or ammonium ions, and the need for further critical studies, the balance of the evidence suggests that in some species these preferences do occur. This

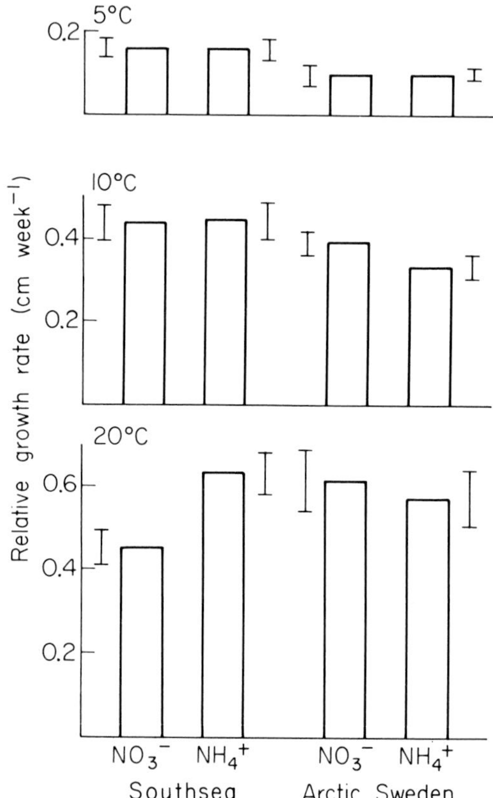

FIG. 5.5. Relative growth rate (week^{-1}) of a temperate and arctic population of *Poa annua* L. grown at different temperatures with nitrogen supplied as either nitrate or ammonium. Standard errors are indicated.

evidence is perhaps strongest in species which lack appreciable nitrate reductase activity and grow poorly on nitrate (Lee & Stewart 1978), demonstrating clearly an ammonium-based nutrition. But what remains to be proved is whether these preferences are simply a reflection of the availability of nitrate and ammonium ions in different habitats or of the varying competitive ability of species for the different ions.

NITROGEN AVAILABILITY AND SUCCESSION

The ability of nitrogen fertilizer addition to produce an increased yield with change in species composition of the plant community is perhaps the simplest example of the potential importance of nitrogen in successional processes. It is apparent that nitrogen supply is likely to be a critical part of the replacement of a smaller, less productive community by a larger, more productive one because of the larger nitrogen demand of the latter. This fact has determined much of the thinking concerning the role of nutrients in primary succession, although apart from the work of Crocker & Major (1955) and Marrs *et al.* (see Chapter 6) there have been very few studies which have provided real data to support the hypothesis that nitrogen availability is a key determinant of successional events. These studies have concentrated on the build-up of nitrogen in the soils as succession proceeds, and have assumed that this reflects an increased availability of nitrogen for plant growth. However, there need be no correlation between soil total and mineralizable nitrogen. But perhaps the most intriguing aspect of nitrogen as a determinant of succession does not involve the absolute rate of supply of the element but instead its form. If species do vary in their ability to utilize nitrate and ammonium, then the form in which nitrogen becomes available may determine which species may grow in a particular soil. Conversely, and controversially, a species may determine which species can coexist with it by affecting not only the absolute amount of nitrogen available but also its form. Considerable attention has been paid to the apparent ability of some plant species to inhibit nitrification, thus causing ammonium to be the nitrogen source. There is evidence to suggest that plant toxins may inhibit the growth of nitrifying bacteria in some veldt and savannah grasslands (e.g. Theron 1951; Stiven 1952; Meiklejohn 1968), but not all the evidence supports this view (Purchase 1974), and the immobilization of nitrogen may be responsible (see Chapter 8). However, the possibility that a dominant species can affect the form of nitrogen available to other species requires investigation as a possible successional process. Preliminary evidence to support this view comes from sand dunes. *Senecio jacobea* L. is widely distributed on British dune systems being a component of the mobile fore-dunes dominated by *Ammophila arenaria* (L.) Link and also of fixed dune grassland dominated by *Festuca rubra* L. The latter habitat has the greater soil organic nitrogen and therefore a presumed greater supply of mineral nitrogen. However, as Table 5.1 shows, *S. jacobea* plants have a lower nitrate reductase activity implying a reduced nitrate supply in the fixed dune habitat. There are, of course, several possible explanations for this, including the simplest which is that *S. jacobea* is a poor competitor for nitrate. However, soils from under *F. rubra* may show appreciable ammonification but low rates of nitrification (Lee & Greenwood 1976) and the possibility remains that at least some populations of this species can inhibit nitrification.

Perhaps the most extravagant claims for the role of nitrogen in successional processes comes from a consideration of secondary succession. Rice & Pancholy (1972) in a study of old field succession in Oklahoma found that

TABLE 5.1. Nitrate reductase activity (μmol NO_2 $h^{-1}g$ fresh wt.$^{-1}$) in plants of *Senecio jacobea* on a sand dune. The figures are the means of at least three determinations.

Dune community	Nitrate reductase activity
Mobile dune with *Ammophila arenaria*	1.00
Fixed dune with *Phleum arenarium*	1.10
Fixed dune with *Festuca rubra*	0.68
LSD ($P < 0.05$)	0.24

quantities of nitrate and numbers of nitrifying microorganisms decreased with successional stage whereas amounts of ammonium increased. These workers suggested that climax ecosystems suppress nitrification, and that this results in the conservation of nitrogen and energy in these communities since nitrate is the form in which nitrogen is most easily lost, and energy need not be expended by plants in reducing nitrate (see also Chapter 8 for further discussion). Other evidence can be cited to support this view: for example, moist evergreen forest soils of Africa show rapid ammonification but only a low rate of nitrification, possibly due to low numbers of nitrifying microorganisms (Nye & Greenland 1960). Similarly, the humus layers of organic soils under spruce and pine in Finland contain ammonium as the predominant form of available nitrogen (Viro 1963). A study of a Venezuelan forest soil showed very high quantities of ammonium compared to nitrate in the surface soil horizon (Jordan, Todd & Escalente 1979). This was attributed to low numbers and activities of nitrifying bacteria caused by low pH and high concentrations of tannins. However, evidence is accumulating that the apparent inhibition of nitrification in climax communities is not universal. Pfadenhauer (1979) showed that nitrate was dominant in acid soils supporting subtropical forest in South Brazil, and Lee & Stewart (1978) showed appreciable nitrate utilization in species from *Quercus robur* L. forest in England demonstrating high nitrate availability. The most convincing evidence against the inhibition of nitrification by climax communities comes from secondary succession to Australian rainforest. Lamb (1980) showed that quantities of ammonium and rates of ammonification were generally low in all successional stages studied whereas the amounts of nitrate were higher and rates of nitrification increased with succession. Recently Robertson & Vitousek (1981) have provided further evidence against the suppression of nitrification by climax communities.

CONCLUSION

There can be little doubt that nitrogen is an important factor limiting plant growth and development, but in ecological terms we are further than ever from broad generalizations. These generalizations can only appear after further studies, for despite the undoubted importance of nitrogen as an ecological factor, too little attention has been paid to it. This has been due not only to the intractable nature of working with an element the supply of which depends largely on the activities of microorganisms, but also to the relative analytical ease of working with other plant nutrients. Recent work has helped to destroy several generalizations which existed, e.g. it has been confirmed that nitrate is utilized in some acid, cold and waterlogged soils (Lee & Stewart 1978) and that nitrification is not inhibited by all climax communities. But the relative importance of nitrate and ammonium as nitrogen sources, and the possible implications of this for plant competition, has not been clearly established. Climax ecosystems may not suppress nitrification, but some species may be able to exert a marked effect on their competitors by influencing the form of available nitrogen and in this way perhaps determine successional events. Nitrogen availability can be shown to limit plant productivity in the field, but the subtleties of the interrelationships between species and the forms of nitrogen availability are largely unknown.

REFERENCES

Bogner W. (1968) Experimentelle Prüfung von Waldbodenpflanzen auf ihre Ansprüche an die Form der Stickstoff-Ernährung. *Mitteilungen des vereins für Forstliche Standortskunde und Forstpflanzenzüchtung*, **18**, 3–45.

Clarkson D.T. & Warner A.J. (1979) Relationships between root temperature and the transport of ammonium and nitrate ions by Italian and Perennial ryegrass (*Lolium multiflorum* and *Lolium perenne*). *Plant Physiology*, **64**, 557–61.

Cox W.J. & Reisenauer H.M. (1973) Growth and ion uptake by wheat supplied nitrogen as nitrate, or ammonium or both. *Plant and Soil*, **38**, 363–80.

Crocker R.L. & Major J. (1955) Soil development in relation to vegetation and surface age at Glacier Bay, Alaska. *Journal of Ecology*, **43**, 427–48.

Davy A.J. & Taylor K. (1974) Seasonal pattern of nitrogen availability in contrasting soils in the Chiltern Hills. *Journal of Ecology*, **62**, 793–807.

Flint P.S. & Gersper P.L. (1974) Nitrogen nutrient levels in arctic tundra soils. In *Soil Organisms and Decomposition in Tundra* (eds A.J. Holding, O.W. Heal, S.F. Maclean & P.W. Flanagan), pp. 375–87. Tundra Biome Steering Committee, Stockholm.

Gigon A. & Rorison I.H. The response of some ecologically distinct plant species to nitrate and to ammonium nitrogen. *Journal of Ecology*, **60**, 93–112.

Grasmanis V.O. & Nicholas D.J.D. (1970) Annual uptake and distribution of N^{15}-labelled ammonia and nitrate in young Jonathan/MM104 apple trees grown in solution cultures. *Plant and Soil*, **35**, 95–112.

Greenland D.J. (1958) Nitrate fluctuations in tropical soils. *Journal of Agricultural Science*, **50**, 82–92.

Grimme K. (1975) Wasser und Nährstoffversrgung von Hangbuchenwäldern auf Kalk in der weiteren Umgebung von Göttingen. Dissertation, Universität der Göttingen.

Haag R.W. (1974) Nutrient limitations to plant production in two tundra communities. *Canadian Journal of Botany*, **52**, 103–16.

Harmsen G.W. & van Schreven D.A. (1955) The mineralization of organic nitrogen in the soil. *Advances in Agronomy*, **7**, 299–398.

Havill D.C., Lee J.A. & Stewart G.R. (1974) Nitrate utilization by species from acidic and calcareous soils. *New Phytologist*, **73**, 1221–31.

Hewitt E.J. (1966) *Sand and Water Culture Methods Used in the Study of Plant Nutrition.* Commonwealth Agricultural Bureau, Farnham Royal, Bucks.

Hiatt A.J. (1978) Critique of "Absorption and utilization of ammonium nitrogen by plants". In *Nitrogen in the Environment* 2. *Soil-Plant-Nitrogen Relationship* (eds D.R. Nielsen & J.G. MacDonald), pp. 157–70. Academic Press, London & New York.

Ignaciuk R. & Lee J.A. (1980) The germination of four annual strand-line species. *New Phytologist*, **84**, 581–94.

Jefferies R.L. & Perkins N. (1977) The effects on the vegetation of the additions of inorganic nutrients to salt marsh soils at Stiffkey, Norfolk. *Journal of Ecology*, **65**, 867–82.

Jordan C.F., Todd R.L. & Escalante G. (1979) Nitrogen conservation in a tropical rain forest. *Oecologia (Berlin)*, **39**, 123–8.

Kirby E.A. (1968) Influence of ammonium and nitrate nutrition on the cation–anion balance and nitrogen and carbohydrate metabolism of white mustard plants grown in dilute nutrient solutions. *Soil Science*, **105**, 133–41.

Kriebitzch, W-U. (1976) Bedingungen der stickstoffmineralisation und nitrifikation in sauren Waldböden Nordwest-Deutschlands. Dissertation, Universität der Göttingen.

Lamb D. (1980) Soil nitrogen mineralisation in a secondary rainforest succession. *Oecologia (Berlin)*, **47**, 257–63.

Lee J.A. & Greenwood B. (1976) The colonisation by plants of calcareous wastes from the salt and alkali industry in Cheshire, England. *Biological Conservation*, **10**, 131–49.

Lee J.A. & Stewart G.R. (1978) Ecological aspects of nitrogen assimilation. *Advances in Botanical Research*, **6**, 1–43.

Lycklama J.C. (1963) Absorption of ammonium and nitrate by perennial ryegrass. *Acta Botanica Neerlandica*, **12**, 361–423.

McCown B.H. (1978) The interactions of organic nutrients, soil nitrogen, and soil temperature and plant growth and survival in the arctic environment. In *Vegetation and Production Ecology of an Alaskan Arctic Tundra* (ed. L.L. Tiezen), pp. 435–56. Springer-Verlag, Berlin.

Meiklejohn J. (1968) Numbers of nitrifying bacteria in some Rhodesian soils under natural grass and improved pastures. *Journal of Applied Ecology*, **5**, 291–300.

Middleton K.R. & Smith G.S. (1979) A comparison of ammoniacal and nitrate nutrition of perennial ryegrass through a thermodynamic model. *Plant and Soil*, **53**, 487–504.

Nye P.H. & Greenland D.J. (1960) *The soil under shifting cultivation.* Technical communication 51, Commonwealth Bureaux of Soils, Harpenden.

Pardo J.H. (1935) Ammonium in the nutrition of higher green plants. *Quarterly Review of Biology*, **10**, 1–31.

Patrick W.H. & Delaune R.D. (1976) Nitrogen and phosphorus utilization by *Spartina alterniflora* in a salt marsh in Barataria Bay, Louisiana. *Estuarine and Coastal Marine Science*, **4**, 59–64.

Pfadenhauer J. (1979) Die stickstoff mineralisation in Böden sub-tropischer Ragenwälder in Südbrasilien. *Oecologia Plantarum*, **14**, 27–40.

Pigott C.D. (1969) Influence of mineral nutrition on the zonation of flowering plants in coastal salt marshes. In *Ecological Aspects of the Mineral Nutrition of Plants* (ed. by I.H. Rorison), pp. 25–35. Blackwell Scientific Publications, Oxford.

Pigott C.D. & Taylor K. (1964) The distribution of some woodland herbs in relation to the supply of nitrogen and phosphorus in the soil. *Journal of Ecology*, **52** (supplement), 175–85.

Purchase B.S. (1974) Evaluation of the claim that grass root exudates inhibit nitrification. *Plant and Soil*, **41**, 527–39.

Raven J.A. (1976) Division of labour between chloroplast and cytoplasm. In *The Intact Chloroplast* (ed. J. Barber), pp. 403–43. Elsevier North Holland Biomedical Press, Amsterdam, New York & Oxford.

Reisenauer H.M. (1978) Absorption and utilization of ammonium nitrogen by plants. In *Nitrogen in the Environment* 2. *Soil-Plant-Nitrogen Relationship*. (eds D.R. Neilsen & J.G. MacDonald), pp. 157–70. Academic Press, London & New York.

Rice E.L. & Pancholy S.K. (1972) Inhibition of nitrification by climax ecosystems. *American Journal of Botany*, **59**, 1033–40.

Robertson G.P. & Vitousek P.M. (1981) Nitrification potentials in primary and secondary succession. *Ecology*, **62**, 376–86.

Runge M. (1974) Die Stickstoff-Mineralisation un Boden einer Sauerhumus-Buchenwaldes Teil 2: Die Nitratproduktion. *Oecologia Plantarum*, **9**, 219–30.

Russell R.S. (1940) Physiological and ecological studies on an arctic vegetation II. The development of vegetation in relation to nitrogen supply and soil microorganisms on Jan Mayen Island. *Journal of Ecology*, **28**, 269–88.

Saebo S. (1970) The autecology of *Rubus chamaemorus* L. II. Nitrogen economy of *Rubus chamaemorus* in an ombrotrophic mire. *Meldinger fra Norges Lanbrukshøgskole*, **49**, 1–37.

Stewart G.R., Lee L.A. & Orebamjo T.O. (1972) Nitrogen metabolism of halophytes I. Nitrate reductase activity in *Suaeda maritima*. *New Phytologist*, **71**, 263–87.

Stewart G.R., Lee J.A. & Orebamjo T.O. (1973) Nitrogen metabolism of halophytes II. Nitrate availability and utilization. *New Phytologist*, **72**, 539–46.

Stewart G.R., Larher F., Ahmad I. & Lee J.A. (1979) Nitrogen metabolism and salt tolerance in higher plant halophytes. In *Ecological Processes in Coastal Environments* (eds R.L. Jefferies & A.J. Davy), pp. 211–27. Blackwell Scientific Publications, Oxford.

Stiven G. (1952) Production of antibiotic substances by roots of a grass, *Trachypogon plumosus* (H.B.K.) Nees, and of *Pentosisia variabilis* (E.Mey) Harv. (Rubiaceae). *Nature*, **170**, 712.

Street H.E. & Sheat D.E.G. (1958) The absorption and availability of nitrate and ammonia. In *Encyclopaedia of Plant Physiology* (ed. W. Ruhland), pp. 150–65. Springer-Verlag, Berlin, Göttingen & Heidelberg.

Sullivan M.J. & Daiber F.C. (1974) Response in production of cord grass, *Spartina alterniflora*, to inorganic nitrogen and phosphorus fertilizer. *Chesapeake Science*, **15**, 121–4.

Summerhayes J.S. & Elton C.S. (1928) Further contributions to the ecology of Spitzbergen. *Journal of Ecology*, **16**, 113–268.

Taylor A.A. (1979) Assimilation of nitrate and ammonium by grassland plants. Ph.D. thesis, University of London.

Theron J.J. (1951) The influence of plants on the mineral nitrogen and the maintenance of organic matter in the soil. *Journal of Agricultural Science*, **41**, 289–96.

Thurston J.M. (1969) The effect of liming and fertilizers on the botanical composition of permanent grassland and on the yield of hay. In *Ecological Aspects of the Mineral Nutrition of Plants* (ed. I.H. Rorison), pp. 3–10. Blackwell Scientific Publications, Oxford.

Tyler G. (1967) On the effect of phosphorus and nitrogen supplied to Baltic Shore-meadow vegetation. *Botoniska Notiser*, **120**, 433–47.

Valiela I. & Teal J.M. (1974) Nutrient limitation in salt marsh vegetation. In *Ecology of*

Halophytes (eds R.J. Reimold & W.H. Queen), pp. 547–63. Academic Press, London & New York.

Van Egmond F. (1978) Nitrogen nutritional aspects of the ionic balance of plants. In *Nitrogen in the Environment 2. Soil-Plant-Nitrogen Relationship* (eds D.R. Nielsen & J.G. MacDonald), pp. 171–89. Academic Press, London & New York.

Viro P.J. (1963) Factorial experiments on forest humus decomposition. *Soil Science*, **95**, 24–30.

Vladimov A.A. (1945) Influence of nitrogen sources in the formation of oxidised and reduced organic compounds in plants. *Soil Science*, **60**, 265–75.

Warren-Wilson, J. (1966) An analysis of plant growth and its control in arctic environments. *Annals of Botany, NS* **30**, 383–402.

Watson R. & Fowden L. (1975) The uptake of phenylalanine and tyrosine by seedling root tips. *Phytochemistry*, **14**, 1181–6.

Wight J.R. & Black A.L. (1979) Range fertilization: plant response and water use. *Journal of Range Management*, **32**, 345–53.

Williams J.T. (1969) Mineral nitrogen in British grassland soils. I. Seasonal patterns in simple models. *Oecologia Plantarum*, **4**, 307–20.

Willis A.J. (1963) Braunton Burrows: the effects of the addition of mineral nutrients to the dune soils. *Journal of Ecology*, **51**, 353–74.

Wright D. (1962) Amino acid uptake by plant roots. *Archives of Biochemistry and Biophysics*, **97**, 174–80.

6. NITROGEN AND THE DEVELOPMENT OF ECOSYSTEMS

R. H. MARRS*, R. D. ROBERTS†, R. A. SKEFFINGTON‡ AND A. D. BRADSHAW

Department of Botany, University of Liverpool, P.O. Box 147, Liverpool L69 3BX

SUMMARY

This paper reviews the importance of nitrogen for the creation of new ecosystems during natural and artificially accelerated primary successions. This will be illustrated by reference to studies made on china clay waste materials and other new environments which are also deficient in nitrogen. Ecosystem development requires the accumulation of nitrogen within the system. The minimum amounts of nitrogen required for *Salix* scrub and *Quercus-Betula* woodland are estimated to be 1000 and 1800 kg N ha^{-1} respectively. Accumulation can be accelerated during reclamation by fertilizer addition and the use of forage legumes. Accumulation is not, however, enough and efficient cycling is also required. This is discussed in relation to the management options available for the successful creation of new ecosystems on these areas.

INTRODUCTION

Nitrogen is the most abundant mineral element in plant tissues derived from the soil. It is usually present in much larger concentrations than all other essential elements, except potassium which may have a similar tissue concentration (Epstein 1972; Hewitt & Smith 1975). However, if these tissue concentrations are corrected for elemental atomic weight, it becomes clear that plant tissues normally contain four times as many nitrogen as potassium atoms, and at least 8–10 times the number of atoms of any other element

* Present address: Institute of Terrestrial Ecology, Monks Wood Experimental Station, Abbots Ripton, Huntingdon, Cambs., PE17 2LS.

† Present address: Department of Biology, University of Essex, Colchester, Essex, CO4 3SA.

‡ Present address: Central Electricity Generating Board, Biology Section, Kelvin Avenue, Leatherhead, Surrey, KT22 7SE.

(Epstein 1972). The concentration of nitrogen found in plant tissue varies between 1 and 4% (Allen *et al.* 1974). It is much higher than the concentration found in rock materials, 0.001–0.2% (Wlotzka 1978). The nitrogen which is present in most surface soils is contained almost entirely in plant residues, and is derived initially from atmospheric sources (Black 1968). If this is the case all soil–plant systems as we understand them, owe their existence almost entirely to the evolution and development of nitrogen-fixing organisms. The only other source is a small amount produced by atmospheric processes, which before industrialization probably resulted in no more than a few kg N ha^{-1} $year^{-1}$.

Studies on the development of ecosystems in new environments focussed attention on the discrepancy in nitrogen requirements by plants and its paucity in rock materials. It is therefore the aim of this paper to examine the importance of nitrogen in the development and functioning of ecosystems, by reference to its role in the development of self-sustaining ecosystems in derelict land environments.

THE IMPORTANCE OF AN ADEQUATE NITROGEN SUPPLY

The importance of a sustained nitrogen supply for derelict land restoration can be clearly demonstrated by simple, nutrient-addition experiments to newly established vegetation. Where vegetation has been initially established in the absence of topsoil with the help of some fertilizer, growth quickly ceases and the vegetation regresses. When nutrients are added to such moribund swards on colliery spoil and china clay waste (Fig. 6.1), growth occurs only where nitrogen is added; continued sward growth obviously requires a sustained nitrogen supply. Since fertilizers were added to these swards at the time of establishment, why should they need more for sustained growth?

One reason for sward regression may be excessive leaching, particularly because derelict land materials, which have a low nutrient retention capacity, are highly porous (Bradshaw & Chadwick 1980). For example, Dancer (1975) demonstrated that 98% of nitrogen applied as fertilizer to raw china clay wastes could be leached beyond the rooting zone by one month of average Cornish summer rainfall. The effect is more marked with nitrate than ammonium nitrogen because of the extremely low anion exchange capacity of these wastes. Fertilizer capture studies and detailed lysimeter experiments have, on the other hand, demonstrated that where nutrients were added to established swards nitrogen capture is extremely efficient (70–95% of applied nitrogen). These results were from an area of high rainfall, and were irrespective of the chemical form of nitrogen used (Dancer, Handley & Bradshaw 1979; Marrs, Roberts & Bradshaw 1980; Marrs & Bradshaw 1980).

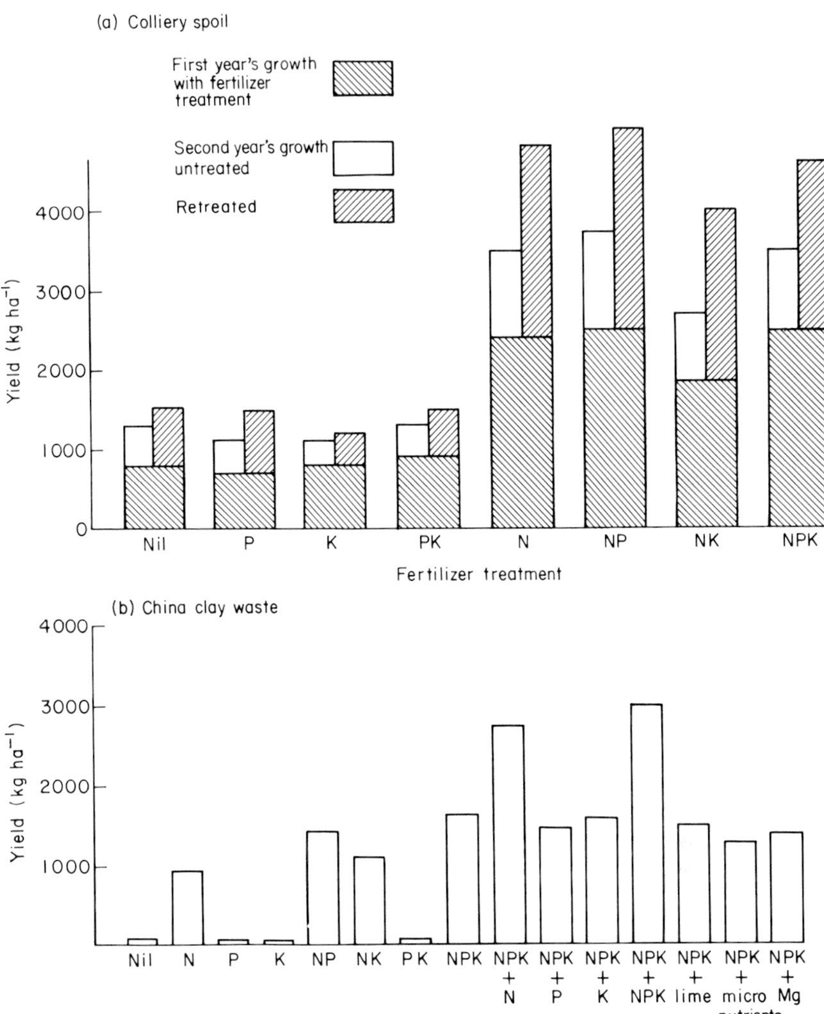

FIG. 6.1. The effect of nutrient amendments to moribund grass swards on (a) colliery spoil, and (b) china clay waste (Bradshaw & Chadwick 1980).

Therefore, although leaching of nitrogen may well be a severe problem where fertilizer is applied in large amounts at the time of seeding, fertilizer applications to developed vegetation are effectively retained and, in spite of the porous nature of these wastes, continued excessive leaching cannot account for sward regression. Similarly, losses of nitrogen from well-developed ecosystems are generally low and appear to be balanced by inputs into

TABLE 6.1. Comparison of major nitrogen pool sizes (kg ha^{-1}), and annual transfers from soil to plant pools (kg ha^{-1} year^{-1}) in a range of mature ecosystems.

Ecosystem	Soil sampling depth (cm)	Soil pool (% of total)	Plant pool	Plant + litter pool	Soil→plant transfer	References
Blanket bog	30	3300 (92)	120	320	60	Martin & Holding (1978)
Tundra:						
subalpine birch	35	11 000 (97)	188	300	81	Wielgolaski, Kjelvik & Kallio (1975)
alpine lichen heath	35	2400 (97)	38	65	21	
alpine dry meadow	35	11 600 (99)	73	144	74	
alpine wet meadow	35	16 900 (97)	199	480	131	
willow thicket	35	10 000 (97)	194	295	112	
Temperate grasslands:						
upland pasture	30	10 460 (98)	123	160	162	Perkins (1978)
short grass prairie	30	5400 (96)	—	200	100	Coupland & Van Dyne (1979)
mixed prairie	36 (Roots 60)	3340 (95)	—	180	80	
Heathland	20	5420 (96)	169	226	—	Robertson & Davies (1965)
Northern hardwood	9	1256 (70)	532	—	80	Bormann & Likens (1979)
Temperate woodland:						
birch	30	1300 (81)	264	311	74–104	Duvigneaud & Denaeyer-de Smet (1970)
oak	30	7476 (94)	393	464	45	
beech	30	6640 (93)	[illegible]	[illegible]	[illegible]	

Tropical grassland:						
semi-arid	?	640 (98)	—	12	30	Singh, Singh & Yadava (1979)
dry subhumid	?	4520 (96)	—	174	256	
moist subhumid	?	4510 (97)	—	121	153	
humid	?	3980 (95)	—	196	209	
Tropical forest:						
Jamaican montane						
Mor ridge	45	9000 (79)	2404	—	—	Estimated from Tanner (1977, 1980)
Mull ridge	40	7000 (52–58)	5023–6553	—	—	
Wet slope	30	3000 (51)	2921	—	—	
Gap	45	9000 (68)	4189	—	—	
Ivory Coast	50	2600–6600 (72–82)	—	1000–1400	—	UNESCO/UNEP/FAO (1978)
Ghana	25	5600 (69)	2010	2045	—	

the system from atmospheric deposition and biological fixation (Bormann & Likens 1979). In order to identify the cause of this problem it is necessary to examine nitrogen stores and fluxes in both new and well-developed ecosystems.

If the nitrogen distribution for a range of established ecosystems is compared (Table 6.1), it is clear that for most systems the largest proportion occurs in the surface soils. Thus, although the nitrogen capital varies considerably between these different ecosystems, usually more that 90% of the nitrogen capital is present in the surface soil in temperate and tundra systems. In some tropical forest ecosystems, however, a larger proportion of the total capital may be stored in the plant pools (Table 6.1), presumably due to faster rates of decomposition and soil cycling processes (Odum 1971). The annual transfer of nitrogen from the soil to plant pools is also variable, and is related to the productivity of the ecosystem, the transfer has been estimated as 160 kg N ha^{-1} year^{-1} for upland pastures, and between 45–104 kg N ha^{-1} year^{-1} for various temperate woodlands (see Table 6.1). A small amount of this may be derived from deposition inputs, but in the absence of biological fixation the majority is from soil organic matter and litter, by decomposition processes.

It is, however, extremely difficult to measure production and decomposition of individual plant organs (especially roots) accurately, and uptake values may be underestimates, although some nitrogen used in growth must be supplied through internal nitrogen cycling processes. Irrespective of these uncertainties it appears that the plant uptake fluxes in well-developed woodland and grassland ecosystems are at least between 45–160 kg N ha^{-1} year^{-1}. The essential validity of this figure is borne out by the increasing number of measurements showing that temperate ecosystems have a total productivity of about 10 000 kg ha^{-1} year^{-1}, which at an average nitrogen concentration of 1.5% implies an annual uptake of 150 kg N ha^{-1} year^{-1} (Whitehead 1970). The crucial point then is that the soil pool and soil decomposition processes must be sufficient to supply a major proportion of this amount, bearing in mind the following: (a) nitrogen differs from other major plant nutrients in being stored mainly in organic matter (which may contain >95% of the soil nitrogen), (b) the soil–plant flux is greater than for any other nutrient, e.g. upland pastures (Table 6.2).

In marked contrast to well-developed ecosystems, the soil nitrogen pools in most derelict land materials are extemely low. Nitrogen contents of less than 200 kg N ha^{-1} in the surface 21 cm are the norm for raw mineral wastes (calculated from data in Hutnik & Davis 1973), and even in subsoils the maximum nitrogen content we have found in a survey of a wide range of materials was 500 kg N ha^{-1} in the surface 21 cm. Even if soil cycling processes operate extremely efficiently, such low amounts of soil nitrogen mean that the

TABLE 6.2. The distribution and transfer of elements between main ecosystem pools and atmospheric inputs in an upland pasture (Perkins 1978).

(a) Pools (kg ha^{-1}).

	Element				
Pool	N	P	K	Ca	Mg
Soil	10 460.0	3020.0	13 610.0	2770.0	11 3400.0
Root	54.3	3.7	3.1	9.7	32.0
Standing crop	43.4	4.0	25.2	5.8	11.3
Standing dead	26.6	3.0	10.0	4.3	15.0
Litter	36.6	2.7	5.0	7.0	21.7

(b) Transfer between pools (kg ha^{-1} year^{-1}).

Transfer	N	P	K	Ca	Mg
Soil→root	162.6	16.0	73.7	30.0	70.0
Root→standing crop	132.3	13.9	72.0	24.6	52.1
Standing crop cycling	113.3	7.7	89.3	10.8	0
Standing crop→standing dead	78.8	8.7	29.6	12.8	44.5
Standing dead→litter	72.7	8.0	27.3	11.3	41.5
Litter→soil	73.1	10.1	30.6	16.3	38.4
Root→soil	30.3	2.1	17.2	5.4	17.9
Through grazers	72.2	7.0	47.7	15.5	18.1

(c) Inputs from rainfall (kg ha^{-1} year^{-1}).

	N	P	K	Ca	Mg
	18.4	1.7	3.0	26.0	11.3

nitrogen and the low nitrogen supply to the plant pool. For most restoration well-developed ecosystems. Normal rates of nitrogen mineralization are about 2% year^{-1}; a capital of 200 kg N ha^{-1} would mean a supply from soil mineralization of only 4 kg N ha^{-1} year^{-1}.

Clearly, therefore, the cause of sward regression is the paucity of soil nitrogen, and the low nitrogen supply to the plant pool. For most restoration programmes fertilizer nitrogen additions of approximately 100 kg N ha^{-1} are made. Whilst this supply is comparable with one year's supply in some well-developed ecosystems, it is immediately locked up in the developing ecosystems (Bradshaw *et al.* 1975). Even if recycling did occur, 100 kg N ha^{-1} is a very small addition to the soil pool (c.f. natural ecosystems; Tables 6.1,

6.2), and would not even be sufficient to maintain upland pasture productivity for one year. However, simply because observed uptake fluxes in well-developed systems are of the order of 100 kg N ha^{-1} year^{-1} this does not necessarily imply that this amount is actually needed to maintain the productivity of the vegetation and the stability of the ecosystem. The first step to unravel this problem is to find out what can be regarded as the minimum amount of nitrogen that specified ecosystems need to maintain themselves. Some information can be obtained from studies of nitrogen accumulation in new and developing ecosystems.

NITROGEN CAPITAL REQUIREMENTS

In the case of china clay extraction, waste heaps have been produced over a long period. They vary in age from 16–116 years since tipping ceased and now support four distinct species assemblages distinguishable by ordination analysis (Table 6.3). Unless colonization by *Lupinus arboreus** occurs most tips appear to remain more or less uncolonized for approximately 30 years until a vegetation consisting of *Calluna vulgaris, Ulex europaeus* (or *U. gallii*) and *Sarothamnus scoparius* develops, with the leguminous species forming a large proportion of the vegetation cover (Roberts *et al.* 1981, 1982). After 50–60 years an intermediate species assemblage containing *Salix atrocinerea* develops, which gives way finally to a mature woodland group containing *Betula pendula, Quercus robur*, and *Rhododendron ponticum* after approximately 75 years.

Throughout this period the nitrogen content of the total ecosystem accumulates at an average annual rate of 10 kg N ha^{-1} year^{-1} (Fig. 6.2; Roberts *et al.* 1981). At this accumulation rate, it takes approximately 100 years to accumulate 1000 kg N ha^{-1} in these new ecosystems. The nitrogen content for each type of species assemblage (Table 6.4) shows that the intermediate *S. atrocinerea* group occurred on sites which have a total nitrogen content of approximately 1000 kg N ha^{-1}, with 700 kg N ha^{-1} in the soil pool (0–21 cm depth). The mature woodland assemblage is found on sites which have 1800 kg N ha^{-1}, with 1200 kg N ha^{-1} (0–21 cm depth) in the soil pool (Marrs *et al.* 1981; Roberts *et al.* 1982).

It may be that the processes of vegetation development and nitrogen accumulation are occurring in parallel without any causal connection. However, although scrub and woodland species can be observed to germinate and become established in earlier stages of the succession, and even to appear as primary colonizers, their subsequent growth and development in the early

* Nomenclature follows Clapham, Tutin & Warburg (1962).

TABLE 6.3. Species assemblages derived from ordination studies of vegetation which has naturally colonized china clay wastes (Roberts *et al.* 1981).

Group	Species
A	*Digitalis purpurea*
	Holcus lanatus
	Lupinus arboreus
	Rumex acetosa
B	*Agrostis tenuis*
	Calluna vulgaris
	Cotoneaster simonsii
	Crataegus monogyna
	Deschampsia flexuosa
	Erica cinerea
	Glechoma hederacea
	Knautia arvensis
	Lamium purpureum
	Osmunda regalis
	Pteridium aquilinum
	Rubus fruticosus agg.
	Sarothamnus scoparius
	Taraxacum officinale
	Teucrium scorodonia
	Ulex europaeus
	Ulex gallii
	Vaccinium myrtillus
C	*Rumex acetosella*
	Salix atrocinerea
D	*Betula pendula*
	Corylus avellana
	Hedera helix
	Lonicera periclymenum
	Quercus robur
	Rhododendron ponticum
	Sorbus aucuparia

stages is negligible. They only contribute a significant proportion to the species cover and standing crop at sites where the nitrogen contents approach 1000 and 1800 kg N ha^{-1} (0–21cm depth). So although an additional nitrogen input could well increase productivity in these ecosystems, these values for nitrogen capital may be regarded as a first approximation to the minimum values required for the full development and maintenance of these non-leguminous shrub and woodland ecosystems.

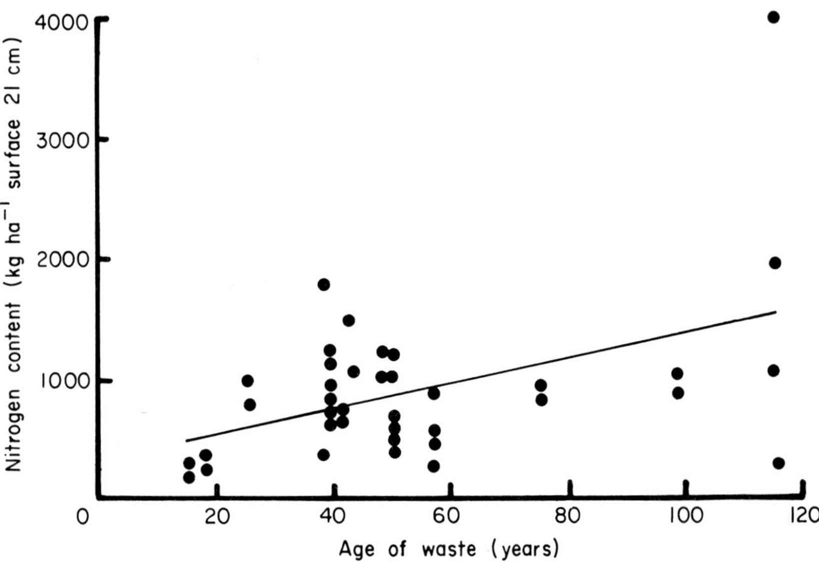

Fig. 6.2. The nitrogen content (kg N ha^{-1} in the surface 21 cm) of ecosystems which have naturally developed on china clay wastes of different age. (Roberts *et al.* 1981.)

Table 6.4. Comparison of nitrogen content and compartmentation between major ecosystem pools in naturally colonized and reclaimed china clay wastes. (n.d. not detectable.)

Ecosystem	Age	Total N (kg ha^{-1})	% Shoots	% Roots	% Litter	% Soil	Reference
(a) Naturally colonized china clay wastes							
Pioneer							
(*Lupinus arboreus*)	16–18	291	37	1	6	56	Marrs *et al.* (1981)
(*Calluna vulgaris*/ *Ulex europaeus*)	30–55	823	13	5	3	79	Marrs *et al.* (1981)
Intermediate							
(*Salix atrocinerea*)	40–76	981	8	18	6	68	Marrs *et al.* (1981)
Woodland							
(*Betula pendula*/ *Rhododendron ponticum*/ *Quercus robur*)	40–116	1770	30	3	0	67	Marrs *et al.* (1981)
(b) Reclaimed china clay wastes							
Sand tips	3–84	211	11	59	n.d.	30	Roberts *et al.* (1980)
Mica dam walls	3–84	441	8	61	n.d.	31	Roberts *et al.* (1980)

The development of ecosystems has been analysed in a similar manner (Fig. 6.3) for glacial moraines in Alaska (Crocker & Major 1955; Crocker & Dickson 1957), ironstone spoil in Minnesota (Leisman 1957), and sand dunes in Michigan (Olson 1958). On glacial moraines, a mature woodland ecosystem develops after approximately 100 years, when the soil nitrogen concentration has reached approximately 1200 kg N ha^{-1} in the surface 30 cm with another 1000 kg N ha^{-1} in the litter. On ironstone spoil, a woodland develops quickly (*c.* 21 years) when the nitrogen content of the soil was calculated to be approximately 600 kg N ha^{-1} in the surface 21 cm. In contrast, in the sand dune ecosystem, where atmospheric and biological nitrogen inputs are low, approximately 100 years is required for the development of a woodland ecosystem which then has 400 kg N ha^{-1} in the surface 10 cm. This apparently lower accumulation value for sand dunes may reflect the fact that the soil was sampled to a lesser depth, but it does raise the possibility that different types of woodland could have different annual, and therefore capital, requirements.

The pool sizes in these developing ecosystems are considerably lower than those in comparable developed ecosystems. The small pool sizes must be related to the age of the systems. Further increases can really only occur in the soil pool and do appear to do this (Marrs *et al.* 1981; Roberts *et al.* 1982). This would then lead to the larger amounts of soil and litter organic matter which are found in developed ecosystems (see Table 6.2). These amounts perhaps represent a greater capital of nitrogen than is absolutely necessary. Present estimates of potential soil to plant transfers (soil nitrogen mineralization) suggest that approximately 2% of the total soil nitrogen capital is mineralized annually (Geist, Reuss & Johnson 1970; Reuss & Innis 1977). If this were the case in these china clay ecosystems, with 700–1200 kg N ha^{-1} in the soil pool, soil mineralization would provide approximately 14–24 kg N ha^{-1} $year^{-1}$. Even if mineralization rates are higher than this value, soil organic matter to plant fluxes must be considerably lower than observed fluxes in well-developed ecosystems.

However, these low mineralization rates, together with annual additions through rainfall (approximately 9 kg N ha^{-1} $year^{-1}$ for china clay wastes), litter decomposition, and plant internal nitrogen cycling processes, must be sufficient to meet the annual requirements of the vegetation. This means that either internal cycling within the plant–root–litter system is extremely efficient, or the greater soil organic nitrogen–plant fluxes in well-developed systems are well in excess of minimal maintenance requirements. Whatever the explanation, it can be concluded that where the restoration of china clay wastes is directed towards a *S. atrocinerea* scrub or woodland vegetation, soil nitrogen pools of 700–1200 kg N ha^{-1} can be regarded as target levels.

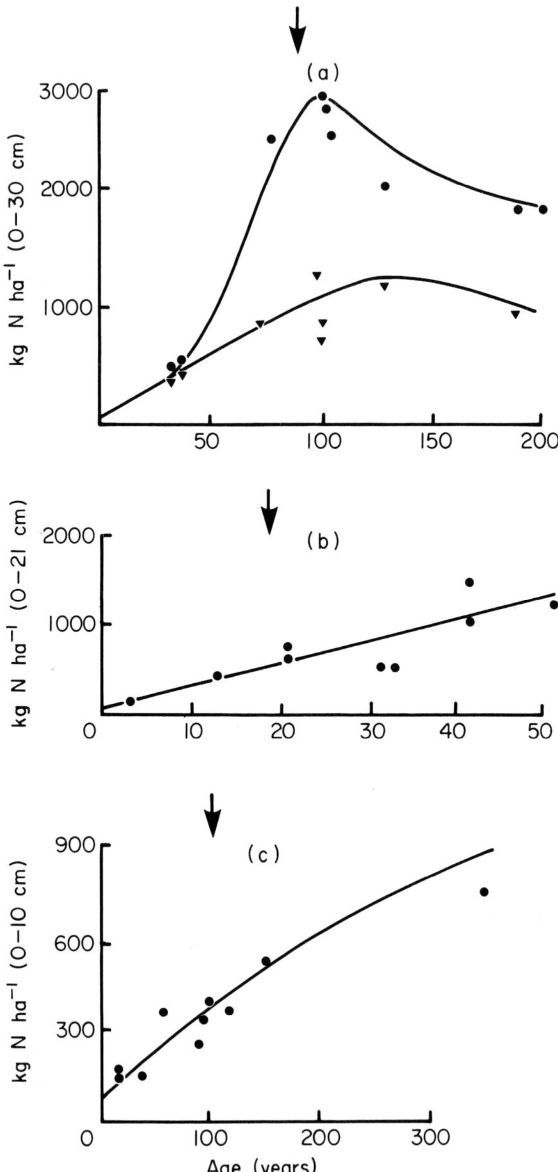

FIG. 6.3. The nitrogen content of soil during primary successions on (a) glacial moraines (Crocker & Major 1955), (b) ironstone spoils (Leisman 1957), and (c) sand dunes (Olson 1958). The age at which non-nitrogen-fixing tree species make a significant contribution to the vegetation is arrowed.

Data from Leisman (1957) calculated using the 1 inch sample for 0–2.5 cm depth, and the 4 inch sample for 2.5–21 cm depth.

These target values may, however, not necessarily apply to other ecosystems (Olson 1963; Swift, Heal & Anderson 1979). If turnover was lower a greater capital would be necessary to provide the same nitrogen supply. This may be true for coniferous woodland, where litter breakdown is slow. If turnover was higher, as in a sward with a high leguminous content, a smaller capital would be adequate. Ecosystems may differ in their nitrogen requirements; heathland communities, for example, certainly have a low productivity (Gimingham 1972), and presumably a low potential requirement. Further research on the determination of the minimum uptake and capital requirement of other communities (including agricultural as well as semi-natural communities) is urgently required.

PRINCIPLES OF NITROGEN ACCUMULATION

If the major aim during land restoration on china clay wastes is to promote nitrogen accumulation, it is necessary to consider the potential sources of inputs and losses of nitrogen. Inputs must essentially be through atmospheric deposition, biological fixation, fertilizers and imports through animal immigration. Losses must be through leaching, surface run-off, denitrification, fire and animal emigration.

Atmospheric deposition of nitrogen in the china clay area was estimated to be 9 kg N ha^{-1} $year^{-1}$ during 1978–9 (Roberts *et al.* 1981). Biological fixation has been estimated in two different ways. Acetylene reduction has been used on naturally colonized and reclaimed sites using incubated plant–soil cores (Skeffington & Bradshaw 1980), and accumulation under field conditions for some legume species in single species field plots (Dancer *et al.* 1977). The values from acetylene reduction demonstrated a low input from free-living microorganisms in soil (<1 kg N ha^{-1} $year^{-1}$), and also from rhizosphere fixation associated with *Calluna vulgaris, Rumex acetosella, Festuca rubra* and *Agrostis tenuis* (<3 kg N ha^{-1} $year^{-1}$). Whilst these non-leguminous fixation rates may be important on an individual plant basis, they are too low to be of significance in whole ecosystem development. In terms of ecosystem development, by far the most significant input is associated with leguminous species (*Ulex europaeus* 27 kg N ha^{-1} $year^{-1}$, *Trifolium repens* 49 kg N ha^{-1} $year^{-1}$, *Lupinus arboreus* 72 kg N ha^{-1} $year^{-1}$).

Accumulation in species trial plots (Table 6.5) gives values ranging from 14 kg N ha^{-1} $year^{-1}$ for *Vicia sativa* to 135 kg N ha^{-1} $year^{-1}$ for *Trifolium repens* and 157 kg N ha^{-1} $year^{-1}$ for *Trifolium pratense* over a two-year period. These data were, however, from plants grown under good conditions with adequate lime and fertilizer. What is perhaps more remarkable is that nitrogen accumulation beneath individual wild plants of *L. arboreus* was found to be as high as 185 kg N ha^{-1} year^{-1} (Palaniappan, Marrs & Bradshaw 1979). Similar values have been obtained for this species in coastal sand dune forestry in New Zealand (Gadgil 1971).

The other major source of nitrogen input in managed systems is through fertilizer addition. The mean annual additions of nitrogen applied to sand tips have been 72 kg N ha^{-1} year^{-1} (Marrs *et al.* 1981). During reclamation up to 200 kg N ha^{-1} year^{-1} may be applied without adverse visual effects on grass swards, but only 50 kg N ha^{-1} year^{-1} can be applied to grass-clover swards without reducing the rates of nitrogen fixation by *T. repens* (Skeffington & Bradshaw 1980). However, as fertilizer is applied to steep slopes by hydraulic means, this is expensive.

Losses of nitrogen through leaching may be excessive on bare soil. Dancer (1975), for example, showed that ammonium and nitrate ions were leached at rates of 2.5 and 4.1 cm depth for each cm of rainfall addition. Vegetation has a major effect on nutrient capture. When fertilizer is applied to established swards leaching losses are low (Fig. 6.4). Lysimeter studies have shown mean losses of 5 kg N ha^{-1} after one year when 95 kg N ha^{-1} year^{-1} as fertilizer and rainfall was added (Marrs & Bradshaw 1980).

Surface run-off from colliery spoil (Collier, Pickering & Musser 1970; Dennington & Chadwick 1978) has been found to vary between 34 and 44% on bare surfaces, but on vegetated plots this loss was generally lower (up to 89% reduction in run-off). If the maximum loss of the nitrogen in rainfall and fertilizers (nitrogen fertilizers are applied hydraulically and would be as susceptible to run-off as rainfall nitrogen) on china clay wastes are assumed to

TABLE 6.5. Net nitrogen fixation by legumes grown on china clay sand waste (Dancer *et al.* 1977).

Legume	kg N ha^{-1} year^{-1}
Trifolium pratense (Early)	151
Trifolium pratense (S123)	141
Trifolium pratense (Altaswede)	119
Trifolium repens (S184)	135
Trifolium repens (S100)	130
Trifolium hybridium (Alsike)	111
Trifolium debium (Suckling)	65
Medicago lupulina (English trefoil)	132
Medicago sativa (Lucerne)	115
Lotus corniculatus (Birdsfoot trefoil)	117
Lupinus angustifolius (Blue lupin)	32
Vicia sativa (Vetch)	14

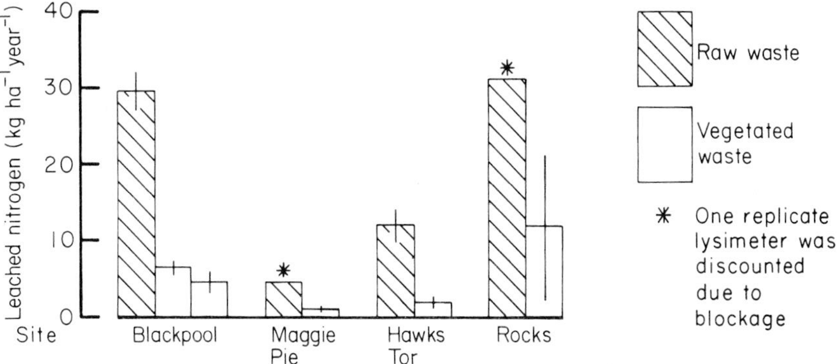

FIG. 6.4. Nitrogen leaching from lysimeters with developing and established swards on different china clay sand wastes which have been treated by standard reclamation techniques (Marrs & Bradshaw 1980).

be 40%, this would mean a loss of 32 kg N ha^{-1} year^{-1} in sites where the mean annual nitrogen input of 81 kg N ha^{-1} year^{-1} is applied (72 kg N ha^{-1} year^{-1} fertilizer and 9 kg N ha^{-1} year^{-1} rainfall). But this loss of nitrogen in surface run-off must represent the maximum potential loss because of the presence of a vegetation cover and the highly porous nature of the china clay wastes resulting in downward rather than surface waterflow. The actual losses from this source are more likely to be < 10 kg N ha^{-1} year^{-1}.

Transfers from the system through grazing import and export, and losses through denitrification have not been quantified for china clay wastes, but are assumed to be relatively insignificant. Denitrification appears to be important in soils only with a high organic matter content, and where there is some degree of anaerobiosis caused by waterlogging (Broadbent & Clark 1965; Black 1968; Russell 1973). China clay wastes have a low organic matter content and are well aerated. The effect of immigration of animals in terms of total ecosystem nitrogen will be very low, and should be offset by emigration, leading to no net nitrogen input from this source.

From these considerations it can be concluded that the major nitrogen input is from atmospheric deposition, legume fixation and fertilizer application. Fertilizer and atmospheric nitrogen will not be efficiently retained during the establishment phase, but capture efficiency increases when the vegetation becomes established. To confirm these conclusions, nitrogen accumulation rates were assessed for all china clay sites which have been reclaimed (Marrs, Roberts & Bradshaw 1980), and were compared with estimated nitrogen inputs. Mean net accumulation rates were calculated for a range of all sites between 3 and 84 months old by measuring the total nitrogen at each site and correcting for the initial nitrogen content of the raw waste. These results were then compared with measured inputs from fertilizer and atmospheric deposition (Table 6.6). The respective values are 108 kg N ha^{-1} year^{-1} actually accumulated in the ecosystem with measured inputs of 81 kg N ha^{-1} year^{-1}. In contrast to nitrogen, all other elements show net annual losses. This discrepancy between the behaviour of nitrogen and other elements reflects the additional inputs through legume fixation, mainly from *Trifolium repens* which had been included as a component of the original seed mixture. Fixation rates must therefore be at least 27 kg N ha^{-1} year^{-1}, assuming no losses from the system; for theoretical considerations, where maximum potential losses of 36 kg N ha^{-1} year^{-1} (32 run-off + 4 leaching) are assumed,

TABLE 6.6. An overall nutrient budget (kg N ha^{-1} year^{-1}) for reclaimed china clay wastes (Marrs *et al.* 1980).

	Element				
	N	P	K	Ca	Mg
Mean net accumulation	+108	+83	+42	+76	+70
Mean input from fertilizers	+72	+87	+62	+344	+73
Input from atmospheric deposition	+9	+0.3	+17	+10	+18
Balance	+27	−4	−37	−178	−21

fixation rates would be a maximum of 63 kg N ha^{-1} year^{-1}. This range of values (27–63 kg N ha^{-1} year^{-1}) compares favourably with estimates of fixation rates estimated for *T. repens* under field conditions by acetylene reduction (49 kg N ha^{-1} year^{-1}); field plot trial values (130–135 kg N ha^{-1} year^{-1}; see Table 6.5) are much higher because of more favourable growing conditions.

Similar calculations can be made for naturally colonized wastes, where nutrient inputs are considerably lower. There are no fertilizer additions to these waste tips, and nitrogen inputs are derived entirely from atmospheric and biological sources. Atmospheric inputs can be asumed to be 9 kg N ha^{-1} year^{-1} over the entire period although this may be an overestimate in the early stages before pollution inputs increased. Maximum losses from these sites during colonization can be calculated using the following assumption that losses of 100% occur during the first 30 years when most tips remain uncolonized (Dancer 1975), but are reduced to 45% when the tips become vegetated (40% surface run-off, 5% leaching).

Nitrogen accumulation on naturally colonized wastes has already been shown to be 10 kg N ha^{-1} year^{-1}. If there were no losses from the system this could almost be contributed by atmospheric deposition and inputs from legume fixation could therefore be negligible (1 kg N ha^{-1} $year^{-1}$). However, from the preceding discussion we can calculate approximately the maximum potential losses of nitrogen. During the first 30 years losses of nitrogen would be equal to 270 kg N ha^{-1} (30×9), and in the following 70 years the maximum loss would be 283 kg N ha^{-1} ($70 \times 9 \times 0.45$), giving a total maximum potential loss of 553 kg N ha^{-1} over a 100-year period. As the total input from atmospheric deposition during this time would be 900 kg N ha^{-1}, this means that 347 kg N ha^{-1} supplied from atmospheric inputs would be retained in the ecosystem. The legume contribution to the observed nitrogen accumulation capital of 1000 kg N ha^{-1} during this period would be 653 kg N ha^{-1}, and as legumes (mainly *Ulex europaeus*) form a large proportion of the vegetation cover and biomass for approximately a 25-year period, this is equivalent to an annual fixation rate of 22 kg N ha^{-1} $year^{-1}$. It is interesting that acetylene reduction studies have estimated nitrogen fixation to be in a similar order of magnitude (26 kg N ha^{-1} $year^{-1}$; Skeffington & Bradshaw 1980).

These calculations are based on gross assumptions, the accuracy of which is impossible to assess. In particular we do not have direct estimates of losses through leaching and surface run-off from these developing and mature ecosystems on china clay wastes. It is very difficult to assess the effect of these processes on the build-up of soil nitrogen in organic matter accurately. Nevertheless, these calculations serve to show the orders of magnitude of the various ways in which nitrogen accumulates, and suggest that ecosystem development can be considerably accelerated by increasing nitrogen inputs. They also emphasize the importance of biological nitrogen fixation during the development of primary successions, and the role that fixation may have as a tool for the successful reclamation of derelict land.

THE DEVELOPMENT OF NITROGEN CYCLES

Accumulation of nitrogen within the ecosystem should not be regarded as the only factor required for successful restoration. Attention must also be paid to the compartmentation between major ecosystem pools, and to efficient cycling processes. Analysis of plant shoot, root, litter (where present), and soil pool sizes in ecosystems developing on naturally colonized and reclaimed china clay wastes shows that accumulation occurs in all pools (Table 6.4). In naturally colonized wastes the major site of accumulation is the soil, whereas in reclaimed wastes the highest accumulation is in the roots.

As we have seen, the transfer of nitrogen from soil organic matter to plant is critical. Modified mineralization tests of Keeney & Bremner (1967) have demonstrated that soil mineralization increases in restored sites with age and quality of the sward (Roberts, Marrs & Bradshaw 1980; Skeffington & Bradshaw, 1981). Mineralization is significantly positively correlated ($P<0.001$) with the total nitrogen content of the vegetation (Fig. 6.5; Skeffington & Bradshaw 1981). The percentage of the total nitrogen mineralized in a two-week incubation, an index of nitrogen turnover, has a slight but non-significant reduction with age (see Fig. 6.5), but after six years is close to values for adjacent rough pasture. This implies that the transfer of soil nitrogen to plants is low in reclaimed sites because of the small nitrogen capital. Whilst the supply from soil mineralization is increasing, these yield little quantitative information on the actual rates of nitrogen supplied from soil organic matter decomposition under field conditions. Indeed, as the highest mean soil nitrogen content in the oldest reclaimed sites was approximately 400 kg ha^{-1} (0–21 cm depth), and if the annual mineralization supply of 2% were assumed, the soil to plant transfer would be only 8 kg N ha^{-1} year^{-1}. Although the major factor limiting the supply from the soil mineralization sources is the low nitrogen capital, an increased efficiency of mineralization per unit of nitrogen capital could also be important. Both soil mineralization and decomposition of dead plant material is faster in substrates with a low carbon/nitrogen ratio. One way of achieving this is to promote the growth of leguminous species, whose tissues have a higher nitrogen concentration than grass species.

A high legume component in the vegetation results in visually attractive swards with a higher mineralization rate than grass swards (Skeffington & Bradshaw 1981), in which the litter decomposes faster (Lanning & Williams 1979). We have no quantitative information on the amounts of nitrogen released for plant uptake by soil mineralization and litter decomposition. It is possible that if fast recycling of litter nitrogen or internal redistribution of plant nitrogen occurs, similar to the situation in shortgrass prairies (Clark 1977), these reclaimed systems could function with low supplies from the mineralization of soil organic nitrogen.

If forage legumes are to be included in the vegetation this will require the addition of suitable amounts of lime and phosphorus fertilizers to maintain healthy growth (Handley *et al.* 1978; Roberts *et al.* 1982). But some leguminous species, particularly *Lupinus arboreus*, can grow vigorously at very low calcium and phosphate levels, and not only increase the plant and soil nitrogen pools, but also considerably increase the amount of mineralizable nitrogen in the soil (Palaniappan, Marrs & Bradshaw 1979).

In addition to soil to plant fluxes, attention should be paid to other transfer processes. Cutting and grazing can be used, for example, to increase the transfers from the above-ground plant to the roots and soil. In upland pastures (Perkins 1978), sheep grazing can account for the transfer of up to

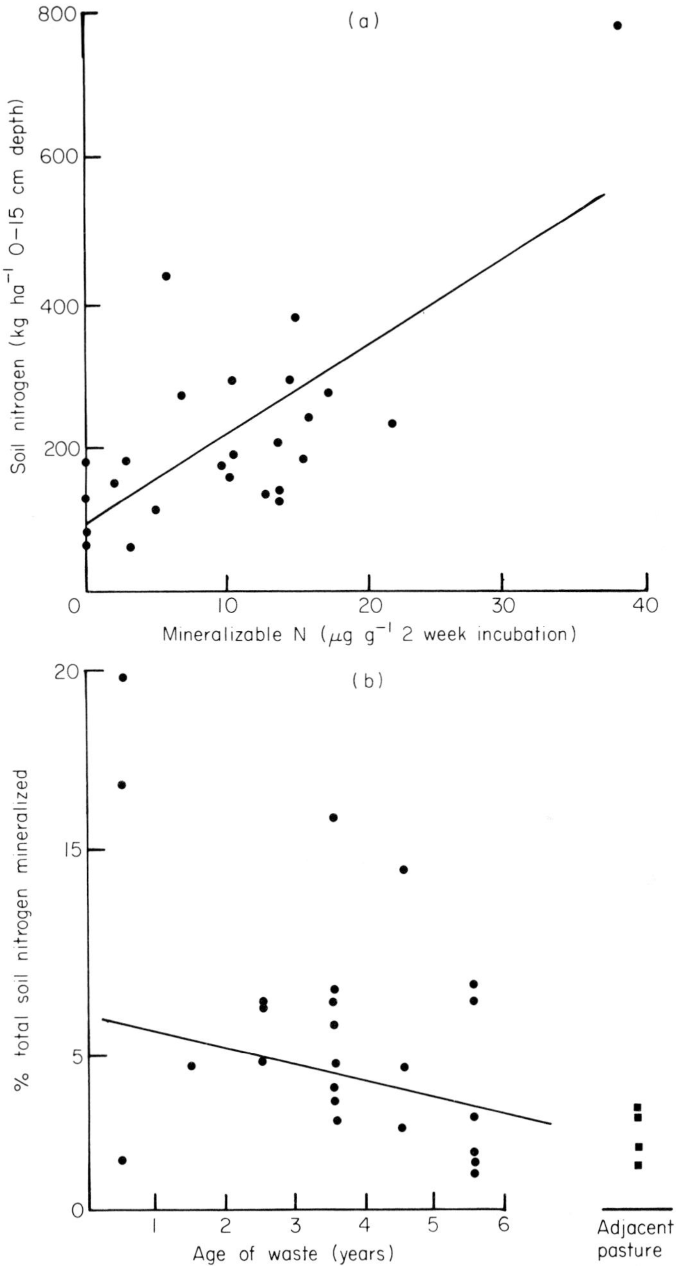

FIG. 6.5. Mineralization of soil nitrogen in reclaimed china clay wastes; (a) mineralization rate ($\mu g\ g^{-1}$ 2 weeks at 30°C) plotted against total soil nitrogen, (b) index of nitrogen turnover (mineralization rates as % of total soil nitrogen) in sites of different age compared to adjacent pasture soils. (Skeffington & Bradshaw 1981.)

18% of the nitrogen transferred annually to the plant pool and, moreover, the nitrogen is returned in urine in a form readily available for plant uptake, or in faeces in a form which is mineralized well at low temperatures (Floate 1970a & b). On reclaimed china clay wastes, grazing has been shown to have a marked effect in reducing the proportion of dead plant material present in the vegetation (Fig. 6.6); 14–27% of the annual uptake of nitrogen into the plant pool could be recycled by sheep grazing (Marrs, Granlund & Bradshaw 1980). This additional transfer is extremely important in the management of these wastes, in view of the low amounts of nitrogen supplied from other cycling sources. The effects of cutting on nutrient cycling is not well understood in reclaimed china clay wastes, but it may provide a means of improved nitrogen utilization in situations where grazing is not possible.

In some derelict land environments where adverse substrate conditions may limit microbial decomposition and transfer processes, the effects of grazing and/or cutting may be even more important. Clearly more detailed work is required on techniques for promoting nutrient cycling in developing ecosystems.

CONCLUSIONS

It is inevitable that there are errors in some of the calculations of nitrogen inputs and outputs, and of rates of nitrogen release by mineralization. Nevertheless the previous discussion here gives evidence of the general magnitude of nitrogen stores and fluxes in developing ecosystems. None of this evidence contradicts the hypothesis that nitrogen supply is a critical factor in the development of ecosystems. It would appear that species distribution and biomass development are both profoundly influenced by nitrogen supply.

This supply comes from the mineralization of the capital of nitrogen accumulated in the soil store. Both the size of the capital and the rate at which it is mineralized are crucial. The available evidence suggests that the situations which occur in fully developed ecosystems will not necessarily be a model for what occurs in developing ecosystems; young ecosystems appear to be able to subsist on a much smaller capital. This is partly because they are composed of nitrogen-fixing species which not only are themselves self-sufficient in nitrogen but also provide an organic matter input which readily decays and releases its nitrogen.

From this we can begin to see three important principles that should be applied to the restoration of land. Firstly, there is a minimum level of soil nitrogen which must be accumulated before scrub ecosystems are likely to be self-sustaining; we have found this to be 700 kg N ha^{-1} for a *Salix* ecosystem. Further research is required to estimate minimum requirements for other

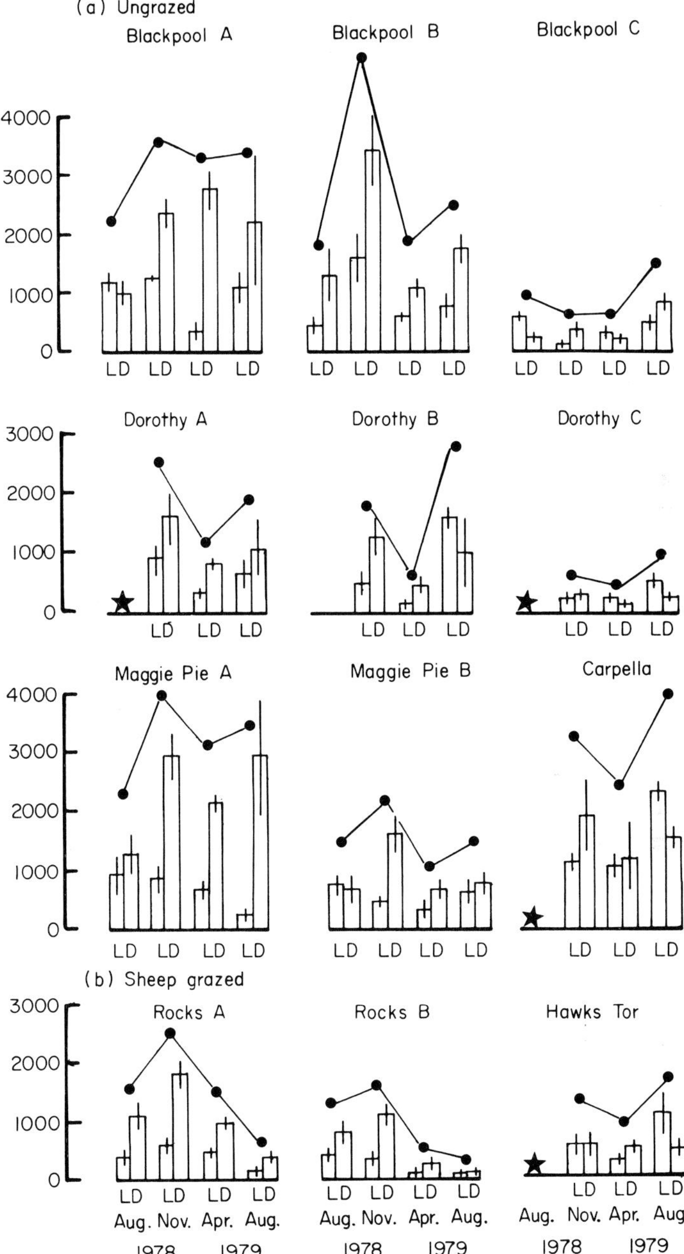

FIG. 6.6. The mean standing crop (L), standing dead (D) and total above-ground plant material (circles) on ungrazed and sheep-grazed reclaimed china clay wastes between 1978 and 1979. Standard errors are presented for standing crop and standing dead. * Samples missing. (Marrs, Granlund & Bradshaw 1980.)

types of ecosystem. Secondly, the only reasonable way to achieve this is by the use of legumes. Thirdly, means must be found of ensuring the rapid return and decomposition of organic matter. Otherwise, heavy and expensive application of fertilizers must be given if the developing vegetation is not to become moribund.

ACKNOWLEDGEMENTS

We would like to thank NERC and English China Clays Ltd for financial support.

REFERENCES

Allen S.E., Grimshaw H.M., Parkinson J.A. & Quarmby C. (1974) *Chemical Analysis of Ecological Materials.* Blackwell Scientific Publications, Oxford.

Black C.A. (1968) *Soil–Plant Relationships.* John Wiley & Sons, New York.

Bormann F.H. & Likens G.E. (1979) *Pattern and Process in a Forested Ecosystem.* Springer-Verlag, New York.

Bradshaw A.D. & Chadwick M.J. (1980) *The Restoration of Land.* Blackwell Scientific Publications, Oxford.

Bradshaw A.D., Dancer W.S., Handley J.F. & Sheldon J.S. (1975) Biology of land revegetation and reclamation of china clay wastes. In *The Ecology of Resource Degredation and Renewal* (eds M.J. Chadwick & G.T. Goodman), pp. 363–84. Blackwell Scientific Publications, Oxford.

Broadbent F.E. & Clark F.E. (1965) Denitrification. *Agronomy*, **10**, 344–59.

Clapham A.R., Tutin T.G. & Warburg E.F. (1962) *Flora of the British Isles*, 2nd edn. Cambridge University Press, London.

Clark F.E. (1977) Internal cycling of ^{15}nitrogen in shortgrass prairie. *Ecology*, **58**, 1322–33.

Collier C.R., Pickering R.J. & Musser J.J. (1970) Influences of strip-mining on the hydrologic environment of parts of Beaver Creek Basin, Kentucky, 1955–1966. *U.S. Geological Survey*, Professional Paper 427c.

Coupland R.T. & Van Dyne G.M. (1979) Natural temperate grasslands: systems synthesis. In *Grassland Ecosystems of the World: Analysis of Grasslands and Their Uses* (ed R.T. Coupland), pp. 97–106. Cambridge University Press, London.

Crocker R.L. & Dickson B.A. (1957) Soil development on the recessional moraines on the Herbert and Mendenhall glaciers, south-eastern Alaska. *Journal of Ecology*, **45**, 169–85.

Crocker R.L. & Major J. (1955) Soil development in relation to vegetation and surface age of Glacier Bay, Alaska. *Journal of Ecology*, **43**, 427–48.

Dancer W.S. (1975) Leaching losses of ammonium and nitrate in the reclamation of sand spoils in Cornwall. *Journal of Environmental Quality*, **4**, 499–504.

Dancer W.S., Handley J.F. & Bradshaw A.D. (1977) Nitrogen accumulation in kaolin mining wastes in Cornwall. II. Forage legumes. *Plant and Soil*, **48**, 303–14.

Dancer W.S., Handley J.F. & Bradshaw A.D. (1979) Nitrogen accumulation in kaolin mining wastes in Cornwall. III. Nitrogenous fertilizers. *Plant and Soil*, **51**, 471–84.

Dennington V.M. & Chadwick M.J. (1978) The nutrient budget of colliery spoil tip sites. I. Nutrient input in rainfall and nutrient losses in surface run-off. *Journal of Applied Ecology*, **15**, 303–16.

Duvigneaud P. & Denaeyer-de Smet S. (1970) Biological cycling of minerals in temperate deciduous forests. In *Ecological Studies. I. Temperate Forests* (ed. D.E. Reichle), pp. 199–225. Springer-Verlag, Berlin.

Epstein E. (1972) *Mineral Nutrition of Plants: Principles and Perspectives,* John Wiley & Sons, New York.

Floate M.J.S. (1970a) Decomposition of organic materials from hill soils and pastures. II. Comparative studies on the mineralization of carbon, nitrogen, and phosphorus from plant materials and sheep faeces. *Soil Biology and Biochemistry*, **2**, 173–86.

Floate M.J.S. (1970b) Decomposition of organic materials from hill soils and pastures. III. The effect of temperature on the mineralization of carbon, nitrogen and phosphorus from plant materials and sheep faeces. *Soil Biology and Biochemistry*, **2**, 187–96.

Gadgil R.L. (1971) The nutritional role of *Lupinus arboreus* in coastal sand dune forestry. 3. Nitrogen distribution in the ecosystem before tree planting. *Plant and Soil*, **35**, 113–26.

Geist J.M., Reuss J.O. & Johnson D.D. (1970) Prediction of nitrogen requirements of field crops. II. Application of theoretical models to malting barley. *Agronomy Journal*, **62**, 385–9.

Gimingham, C.H. (1972) *Ecology of Heathlands*. Chapman & Hall, London.

Handley J.F., Dancer W.S., Sheldon J.C. & Bradshaw A.D. (1978) The nitrogen problem in derelict land reclamation with special reference to the British china clay industry. In *Environmental Management of Mineral Wastes* (eds G.T. Goodman & M.J. Chadwick), pp. 215–36. Sijthoff & Noordhoff, Alphen aan den Rijn.

Hewitt E.J. & Smith T.A. (1975) *Plant Mineral Nutrition*. English Universities Press, London.

Hutnik R.J. & Davis G. (1973) *Ecology and Reclamation of Devastated Land.* Gordon & Breach, New York.

Keeney D.R. & Bremner J.M. (1967) Determination and isotope-ratio analysis of different forms of nitrogen in soils. 6. Mineralizable nitrogen. *Proceedings of the Soil Science Society of America*, **31**, 34–9.

Lanning S. & Williams S.T. (1979) Nitrogen in revegetated china clay sand waste I. Decomposition of plant material. *Environmental Pollution*, **20**, 147–59.

Leisman G.A. (1957) A vegetation and soil chronosequence on the Mesabi iron range spoil banks, Minnesota. *Ecological Monographs*, **27**, 221–45.

Marrs R.H. & Bradshaw A.D. (1980) Ecosystem development on reclaimed china clay wastes. III. Leaching of nutrients. *Journal of Applied Ecology*, **17**, 727–36.

Marrs R.H., Granlund I.H. & Bradshaw A.D. (1980) Ecosystem development on reclaimed china clay wastes. IV. Recycling of above-ground plant nutrients. *Journal of Applied Ecology*, **17**, 803–13.

Marrs R.H., Roberts R.D. & Bradshaw A.D. (1980) Ecosystem development on reclaimed china clay wastes. I. Assessment of vegetation and capture of nutrients. *Journal of Applied Ecology*, **17**, 709–17.

Marrs R.H., Roberts R.D., Skeffington R.A. & Bradshaw A.D. (1981) Ecosystem development on naturally colonized china clay wastes. II. Nutrient compartmentation. *Journal of Ecology*, **69**, 163–9.

Martin N.J. & Holding A.J. (1978) Nutrient availability and other factors limiting microbial activity in the Blanket Peat. In *Production Ecology of British Moors and Montane Grasslands* (eds O.W. Heal & D.F. Perkins), pp. 113–35. Springer-Verlag, Berlin.

Odum E.P. (1971) *Fundamentals of Ecology*. W.B. Saunders, Philadelphia.

Olson J.S. (1958) Rates of succession and soil changes on southern Lake Michigan sand dunes. *Botanical Gazette*, **119**, 125–70.

Olson J.S. (1963) Energy storage and the balance of producers and decomposers in ecological systems. *Ecology*, **44**, 322–31.

Palaniappan V.M., Marrs R.H. & Bradshaw A.D. (1979) The effect of *Lupinus arboreus* on the nitrogen status of china clay wastes. *Journal of Applied Ecology*, **16**, 825–31.

Perkins D.F. (1978) The distribution and transfer of nutrients in the *Agrostis–Festuca* grassland ecosystem. In *Production Ecology of British Moors and Montane Grasslands* (eds O.W. Heal & D.F. Perkins), pp. 375–96. Springer-Verlag, Berlin.

Reuss J.O. & Innis G.S. (1977) A grassland nitrogen flow simulation model. *Ecology*, **58**, 379–88.

Roberts R.D., Marrs R.H. & Bradshaw A.D. (1980) Ecosystem development on reclaimed china clay wastes. II. Nutrient compartmentation and nitrogen mineralization. *Journal of Applied Ecology*, **17**, 719–25.

Roberts R.D., Marrs R.H., Skeffington R.A. & Bradshaw A.D. (1981) Ecosystem development on naturally colonised china clay wastes. I. Vegetation changes and overall accumulation of organic matter and nutrients. *Journal of Ecology*, **69**, 153–61.

Roberts R.D., Marrs R.H., Skeffington R.A. & Bradshaw A.D. (1982) The importance of plant nutrients in the restoration of derelict land and mine wastes. *Transactions of the Institute of Mining and Metallurgy*, **91A**, 42–4.

Robertson R.A. & Davies G.E. (1965) Quantities of plant nutrients in heather ecosystems. *Journal of Applied Ecology*, **2**, 211–19.

Russell E.W. (1973) *Soil Conditions and Plant Growth*. 9th edn. Longmans, London.

Singh J.S., Singh K.P. & Yadava P.S. (1979) Tropical grasslands: ecosystem synthesis. In *Grassland Ecosystems of the World: Analysis of Grasslands and Their Uses* (ed R.T. Coupland), pp. 231–40. Cambridge University Press, London.

Skeffington R.A. & Bradshaw A.D. (1980) Nitrogen fixation by plants grown on reclaimed china clay waste. *Journal of Applied Ecology*, **17**, 469–77.

Skeffington R.A. & Bradshaw A.D. (1981) Nitrogen accumulation in kaolin mining wastes in Cornwall. IV. Sward quality and the development of a nitrogen cycle. *Plant and Soil*, **62**, 439–51.

Swift M.J., Heal O.W. & Anderson J.M. (1979) *Decomposition in Terrestrial Ecosystems*. Blackwell Scientific Publications, Oxford.

Tanner E. (1977) Four montane rain forests of Jamaica: a quantitative characterization of the floristics, the soils and the foliar mineral levels, and a discussion of their interrelations. *Journal of Ecology*, **65**, 883–918.

Tanner E. (1980) Studies on the biomass and productivity in a series of montane rain forests in Jamaica. *Journal of Ecology*, **68**, 573–98.

UNESCO/UNEP/FAO (1978) *Tropical Forest Ecosystems*, Paris.

Whitehead D.C. (1970) *The Role of Nitrogen in Grassland Productivity*. Bulletin 48, Commonwealth Agricultural Bureau, Farnham Royal, Bucks.

Wielgolaski S.E., Kjelvik S. & Kallio P. (1975) Mineral content of tundra and forest tundra plants in Fennoscandia. In *Fennoscandian Tundra Ecosystems, Part I. Plants and Micro-organisms* (ed. S.E. Wielgolaski), pp. 316–32. Springer-Verlag, Berlin.

Wlotzka F. (1978) Nitrogen. In *Handbook of Geochemistry* (ed K.H. Wedepohl), Vol. II, 7B–M. Springer-Verlag, Berlin.

7. THE RELATION BETWEEN THE NITROGEN METABOLISM OF *PLANTAGO* SPECIES AND THE CHARACTERISTICS OF THE ENVIRONMENT

J.W.WOLDENDORP
Institut voor Oecologisch Onderzoek,
Boterhoeksestraat 22,
6666 GA Heteren,
The Netherlands

SUMMARY

As part of a study of the demographic, physiological and genetic adaptations of grassland species to their environment, special attention was paid to the nitrogen metabolism of the genus *Plantago*. In an old, established semi-natural dune grassland the capacity of the soil to provide three *Plantago* species with ammonium and nitrate was investigated. Nitrogen mineralization established using an incubation technique *in situ*, the numbers of nitrifying bacteria in the rhizosphere, the ionic balance and the level of nitrate reductase of the *Plantago* species all pointed to nitrate formation and uptake. A considerable small-scale spatial variability in nitrogen mineralization was observed and a close correlation was found to exist between numbers of nitrifying bacteria in the rhizosphere and nitrate reductase activity in the leaves. In addition, nitrogen availability was highly influenced by the moisture content of the soil and it was concluded that *Plantago major* L. occurred in places with a more continuous nitrogen supply than *Plantago lanceolata* L. The latter species is found over a wide range of habitats to which it is adapted by the high metabolic flexibility of its individuals.

In accordance with these conclusions it was found in laboratory studies that *P. lanceolata* was better adapted to shifts in the nutrient level than *P. major*. The former species rapidly adjusted its growth rate, organic nitrogen content and the activities of nitrate reductase, glutamate dehydrogenase and glutamine synthetase to a changed nutrient level; in *P. major* these parameters were largely independent of the nutrient level. Using a new technique to attain low levels of nutrient supply, indications were obtained that the kinetics (K_m and V_{max}) of nitrate uptake were related to the nitrogen availability of the environment, i.e. species of poor or rich environments possessing a low K_m and a high V_{max} at low or high nitrate levels respectively.

The results obtained for the nitrogen metabolism of *P. major* and *P. lanceolata* were shown to be in accordance with the genetic structure of their populations.

INTRODUCTION

Some years ago an integrated project was started to study the demographic physiological and genetic adaptations of grassland species to their environment. A basic assumption in this project is that differences between closely related species may point to specific adaptations to their environment. For this comparison mainly species of the genus *Plantago* were chosen. In this study special attention is being paid to the nitrogen nutrition of these *Plantago* species.

In the present paper the results of this study on the nitrogen nutrition of *Plantago* species are summarized in a preliminary way. It aims at evaluating the prospects of a concerted action of soil chemists, soil microbiologists and plant physiologists in explaining differences in the metabolism of plants in an ecologically relevant way. No attempt is made to compile the vast literature on soil nitrogen and on the nitrogen metabolism of plants, and it is assumed that the reader is familiar with its principles, which are summarized in recent reviews and books on nutrients in the soil (Nye & Tinker 1977; Rorison 1980), nitrogen in the environment (Nielsen & MacDonald 1978a, b), nitrogen transformations in the rhizosphere (Woldendorp 1975, 1978, 1981), nitrogen uptake and metabolism of plants (Beevers & Hageman 1969; Haynes & Goh 1978; Hewitt & Cutting 1979; Clarkson & Hanson 1980), and the mineral nutrition of wild plants (Chapin 1980).

DESCRIPTION OF *PLANTAGO* SPECIES AND THEIR HABITATS

The genus *Plantago* is represented in the Netherlands by five species, i.e. *P. lanceolata* L., *P. major* L., *P. media* L., *P. coronopus* L., and *P. maritima* L. A phylogenetic investigation using the allozyme technique indicated that the genetic distances between *P. lanceolata* and the four other species were considerably higher than those between these latter species, that between *P. major* and *P. media* being the smallest (E. Boekema, unpublished data).

P. lanceolata can be found in a wide range of low vegetation types, such as dry and wet hayfields, meadows, and road verges. *P. major* is represented by two subspecies, i.e. *P. major* spp. *major* and *P. major* spp. *pleiosperma*. The former occurs in open, trampled places with a high degree of soil compaction, such as paths and grazed meadows; *P. major* spp. *pleiosperma* is found in places where the soil is compacted by changes in the water table, e.g. river banks. *P. media* is restricted to soils with a high content of calcium carbonate, such as chalk grasslands. *P. maritima* and *P. coronopus* are found along the coasts, the former generally in locations which are exposed to the sea. For *P. coronopus* a few inland populations are known, but these are much more rare than in some other West European countries. Because of its wide range of occurrence, *P. lanceolata* is often found together with one of the other *Plantago* species (except *P. maritima*), which facilitates comparison of their populations.

The range of values of some relevant soil parameters for locations of the *Plantago* species is given in Table 7.1. *P. lanceolata* is found over a wide range of soil conditions, particularly as regards pH and organic matter content, and only at the salinities at which *P. maritima* is found is the species absent.

Although the soil chemical characteristics of *P. major* locations overlap nearly completely with those of *P. lanceolata* (as well as in other parameters not shown in Table 7.1), this does not mean that the nutrient supply of both species is the same. Due to its occurrence in open positions with no competition and on compacted soils with more favourable moisture conditions, the nutrient supply of *P. major* may be somewhat higher and more continuous than that of *P. lanceolata.*

In the present paper it is the nitrogen nutrition of *P. major* spp. *major* and *P. lanceolata* which is mainly discussed. In addition, a comparison is made with the other *Plantago* species, with *Hypochaeris radicata* L. a species of poor, acid soils (see Table 7.1), and with *Urtica dioica* L. a competitive species from nitrogen-rich habitats.

For the interpretation of differences in the nitrogen metabolism of *P.*

TABLE 7.1. The range of soil chemical characteristics of the 0–10 cm layer of locations of *Plantago* species and *Hypochaeris radicata* L.

Species	pH–H_2O	$CaCO_3$ (%)	Organic matter (%)	Total N ($\mu g\ g^{-1}$)	Cl^- (mg 100 g^{-1})
P. media	5.8–7.8	0–53	2.8–13	750–4000	0–3.6
P. major	5.3–8.4	0–10	0.9–33	280–4400	0–5.0
P. lanceolata	4.3–7.8	0–53	1.3–74	350–24 000	0–9.6
P. coronopus	5.1–8.4	0–5	4.3–13	1500–4500	0–110
P. maritima	7.8	5–10	13–14	4500–5000	110–350
H. radicata L.	4.5–5.9	0	1.4–5.7	390–1800	0–0.4

major and *P. lanceolata* it should be known that the population characteristics of both species are different (van de Dijk 1981b). In *P. major*, reproduction takes place mainly by selfing, in contrast to *P. lanceolata* which is a self-incompatible species, its pollen being distributed by wind (Table 7.2). The degree of polymorphism (the fraction of enzyme loci for which variation has been observed) is about the same in both species, but the degree of heterozygosity (the average fraction of loci for which individual plants are heterozygous) in *P. lanceolata* is ten times as high as in *P. major*. Populations of the latter species can be subdivided at short distance (1 m) into small

TABLE 7.2. Population genetical characteristics of *P. major* L. and *P. lanceolata* L.

Species	Polymorphism	Heterozygosity	Cross-fertilization (%)
P. major L.	0.25	0.010	5–45
P. lanceolata L.	0.39	0.100	100

subpopulations with different allele frequencies. On the contrary, notwithstanding the high degree of variation present in *P. lanceolata*, such differentiation at short distances is not found, which is not suprising for a species reproducing by cross-fertilization. Not only within, but also between populations the allele frequencies are the same. The differences between both species can be explained by assuming that *P. major* copes with environmental variation by means of genetically different subpopulations of which the individuals have similar properties. The between-population variation in *P. lanceolata* is much smaller, and the species can adapt to the wide range of habitats in which it is found by the high phenotypic plasticity of its individuals.

PRODUCTION OF MINERAL NITROGEN IN THE SOIL

Plant roots have to compete for mineralized ammonium with heterotrophic bacteria which immobilize it into organic compounds and with the nitrifying bacteria which oxidize it to nitrate. In studying specific adaptations of the nitrogen metabolism of plants, it is of paramount importance to know in what quantities and proportions ammonium and nitrate become available to the plants. The latter is determined not only by microbial activity but also by the physical characteristics of the soil which influence the transport to the root surface.

In grassland soils the steady state concentrations of mineral nitrogen are nearly always low, particularly those of nitrate. Such low nitrate levels have been attributed to a suppression of nitrifying bacteria by allelopathic compounds excreted by plant roots (Rice 1974). But, more likely, the low nitrification rates are due to the absence of sufficient ammonium ions which are thought to be used preferentially in nitrogen immobilization and uptake by plant roots (Woldendorp 1975, 1978). In the highly fertilized pastures of the Netherlands numbers of nitrifying bacteria are low (Woldendorp 1963), but no data are available for more natural grasslands. Therefore, the production of ammonium and nitrate was studied in a natural dune grassland. It was not considered feasible to follow the nitrogen turnover processes in this soil using ^{15}N-labelled compounds, since these may cause a shift in nitrogen transformations. As an alternative procedure, the spatial and temporal variability in the distribution of ammonium and nitrate was established and related to their production which was determined using a field incubation technique. In addition, some plant parameters, i.e. nitrate reductase activities and the differences between accumulated amounts of inorganic cations and anions, were used to gain an impression of the contribution of nitrate and ammonium to nitrogen uptake by the *Plantago* species.

Description of the study area

The study was carried out at the Westduinen, an old dune grassland on the island of Goeree (the Netherlands) which has been extensively grazed by horses and cows for several centuries. It has a rolling relief with maximal differences in height of about 2–4 metres. The topography exerts a major influence on soil profile development via its impact on the microclimate and hydrological regime (slope aspect, distance to the ground water; some depressions are flooded during the winter and early spring).

Two forms of *P. lanceolata* occurred in the study area; one form of large individuals on wet sites and the other of small individuals on drier parts of the area (Blom *et al.* 1979; Noë & Blom 1981). *P. major* was found in relatively wet sites with a low percentage of air-filled pore volume and a relatively high organic matter content. *P. coronopus* was observed in spots with a relatively low total pore volume.

Spatial and temporal variability of soil chemical properties and mineral nitrogen

To investigate the small-scale variability within a topographically uniform unit (no slope effects), a 60 cm^2 grid, was subdivided into 16 units of 15 cm^2 at

19 different locations. Eight units of the grid were sampled separately and the other eight were combined as a bulk sample. By analysis of these samples an indication of the small-scale variability could be obtained, while a combination of grid and transect analyses yielded information of the topographic variation (Troelstra 1977). In Table 7.3 an example is given of such small-scale variability in some soil factors at one of these locations. The organic matter content and the factors related to it (e.g. organic phosphorus and nitrogen, exchangeable cations, moisture) showed considerable variation. Therefore, the determination of only a mean value of a bulk sample may involve the loss

TABLE 7.3. Spatial variability shown as the range of some soil chemical factors in the 0–10 cm layer in the Westduinen. (For details, see text)

	Small-scale variability (0.36 m^2)	Topographic variation
pH–H_2O	4.7–5.5	4.2–6.4
Organic matter (%)	10.9–15.7	0.9–33.0
Total phosphorus (mg 100 g^{-1})	16.0–20.2	7.4–51.0
Exchangeable magnesium (m-equiv. 100 g^{-1})	0.84–1.20	0.08–2.5
Exchangeable calcium (m-equiv. 100 g^{-1})	5.43–6.38	0.12–11.5
Exchangeable potassium (m-equiv. 100 g^{-1})	0.22–0.32	0.03–0.67

of ecologically important information related to the small-scale variability of the plots.

A considerable variation was also found to be present in the mineral nitrogen at the six locations which were analysed seasonally (Table 7.4). Not only did considerable differences between these locations exist but also over the year the quantities of ammonium and nitrate varied appreciably. Towards the end of the growing season they tended to increase. The quantities of ammonium, which occur mainly in the exchangeable form, were nearly always distinctly higher than those of nitrate. The nitrate contents were expressed as concentration in the soil solution to allow comparison with concentrations used in laboratory experiments. High nitrate concentrations were generally found to occur at low moisture contents of the soil; low values were obtained in moist soils. The nitrate values presented in Table 7.4 therefore do not necessarily give information on nitrate availability to the plant. Transport of nitrate to the roots, which at these low concentrations (20–30 mM is the normal concentration in fertilized agricultural soils) is mainly by diffusion, is

TABLE 7.4. The range and seasonal variation in ammonium and nitrate levels at six sites in the Westduinen during 1978. NO_3^- concentration was determined in the expressed soil solution. NH_4^+ was determined from a 1M KCl extract.

Date	NO_3^- (mM)	NH_4^+ (μg g^{-1} N)
3rd March	0.10–0.78	0.59–9.47
9th May	0.09–0.68	1.22–7.65
5th July	0.23–0.66	1.43–15.66
5th September	0.14–3.16	0.87–13.39
7th November	0.10–1.63	1.55–23.33

considerably influenced by the moisture content of the soil (Nye & Tinker 1977; Chapin 1980).

The mineralization of organic nitrogen was followed by incubating undisturbed soil cores for 4–5 weeks *in situ* in the field (Troelstra & Wagenaar 1979, 1980). The cores were taken each month at four sites. Although the value of incubation techniques in establishing nitrogen availability to plant roots may be questioned, the present method can be used in comparing different sites in the same area and in establishing flushes in mineral nitrogen over the year. The quantities of nitrogen which were found to be mineralized in this way from March to November 1979 agreed fairly well with those taken up by the vegetation during this period. The mineralization rate varied considerably over the year (Table 7.5). The highest rates occurred in early summer, but during a dry spell mineralization came to a standstill. When calculated over the total period, in the wet sites colonized by *P. major* the ammonium/nitrate ratio was much higher than in the dry sites supporting *P. lanceolata* and *P. coronopus*. This does not mean that both ions are taken up in the same proportions by the plant roots. On the contrary, data on the nutrient content of the plants (see below) suggest that nitrogen uptake in the wet sites was proportionally more in the nitrate form than in the dry sites.

Production of nitrate in the rhizosphere

To obtain more insight into the scale at which nitrogen mineralization is of relevance to the individual plant, and to establish a possible suppression of nitrification in grassland soils, the nitrate production in the rhizosphere of individual plants of *P. lanceolata*, *P. coronopus* and *P. major* from the Westduinen was determined (Smit & Woldendorp 1981).

The absence of nitrate in the rhizosphere is not necessarily an indication of the absence of nitrification and uptake by plant roots: nitrate has been found

TABLE 7.5. The range of nitrogen mineralization rates at four sites (0–10 cm) in the Westduinen during the period March 1978 to November 1979.

Site	$g\,N\,m^{-2}\,week^{-1}$ NO_3^--N	NH_4^+-N	Mean ratio NH_4^+/NO_3^-
1 (wet)	−0.01–0.11	−0.09–0.99	16.0
2 (dry)	−0.02–0.21	0.02–0.64	2.0
3 (wet)	−0.02–0.08	−0.03–0.40	14.5
4 (dry)	−0.01–0.20	−0.01–0.51	2.8

in grass tops even though the compound could not be detected in the soil. A good impression of nitrate production in the rhizosphere can be obtained from counts of nitrifying bacteria, as these chemolithotrophic organisms are completely dependent for growth and maintenance on the oxidation of ammonium or nitrite. Counts of nitrifying bacteria, however, give little information on nitrate uptake by plants.

Besides nitrate accumulation in the above-ground parts, several other plant parameters can provide information on nitrate utilization. The differences between accumulated amounts of inorganic cations and anions can be related to nitrate uptake (see below). Similarly, the presence of nitrate reductase which is an inducible enzyme is an indication of nitrate uptake. From the levels of this enzyme it was concluded that in some acid and calcareous grassland nitrate could be used as a nitrogen source (Havill *et al.* 1974, 1977).

For the interpretation of levels of nitrate reductase in the field, its basal levels in the absence of nitrate have to be known. To establish such basal nitrate reductase levels, *P. major*, *P. media*, *P. coronopus* and *P. lanceolata* were precultivated for six weeks on a modified Hoagland solution with ammonium as the nitrogen source. The plants were transferred to the same solution without ammonium 145 hours before nitrate reductase activity was measured. 20 hours before the measurement of nitrate reductase activity the plants were transferred to solutions with various nitrogen sources (Table 7.6). Nitrate reductase was determined with the intact tissue assay (*in vivo*) of Jaworski (1971) which gave a good estimate of the real nitrate reduction (Stulen *et al.* 1981b). The basal levels were low in *P. lanceolata* and *P. coronopus* but significantly higher in *P. major* and *P. media*. In all species nitrate reductase was induced by nitrate but when ammonium was also present the activity of the enzyme was suppressed. The high basal levels of *P. major* and *P. media* point to a specific adaptation to environments with a continuous nitrate supply, since the maintenance of superfluous levels of nitrate reductase in the absence of nitrate can be considered a waste of energy.

The nitrate reductase levels found in the laboratory were compared with those in the field (Table 7.7). It should be kept in mind that the plants in the field differed from those in the laboratory in terms of light intensity, age, growth stage, growing season, and the temperature and moisture levels in the environment. Since most of these factors probably had no effect, or only negative effects, on nitrate reductase levels in the field, it may be concluded from the nitrate reductase levels in the field material of the three *Plantago* species, which were higher than the basal ones, that at least part of the nitrogen was taken up in the nitrate form. Furthermore, the numbers of nitrifying bacteria in the rhizosphere, which are in good agreement with the nitrate reductase levels, point to the formation of nitrate. Thus, in the present study no indications were obtained that nitrifying bacteria are inhibited in the rhizosphere. On the contrary, nitrogen transformations in the rhizosphere seem to be highly important for the nitrogen nutrition of the individual plants.

A similar correlation between numbers of nitrifying microorganisms in the

TABLE 7.6. Nitrate reductase activity (μmol NO_2^- h^{-1} g dry wt.$^{-1}$) in the leaves of four *Plantago* species pregrown on solutions containing ammonium as the sole nitrogen source before transfer to the experimental treatments. (For details, see text.)

Treatment	No nitrogen	NH_4^+	NO_3^-	NH_4NO_3
P. lanceolata	0.05	0.04	2.22	1.59
P. coronopus	0.03	0.02	3.96	0.97
P. major	0.55	0.29	4.67	3.60
P. media	0.42	0.20	4.36	0.38

TABLE 7.7. Some rhizosphere characteristics and nitrate reductase activities (NR) of *Plantago* species in the Westduinen. All figures are means.

				Numbers of nitrifiers g dry soil^{-1}		NR
Species	pH–H_2O	NH_4^+-N (μg g^{-1})	NO_3^--N (μg g^{-1})	NH_4^+-oxidizers	NO_2^--oxidizers	μmol $NO_2^-\cdot h^{-1}$ g fresh wt.$^{-1}$
P. lanceolata	4.4	6.2	0.5	5500	4600	0.38
P. coronopus	6.6	4.2	0.5	11 900	6600	0.88
P. major	7.6	17.5	1.2	64 700	44 000	1.02

N.B. NO_3^- content of the shoots was always <20 mmol kg dry wt.$^{-1}$. NH_4^+-N and NO_3^--N data are for soils extracted with 1M KCl.

rhizosphere and nitrate reductase levels in the leaves was also found when individual plants of *P. lanceolata* were compared (Table 7.8). This demonstrated that the nitrogen nutrition of an individual plant is influenced by the processes going on in its rhizosphere. Surprisingly, no correlation existed between the amounts of ammonium in the root environment and the numbers of nitrifying bacteria. This result may be due to suppression of nitrification by root compounds, but is more likely the result of differences in the distribution of ammonium and nitrifying bacteria over the microsites in the soil.

TABLE 7.8. Correlation between nitrate reductase activities (NR) in 13 plants of *Plantago lanceolata*, the number of nitrifiers and the ammonium content of the soil. NS = not significant.

		r_s	Significance
Soil-NH_4^-	NH_4^+-oxidizers	0.289	NS
Soil-NH_4^-	NO_2^--oxidizers	0.180	NS
NH_4^+-oxidizers	NO_2^--oxidizers	0.872	***
NH_4^+-oxidizers	NR-leaves	0.807	***
NO_2^--oxidizers	NR-leaves	0.794	**

The ionic balance of Plantago *species and its relation to nitrogen uptake*

An impression of the nature of the nitrogen nutrition of plants can be obtained by analysing the amounts of cations (Na^+, K^+, Mg^{2+}, Ca^{2+}, NH_4^+) and inorganic anions (Cl^-, SO_4^{2-}, $H_2PO_4^-$, NO_3^-) in the plant material. Generally, the difference between cations and inorganic anions (C – A) has a positive value and is a measure of the amounts of organic anions (malate, oxalate, citrate, etc.) which are accumulated in the plant tissue to maintain electroneutrality. The size of this carboxylate pool is influenced by the proportions of cations and anions taken up, by the reduction of nitrate and sulphate, and by the incorporation of ammonium into amino acids (Breteler 1975; van Egmond 1978). In the absence of nitrate, roots always absorb more cations than anions, and excrete more H^+ than HCO_3^-, the cationic excess inside the plant being compensated by HCO_3^- (i.e. the synthesis of organic anions) as described by Dejaegere *et al.* (1980). Ammonium uptake is therefore attended by a decrease of the rhizosphere pH, low positive C – A values, and a low ratio between C – A and organic nitrogen.

Upon reduction, nitrate is replaced by organic anions, which leads to higher C – A values than with ammonium as the nitrogen source. In principle, the level of C – A should be about equal to the organic nitrogen content (C – A/organic-N ratio = 1), but ratios < 1 (decarboxylation) or > 1 (carboxylate formation in excess of nitrate and sulphate reduction) are commonly found, with resulting rhizosphere pH effects of an increase or a decrease respectively. At high nitrate levels commonly low C – A/organic ratios are found, but at low levels the ratio is higher.

The levels of the C – A values and the C – A/organic-N ratio are dependent on factors such as the plant species, the part of the plant analysed, and the age of the plant, as has been shown in many analyses of crop plants. Typical data on the *Plantago* species, which are derived from a large number of water culture experiments and plants from a wide range of natural habitats, are given in Table 7.9 (Troelstra & Smant 1980). Under identical growth conditions with nitrate as the nitrogen source, the C – A values of *P. media* and *P. major* are > 1000 and those of *P. lanceolata* and *P. coronopus* are around 1000; in the case of *P. maritima* the values (not recorded in Table 7.9) are considerably below 1000. With ammonium, in waterculture, much lower C – A values are found and also the C – A/organic-N ratios (low with ammonium, higher with nitrate) are typical for both nitrogen sources. The C – A values and C – A/organic-N ratios of the field material are higher than those of ammonium plants. This suggests that, under field conditions, at least

TABLE 7.9. Comparison of ionic balances of *Plantago* species grown in watercultures with ammonium or nitrate as the nitrogen source, with those of plants from natural habitats.

Species	Growth conditions	C – A* (m-equiv. kg dry wt.$^{-1}$)	C – A/organic N	NO_3^- (mmol kg dry wt.$^{-1}$)
P. media	Waterculture NO_3^-	1680–1820	0.66–0.68	1100–1440
	Waterculture NH_4^+	620	0.15	< 20
	Natural habitats	1460–2490	1.24–3.01	0–30
	Westduinen	—	—	—
P. major	Waterculture NO_3^-	980–2300	0.40–1.84	200–1100
	Waterculture NH_4^+	680–1050	0.26–0.35	< 20
	Natural habitats	1140–1970	0.60–2.60	0–120
	Westduinen	1200–1300	0.75–0.87	0–40
P. lanceolata	Waterculture NO_3^-	890–2240	0.38–1.31	200–1850
	Waterculture NH_4^+	320–590	0.13–0.25	< 20
	Natural habitats	650–1590	0.45–1.58	00–260
	Westduinen	710–910	0.49–0.61	0–12
P. coronopus	Waterculture NO_3^-	1030–1520	0.44–0.67	1400–1930
	Waterculture NH_4^+	600–640	0.14–0.19	< 20
	Natural habitats	920–1310	0.64–1.34	5–40
	Westduinen	—	—	—

* For details, see text.

part of the nitrogen had been taken up as nitrate. This also applies to *P. major* and *P. lanceolata* plants from the Westduinen, which is in agreement with the data on nitrate reductase levels of these plants.

In Table 7.5 it was recorded that the highest ratios of mineralized ammonium and nitrate were found in wet sites. Nevertheless, the C – A/organic-N ratios of the vegetation on these sites were higher than those on dry sites. This suggests that on the wet sites proportionally more nitrate was taken up, which was in agreement with the higher ammonium concentrations in the vegetation on the wet sites. These results demonstrate that the proportions in which nitrate and ammonium are taken up from the soil can be different from those obtained in mineralization experiments.

Although high nitrate levels were sometimes found in plants from natural habitats, the levels were generally very low and below the limit which could be accurately established with an ion-specific nitrate electrode. They were much lower than those in waterculture experiments, where it was found that under identical conditions *P. lanceolata* accumulated more nitrate than *P. major*.

General discussion of the field data

Soil analysis of a large number of locations of *Plantago* species revealed that *P. lanceolata* occurred over a wider range of soil chemical conditions than the other species. It can be found in acidic and alkaline soils, dry and wet situations as well as in soils low or rich in nutrients; however, it is excluded from the compacted, trampled sites in which *P. major* is commonly found (Blom 1979). Population genetic studies suggest that *P. lanceolata* can cope with this wide range of conditions by the phenotypic plasticity of its individuals; in *P. major* adaptation at the subpopulation level may occur to soil conditions which are physically more favourable (a more continuous water supply) than those of *P. lanceolata*.

Nitrate reductase activities, C – A levels and C – A/organic-N ratios in field material, and the numbers of nitrifying bacteria in the rhizosphere, all point to nitrate uptake under field conditions by the *Plantago* species. However, at present no exact information is available on the proportions in which ammonium and nitrate are taken up. This proportion is certainly different from that found in nitrogen mineralization experiments under field conditions. The latter ratios may underestimate nitrate uptake by the roots, as the mobility of nitrate in the soil is higher than that of ammonium, and additional transport from the subsoil may occur. Also, in relatively poor soils the nitrogen turnover processes actually occurring in the rhizosphere may be important in uptake by the plant; mineralization experiments with large samples cannot be expected to give reliable information on such processes at the microhabitat level. All of these factors may have contributed to the much higher ammonium/nitrate ratios observed in the wet than in the dry sites of the Westduinen. Nevertheless, proportionally more nitrate was taken up by the vegetation in these wet sites than in the dry ones.

In the Westduinen in topographically uniform sites a considerable small-scale variability in soil chemical factors was found, particularly in contents of organic matter and related factors such as soil moisture. Also, at the level of the rhizosphere of individual plants such small-scale variability occurred. Information on small-scale variability is important when the performances of individual plants or small subpopulations are to be studied, which is the case in plant demographic investigations. Demographic parameters, such as longevity of the individuals, are highly influenced by their nutritional state (Harper 1978).

In natural habitats the availability of nutrients which reach the root surface mainly by diffusion rather than by mass flow (Nye & Tinker 1977; Chapin 1980) will be governed by the moisture content of the soil. In dry soils the diffusion coefficients will decrease, mass flow will be lower and dry zones around the roots may arise. The absorbing power of the roots for nutrients may be reduced by an increased water potential. All of these unfavourable factors may not be compensated for by the higher concentrations of nutrients in dry soils (Table 7.4). Consequently, low values of nitrate reductase in plants (a measure of nitrate supply) were found to occur under conditions of moisture shortage (Havill *et al.* 1977; Aparacio-Tejo *et al.* 1980; Meguro & Magalhaes 1980).

In the field, the moisture conditions are changing continuously, due to topographic and small-scale variability and to irregularities in the rainfall. As a result, nutrient availability may also vary continuously. As has been discussed above, the moisture conditions of the habitat are expected to fluctuate more with *P. lanceolata* than with *P. major*. It was decided, therefore, to investigate in the laboratory the effects of changes in the nutrient level on the performances of both species in more detail. The results of these experiments are discussed below.

LABORATORY EXPERIMENTS WITH *PLANTAGO MAJOR* AND *P. LANCEOLATA*

Nitrate uptake and growth from continuous and intermittent supply of nutrients

To test the hypothesis derived from the field work that *P. lanceolata* may be better adapted to a fluctuating nutrient supply than *P. major* (see above), both species were grown at divergent and varying availabilities of nitrate (Freijsen & Otten 1979, 1981). These were realized by the application of culture solutions with constant nitrate concentration (750 and 25 μM) and were supplied either continuously or intermittently. The high concentration was sufficient to ensure optimal growth conditions and the low concentration was within the range of apparent K_m values recorded in a flowing culture system. This technique has already been used by other authors (Asher & Edwards 1978).

At 750 μM *P. lanceolata* and *P. major* attained average relative growth rates of 0.21 and 0.23 $g.g^{-1}$ day^{-1}, respectively (Table 7.10). The initial seedling weights of *P. lanceolata* were higher than those of *P. major* but the 10% higher growth rate of juvenile *P. major* plants meant that after 3–4 weeks the same weight was attained as with *P. lanceolata*. At the end of the experiment *P. lanceolata* had a higher shoot/root ratio and a higher nitrate content of the leaves than did *P. major*.

A low nitrate concentration (25 μM) gave a small growth reduction with *P. lanceolata* and no reduction at all with *P. major* (Table 7.11). These results agree with data of Clement *et al.* (1978), who found that shoot growth of *Lolium perenne* L. was reduced only slightly when this species was grown in a flowing culture system at the low nitrate concentrations of 14 and 1.4 μM.

TABLE 7.10. Properties of juvenile *Plantago* species grown for 24 days at 750 μM nitrate.

Property	*P. lanceolata*	*P. major*
Dry weight (mg $plant^{-1}$)	224	93
Relative growth rate ($g.g^{-1}$ day^{-1})	0.21	0.23
Shoot/root ratio	2.7	2.3
NO_3^- content of shoot (mmol kg dry $wt.^{-1}$)	1488	1027
Organic-N content of shoot (mmol kg dry $wt.^{-1}$)	2960	2800

TABLE 7.11. Properties of juvenile *Plantago* species grown at 750 and 25 μM nitrate, respectively.

Property	*P. lanceolata*		*P. major*	
	μM NO_3			
	750	25	750	25
Dry weight (mg $plant^{-1}$)	87	77	48	47
Relative growth rate ($g.g^{-1}$ day^{-1})	0.23	0.22	0.18	0.18
Shoot/root ratio	2.3	2.1	2.8	2.3
NO_3^- content of shoot (mmol kg dry $wt.^{-1}$)	895	293	892	975
Organic-N content of shoot (mmol kg dry $wt.^{-1}$)	3048	2789	3940	4025

Three factors seem to be involved which can explain why the effects of low NO_3^- concentrations on the dry weight yield were so small. In *P. lanceolata* a lower content of organic nitrogen was found in the plants of the 25 μM nitrate treatment, whereas the dry matter content stayed at the same level. Apparently, this species used the available nitrogen in a more economical way. Furthermore, both species exhibited a somewhat lower shoot/root ratio, so that the larger root system could withdraw more nitrate from the culture solution. The third factor, possibly the most important one, resulted from the relatively high uptake rates of those at 750 μM which in *P. lanceolata* and *P. major* were 70% and 87% respectively. These results clearly disagree with the supposition that the uptake rate at the lower concentration would be twice as low as at the higher one. The most reasonable explanation is that an uptake rate equal to V_{max} was not attained, because the demand for nitrate was already satisfied at a lower average rate (Clarkson & Hanson 1980). In this view the nitrate influx was not dependent on the external concentration but regulated by some other factor, e.g. intermediates of the nitrogen metabolism (see below).

It should be noted that in the field, bulk nitrate concentrations of 90–3500 μM were found (see Table 7.4). At these concentrations plants developed at a suboptimal rate. Apparently, there is no direct relation between conditions of nutrient supply in water cultures and those in soil solutions. This is underlined by the negligibly low nitrate levels in the field material (see Table 7.9) which contrasted with the nitrate contents of the shoots of plants grown at 25 μM nitrate in the water cultures. In view of this discrepancy the preliminary conclusion can be drawn that nutrient conditions in the field are badly simulated by concentration experiments. We agree with Ingestad & Lund (1979) who showed that the nutrient concentration *per se* is not as important as the quantity which is supplied per unit of time.

In a subsequent experiment a continuous supply at two nitrate levels was compared with an intermittent one, in which plants were provided with the nitrate concentrations on alternate days. *P. major* appeared to be much more susceptible to the intermittent supply of nitrate than to the influence of a low concentration of this ion (Table 7.12). It should be noted that with the interrupted supply at 750 μM much more nitrate was available to the plants than with a continuous concentration of 25 μM. The interrupted supply resulted in a distinct growth reduction, and the shoot/root ratio and the total nitrogen contents were also decreased. The interruptions in the supply were not compensated for by an increased uptake on days when nitrate was supplied, which was in contrast to the results of Clement *et al.* (1979), who found such a compensatory uptake in *Lolium perenne*. In a similar experiment it was found that *P. lanceolata*, too, was more affected by intermittent supply

TABLE 7.12. Properties of juvenile plants of *Plantago major* grown with continuous or intermittent supply of 750 and 25 μM nitrate. n.d. = not detectable

Property	750 μM NO_3^- Continuous	750 μM NO_3^- Intermittent	25 μM NO_3^- Continuous	25 μM NO_3^- Intermittent
Dry weight (mg plant^{-1})	48	20	47	15
Relative growth rate (g.g^{-1} day^{-1})	0.18	0.14	0.18	0.13
Shoot/root ratio	2.8	1.6	2.3	1.7
NO_3^- content of shoot (mmol kg dry wt.$^{-1}$)	892	n.d.	975	n.d.
Organic-N content of shoot (mmol kg dry wt.$^{-1}$)	3940	3214	4025	3000

than by exposure to a low concentration of nitrate (Freijsen & Otten 1979). When a concentration of 25 μM nitrate was supplied on alternate days, the growth and the levels of organic nitrogen were reduced more in *P. major* than in *P. lanceolata* (Table 7.13). When 25 μM nitrate was given only on one out of three days, the properties of both species altered still more. Again, the growth of *P. major* was inhibited more than that of *P. lanceolata*.

The results so far obtained are in agreement with the hypothesis that individuals of *P. lanceolata* are phenotypically more flexible than those of *P. major*. To give further support to this hypothesis, the changes which occur in

TABLE 7.13. Properties of juvenile *Plantago* species grown with either a continuous supply of 25 μM nitrate or with an intermittent supply. The latter was a cycle of either one day nitrogen and one day no nitrogen or one day nitrogen and three days without. n.e. = not examined.

Property	Cycle (days)	*P. lanceolata* Continuous	*P. lanceolata* Intermittent	*P. major* Continuous	*P. major* Intermittent
Dry weight (mg plant^{-1})	1	77.6	37.8	72.6	18.3
	3	78.4	19.6	65.4	8.6
Relative growth rate (g.g^{-1} day^{-1})	1	0.21	0.17	0.23	0.17
	3	0.21	0.15	0.25	0.15
Shoot/root ratio	1	3.3	2.2	2.8	1.8
	3	3.1	1.4	2.2	1.0
NO_3^- content of shoot	1	1236	495	953	178
(mmol kg dry wt.$^{-1}$)	3	1008	15	829	n.e.
Organic-N content of shoot	1	2448	2316	2714	1822
(mmol kg dry wt.$^{-1}$)	3	2806	1328	2739	1602

the nitrogen metabolism after a sudden shift in the levels of nutrients were investigated in more detail.

The effects of changes in the nutrient level on the nitrogen metabolism of P. lanceolata *and* P. major

In these experiments plants of *P. lanceolata* (12 days old) and *P. major* (18 days old) were grown either on a high nutrient level ($\frac{1}{4}$ strength Hoagland; 3750 μM nitrate = 100%) or on a diluted nutrient solution (2% of $\frac{1}{4}$ strength Hoagland; 75 μM nitrate = 2%). After another 16 days, half of the plants grown in full nutrient solution were transferred to the diluted nutrient solution (100%→2% plants). The other half were kept in the full nutrient solution (100% plants). In a similar way a series of 2%→100% and 2% plants were created. A number of physiological processes were followed before and after the switch. The experiments produced a wealth of data parts of which are summarized below. For more detailed information the reader is referred to Lambers *et al.* (1981a, b) and to Stulen *et al.* (1981a, b).

Upon transfer from the concentrated to the dilute solution (100%→2% shift), shoot growth in *P. lanceolata* stopped immediately, but growth of the roots was not influenced and continued at the same rate. In *P. major* growth of the shoots and the roots both decreased to the same degree. As a result the shoot/root ratio in *P. lanceolata* decreased but in *P. major* it changed only slowly (Fig. 7.1). In the 2%→100% shift the reverse picture was observed: in *P. lanceolata* the shoot/root ratio increased immediately, due to a decreased root growth, but in *P. major* no change occurred at first.

Root respiration in *P. lanceolata* was more efficient than in *P. major* and the contribution of the alternative non-phosphorylating respiration pathway was more flexible. This alternative respiration pathway has an 'overflow' function in which assimilates are 'wastefully' oxidized and it is of significance in the removal of carbohydrates, which cannot be utilized for growth, energy production, storage of sugars, etc. (Lambers 1979; Lambers & Steingröver 1978). In *P. major* total root respiration and the activity of the alternative respiratory chain both decreased during the ageing of the plants and reacted less flexibly to changes in the nutrient supply than *P. lanceolata.*

In *P. lanceolata* the activities of nitrate reductase, glutamate dehydrogenase and glutamine synthetase in the roots showed a good correlation with the level of the nutrient supply. Upon shifts to a different nutrient supply the enzyme contents were rapidly adjusted to the appropriate levels. The reduced nitrogen content of both roots and shoots, too, was correlated with the nutrient supply and changed rapidly after a switch. On the other hand, in *P. major* levels of the above enzymes and also the reduced nitrogen content were

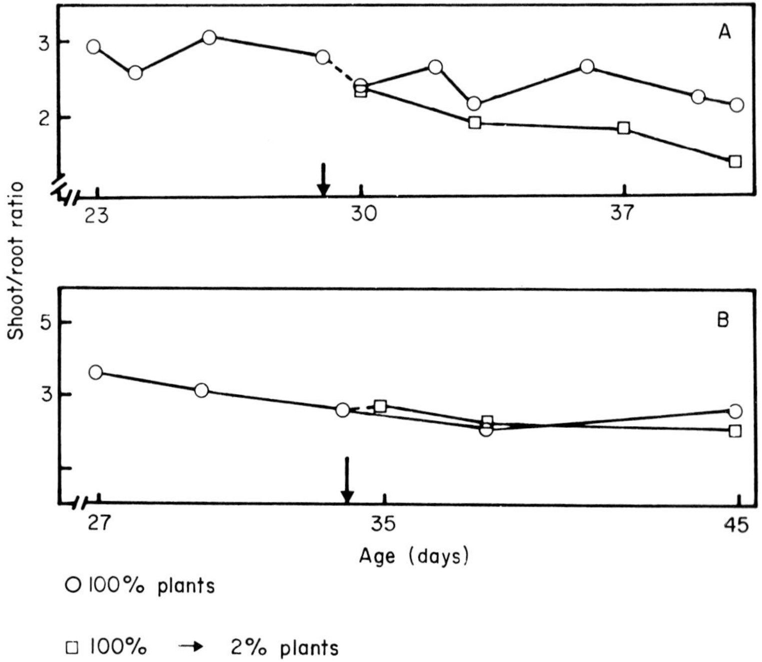

FIG. 7.1. Time course of the shoot/root ratio of *Plantago lanceolata* (A) and *Plantago major* (B) as affected by the level and change in the level of nutrients. Each symbol represents the mean of at least three independent determinations with six plants each. Standard deviation was *c* 15%. Time of the switch is indicated by the arrow. (For details, see text.)

all rather independent of the nutrient supply. Some representative examples which illustrate the differences in the nitrogen metabolism of *P. lanceolata* and *P. major* are given in Fig. 7.2 (reduced nitrogen in the shoots), Fig. 7.3 (nitrate reductase in the shoots) and Fig. 7.4 (glutamine synthetase in the roots).

The changes which take place in the energy supply and the nitrogen metabolism of *P. lanceolata* and *P. major* upon a shift to a different nutrient level are fully in accordance with the picture of the species already formed. *P. lanceolata* was able to adjust its enzyme levels rapidly to the new conditions, but in *P. major* these levels were rather independent of the nutrient supply. In separate experiments it was found that *P. lanceolata* also rapidly adapted its ATPase levels and the lipid composition of its membranes to changed environmental conditions, while *P. major* did not (Kuiper & Kuiper 1978, 1979a, b).

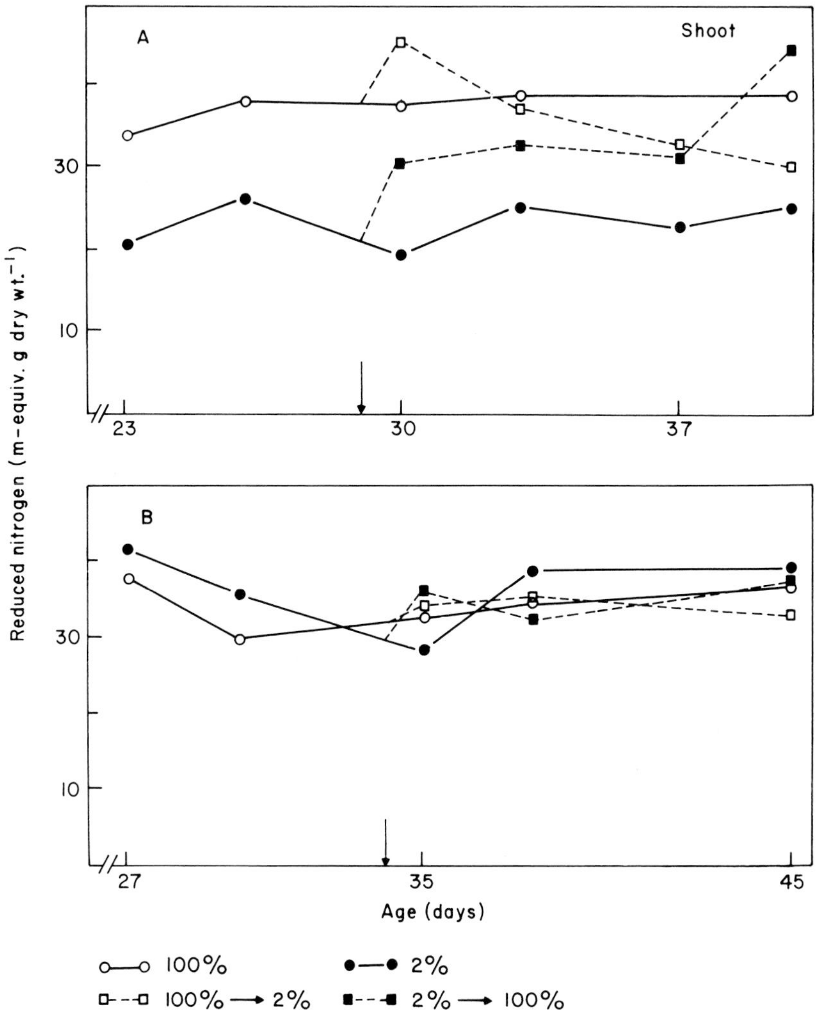

FIG.7.2. Reduced nitrogen content of shoots of *Plantago lanceolata* (A) and *Plantago major* (B) as affected by the level and change in the level of nutrients. (For further explanation, see legend to Fig. 7.1.)

Nitrate uptake kinetics

The factors which influence the uptake of nutrients in the root zone have recently been discussed by Clarkson & Hanson (1980) and Chapin (1980). In both reviews it was argued that under the poor or moderately rich soil conditions in which most plant species have evolved, the diffusion rates to the

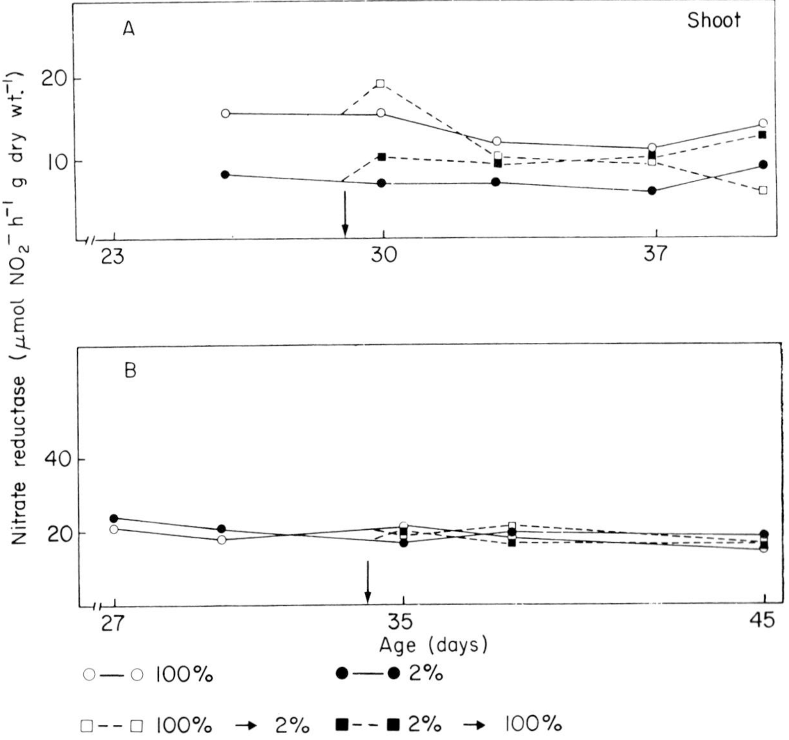

FIG. 7.3. Nitrate reductase in the shoots of *Plantago lanceolata* (A) and *Plantago major* (B) as affected by the level and changes in the level of nutrients. (For further explanation, see the legend to Fig. 7.1.)

root surface are much lower than the uptake capacity of the roots. As was stated by Chapin, 'The kinetics of nutrient absorption by roots should exert only minor influence over nutrient acquisition by plants, except under conditions of luxury consumption, i.e. under even mild nutrient limitation, soil processes are the primary control over nutrient absorption by each unit of root'. Therefore, factors such as the surface area of the root system relative to its weight, the number of root hairs, the presence of mycorrhizas, and other factors which also determine the amount of soil accessible to the plant root, are thought to be particularly important in plant nutrient acquisition.

The ability of plant species to absorb nutrients has often been expressed as the root absorption capacity (RAC), which is the absorption rate per unit of root measured under standardized conditions. Species which are characteristic of rich soils, such as most agricultural crops, have a high RAC which is positively correlated with the growth rate and the shoot/root ratio. Species

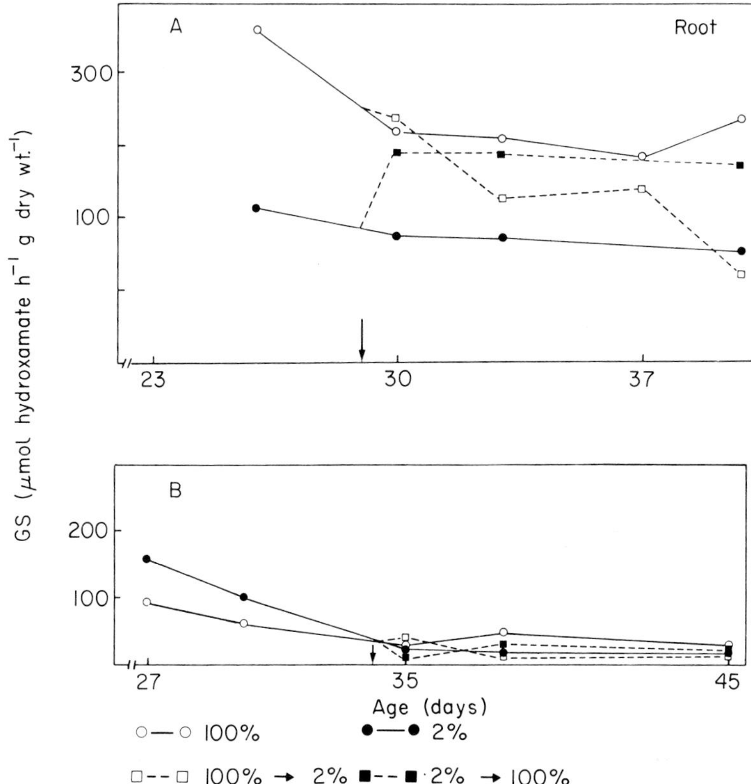

FIG. 7.4. Glutamine synthetase transferase activity (GS) in the roots of *Plantago lanceolata* (A) and *Plantago major* (B) as affected by the level and changes in the level of nutrients. (For further explanation, see legend to Fig. 7.1.)

from infertile soils absorb considerably less nutrients under high nutrient conditions but similar quantities at low nutrient levels. In summarizing the literature, Chapin states that the RAC is usually higher in rapidly growing species from fertile habitats than in plants from infertile habitats under all growth conditions, and in most cases plants have not adapted to nutrient stress through the evolution of an enhanced capacity to extract minerals from the soil (Chapin 1980).

Over the years the uptake kinetics, expressed in terms of K_m and V_{max}, have been established for a large number of plant species grown under a wide range of experimental conditions. These experiments have sometimes yielded conflicting results on both uptake parameters. In the present state of knowledge K_m and V_{max} are thought to be both considerably influenced by metabolic compounds. For instance, in the case of nitrate uptake by *Arabidopsis thaliana* (L.) Heynh. it was shown by Doddema (1978), working with a mutant blocked in its nitrate reductase activity in which no nitrogen could be assimilated, that regulation of the nitrate uptake by system 1 (Epstein 1976) may be exercised in at least three different ways:

1. An activation of the carrier by nitrate, leading to an increased affinity (low K_m) and/or V_{max}.
2. An inhibition by certain amino acids which is counteracted by nitrate and which can be reversed by starvation for nitrogen.
3. Very long starvation times, which may lead to a gradual degradation of the carrier protein, which is expressed as a decrease of the V_{max} value after several days of starvation.

Therefore, under different experimental conditions varying values of K_m and V_{max} can be found. The picture of varying levels of K_m and V_{max} makes the conclusion that the K_m values play a part in the distribution of plants in the field doubtful (Lancaster, cited by Huffaker & Rains 1978); with three grass species this author found a relation between K_m values for nitrate and the nutrient status of their soils. The K_m and V_{max} values for nitrate uptake which have been recorded in the literature were mostly obtained with plants grown in liquid media at relatively high nitrogen levels. Under such conditions nitrate reductase is limiting nitrate assimilation, and the accumulation of nitrate in the plant tissues is a common phenomenon. Such nitrate accumulation even took place in flowing nutrient cultures at the low concentration of 25 μM (see section above). Therefore, it may be questioned whether K_m and V_{max} have ever been determined under conditions where no allosteric inhibition by compounds of nitrogen metabolism was involved or where the plants were not completely starved.

It is by no means established beyond any doubt that specific differences in uptake parameters do not exist under field conditions, i.e. under such conditions that nitrate levels in the plant are negligibly small and nitrogen metabolism is determined by uptake rather than by nitrate reductase activity, as is the case in waterculture experiments. In the field, diffusion of nitrate may be the step which is limiting the uptake rate, when roots of different species are not intermingled. However, diffusion zones of the relatively mobile nitrate ion may easily overlap each other in neighbouring roots (Nye & Tinker 1977). In the production of nitrate small-scale variability has to be taken into account (see above) and, locally, higher nitrogen levels may occur for which the root has to compete at short distances with other roots and with rhizosphere microorganisms. Under such conditions of competition, roots may benefit from particular kinetic features in two ways. Firstly, a low K_m value results in a more efficient use of carriers. Secondly, a high V_{max} of the nitrate uptake system is always favourable: the more carrier, the higher the uptake rate, even at low concentrations. Finally, where two individuals have to compete for the same limiting amount of nitrate, the larger root system will result in a higher uptake even when the K_m and V_{max} are the same. The effects of these factors were tested in a number of experiments, which are described below.

In plants of *P. major* and *P. lanceolata* grown at high nutrient level (3750 μM NO_3^-, 100%) and low nutrient level (75 μM NO_3^-, 2%), K_m and V_{max} values were determined. Average values are recorded in Tables 7.14 and 7.15 (van de Dijk 1981a). As a comparison, the same parameters were established in *Urtica dioica*, a species from nutrient-rich habitats, and *Hypochaeris radicata*, a species from nutrient-poor soils (see Table 7.1). Both species were grown in the same way as the two *Plantago* species. In all four species the nitrate uptake pattern up to 200 μM NO_3^- showed the characteristic features of Michaelis–Menten kinetics. No significant differences in apparent K_m values were observed between the species, neither at a high nutrient level nor at a low level of nutrients (Table 7.14). The values were of the same order of magnitude as those recorded in the literature (Nye & Tinker 1977; Doddema 1978). No correlation was found with the nutrient richness of the habitat as was found for the three grass species by Lancaster (cited by Huffaker & Rains 1978). The average V_{max} values of *U. dioica* were higher than those of the other species, even at a low nutrient level (Table 7.15), where they were considerably lower than at the 100% level. In this species the root absorption capacity was clearly related to the growth rate, as has been found for many other species from nutrient-rich environments (Chapin 1980). When *U. dioica* plants were switched from high nutrient to low nutrient conditions (100%→2%) or in the reverse direction (2%→100%) the V_{max} values were rapidly adjusted to the appropriate levels. In contrast to *U. dioica*, both subspecies of *H. radicata* which were tested, had significantly higher V_{max} values at the 2% nutrient level than at 100% nutrients. Nevertheless, at the 2% level the values were lower

TABLE 7.14. Apparent K_m nitrate values (μM) followed by standard errors of four species cultivated at two levels of nutrients (either 2% or 100%). (For details, see text.)

Species	K_m NO_3^- 2%	K_m NO_3^- 100%
Plantago lanceolata	20±4	23±10
Plantago major	15±5	15±6
Hypochaeris radicata	13±6	13±4
Urtica dioica	11±3	9±4

TABLE 7.15. Average V_{max} values (μmol NO_3 h^{-1} g dry wt.$^{-1}$) of roots of four species at two levels of nutrients (either 2% or 100%). (For details, see text.)

Species	V_{max} 2%	V_{max} 100%
Plantago lanceolata	60	–*
Plantago major	39	43
Hypochaeris radicata†	90	75
Hypochaeris radicata‡	82	70
Urtica dioica	94	150

* V_{max} declining during experiment
† ssp. *ericitorum*
‡ ssp. *radicata*

than those of *U. dioica*. The V_{max} values for *P. major* were about the same at both nutrient levels and did not vary much over the experimental period. This result is in accordance with those above where it was shown that other enzymes involved in nitrogen metabolism of *P. major* were also little influenced by the nutrient level. Even basal levels of nitrate reductase in nitrogen-starved plants were still high (see Table 7.6). Therefore, it can be concluded that the rates of the processes involved in uptake and metabolism of nitrate in *P. major* are not very liable to modification under the influence of the external nutrient levels. In *P. lanceolata* the results were difficult to interpret. With low nutrients the V_{max} did not change much during the experiment but at 100% nutrients the level was considerably higher at the start of the experiment, but much lower at the end of it. In both *Plantago* species no indications were obtained of a higher V_{max} value at a low nutrient level than one at a high nutrient level, such as was found in *H. radicata*.

To overcome the problems of the flowing culture experiments in which it was found that at the low, continuous supply of 25 μM nitrate hardly any growth reduction could be detected, a new technique was developed to grow plants at much lower nitrogen levels (van de Dijk 1981b). Using this technique, seedlings were grown either in monoculture (72 seedlings per culture vessel) or in mixed culture (36 seedlings of each species in a culture vessel). With the aid of an infusion pump, on the first day 34 μM nitrate was added to each culture vessel. In preliminary experiments this quantity was shown to limit plant growth and nitrate uptake. On subsequent days the nitrate quantity was increased exponentially 5% per day, so that after 23 days 100 μM was given per culture vessel. With this technique, steady state nitrate concentrations of 1 μM or lower were realized, which were below those used in the flowing culture systems described by Clement, Hopper & Jones (1978), Ingestad & Lund (1979) and Freijsen & Otten (1981). The plants which were grown in this way contained at the end of the experimental period less than 20 mmol nitrate per kg dry weight. Such low nitrate concentrations are also commonly found in field material. Therefore, it may be concluded that the present technique yielded plants which were more like those in their natural habitat than those grown using other techniques.

In the initial experiment the hypothesis was tested that at nitrate concentrations limiting the uptake rate, plants with a large root system at the start of the experiment will possess a higher uptake rate than plants with a small root system, even when the K_m and V_{max} of the two sets of plants are the same. Seedlings of *H. radicata* which differed one week in age, were grown separately or in mixed culture (Fig. 7.5). As can be seen, the larger plants obtained most of the nitrogen. This result can only be attributed to different amounts of carrier per plant. It illustrates that in the field small differences in age between seedlings may be very important for the outcome of intraspecific competition.

In a second experiment, *H. radicata* and *U. dioica* were grown in monoculture and in mixed culture. In this experiment the initial root dry

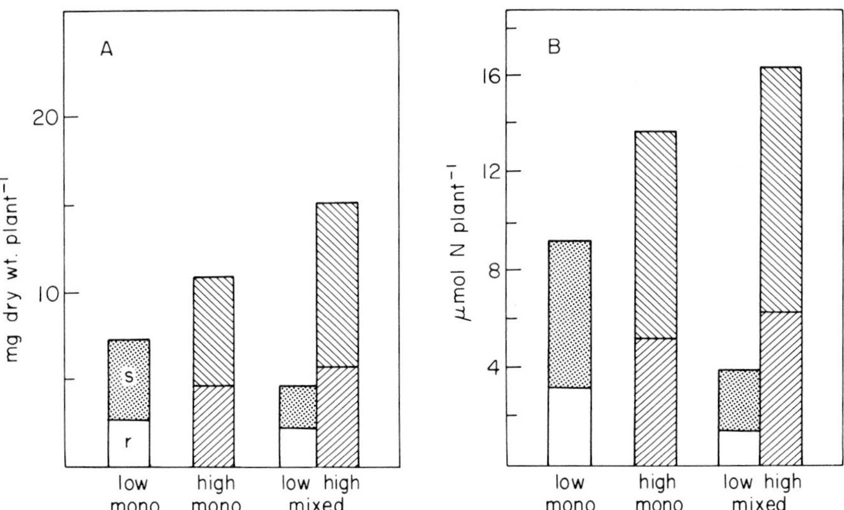

FIG. 7.5. Dry weight (A) and nitrogen accumulation (B) per plant of mono- and mixed cultures of seedlings of *Hypochaeris radicata* with high and low root weight at the start of the experiment. (For details, see text.)
Lower part of each column: root weight (r).
Upper part of each column: shoot weight (s).

weight of *U. dioica* was considerably higher than that of *H. radicata*. In addition, the initial shoot/root ratio was higher. Moreover, it has been found that V_{max} of *U. dioica* was always higher than that of *H. radicata* (see Table 7.15). Nevertheless, dry weight and nitrogen accumulation of *U. dioica* in mixed culture were significantly lower than in the monoculture (Fig. 7.6). However, with *H. radicata* both parameters were significantly higher in the

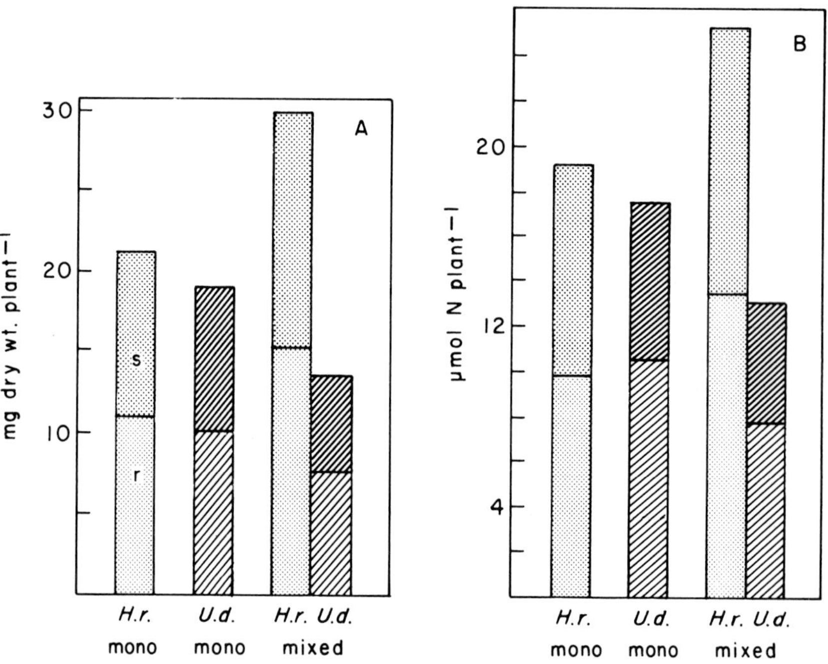

FIG. 7.6. Dry weight (A) and nitrogen accumulation (B) per plant of mono- and mixed cultures of seedlings of *Hypochaeris radicata* (*H.r.*) and *Urtica dioica* (*U.d.*). (For further explanation, see legend to Fig. 7.5 and text.)

mixed culture. This result can only be attributed to a difference in K_m or, more likely, in V_{max} between both species at these low nitrate levels. It is worth noticing that under high nutrient conditions the nitrate absorption rate of *U. dioica* is much higher than that of *H. radicata*. Apparently, both species are well adapted to their particular natural environment. The present technique thus seems to offer new possibilities in cultivating plants in the laboratory under conditions which simulate the natural availability of nutrients in a more satisfactory way than has been achieved by other techniques.

CONCLUSIONS

In the present paper attempts to correlate adaptation in the nitrogen metabolism of a number of *Plantago* species with the nitrogen supply in their natural environment are described. On the one hand, it proved difficult to describe the capacity of the environment to provide the plant with nitrate and ammonium in a satisfactory way. Various techniques, e.g. nitrogen mineralization experiments, counts of nitrifying bacteria, determinations of nitrate reductase levels in the shoots of field material, and the ionic balance of plants were applied to give such a description. The small-scale variability in soil chemical conditions and the moisture content of the soil proved to be important parameters in the nutrient supply to plants growing in natural habitats. On the other hand, we were faced with many problems in designing laboratory experiments which more or less simulated natural conditions and which yielded plants with the characteristics of field-grown ones. The capacity for adaptation to changes in the external nitrogen level appeared to be an important characteristic in the specific adaptation to the environment.

The physiological properties of two *Plantago* species (*P. major* and *P. lanceolata*) which were investigated in detail were in accordance with the characteristics of their habitat and with the genetic structure of their populations. *P. major* occurs in open, trampled locations with a good soil water status and a low level of competition from other plants. The growth conditions do not therefore vary to a large extent, and a continuous supply of nutrients is assured. Within the small subpopulations there is not much genetic variation present and the flexibility in the metabolism of the individuals is low. Under constant environmental conditions this is a satisfactory strategy, as it saves investment in energetically expensive regulation mechanisms. *P. lanceolata* occurs in a wide range of habitats with large variation in growth conditions. There are no differences in genetic composition between the populations occurring in these habitats, and the species has opted for a high phenotypic plasticity of its individuals. This plasticity is reflected in the flexibility of its metabolic processes, e.g. energy supply, nitrogen metabolism, membrane composition and ATPase levels.

REFERENCES

Aparicio-Tejo P., Barcala N. & Sanchez-Diaz M. (1980) Water stress and nitrate reduction activity in legumes. Second Congress of the Federation of European Societies of Plant Physiology, Santiago de Compostela. Abstracts of lectures and posters, pp. 184–5.

Asher C.J. & Edwards D.G. (1978) Relevance of dilute solution culture studies to problems of low fertility tropical soils. In *Mineral Nutrition of Legumes in Tropical and Subtropical Soils* (eds C.S. Andrew & E.J. Kamprath), pp. 130–52. CSIRO, Melbourne.

Beevers L. & Hageman R.H. (1969) Nitrate reduction in higher plants. *Annual Review of Plant Physiology*, **20**, 495–522.

Blom C.W.P.M. (1979) Effect of trampling and soil compaction on the occurrence of some *Plantago* species in coastal sand dunes. Ph.D. thesis, Nijmegen.

Blom C.W.P.M., Husson L.M.F. & Westhoff V. (1979) Effects of trampling and soil compaction on the occurrence of some *Plantago* species in coastal sand dunes. IV. The vegetation of two dune grasslands in relation to physical soil factors. *Proceedings of the Koninklijke Nederlandse Akademie van Wetenschappen, (Series C)*, **82**, 245–73.

Breteler H. (1975) Carboxylates and the uptake of ammonium by excised maize roots. Ph.D. thesis, Wageningen.

Chapin F.S. (1980) The mineral nutrition of wild plants. *Annual Review of Ecology Systematics*, **11**, 233–60.

Clarkson D.T. & Hanson J.B. (1980) The mineral nutrition of higher plants. *Annual Review of Plant Physiology*, **31**, 239–98.

Clement C.R., Hopper M.J. & Jones L.H.P. (1978) The uptake of nitrate by *Lolium perenne* from flowing nutrient solution. I. Effect of NO_3^- concentration. *Journal of Experimental Botany*, **29**, 453–64.

Clement C.R., Jones L.H.P. & Hopper M.J. (1979) Uptake of nitrogen from flowing nutrient solution: effect of terminated and intermittent nitrate supplies. In *Nitrogen Assimilation of Plants* (eds E.J. Hewitt & C.V. Cutting), pp. 123–33. Academic Press, London & New York.

Dejaegere D., Neirinchx L., Delegher V. & Francois G. (1980) Proton pumps and ion uptake in barley seedlings. Second Congress of the Federation of European Societies of Plant Physiology, Santiago de Compostela. Abstracts of lectures and posters, pp. 749–50.

Doddema H. (1978) Uptake of nitrate by chlorate resistant mutants of *Arabidopsis thaliana* (L.) Heynh. Ph.D. thesis, Groningen.

Epstein E. (1976) Kinetics of ion transport and the carrier concept. In *Encyclopaedia of Plant Physiology, New Series Vol. 2, Part B. Transport in Plants. II. Tissues and Organs* (eds U. Lüttge & M.G. Pitman), pp. 70–94. Springer-Verlag, Berlin.

Freijsen A.H.J. & Otten H. (1979) The growth and nitrate uptake of *Plantago lanceolata* and *P. major* in experiments with continuously and intermittently supplied nitrate concentrations of 750 and 25 μM. *Verhandelingen Koninklijke Nederlandse Akademie Wetenschappen, Afdeling Natuurkunde, 2e Reeks*, **73**, 364–7.

Harper J.L. (1978) The demography of plants with clonal growth. In *Structure and Functioning of Plant Populations* (eds A.H.J. Freijsen & J.W. Woldendorp), pp. 27–48. North Holland Publishing Company, Amsterdam.

Havill D.C., Lee J.A. & De-Felice J. (1977) Some factors limiting nitrate utilization in acidic and calcareous grasslands. *New Phytologist*, **78**, 649–59.

Havill D.C., Lee J.A. & Stewart G.R. (1974) Nitrate utilization by species from acidic and calcareous soils. *New Phytologist*, **73**, 1221–31.

Haynes R.J. & Goh K.M. (1978) Ammonium and nitrate nutrition of plants. *Biological Reviews*, **53**, 465–510.

Hewitt E.J. & Cutting C.V. (1979) *Nitrogen Assimilation of Plants*. Academic Press, London & New York.

Huffaker R.C. & Rains D.W. (1978) Factors influencing nitrate acquisiton by plants; assimilation and fate of reduced nitrogen. In *Nitrogen in the Environment, 2. Soil–Plant–Nitrogen Relationship* (eds D.R. Nielsen & J.G. MacDonald), pp. 1–43. Academic Press, London & New York.

Ingestad T. & Lund A.B. (1979) Nitrogen stress in birch seedlings. *Physiologia Plantarum*, **45**, 137–48.

Jaworski E.G. (1971) Nitrate reductase assay in intact plant tissues. *Biochemical Biophysical Research Communications*, **43**, 1106.

Kuiper D. & Kuiper P.J.C. (1978) Lipid composition of the roots of *Plantago* species: response to alteration of the level of mineral nutrition and ecological significance. *Physiologia Plantarum*, **44**, 81–6.

Kuiper D. & Kuiper P.J.C. (1979a) Ca^{2+} and Mg^{2+} stimulated ATPases from roots of *Plantago major* and *Plantago maritima*: response to alterations of the level of mineral nutrition and ecological significance. *Physiologia Plantarum*, **45**, 1–6.

Kuiper D. & Kuiper P.J.C. (1979b) Ca^{2+} and Mg^{2+} stimulated ATPases from roots of *Plantago lanceolata*, *P. media* and *P. coronopus*: response to alterations of the level of mineral nutrition and ecological significance. *Physiologia Plantarum*, **45**, 240–4.

Lambers H. (1979) Energy metabolism in higher plants in different environments. Ph.D. thesis, Groningen.

Lambers H., Posthumus F., Stulen I., Lanting L., van de Dijk S.J. & Hofstra R. (1981a) Energy metabolism of *Plantago lanceolata* as dependent on the supply of mineral nutrients. *Physiologia Plantarum*, **51**, 85–92.

Lambers H., Posthumus F., Stulen I., Lanting L., van de Dijk S.J. & Hofstra R. (1981b) Energy metabolism of *Plantago major* L. ssp. *major* as dependent on the supply of mineral nutrients. *Physiologia Plantarum*, **51,** 245–52.

Lambers H. & Steingröver R. (1978) Efficiency of root respiration of a flood-tolerant *Senecio* species as affected by low oxygen tension. *Physiologia Plantarum*, **42**, 179–84.

Meguro N.E. & Magalhaes A.C. (1980) Water stress affecting nitrogen reduction and diffusive resistance in coffee leaves. Second Congress of the Federation of European Societies Plant Physiology Santiago de Compostela, Abstracts of lectures and posters, pp. 493–4.

Nielsen D.R. & MacDonald J.G. (eds) (1978a) *Nitrogen in the Environment 1. Nitrogen Behaviour in Field Soil.* Academic Press, London & New York.

Nielsen D.R. & MacDonald J.G. (eds) (1979b) *Nitrogen in the Environment 2. Soil–Plant–Nitrogen Relationship.* Academic Press, London & New York.

Noë R. & Blom C.W.P.M. (1981) Occurrence of three *Plantago* species in coastal dune grasslands in relation to pore volume and organic matter content of the soil. *Journal of Applied Ecology*, **19,** 177–82.

Nye P.H. & Tinker P.B. (1977) *Solute Movement in the Soil–Root System.* Blackwell Scientific Publications, Oxford.

Rice E.L. (1974) *Allelopathy*. Academic Press, London & New York.

Rorison I.H. (1980) The effects of soil acidity on nutrient availability and plant response. In *Effects of Acid Precipitation on Terrestrial Ecosystems* (eds T.C. Hutchinson & M. Havas), pp. 283–304. Plenum Publishing Corporation, New York.

Smit A.J. & Woldendorp J.W. (1981) Nitrate production in the rhizosphere of *Plantago* species. *Plant and Soil*, **61**, 43–52.

Stulen I., Lanting L., Lambers H., Posthumus F., van de Dijk S.J. & Hofstra R. (1981a) Nitrogen metabolism of *Plantago lanceolata* as dependent on the supply of mineral nutrients. *Physiologia Plantarum*, **51**, 93–8.

Stulen I., Lanting L., Lambers H., Posthumus F., van de Dijk S.J. & Hofstra R. (1981b) Nitrogen metabolism of *Plantago major* L. ssp. *major* as dependent on the supply of mineral nutrients. *Physiologia Plantarum*, **52,** 108–14.

Troelstra S.R. (1977) Spatial variability of soil chemical properties in the older dune area (Westduinen) of Goeree. *Verhandelingen Koninklijke Nederlandse Akademie Wetenschappen, Afdeling Natuurkunde*, 2e *Reeks*, **71,** 321–7.

Troelstra S.R. & Smant W. (1979) The ionic balance of some plant species from natural vegetations and its relation to nitrogen uptake and salt tolerance. *Verhandelingen Koninklijke Nederlandse Akademie Wetenschappen, Afdeling Natuurkunde, 2e Reeks*, **73**, 381–9.

Troelstra S.R. & Smant W. (1980) The ionic balance of some plant species from natural vegetations: comparison of plants grown in the greenhouse and in the field. *Verhandelingen*

Koninklijke Nederlandse Akademie Wetenschappen, Afdeling Natuurkunde, 2e Reeks, **75**, 46–51.

Troelstra S.R. & Wagenaar R. (1979) Mineral nitrogen and nitrogen net-mineralization in an older dune area (Westduinen) on the island of Goeree. *Verhandelingen Koninklijke Nederlandse Akademie Wetenschappen, Afdeling Natuurkunde, 2e Reeks*, **73**, 304–91.

Troelstra S.R. & Wagenaar R. (1980) Seasonal patterns in a relatively old dune area (Westduinen) on the island of Goeree. *Verhandelingen Koninklijke Nederlandse Akademie Wetenschappen, Afdeling Natuurkunde, 2e Reeks*, **75**, 37–41.

van de Dijk S.J. (1981a) Kinetics of nitrate uptake by different species from nutrient-rich and nutrient-poor habitats as affected by the nutrient supply. *Physiologia Plantarum*, **55,** 103–10.

van de Dijk, S.J. (1981b) Differences in nitrate uptake of species from habitats rich or poor in nutrients when grown at low nitrate concentrations, using a new growth technique. *Plant and Soil*, **62,** 262–78.

van Egmond F. (1978) Nitrogen nutritional aspects of the ionic balance of plants. In *Nitrogen in the Environment 2: Soil–Plant–Nitrogen Relationship* (eds D.R. Nielsen & J.G. MacDonald), pp. 169–89. Academic Press, London & New York.

Woldendorp J.W. (1963) The influence of living plants on denitrification. *Mededelingen Landbouwhogeschool, Wageningen*, **63**, 1–100.

Woldendorp J.W. (1975) Nitrification and denitrification in the rhizosphere. *Société Botanique de France*, Colloque 'La Rhizosphère', **112**, 89–107.

Woldendorp J.W. (1978) The rhizosphere as part of the plant–soil system. In *Structure and Functioning of Plant Populations* (eds A.H.J. Frijsen & J.W. Woldendorp), pp. 237–67. North Holland Publishing Company, Amsterdam.

Woldendorp J.W. (1981) Nutrients in the rhizosphere. In *Agricultural Yield Potentials in Continental Climates*. pp. 99–125. Proceedings of the 16th Colloquium of the Potash Institute, Berne.

8. STUDIES OF NITRATE UTILIZATION BY THE DOMINANT SPECIES OF REGROWTH VEGETATION OF TROPICAL WEST AFRICA: A NIGERIAN EXAMPLE

G. R. STEWART[1] AND T. O. OREBAMJO[2]
[1]*Department of Botany, Birkbeck College, University of London, Malet Street, London, WC1E 7HX* and [2]*School of Biological Sciences, University of Lagos, Lagos, Nigeria*

SUMMARY

Different plant communities of West Africa were found to show differences in nitrate utilization. Species of forb-rich fallow, thicket and secondary deciduous forest were found to have large activities of nitrate reductase while those of savanna fallow and woodland had small nitrate reductase activities. Plants of freshwater swamp had high nitrate reductase activities while those of mangrove swamp had low activities. Feeding nitrate had little effect on the nitrate reductase activities of plants in forb-rich fallow sites but markedly increased those of plants of savanna fallow and woodland, and mangrove swamp. There is no evidence that species from communities in which nitrate is not the available nitrogen source lack the enzymic potential to utilize it. None of those species from nitrate-poor sites exhibit any feature indicative of metabolic adaptation to the utilization of ammonia as a nitrogen source.

INTRODUCTION

In a symposium entitled *Nitrogen as an Ecological Factor* it is perhaps unnecessary to point out that the growth of plants, and consequently that of their attendant populations of herbivores, is in many ecosystems restricted by the availability of nitrogen, more than any other mineral nutrient. Except for the relatively small number of plants which can, by means of some specific symbiotic association with a nitrogen-fixing prokaryote, utilize molecular nitrogen, and the even more restricted range of carnivorous and parasitic plants which have access to reduced organic sources of nitrogen, for most plants soil inorganic nitrogen compounds are the available source for growth.

It is not simply the availability of nitrogen which may influence plant growth; a number of studies indicate that the form in which nitrogen is present may also be of importance. It has been reported that some species of Baltic coastal marshes respond to the addition of nitrogen in the ammonia rather than nitrate form and it has been suggested that such plants are specifically adapted to ammonia utilization (Tyler 1967). It is often assumed that nitrification proceeds at low rates in soils of low pH (Raven, Smith & Smith 1978) although there are several studies demonstrating appreciable rates of nitrification in acid soils (Lee & Stewart 1978). Nevertheless, a number of plants characteristic of acid soils are either unable to utilize nitrate or grow better on ammonia (Cain 1952; Colgrove & Roberts 1956; Pharis, Barnes & Naylor 1964; McFee & Stone 1968; Gigon & Rorison 1972). Townsend (1970) has proposed that this inability to use nitrate is associated with a lack of the enzyme nitrate reductase and many species in the Ericaceae have very low activities of this enzyme (Dirr, Barker & Maynard 1972).

Many grassland and other 'climax' communities are characterized by low rates of nitrification. Grasslands such as the Guinea Savanna of West Africa (Greenland 1958), New Zealand tussock (Robinson 1963), Veldt of Southern Africa (Stiven 1952; Mills 1953), the Patana of Ceylon (Eden 1951) and old-field succession to prairie in Oklahoma (Rice 1974) have little nitrate in the soil and rates of nitrification are low. Wiltshire (1973) has shown climax perennial grass species from Zimbabwe to exhibit greater growth on ammonia compared with nitrate while earlier seral species grow better on nitrate. Bate & Heelas (1975) have shown in a comparison of the climax species *Hyparrhenia filipendula* and the pioneer species, *Sporobolus pyramidalis*, that the latter has a much greater capacity to reduce nitrate.

Evidence for the inhibition of nitrification in other climax communities is less well documented. The work of Dommergues (1954, 1956) in Senegal and Madagascar suggests nitrification is low and ammonification high in humid forest soils while nitrification is greater in dry forest soils. Similar low rates of nitrification have been reported for forest soils of the Ivory Coast (Jacquemin & Berlier 1956; Berlier, Dabin & Leneuf 1956).

The basis of these low rates of nitrification is uncertain. Rice (1974) has argued that they result from an inhibition by polyphenolic toxins in the litter or root exudates of some species. Theron (1951) has suggested that in soils supporting *Hyparrhenia* nitrification is inhibited by root exudates and Meiklejohn (1968) has shown that there are few nitrifying bacteria under *Hyparrhenia* grasslands in Zimbabwe. Allelopathic factors in interactions between other savanna species have been implicated by Boughey *et al.* (1964). Purchase (1974) has, however, argued that low rates of nitrification result from immobilization of nitrogen while Roux (1954) claims that substrate availability is the main factor limiting nitrification.

Rice (1974) has further argued that there are a number of ecological implications in the inhibition of nitrification in climax communities. Reduced nitrification is suggested to be a means of conserving nitrogen since the ammonium ion is less readily leached from soil than the nitrate ion. Ammonia is also postulated to be a less energy-demanding nitrogen source than nitrate, and reduced nitrification is to be regarded as energy conserving. In addition, plants which are able to utilize ammonia efficiently are considered to have a selective advantage in climax vegetation.

Possible differences in the form of available nitrogen in different plant communities indicate the possibility of different physiological types adapted to the assimilation of ammonia or nitrate. The assimilation of nitrate can usefully be divided into three components: the uptake of nitrate ions from the soil solution; its reduction via nitrite to ammonia; and the assimilation of ammonia into organic nitrogen compounds. The uptake of nitrate occurs via an energy-dependent, adaptive transport system (Butz & Jackson 1977). The enzymes of nitrate reduction comprise, in angiosperms, a pyridine nucleotide-dependent nitrate reductase and a ferredoxin-dependent nitrite reductase (Hewitt 1975; Beevers & Hageman 1980). Ammonia assimilation, including ammonium ions absorbed from the soil solution, is now generally thought to take place via the glutamate synthase cycle (Miflin & Lea 1980). In this cycle two enzymes, glutamine synthetase and glutamate synthase, catalyse on ATP-dependent synthesis of glutamate. There is little evidence that exogenous ammonia is assimilated via glutamate dehydrogenase, as was thought. Most recent evidence supports the view that the glutamate synthase cycle is the sole route of ammonia assimilation in angiosperms (Rhodes, Sims & Folkes 1980). A major difference between nitrate- and ammonia-assimilating plants is in the site of assimilation. Ammonia assimilation occurs more or less exclusively in the root system. This restriction of ammonia assimilation to the root results from the necessity to regulate pH, as large amounts of hydrogen ions are produced during ammonia assimilation and plant leaf cells lack any great capacity to store hydrogen ions or biochemically neutralize them. Consequently, the release of hydrogen ions from root cells to the soil solution represents the only feasible means of pH control (Raven & Smith 1976). In contrast, the reduction of nitrate can be biochemically pH-regulated and the shoot system is the major site of nitrate reduction (Pate 1971). Nitrate reduction can, in the shoot system, be directly light-driven through its coupling to the light reaction of photosynthesis (Beevers & Hageman 1980).

In this contribution we examine the hypothesis that the form of available nitrogen varies in different plant communities of West Africa and consider some of the physiological characteristics associated with ammonia and nitrate assimilation.

MATERIALS AND METHODS

Sampling sites

Secondary succession leading to deciduous rain forest

One of the first kinds of community recognizable once a farm has been abandoned is that known as forb regrowth (Ahn 1958) and it consists largely of fast-growing herbs and forbs. It can contain crops and trees left from the cultivation period and coppiced shrubs and trees left from the original vegetation. Sites corresponding to this type of community were sampled on the University of Lagos campus.

Thicket succeeds forb regrowth and is characterized by the presence of woody-stemmed climbers, coppiced shoots, tree seedlings, small trees and shrubs which replace the herb and forb species. Again sites corresponding to this stage were sampled on or close to the University of Lagos campus.

A second type of early regrowth community was apparent in several areas; this we refer to as grass-rich fallow since it was dominated by the grass *Panicum maximum**.

The area of mature deciduous rain forest selected for sampling was in the Gambari Forest Reserve, some 50 km south of Ibadan (see Keay 1962 for details).

Savanna vegetation

The majority of the savanna sampling sites were located in what is generally referred to as Guinea Savanna (Keay 1959). Two types of savanna woodland were sampled, the first being protected from burning and the second subject to annual burning. One of these protected woodland sites was located near to Ilorin airport and the other in the Samaru forest reserve, Zaria. At Ilorin the grass layer consisted mainly of *Andropogon gayanus* while at Samaru, *Hyparrhenia* and *Pennisetum* were abundant. At both, *Daniella oliveri* and *Parkia clappertoniana* were the main tree species.

The burnt savanna woodland sites were located near to Ilorin University and Ahmadu Bello University, Zaria. At Ilorin the grass layer was dominated by *A. gayanus* and the main trees were *Butyrospermum paradoxum*, coppices of *D. oliveri* and *P. clappertoniana* which formed a discontinuous canopy. At

* Nomenclature follows Hutchinson & Dalziel (1954–1972).

the Ahmadu Bello sites the vegetation resembled that of the Sudan Savanna, rather than the Guinea Savanna, with the shorter feathery grasses rather than the tall grasses being abundant. The main tree species were *B. paradoxum* and *Isoberlina doka*.

Sampling was also carried out at a number of 'damp' savanna sites which consisted of patches of vegetation in or close to river beds or dam margins.

Mangrove vegetation

The sampling sites for mangrove were located in creeks of the River Ogun estuary in Lagos Lagoon. The general features of the vegetation in this area have been described by Jackson (1964) and Orebamjo & Njoku (1970).

Sampling was carried out over the period 1978 (January–May) to 1981.

Nitrate reductase measurements

Nitrate reductase activity was determined essentially as described by Stewart & Orebamjo (1980). In this assay leaf fragments are infiltrated with assay solution and incubated in the dark. The amount of nitrite produced during the incubation period is then a measure of *in vivo* nitrate reductase activity. Normally 1 g fresh weight of leaf tissue was sliced into fragments and from this subsamples of 100–200 mg were taken for assay. Two or three samples were normally processed in this way.

Total nitrogen

Total nitrogen was determined as ammonia (McCullough 1967) after digestion of dried leaf samples.

Ureide determination

Allantoin and allantoic acid were determined as described by Vogel & Van der Drift (1970).

RESULTS

Nitrate utilization in the succession from fallow to secondary deciduous rain forest.

The results in Table 8.1 show the levels of nitrate reductase present in leaf tissue of plants growing in communities representative of seral stages in the

TABLE 8.1. Nitrate reductase activity in plants of communities representing different seral stages in the regrowth of deciduous rain forest.

Species	Nitrate reductase activity (μmol h^{-1} g fresh wt.$^{-1}$) Forb-rich	Grass-rich	Early thicket	Thicket	Rain forest
Ageratum conyzoides	1.3		0.3		
Albizia agg.					3.6
Alchornea cordifolia			3.2	5.0	
Alstonia congensis				4.6	
Alternanthera repens	5.3		2.4		
Amaranthus viridis	3.9				
Andropogon tectorum	1.6				
Angylocalyx oligophyllus					0.2
Anthocleista vogelli				0.1	
Anthonotha macrophylla				0.2	
Antiaris africana			8.9	6.0	4.8
Aspilia africana	1.2	0.7	0.7		
Axonopus flexuosus		0.9			
Baphia nitida				0.4	0.2
Blighia sapida			0.3	0.3	0.2
Blighia unijugata			0.8		
Boerhavia diffusa	3.5		1.6		
Bridelia atroviridis				4.3	
Bridelia micrantha				4.5	1.3
Bombax bunopozense					0.4
Brachiaria deflexa	1.4	0.8			
Brachiaria distichophylla		2.7			
Byrsocarpus coccineus	3.7	2.3		2.2	
Cassia hirsuta	4.2		2.6		
Celosia argenta	3.0				
Celtis zenkeri					0.5
Celtis sp.					1.2
Chlorophora excelsa					3.7
Cleome ciliata	4.4		2.2		
Cnestis ferruginea				0.6	
Cola cordifolia					0.5
Cola gigantea					0.7
Cola millenii					0.4
Combretum smeathmannii			0.8	0.6	1.2
Commelina benghalensis	2.2				
Commelina vogelli	4.4	1.5	2.1	0.6	
Crotolaria govensis		0.6			
Croton lobatus	5.1		3.0		
Culcasia scandens					1.9
Cynodon dactylon		1.1	1.1		
Dialium guineense				0.1	
Digitaria horizontalis	2.0	0.8			
Dissotis rotundifolia			1.6		

TABLE 8.1. (*cont.*)

Species	Nitrate reductase activity (μmol h^{-1} g fresh wt.$^{-1}$) Forb-rich	Grass-rich	Early thicket	Thicket	Rain forest
Dissotis segregata			2.3		
Dracaena manni				0.2	
Duparquetia orchidacea			0.4		
Elaeis guineensis	1.0				
Eleusine indica	1.0				
Emila sonchifolia		0.6	0.2		
Eupatorium odoratus	3.6	1.0	0.9	0.8	
Euphorbia glomerifera	5.8	1.1	1.6		
Euphorbia heterophylla	5.4	2.0	1.3		
Euphorbia hirta	3.3	1.2	1.2		
Fagara zanthoxyloides				1.4	
Ficus asperifolia			3.8	2.2	
Gliricidia maculata				0.2	
Harungana madagascarensis			1.2	0.7	
Hibiscus surattensis		0.8			
Hibiscus vitifolia	2.6				
Holarrhena floribunda			3.9	2.3	
Hyptis suaveolens		0.3			
Icacina trichantha					1.2
Ipomea involucrata	2.4	0.8	0.4	0.1	
Jussiaeu suffruticosa			4.0		
Lantana eamara	2.2		1.6		
Lecanodiscus cupanoides			4.0		
Luffa aegyptica			3.0	6.4	
Mariscus alternifolius		0.4	0.1		
Monodora tenuifolia				0.7	1.2
Morinda lucida			1.9	1.3	
Musanga cecropioides	0.8		1.4	0.9	
Newbouldia laevis				6.0	5.0
Oldenlandia affinis	4.6	2.5			
Olyra latifolia				0.6	
Oplismenus burmanni			2.6	1.0	
Panicum maximum	2.5	0.7	0.9		
Panicum nigrense		0.4			
Paspalum commersonii	1.4		0.4		
Pennisetum polystachion		0.3			
Phyllanthus meullerianus				0.2	
Rauwolfia vomitoria				1.7	0.4
Schwenkia americana	2.9	0.9	1.0		
Sida acuta	11.5	3.2	4.8	3.2	
Solanum torvum	12.3				
Solanum verbascifolium					5.3
Spathodea companulata				2.1	

TABLE 8.1. (*cont.*)

Species	Nitrate reductase activity (μmol h^{-1} g fresh wt.$^{-1}$) Forb-rich	Grass-rich	Early thicket	Thicket	Rain forest
Sphenocentrum jollyanum					0.4
Spigelia anthelmia	2.4	0.9	0.9		
Spondias mombin				0.7	
Sporobolus pyramidalis	1.2	0.8	0.3		
Stachytarpheta jamaicensis	2.9	0.8	1.7		
Synedrella nodiflora	4.2	1.9	0.3		
Talinum triangulare	1.5				
Terminalia superba					0.5
Trema guineensis	8.2	1.2	2.3	2.8	
Tridax procumbens	1.5		0.2		
Triplochiton scleroxylon					3.1
Triumfetta rhomboidea	3.9		3.1	1.5	
Uraria picta			0.1		
Urena lobata	6.0	2.1	2.0		
Usteria guineensis			0.2	0.2	
Vernonia cinerea	3.7		1.0		
Xanthosoma mafaffa			3.9		

regrowth of deciduous rain forest following the abandonment of farmland. All of the species examined had measurable nitrate reductase activities and in the earliest forb-rich stage large activities were measurable in many of the forbs and coppiced trees and shrubs. Some of the forb species such as *Alternanthera repens*, *Croton lobatus*, *Euphorbia glomerifora*, *E. hirta*, *Sida actua* and *Urena lobata* and several of the young trees of the thicket stage of regrowth such as *Antiaris africana*, *Newbouldia laevis* and *Trema guineensis* had levels of nitrate reductase comparable with those found in temperate nitrophilous species (Stewart *et al.* 1974; Lee & Stewart 1978).

The activities of nitrate reductase found in plants of the forb-rich sites were considerably greater than those of plants in the grass-rich sites dominated by *Panicum maximum*. This is seen clearly in Table 8.2. The activities in forb, grass shrub and tree species found in these forb-rich sites are 2–4 times those of the corresponding life-form found in the grass-rich fallows. The low levels of nitrate reductase in plants growing in grass-rich fallow is interesting in relation to reports referred to earlier of low rates of nitrification in grassland. Low rates of nitrification are for the most part associated with grasses in the Andropoganeae (Nye & Greenland 1960). Agboola (1971) has, however, reported the suppression of nitrification in soil planted with two ecotypes of *P.*

TABLE 8.2. Nitrate utilization in seral communities leading to deciduous rain forest. (Figures in parentheses are the number of species sampled.)

Site	Nitrate reductase activity (μmol h^{-1} g fresh wt.$^{-1}$)			
	Forb	Grass	Shrub/Climber	Tree
Forb-rich fallow	3.8 (25)	1.7 (8)	4.7 (5)	4.5 (2)
Grass-rich fallow	1.3 (17)	0.7 (7)	2.3 (1)	1.2 (1)
Early thicket	1.7 (24)	0.9 (6)	2.1 (7)	2.6 (8)
Thicket	2.1 (6)	0.8 (2)	1.4 (12)	1.8 (17)
Deciduous forest	1.9 (1)	—	0.7 (7)	2.0 (16)

maximum from Nigeria, although not with Kenyan ecotypes. Except in the case of these grass-rich fallows there appears to be a general decline in nitrate reductase levels passing from forb to thicket. This decline is seen in many of the forb species, including *Aspilia africana*, *Commelina vogelli*, *Eupatorium odoratum*, *Ipomoea involucrata* and *S. actua*, which persist into the thicket stage.

Although the average nitrate reductase activity in the three species of the forb-rich stage is 2–3 times greater than that of the deciduous rain forest tree species, a number of the latter do exhibit quite large nitrate reductase activities. In particular, *Antiaris africana*, *Chlorophora excelsa*, *N. laevis*, *Solanum verbascifolium* and *Triplochiton scleroxylon* have activities in excess of 3.0 μmol h^{-1} g fresh wt.$^{-1}$ suggesting these rain forest species are very active in nitrate assimilation.

Nitrate utilization in savanna communities

The nitrate reductase activities found in leaf tissue of plants growing in different savanna communities (Tables 8.3 and 8.4) are in general considerably lower than those found in plants of the seral stages leading to secondary deciduous rain forest. In the fallow sites most species were found to have nitrate reductase activities less than 1.0 μmol h^{-1} g fresh wt^{-1}. Exceptions to this were *Acalypha crenata*, *Nauclea latifolia* and *Nelsonia canescens*, which had activities in excess of 2.0 μmol h^{-1} g fresh wt.$^{-1}$ (see Table 8.3). In the damp savanna sites the nitrate reductase activities of all of the life-forms examined were greater than at any of the other savanna sites (see Table 8.4). The activities of the forb and grass species of savanna woodland are small, lower than the activities found in forb and grass species characteristic of temperate acidic grassland soils (Havill, Lee & Stewart 1974; Lee & Stewart 1978). The shrub and trees species of savanna woodland exhibit greater nitrate

TABLE 8.3. Nitrate reductase activity in plants of savanna vegetation.

Species	Nitrate reductase activity (μmol h^{-1} g fresh wt.$^{-1}$) Abandoned farmland	'Damp' savanna	Woodland	Woodland (burnt)
Acalypha crenata	2.1			
Acanthospermum hispidum	0.8			
Afrormosia laxiflora			0.2	
Ageratum conyzoides	0.5			
Amaranthus viridis	1.2	2.7		
Andropogon gayanus	0.2	0.4	0.2	0.2
Annona senegalensis	0.1	0.4	0.1	
Anogeissus leiocarpus			0.9	
Arthrosolen chrysanthus	0.3			
Aspilia africana	0.2			0.2
Azadirachta indica	0.5	3.2	0.4	2.0
Balanites aegyptica			3.3	
Bauhinia rufescens	0.4			
Berlinia confusa			0.4	
Bidens pilosa	0.2			
Blighia sapida			0.2	
Blumea aurita	0.1			
Boerhavia diffusa	0.2			
Borreria filiformis	0.2	1.2		
Brachiaria deflexa	0.2			
Bridelia ferruginea			0.1	
Butyrospermum paradoxum	0.1			0.1
Byrsocarpus coccineus				2.2
Calotropis procera			0.9	
Carissa edulis	0.4		0.6	
Cassia rotundifolia			0.2	0.2
Cassia siamea	1.0		0.8	
Ceiba pentandra			0.8	
Celosia argenta	0.9			
Celtis integrifolia	0.9		0.4	
Combretum ghasalense		0.4	0.1	
Combretum nigricans	0.1	0.4		
Crotalaria lachnosema	0.1			
Crotolaria retusa		0.3	0.2	
Croton lobatus	0.7			0.2
Dalbergia sisoo	0.3			
Daniellia oliveri	0.1		0.2	0.1
Delonix regia	0.5			
Desmodium gangeticum			1.1	
Detarium microcarpum			0.7	
Dichrostachys glomerata	0.4		0.2	
Diodia scandens		0.4		

TABLE 8.3. (*cont.*)

Species	Nitrate reductase activity (μmol h^{-1} g fresh wt.$^{-1}$) Abandoned farmland	'Damp' savanna	Woodland	Woodland (burnt)
Diospyros mespiliformis			0.1	
Dodonaea viscosa	0.3			
Ekebergia senegalensis	0.2		0.3	
Eleusine indica	0.4			
Entada africana				0.3
Eragrostis atrovirens		0.9		
Eragrostis tenella	0.6	0.8		
Eriosema andohii	0.1			
Eriosema psoraleoides			0.2	
Euphorbia heterophylla	0.8			
Ficus asperifolia		0.9		
Ficus ingens		5.5	0.6	
Ficus iteophylla	0.9	3.0		
Gardenia aqualla			0.2	
Guiera senegalensis		1.0	0.5	
Heeria insignis	0.4			
Hibiscus sidiformis	0.4			
Hymanocardia acida				0.4
Hyparrhenia cyanescens	0.1	0.3		0.1
Hyparrhenia rufa	0.1		0.2	
Imperata cylindrica	0.3		0.1	0.1
Indigofera congesta			0.2	
Ipomoea aquatica		1.0		
Ipomoea repens		1.5		0.1
Isoberlinia doka	0.4		0.2	
Khaya senegalensis	<0.1			
Lactuca capensis	0.2			
Lawsonia inermis	0.6			
Lepidagathis collina	0.2			
Lonchoarpus cyanescens				0.5
Luffa aegyptica		3.7		
Macrosphyra longistyla			0.2	
Maerua crassifolia			2.0	
Malacantha alnifolia				0.3
Microtrichia perrottetii			0.3	
Mitragyna inermis			1.9	
Monotes kerstingii			0.3	
Nauclea latifolia	2.2		2.0	
Nelsonia canescens	2.6			
Ochna kibbiensis	<0.1			
Olax gambecola	0.2		<0.1	
Oncoba spinosa				0.6
Panicum maximum		0.6		
Parinari curatellifolia	0.3	1.9		1.6

TABLE 8.3. (*cont.*)

Species	Nitrate reductase activity (μmol h^{-1} g fresh wt.$^{-1}$) Abandoned farmland	'Damp' savanna	Woodland	Woodland (burnt)
Parkia clappertoniana			0.9	<0.1
Pennisetum polystachion	<0.1		0.2	0.2
Physalis angulata	2.1			
Piliostigma thonningii	0.7	1.4	0.2	0.3
Polygonum limbatum	4.6			
Prosopis africana	<0.1		0.2	
Psidium guajava	0.2			
Saccharum spontaneum		1.3		
Sclerocarpus africanus				0.2
Setaria barbata	0.2	0.8		
Sphaeranthus senegalensis			0.2	
Spigelia anthelmia	0.2			
Sporobolus pyramidalis	0.2			
Stereospermum kunthianum			0.4	
Synedrella nodiflora	1.8			
Tephrosia vogelii				0.5
Terminalia avicennioides	<0.1	1.7		
Trema guineensis			0.7	
Trianthema portulacastrum		1.4		
Tridax procumbens	0.1			
Urena lobata	0.9			
Vernonia amygdalina	0.8			
Vitex doniana	0.1		0.3	0.6
Waltheria indica			1.1	
Ximenia americana			<0.1	

TABLE 8.4. Nitrate utilization in communities of savanna vegetation. (Figures in parentheses are the number of species sampled.)

Site/Life form	Nitrate reductase activity (μmol h^{-1}g fresh wt.$^{-1}$) Forb	Grass	Shrub	Tree
Fallow	0.6 (21)	0.2 (10)	0.5 (17)	0.4 (18)
Damp	1.9 (10)	0.7 (7)	1.8 (5)	1.8 (6)
Woodland	0.2 (5)	0.2 (4)	0.6 (14)	0.6 (25)
Woodland (burnt)	0.2 (5)	0.1 (4)	0.8 (5)	0.6 (9)

reductase activity than the forb and grass species, although even in the former enzyme activities are considerably lower than those of shrub and tree species of secondary deciduous forest. Two tree species of savanna woodland, *Balanites aegyptica* and *Maerua crassifolia*, had activities in excess of 2.0 μmol h^{-1} g fresh wt.$^{-1}$, suggesting these savanna woodland species were active in nitrate utilization.

Nitrate utilization in swamp vegetation

There was very little nitrate reductase activity detectable in any of the species growing in mangrove vegetation (Table 8.5). The greatest nitrate reductase activity was 0.1 μmol h^{-1} g fresh wt.$^{-1}$, found in *Cyperus articulatus* and *Paspalum vaginatum*. The activities in these mangrove species are considerably lower than those found in plants of another saline habitat, temperate salt marshes (Stewart, Lee & Orebamjo 1973). These results suggest little or no nitrate assimilation by the leaves of these mangrove species.

In contrast, most of the species of freshwater swamp which were examined had activities in excess of 1.0 μmol h^{-1} g fresh wt.$^{-1}$, that is at least ten times those of the activities in plants of the mangrove (Table 8.6).

Enzyme induction in situ

Nitrate reductase is a substrate-inducible enzyme so that by the addition of nitrate fertilizer to vegetation it should be possible, in sites where there is a limited supply of nitrate, to increase nitrate reductase activity. This is illustrated by the results in Table 8.7 which show the increase in nitrate reductase following the addition of nitrate to a fallow site dominated by the grass *P. maximum*. All of the species exhibited 3–5 fold increase in nitrate reductase activity. In contrast, the addition of nitrate to a forb-rich fallow site had little effect on nitrate reductase levels, the maximum increase being 1.5 fold (Table 8.8). These results suggest that in the forb-rich fallow there is sufficient nitrate available to maintain the plants at their maximum nitrate-utilizing capacity. In contrast, in the *Panicum*-dominated fallow less than 30% of the maximum nitrate utilizing capacity is being expressed.

This approach has been applied to several sites in the savanna and seral stages of secondary deciduous rain forest (Table 8.9). It is evident from these results that the herb and forb species of savanna woodland and mangrove swamp exhibit very little of their potential to assimilate nitrate, the control levels of nitrate reductase being less than 10% of those obtained after induction with nitrate fertilizer.

It is interesting that the average induced level of nitrate reductase appears

TABLE 8.5. Nitrate reductase activities in species of the mangrove swamp.

Species	Nitrate reductase activity (μmol h^{-1}g fresh wt.$^{-1}$)
Acrostichum aureum	0.02
Avicennia nitida	0.01
Cyperus articulatus	0.10
Dalbergia ecastaphylum	0.03
Drepanocarpus lunatus	0.01
Hibiscus tiliaceus	0.04
Paspalum vaginatum	0.10
Rhizophora racemosa	0.02

TABLE 8.6. Nitrate reductase activities in species of freshwater swamp.

Species	Nitrate reductase activity (μmol h^{-1}g fresh wt.$^{-1}$)
Alchornea cordifolia	4.5
Alstonia congensis	4.6
Anthocleista vogelii	1.1
Elaeis guineensis	1.3
Fibristylis sp.	1.3
Musanga cecropioides	0.9
Napoleona vogelli	2.8

TABLE 8.7. Induction *in situ* of nitrate reductase in grass-rich fallow.

Species	Nitrate reductase activity (μmol h^{-1}g fresh wt.$^{-1}$) Control	Induced
Aspilia africana	0.6	1.8
Axonopus flexuosus	0.9	2.8
Brachiaria deflexa	0.8	2.5
Byrsocarpus coccineus	2.0	6.5
Crotolaria govensis	0.7	2.4
Digitaria horizontalis	0.7	2.5
Euphorbia glomerifera	1.0	5.1
Euphorbia hirta	1.2	4.8
Panicum maximum	0.5	3.3
Schwenkia americana	1.0	4.2
Spigelia anthelmia	0.7	3.7
Sporobolus pyramidalis	0.5	2.0

TABLE 8.8. Nitrate reductase induction *in situ* in forb-rich fallow.

Species	Nitrate reductase activity (μmol h^{-1} g fresh wt.$^{-1}$) Control	Induced
Celosia argenta	3.1	3.3
Cleome ciliata	2.2	2.6
Comellina vogelli	4.4	4.3
Eleusine indica	1.5	1.6
Euphorbia heterophylla	5.8	6.5
Euphorbia hirta	3.9	4.6
Ipomoea involucrata	2.8	3.1
Sida acuta	11.8	12.6
Spigelia anthelmia	1.9	3.1
Talinum triangulare	1.8	2.3
Tridax procumbens	1.6	2.0
Urena lobata	8.3	9.0

TABLE 8.9. Nitrate-reducing potential of herbs and forbs in different communities.

Site	Nitrate reductase activity (μmol h^{-1}g fresh wt.$^{-1}$) Control	Induced	% Maximum potential being expressed
Forb-rich fallow	4.1	4.4	89
Grass-rich fallow	0.9	3.5	26
Thicket	1.6	4.8	33
Savanna fallow	0.3	1.5	20
Savanna woodland	0.1	1.8	6
Mangrove swamp	0.1	2.1	5

to be lower in the herbs and grasses of savanna than in those of fallow or thicket leading to secondary forest. This is similar in certain respects to the findings of Bate & Heelas (1975) who showed a climax grass to have a lower nitrate-reducing capacity than a pioneer grass species.

The extent to which the capacity of forbs and grasses to assimilate nitrate is expressed appears considerably lower in thicket than in forb-rich fallow, suggesting a decline in nitrate availability. It may be, however, that these forb and grass species are less able to compete in thicket with shrub and tree species for the available nitrate.

There is some indication, however, of a general decline in nitrogen availability going from fallow to secondary forest in so far as plant nitrogen contents decline from 3.9% to 2.7% (Table 8.10).

It is noteworthy that plants of mangrove swamp vegetation which like those of savanna fallow and woodland had small nitrate reductase activities, have nitrogen contents twice those of the savanna plants. This suggests that the savanna fallow and woodland can be characterized as sites low in available nitrogen, irrespective of what form of nitrogen may be available. These results are consistent with studies of savanna soils which indicate small amounts of available nitrogen to be present (du Rham 1970).

The nitrogen status of the mangrove swamp appears somewhat different from that of savanna. Although the species of this community appear not to be active in nitrate assimilation, they have nitrogen contents comparable with those of species in secondary deciduous forest which are active in nitrate assimilation. It would seem likely then that these species are assimilating some nitrogen source other than nitrate.

TABLE 8.10. Nitrogen content of plants growing in different communities.

Site	Nitrogen content mg g dry wt.$^{-1}$
Mangrove	24.5
Thicket	32.6
Forb-rich fallow	38.6
Secondary forest	27.0
Savanna fallow	10.4
Savanna woodland	10.9

DISCUSSION

The results presented here indicate quite clearly that there are very marked differences in nitrate utilization in the different plant communities studied. In recently abandoned farmland characterized by a forb-rich community the occurrence of high levels of nitrate reductase, the small increase in enzyme level observed after nitrate fertilization and the high plant nitrogen contents, all strongly suggest that such sites are rich in nitrate and that nitrate is likely to be the main form of nitrogen utilized by species growing in these communities. These observations reinforce the conclusions of other workers (Nye & Greenland 1960; du Rham 1970) that the abandonment of farmland in such areas does not stem from declining yields which are the result of soil minerals becoming depleted.

Our results suggest that while there may be some decline in nitrate availability as the succession proceeds from fallow to secondary deciduous rain forest there is no major change in the form of available nitrogen. Many of the rain forest species examined were found to be active in nitrate reduction and while no induction experiments were carried out *in situ* with these species (for obvious technical reasons), several of them have been grown on nitrate in the laboratory. The results in Table 8.11 indicate that the field levels of nitrate reductase in some of these rain forest trees are not markedly less than those found when they are grown on nutrient solution containing 5 mM nitrate.

Plants of savanna woodland and fallow are characterized by low nitrogen contents (Nye 1958) and low nitrate reductase activities which increase some 10–20 fold following nitrate fertilization. This suggests savanna soils to be low in nitrogen, particularly in the nitrate form. In many species nitrate reductase activities were on the limits of detection, the level being similar to that of plants grown in nutrient solution containing ammonia as the sole nitrogen source (Orebamjo & Stewart, unpublished results). These results suggest that little nitrate utilization occurs in savanna fallow or woodland and imply that abandonment of farmland in such areas could well be associated with the depletion of soil nitrogen.

The apparent lack of nitrate indicates that alternative sources of nitrogen may be used, although seasonal variation in nitrate availability may complicate such a simplistic view (Birch 1958). Many of the woody species

TABLE 8.11. Nitrate reductase activities in leaf tissue of woody species grown on nitrate.

Species	Nitrate reductase activity (μmol hg^{-1} fresh wt.$^{-1}$) Grown on 5 mM NO_3	Field activity
Avicennia nitida	0.8	<0.1
Cassia hirsuta	7.0	2.5–5.5
Dalbergia ecastaphyllum	2.4	<0.1
Delonix regia	2.0	0.3–0.5
Drepanocarpus lunatus	1.5	<0.1
Guiera senegalensis	1.4	0.1–0.6
Harungana madagascariensis	2.0	0.5–1.5
Newbouldia laevis	5.5	4.1–6.0
Parkia clappertoniana	5.0	<0.1–1.0
Prosopis africana	2.0	0.1–0.3
Rauwolfia vomitoria	3.5	0.2–2.2
Trema guineensis	10.0	1.0–8.6

belong to the Caesalpinaceae and some may be active in nitrogen-fixing association with *Rhizobium* (Allen & Allen 1976). Another possibility which must be considered is molecular nitrogen fixed by rhizosphere bacteria. Many tropical plants, grasses in particular are thought to participate in diazotrophic rhizocoenoses with the bacterium *Azospirillum* (Döbereiner & De-Polli 1980). In Nigeria the roots of 67% of the grass species and 84% of the soils examined were found to have *Azospirillum* associated with them (Döbereiner, Marriel & Nery 1976). The rates of nitrogen fixation are comparable in some species with those observed for nitrate reduction (Table 8.12). Measurements of nitrogen

TABLE 8.12. A comparison of nitrate reduction and nitrogen fixation in some savanna grasses.

Species	Shoot nitrate reduction nmol $h^{-1}g$ fresh wt.$^{-1}$	Root nitrogen fixation* N_2 nmol h^{-1} g fresh wt.$^{-1}$
Andropogon gayanus	100–300	10–180
Hyparrhenia rufa	50–400	20–93
Panicum maximum	100–1800	50
Pennisetum purpureum	100–1050	40

* Data calculated from Day *et al.* (1975).

fixation, whether by *Rhizobium*-root nodule association or diozotrophic rhizocoenoses in savanna woodlands, would be extremely useful in studies of their nitrogen economy.

According to Rice (1974) ammonium ions are the available source in communities where nitrification in inhibited. One community examined here in which ammonia could be the available form of nitrogen is mangrove. Plants of this vegetation type were found to have extremely low levels of nitrate reductase but relatively large nitrogen contents. It would seem likely that in the highly reducing environment of the mangrove (Orebamjo & Njoku 1970) ammonia is the main form of available nitrogen and measurements of soil water ammonia have given concentrations as high as 0.5 μmol cm^{-3}. It is clear, however, from the results in Table 8.11 that some of the woody species have considerable potential to utilize nitrate, and growth studies (Ahmad, Orebamjo & Stewart, unpublished results) confirm that species such as *Avicennia nitida*, *Dalbergia ecastaphyllum*, *Drepanocarpus lunatus* and *Hibiscus tiliaceaus* grow equally well on nitrate or ammonium ions.

Any adaptation to ammonia utilization in the plants of the mangrove does not then appear to be associated with a reduction in capacity to utilize nitrate.

The levels of nitrate reductase present in some of the savanna trees when they are grown on nitrate are also large (see Table 8.11) even though nitrate would not seem to be an important nitrogen source in their natural habitat.

It may be, however, that ammonia is not such an advanced nitrogen source as suggested by Rice. The reduction of nitrate to ammonia requires eight electrons and it is this additional requirement which is the basis for the claim (Rice 1974) that inhibition of soil nitrification and the consequent utilization of ammonia conserves energy. As discussed earlier, ammonia assimilation occurs in the root system and consequently requires the transport of photosynthate to provide carbon skeletons, ATP and reductant to support its assimilation. Reduced nitrogen must, of course, be transported from root to shoot in the xylem. The most frequently found form of transport nitrogen is the amide glutamine; some species, however, employ asparagine or ureides such as allantoin and allantoic acid to transport root-assimilated nitrogen (Pate 1971; Lea & Miflin 1980). Glutamine is the only one of these compounds which can be directly metabolized to glutamate and hence other amino acids. Asparagine and the ureides have to be degraded to ammonia which is then re-assimilated into glutamine (Lea & Miflin 1980). Species transporting asparagine or ureides would appear to assimilate ammonia twice and this may well reduce the apparent energy conservation in ammonia utilization, since for every nitrogen molecule transported as allantoin, for example, 3.25 molecules of ATP must be used in its synthesis and catabolism. A feature of the physiological adaptation of plants to ammonia utilization could therefore be the use of a low energy cost transport form of nitrogen such as glutamine.

It is noteworthy then that several of the woody species examined here transport much of their nitrogen in the form of allantoin and allantoic acid (Table 8.13). Species found in mangrove swamp, savanna woodland, rain

TABLE 8.13. Occurrence of ureide nitrogen in the xylem sap of some woody species. Plants were grown on a nutrient solution containing nitrogen in the form of ammonia ions (0.5 mM).

	μmol cm^{-3}		
Species	Amide N	Amino N	Ureide N
Avicennia nitida	16	0.4	15
Drepanocarpus lunatus	10	0.4	12
Guiera senegalensis	8	0.3	8
Harungana madagascariensis	6	0.2	16
Newbouldia laevis	7	0.2	24
Prosopis africana	12	0.5	16
Trema guineensis	5	0.3	19

forest and pioneer species such as *Trema guineensis* transport ureides when grown on ammonia. The employment of ureides as nitrogen transport compounds, which is costly in energy, suggests that the species of savanna woodland and mangrove are not specifically adapted to ammonia utilization.

Although differences in nitrate availability and utilization have been demonstrated in different West African plant communities there is no evidence that species of communities in which alternative nitrogen sources are utilized lack the potential to assimilate nitrate. Furthermore, none of these species exhibit any features indicative of metabolic adaptation to the utilization of ammonia as nitrogen source.

ACKNOWLEDGEMENTS

The technical assistance of Ann Butterworth, Janice Stewart and Patricia Todd is gratefully acknowledged. T.O. Orebamjo thanks the Royal Society for the award of a Commonwealth Bursary and the University of Lagos for a scientific investigations grant. G.R. Stewart acknowledges financial support from the Royal Society and the Department of Botany, The University of Manchester, in which part of the work was carried out.

REFERENCES

Agboola S.D. (1971) A note on the apparent suppression of nitrification in grassland soils. *Journal of the West African Science Association*, **16**, 171–8.

Ahn P. (1958) Regrowth and swamp vegetation in the western forest areas of Ghana. *Journal of the West African Science Association*, **4**, 163–73.

Allen E.K. & Allen O.N. (1961) The scope of nodulation in the Leguminosae. *Recent Advances in Botany*, **1**, 585–8.

Bate G.C. & Heelas B.V. (1975) Studies on the nitrate nutrition of two indigenous Rhodesian grasses. *Journal of Applied Ecology*, **12**, 941–52.

Beevers L. & Hageman R.H. (1980) Nitrate and nitrite reduction. In *Biochemistry of Plants* (ed. B.J. Miflin), Vol. 5, pp. 116–68. Academic Press, London & New York.

Berlier Y., Dabin B. & Leneuf N. (1956) Comparison physique, chimique et microbiologique entre les sols de forêt et de savanne sur les sables tertiaires de la Basse Côte d'lvoire. *Transactions of the 6th International Congress of Soil Science*, pp. 499–502.

Birch H.F. (1958) The effect of soil drying on humus decomposition and nitrogen availability. *Plant and Soil*, **10**, 9–31.

Boughey A.S., Munro P.E., Meiklejohn J., Strang R.M. & Swift M.J. (1964) Antibiotic reactions between African savanna species. *Nature*, **203**, 1302–3.

Butz R.G. & Jackson W.A. (1977) A mechanism for nitrate transport and reduction. *Phytochemistry*, **16**, 409–17.

Cain J.C. (1952) A comparison of ammonium and nitrate nitrogen for blueberries. *Proceedings of the American Society for Agricultural Science*, **59**, 101–66.

Colgrove M.S. & Roberts A.N. (1956) Growth of the Azalea as influenced by ammonium and nitrate nitrogen. *Proceedings of the American Society for Horticultural Science*, **68,** 522–36.

Day J.M., Harris D., Dart P.J. & Van Berkum P. (1975) The Broadbalk Experiment—an investigation of nitrogen gains from non-symbiotic nitrogen fixation. In *Nitrogen Fixation by Free-living Micro-organisms* (ed. W.D.P. Stewart), pp. 71–84. Cambridge University Press, Cambridge.

Dirr M.A., Barker A.V. & Maynard D.N. (1972) Nitrate reductase activity in the leaves of the high bush blueberry and other plants. *Journal of the American Society for Horticultural Science*, **97**, 329–31.

Döbereiner J. & De-Polli H. (1980) Diazotrophic rhizocoenoses. In *Nitrogen Fixation* (eds W.D.P. Stewart & J.R. Gallon), pp. 301–33. Academic Press, London & New York.

Döbereiner J., Marriel I.E. & Nery M. (1976) Ecological distribution of *Spirillum lipoforum* Berjerinch. *Canadian Journal of Microbiology*, **22**, 1464–73.

Dommergues Y. (1954) Biology of forest soils of central and eastern Madagascar. *Transactions of the 5th International Congress of Soil Science*, **3**, 24–8.

Dommergues Y. (1956) Study of the biology of soils of dry tropical forests and their evolution after clearing. *Transactions of the 6th International Congress of Soil Sciences*, 1956 E, pp. 605–10.

du Rham P. (1970) L'azote dans quelques forêts, savanes et terrains de culture d'Afrique tropicale humide (Côte d'Ivoire). *Veroffentlichungen Geobotanischen Institutes Rübel, Zurich*, **45**, 1–124.

Eden T. (1951) Some agricultural properties of Ceylon montane tea soils. *Journal of Soil Science*, **2**, 43–9.

Gigon A. & Rorison I.H. (1972) The response of some ecologically distinct plant species to nitrate and ammonium-nitrogen. *Journal of Ecology*, **60**, 93–102.

Greenland D.J. (1958) Nitrate fluctuations in tropical soils, *Journal of Agricultural Science*, **50**, 82–90.

Havill D.C., Lee J.A. & Stewart G.R. (1974) Nitrate utilization by species from acidic and calcareous soils. *New Phytologist*, **73**, 1221–31.

Hewitt E.J. (1975) Assimilatory nitrate-nitrite reduction. *Annual Review of Plant Physiology*, **26**, 73–100.

Hutchinson J. & Dalziel J.M. (1954–1972) *Flora of West Tropical Africa*, 2nd edn. Crown Agents, London.

Jackson G. (1964) Notes on West African vegetation–1. Mangrove vegetation at Ikorodu, West Nigeria. *Journal of the West African Science Association*, **9**, 98–110.

Jacquemin H. & Berlier Y. (1956) Évolution du pouvoir nitrificant d'un sol de bosse 6th Côte d'Ivoire sous l'action du climat et de la végétation. *Transactions of the 6th International Congress of Soil Science*, 1956 C, pp. 343–7.

Keay R.W.J. (1959) *Outline of Nigerian Vegetation*, 3rd edn. Government printer, Nigeria.

Keay R.W.J. (1962) Natural vegetation. In *Soils and Land Use in Central Western Nigeria* (eds R.W.J. Keay & F.N. Hepper), pp. 170–6. Ministry of Agriculture and Natural Resources, Ibadan.

Lea P.J. & Miflin B.J. (1980) Transport and metabolism of asparagine and other nitrogen compounds within the plant. In *Biochemistry of Plants* (ed. B.J. Miflin), Vol. 5, pp. 569–608. Academic Press, London & New York.

Lee J.A. & Stewart G.R. (1978) Ecological aspects of nitrogen assimilation. *Advances in Botanical Research* **6**, 1–43.

McCullough H. (1967) The determination of ammonia in whole blood by a direct colorimetric method. *Clinica Chimic Acta*, **17**, 297–304.

McFee W.W. & Stone E.L. (1968) Ammonium and nitrate as nitrogen sources for *Pinus radiata* and *Picea glauca. Proceedings of the Soil Science Society of America*, **32**, 879–84.

Meiklejohn J. (1968) Numbers of nitrifying bacteria in some Rhodesian soils under natural grass and improved pasture. *Journal of Applied Ecology*, **5**, 291–300.

Miflin B.J. & Lea P.J. (1980) Ammonia assimilation. In *Biochemistry of Plants* (ed. B.J. Miflin), Vol. 5, pp. 169–202. Academic Press, London & New York.

Mills W.R. (1953) Nitrate accumulation in Uganda soils. *East Africa Agricultural Journal*, **19**, 53–4.

Nye P.H. (1958) The relative importance of fallows and soils in storing plant nutrients in Ghana. *Journal of the West African Science Association*, **4**, 31–49.

Nye P.H. & Greenland D.J. (1960) *The Soil under Shifting Cultivation*. Technical Communication No. 51, Commonwealth Bureau of Soils, Harpenden.

Orebamjo T.O. & Njoku E. (1970) Ecological notes on the vegetation of the Lagos University site at the time of acquisition. *Journal of the West African Science Association*, **15**, 35–55.

Pate J.S. (1971) Movement of nitrogenous solutes in plants. In *Nitrogen-15 in Soil Plant Studies*, pp. 165–87. International Atomic Energy Agency, Vienna.

Pharis R.P., Barnes R.L. & Naylor A.W. (1964) Effects of nitrogen level, calcium level and nitrogen source upon growth and composition of *Pinus taeda* L. *Physiologia Plantarum*, **17**, 560–72.

Purchase B.S. (1974) Evaluation of the claim that grass root exudates inhibit nitrification. *Plant and Soil*, **41**, 527–39.

Raven J.A. & Smith F.A. (1976) Nitrogen assimilation and transport in vascular land plants in relation to intracellular pH regulation. *New Phytologist*, **76**, 415–31.

Raven J.A., Smith S.E. & Smith F.A. (1978) Ammonia assimilation and the role of mychorrhizas in climax communities in Scotland. *Transactions of the Botanical Society of Edinburgh*, **43**, 47–55.

Rhodes D., Sims A.P. & Folkes B.F. (1980) Pathway of ammonia assimilation in illuminated *Lemna minor*. *Phytochemistry*, **19**, 357–65.

Rice E.L. (1974) *Allelopathy*. Academic Press, London & New York.

Robinson J.B. (1963) Nitrification in a New Zealand grassland soil. *Plant and Soil*, **19**, 173–83.

Roux E. (1954) The nitrogen sensitivity of *Eragrostis curvula* and *Trachypogon plumosus* in relation to grassland succession. *South African Journal of Science*, **50**, 173–6.

Stewart G.R. & Orebamjo T.O. (1980) Some unusual characteristics of nitrate reduction in *Erythrina senegalensis* DC. *New Phytologist*, **83**, 311–19.

Stewart G.R., Lee J.A. & Orebamjo T.O. (1973) Nitrogen metabolism of halophytes II. Nitrate availability and utilization. *New Phytologist*, **72**, 539–46.

Stewart G.R., Lee, J.A., Havill D.C. & Orebamjo T.O. (1974) Ecological aspects of nitrogen metabolism. In *Mechanisms of Regulation of Plant Growth* (eds R.L. Bieleski, A.R. Ferguson & M.M. Creswell), pp. 41–7. The Royal Society of New Zealand, Wellington.

Stiven G. (1952) Production of antibiotic substances by roots of a grass *Trachypogon plumosus* (H.B.K. Nees) and *Pentasia varibilis* (E. May.) Harv. Rubiaceae. *Nature*, **170**, 712–13.

Theron J.J. (1951) The influence of plants on the mineralization of nitrogen and the maintenance of organic matter in soil. *Journal of Agricultural Science*, **41**, 289–96.

Townsend L.R. (1970) Effect of form of nitrogen and pH on nitrate reductase activity in low bush blueberry leaves and roots. *Canadian Journal of Plant Science*, **50**, 603–5.

Tyler G. (1967) On the effect of phosphorus and nitrogen applied to Baltic shore meadow vegetation. *Botaniska Notiser*, **120**, 433–48.

Vogel G.D. & Van der Drift C. (1970) Differential analyses of glyoxalate derivatives. *Analytical Biochemistry*, **33**, 143–57.

Wiltshire G.H. (1973) Response of grasses to nitrogen source. *Journal of Applied Ecology*, **10**, 429–35.

9. NITROGEN SOURCE, TEMPERATURE AND THE GROWTH OF HERBACEOUS PLANTS

I. H. RORISON[1], J. H. PETERKIN[2*] AND
D. T. CLARKSON[3]
[1,2]*Unit of Comparative Plant Ecology (NERC) Department of Botany, The University, Sheffield S10 2TN*, [3]*ARC Letcombe Laboratory, Wantage, OX12 9JT*

SUMMARY

The form of nitrogen available to plants is affected by soil reaction and temperature. Uptake of nitrogen by plants is also affected by temperature, and the responses of species and populations to the form of available nitrogen indicate some adaptation. Effects of temperature on nitrogen supply, on growth, and on uptake and utilization of nitrogen by the plant are considered with particular reference to two species. They are represented by populations of *Arrhenatherum elatius* (L.)† Beav. ex J. & C. Presl and *Festuca ovina* L. which coexist in an unmanaged infertile limestone grassland and show marked differences in their potential growth rate and demand for nutrients and in their phenology. *F. ovina* was shown to be capable of greater absolute growth than *A. elatius* over one year outdoors despite the higher potential relative growth rate of *A. elatius*. This was due to the maintenance of a higher level of activity in *F. ovina* during the winter. In laboratory experiments *A. elatius* showed a higher growth rate and demand for nitrogen than *F. ovina* at near optimal temperatures. At suboptimal temperature, however, *F. ovina* grew faster and took up nitrogen at a higher rate than did *A. elatius*. Proportions taken up of ammonium to nitrate also varied with temperature. The possible importance of these responses in relation to seasonal growth and storage and utilization of nitrogen is discussed.

The observed differences in response of the two species to temperature and in seasonal patterns of growth and nitrogen uptake are considered as part of the explanation of their coexistence in the field. It is suggested that *A. elatius* has as its prime competitive attribute rapid spring and summer growth while for *F. ovina* it is continuous winter growth that is vital.

* Present address: Department of Botany, Birkbeck College, Malet Street, London WC1E 7HX.
† Nomenclature follows Clapham, Tutin & Warburg (1962).

INTRODUCTION

Nitrogen together with phosphorus is the nutrient most limiting to growth and variable in supply in the uncultivated soils of the UK. Plant species may differ in their response to nitrogen when it is supplied in either the nitrate or ammonium form and the availability of the two forms varies with soil reaction (Hewitt 1966; Lee & Stewart 1978). The availability (Williams 1969) and rate of uptake of these ions by plants also varies with shifts in temperature (Williams & Vlamis 1962; Moraghan & Porter 1975; McCown 1978; Clarkson & Warner 1979).

It is therefore of interest to consider, with respect to nitrogen, the relationships between the factors external to the plant which affect availability and the internal factors which affect growth, uptake, utilization and compartmentation by the plant. Of particular interest is the degree to which the distribution and survival of different species may be affected by these factors and what contribution both field and laboratory studies can make to our understanding of the relationships.

In what follows no mention is made of the contribution of the endomycorrhizal associations of which the herbaceous plants under discussion form a part (Read, Koucheki & Hodgson 1976). This is because there have been no studies that demonstrate unequivocally a primary role of vesicular-arbuscular mycorrhizas in augmenting nitrogen uptake by these plants (Bowen & Smith 1981). When an increase in nitrogen has been measured it is usually accompanied by an increase in the uptake of phosphorus and of other nutrient elements (Smith 1980) to which the plant might also respond.

FACTORS EXTERNAL TO THE PLANT

Nitrogen source and plant distribution

Source

In general, there is a preponderance of ammonium in acidic soils and of nitrate elsewhere (Ellenberg 1964). However, there are reports (Runge 1974; van de Dijk & Troelstra 1980) of heterotrophic nitrification taking place in very acidic soils, which could account for the small amounts of nitrate detected.

Levels of total nitrogen, ammonium and nitrate decrease with depth in the profile according to texture and composition. Nitrification is also inhibited in waterlogged and cold soils (Haynes & Goh 1978).

Plant distribution

Groups of plant species may be considered in relation to nitrogen source and soil reaction. Calcifuge plants, restricted largely to acidic soils with predominantly ammonium, have been shown to be tolerant of ammonium but are also (with the exception of some ericaceous species which exhibit minimal nitrate reductase activity (Havill, Lee & Stewart 1974)) able to utilize nitrate.

Calcicole plants are found predominantly in circum-neutral soils, suffer ammonium toxicity under acidic conditions and only thrive in the presence of adequate nitrate (Gigon & Rorison 1972).

In between these two extremes there are species and populations showing intermediate tolerances, e.g. *Holcus lanatus* L. (McGrath 1979).

Seasonal fluctuations in nitrogen supply

Availability in the soil

Nitrogen supply in soils can be heterogeneously distributed and can vary with time (Davy & Taylor 1974). Spring 'flushes' are widely reported in agricultural literature but the timing and intensity of them varies from year to year.

Measurements from brown podzolic soil horizons (Gupta & Rorison 1975) and a rendzina (Fig. 9.1) confirm similar occurrences in uncultivated soils. In the rendzina the level of nitrate falls in the presence of plants, indicating that it is the major source of nitrogen throughout the growing season. In contrast, the ammonium curve fluctuates, falling markedly during the growing season in the presence or absence of plants. This may be largely accounted for by a high rate of nitrification. The accelerated fall in winter indicates ammonium absorption by *Arrhenatherum elatius*.

Effects of temperature on uptake

Some plants are known to take up relatively more nitrogen as ammonium than as nitrate at low temperatures (e.g. *Lactuca sativa* L. at 8°C or below (Frota & Tucker 1972)). Their translocation pattern is also affected.

Seasonal patterns in root and shoot temperatures

In order to test plants under realistic conditions it is important to know the range of root and shoot temperatures which are experienced in the field throughout the year. Potentially, there are large and variable diurnal and

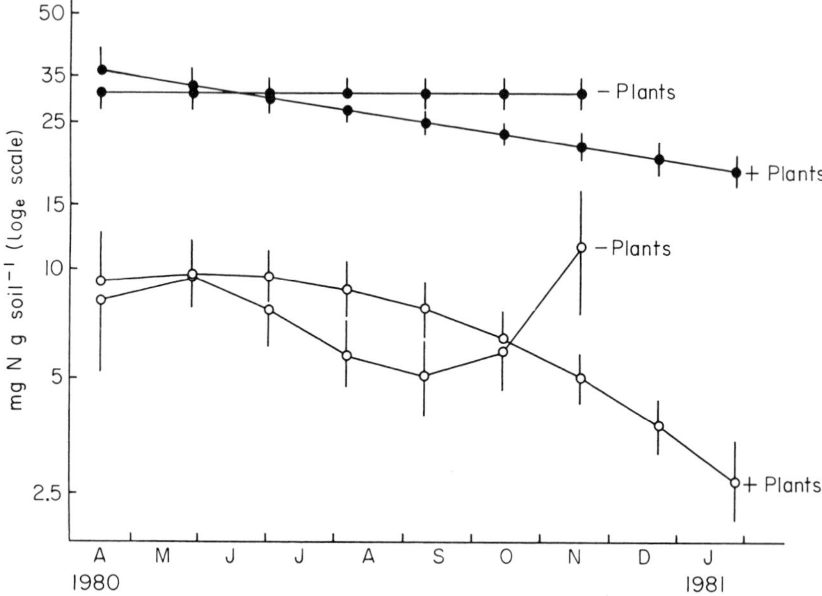

FIG. 9.1. Nitrate-N (●) and ammonium-N (○) extractable (in 0.01 M $CuSO_4$) from a rendzina rendzina soil either in the presence or the absence of seedlings of *Arrhenatherum elatius* during the period April 1980 to January 1981. Note the log scale on the ordinate. Vertical bars represent 95% confidence limits.

seasonal fluctuations of temperature in the field which vary in amplitude according to the distance above and below ground where the measurements are made (Rorison 1981). The aerial parts of plants may reach temperatures which are either considerably above or below those of the surrounding air depending on radiation load, life-form, leafiness, rate of air movement and transpiration rate (Gates 1968). Root temperatures rarely coincide with those of the shoot but it should be remembered that temperatures at or near the soil surface are much closer to those of the air and are liable to fluctuate more than those deep in the soil (Fig. 9.2). In this context it is important to consider the spatial distribution and function of root systems (Caldwell 1979). At one extreme are herbaceous species with adventitious roots, the majority of whose 'feeders' are relatively close to the soil surface (particularly in podzols and other soils with distinct horizons). At the other extreme are the tap-rooted species which reach and exploit water and nutrients to a greater depth, particularly when conditions at the soil surface deteriorate (Newbould 1969).

Thus the form and availability of nitrogen can be shown to vary with time

and in space in a range of soil types. In relating this knowledge to the distribution of plants it is important to know whether:
(a) plants grow and take up nitrogen throughout the year;
(b) given the choice, they exhibit a preference for nitrate or ammonium;
(c) any preference is temperature dependent;
(d) when there is growth and/or uptake and utilization in winter, the response of different species indicates a possible strategy in their survival.

These points will now be discussed in the light of experimental evidence.

RESPONSE OF THE PLANT

Seasonal patterns of growth and development

Information on phenology is based largely upon measurements of above-ground plant material presented on a biomass per unit area basis. It shows that there are species exhibiting quite clearly defined and pronounced seasonal peaks of growth, e.g. *A. elatius*, while in others such as *F. ovina* L. and *Deschampsia flexuosa* (L.) Trin., the peaks are less pronounced (Al-Mufti *et al.* 1977). There are also winter annual species, e.g. *Catapodium rigidum* (L.) C. E. Hubbard (Clark 1974) and *Cardamine hirsuta* L. (Ratcliffe 1961) persisting through winter and flowering in spring/early summer, and annuals, e.g. *Poa annua* L., maturing at any time of year, usually in summer.

A detailed investigation of the phenology of a number of grasses was carried out in order to provide information about changes in individual plant components.

In the present paper we give results for two perennials of contrasted ecology, *A. elatius* and *F. ovina*, which differ in phenology, potential growth rate (Grime & Hunt 1975) and nutrient demand.

Seeds were sown in soil in individual containers to grow outdoors for one year (April 1980–1981). Harvests were taken at five-weekly intervals and dead and living leaves and tillers counted. Dry weights of plant parts, including roots, were determined. In both species, growth, as measured by length of individually tagged leaves, continued throughout winter.

During November, December and January, there was a large increase in the number of dead leaves of *A. elatius* and few new leaves were produced. With *F. ovina* the number of dead leaves, relative to living ones, was very much smaller, due to both the addition of new leaves and to their greater longevity (Sydes 1980). In *A. elatius* there was a proportional increase in root dry weight between October and January and a drop in dry weight of above-ground living material (Fig. 9.3). The relative amount of root also increased in *F. ovina* but the effect was less marked.

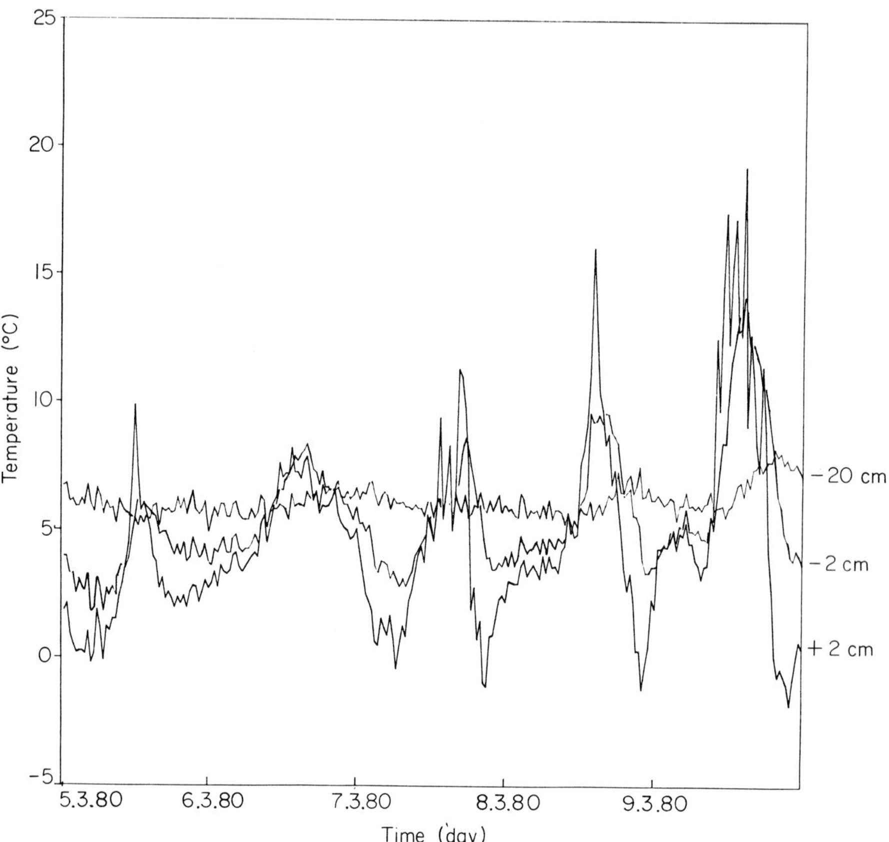

FIG. 9.2. Temperature (°C) 2 cm above the soil surface and 2 and 20 cm below on five consecutive days of increasing solar radiation 5th–9th March 1980. The soil is a rendzina from the same source as that referred to in Figs 9.1 and 9.3.

The evidence shows that *A. elatius* and *F. ovina* differ markedly in the extent to which they grow at different times of year. During the summer (April to October) of 1980 the growth (increase in dry weight) of *A. elatius* was appreciably faster than that of *F. ovina*. During the winter (November to March), however, *A. elatius* stopped growing and lost dry weight of both roots and living leaves. In contrast, *F. ovina* increased in dry weight by continued production of leaves and roots.

These results led to the prediction that temperature, rather than short days, was the important factor in determining the observed phenological differences between the two species. In order to test this prediction a series of experiments was carried out to investigate the physiological relationships

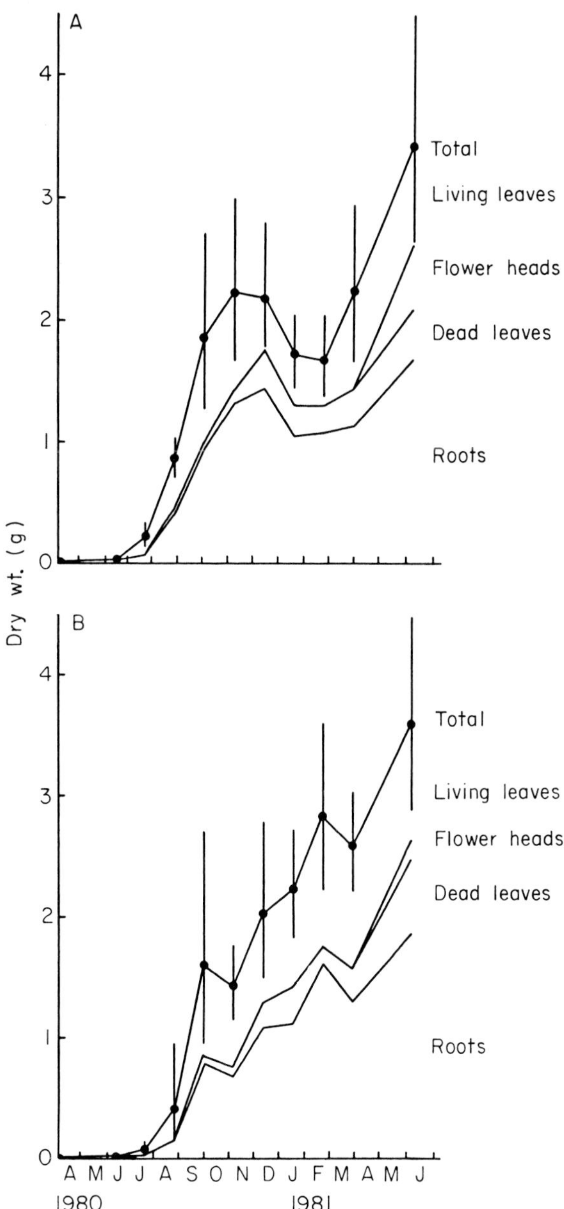

FIG. 9.3. Dry weight of *Arrhenatherum elatius* (A) and *Festuca ovina* (B) grown outdoors from seed for one year showing contribution of parts to total plant weight. Vertical lines indicate 95% confidence limits (after logarithmic transformation and back transformation to arithmetic values).

between growth, nitrogen source, nitrogen uptake and utilization, and temperature with particular reference to species of contrasted phenology.

Basic physiology

There is a wealth of literature on the effects of temperature on various aspects of plant activity, including growth and nutrient uptake (Went 1953; Cooper 1973; Berry & Björkmann 1980).

At low temperatures both nitrogen uptake and growth are reduced (Ingestad 1979) but this tendency is greater for nitrate than for ammonium if the plants are presented with a mixed supply. Integrated approaches to the effects of temperature upon growth *and* nutrient uptake in native species are rare, and the answers to the following questions are still awaited:

1. To what extent can the processes of growth and nutrient absorption be separated temporally or proceed at different rates?
2. Can plants absorb nitrogen in excess of growth requirements, thus building up an internal 'store' to be utilized when growth conditions are more favourable?

These questions are of particular interest in the comparative study of species which survive under suboptimal conditions as, for example, on infertile upland soils. Plants able to maintain growth and to accumulate stores of nutrients in excess of immediate requirements at low temperatures would have a possible advantage over those which are unable to do either or both of these things.

In the knowledge that species vary in phenology and seasonal patterns of dry matter partitioning, one might predict that this variation would be reflected in the equilibrated response of different species to a drop in temperature.

Species responses to a drop in temperature

Nitrogen supplied solely as nitrate

A number of herbaceous species were grown in solution culture for four weeks in a controlled environment at 20°C day, 15°C night followed by four weeks at 10/5°C (a simulated 'autumn'). Weekly harvests were taken and dry weights and nitrogen content of roots and shoots determined. Fig. 9.4a shows that the results obtained for *A. elatius* and *F. ovina* follow the same trend as the curves obtained in the previous experiment (see Fig. 9.3).

For *A. elatius* the relative growth rate (RGR) (Fig. 9.4b) was much reduced at the lower temperature, root/shoot dry weight ratio (Fig. 9.4c) increased, and specific absorption rate (SAR-N) (Fig. 9.5a) was reduced.

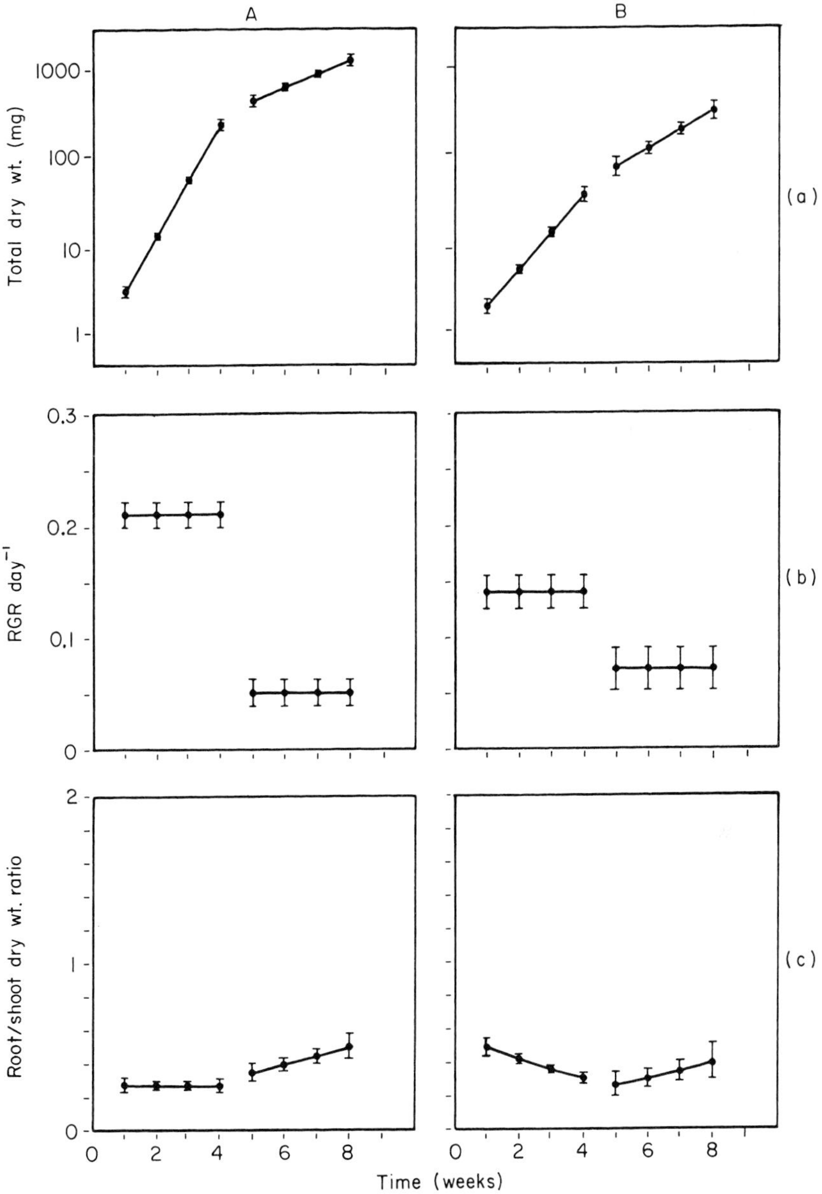

FIG. 9.4. (a) Total dry weight (log$_e$ scale), (b) relative growth rate (RGR) and (c) root/shoot dry weight ratio of *Arrhenatherum elatius* (A) and *Festuca ovina* (B) grown for four weeks at 20°C day/15°C night and a further four weeks at 10°C/5°C. Curves were fitted by the Hunt & Parsons (1974) growth analysis computer program. Vertical lines indicate 95% confidence limits.

Reduction in shoot nitrogen concentration (Fig. 9.5c) may have been due to decreased input from roots or retranslocation of labile nitrogen from the shoots to the roots.

For *F. ovina* the RGR was also reduced by the drop in temperature, though much less than in *A. elatius*; proportionally more root was produced at the lower temperature. The SAR-N was reduced much less than in *A. elatius*. These changes led to increased tissue concentration of nitrogen, especially in the root (Fig. 9.5b).

Therefore, as predicted from phenological measurements (see p. 194), *F. ovina* was generally less adversely affected by the drop in temperature than *A. elatius*.

Ammonium and nitrate supplied together

In the experiment just described, nitrogen was supplied solely as nitrate because of the known susceptibility to ammonium sources of the calcicole species which were also being examined.

Amounts of ammonium and nitrate available to plants in soil vary with pH and season, but it is unusual for either nitrate or ammonium to be the exclusive nitrogen source in soils. Supplying nitrogen solely as nitrate as above or solely as ammonium (which is relevant to only a limited number of very acidic soils) is therefore somewhat unrealistic. Proportionally more of the available nitrogen in soils in autumn and winter is in the ammonium form (Williams 1969; Davy & Taylor 1974). It has also been shown that *Lolium multiflorum* Lam. takes up an increasing amount of ammonium relative to nitrate as the root temperature is lowered from 20°C to 3°C (Clarkson & Warner 1979).

In winter the mean air temperature tends to drop below mean soil temperature (Rorison & Sutton 1976), the soil being more buffered against seasonal changes. Thus, in general, the roots may experience a higher temperature than the shoots during autumn and winter.

With these factors in mind a series of experiments was carried out to investigate the effects of temperature on the uptake of nitrate and ammonium from a solution containing ammonium nitrate. The effect of decreased shoot temperature relative to root temperature was also investigated. Plants were raised in solution culture in a glasshouse at *c.* 20/15°C and later transferred to growth cabinets at 15°C. After seven days plants were transferred to one of a series of lower temperatures in which root temperature was either equal to, or more than, shoot temperature (Table 9.1). When required, root temperatures were maintained higher than shoot temperatures by the use of heated water baths.

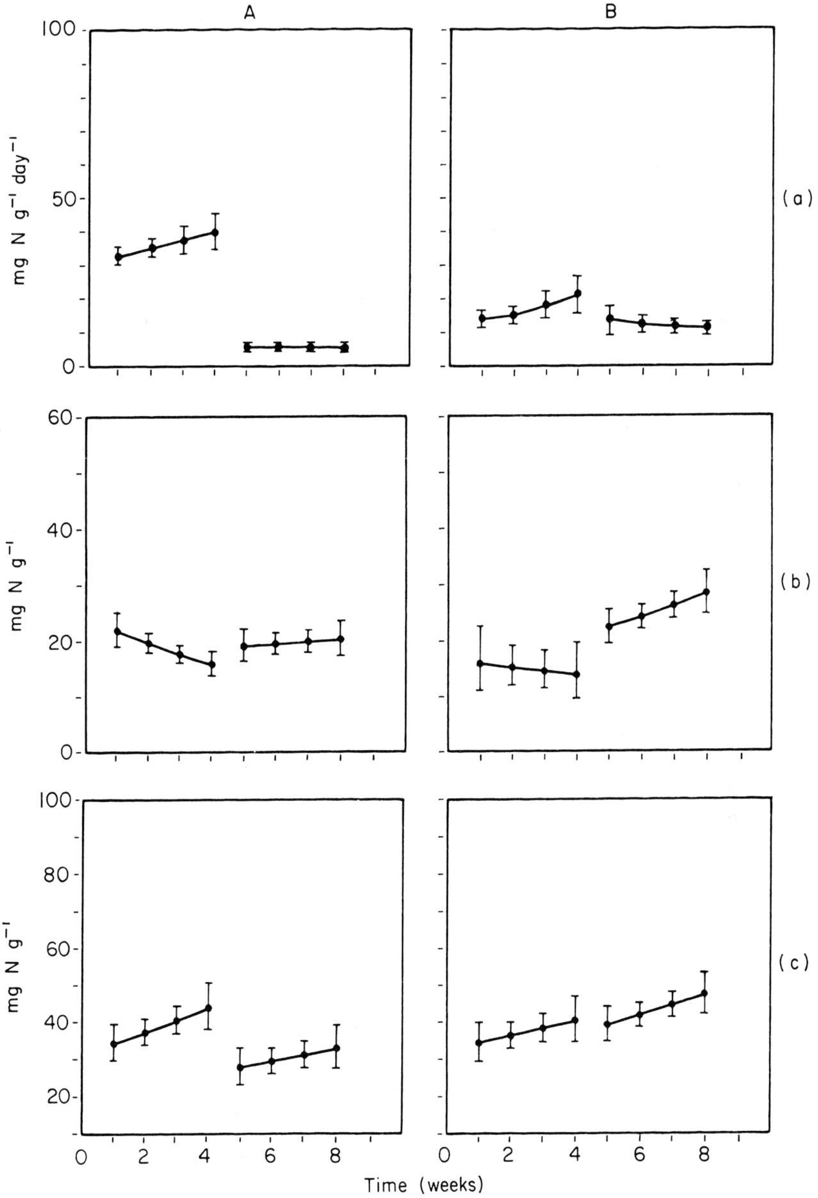

FIG. 9.5. (a) Specific absorption rate for nitrogen (nitrate-N), (b) root and (c) shoot concentration of nitrogen of (A) *Arrhenatherum elatius* and (B) *Festuca ovina* grown for four weeks at 20°C day/15°C night and a further four weeks at 10°C/5°C. Curves were fitted by the Hunt & Parsons (1974) growth analysis computer program. Vertical lines indicate 95% confidence limits.

For the last 24 hours of the experiment plants were placed in solutions where either the ammonium or nitrate was labelled with ^{15}N. They were then harvested and root and shoot dry weight, total nitrogen content and atoms % ^{15}N determined. It was thus possible to calculate how much ammonium and nitrate was taken up by the plants from a solution containing both forms.

Growth. Predictably *A. elatius* showed a significant increase in final dry weight with increasing temperature (Table 9.2). The root/shoot ratio increased significantly with reduced root temperature. Temperature did not significantly affect the final dry weight of *F. ovina*. Lower root temperature

TABLE 9.1. Experimental procedure, Days 1 to 25 after initial seedling establishment (ammonium and nitrate supplied together).

Time (Days)	Temperature					Nitrogen source
1–6	Acclimatization to experimental temperatures, attained by step-wise drops from 15°C					NH_4NO_3
	Treatments					
7–24		1	2	3	4	NH_4NO_3
	Root	5°	10°	10°	15°C	
	Shoot	5°	5°	10°	10°C	
25	(Same treatments as above)					$^{15}NH_4+NO_3$ *or* $NH_4+{}^{15}NO_3$

TABLE 9.2. Total plant dry weight and root/shoot dry weight ratio at Day 25. Values not linked (└─┘) are significantly different at $P=0.05$. Species were grown sequentially such that at Day 25 plants of *Arrhenatherum elatius* were *c.* 9 weeks old and those of *Festuca ovina, c.* 15 weeks old.

	Treatments 1	2	3	4
Root	5°	10°	10°	15°C
Shoot	5°	5°	10°	10°C
A. elatius				
Log_e final dry wt. (mg)	0.46	0.67	0.88	1.10
R/S ratio	1.00	0.82	0.82	0.52
F. ovina				
Log_e final dry wt. (mg)	2.74	2.70	2.72	2.65
R/S ratio	0.36	0.26	0.32	0.30

tended to result in a higher root/shoot ratio; that in Treatment 1 was significantly higher than that in Treatment 2.

Uptake. SAR-N was calculated using the formula of Welbank (1962) after Williams (1946). As expected, higher temperatures, especially of the root, resulted in higher SAR-N (Table 9.3).

In *A. elatius* the SAR for both ammonium and nitrate was significantly depressed at the three lowest temperatures. The ratio of nitrate to ammonium uptake was higher at higher temperatures as reported for lettuce (Frota &

TABLE 9.3. Specific absorption rate (mg N g root dry wt.$^{-1}$ day^{-1}) for ammonium and nitrate during Day 25. Values not linked (└──┘) are significantly different at $P=0.05$.

		Treatments			
		1	2	3	4
	Root	5°	10°	10°	15°C
	Shoot	5°	5°	10°	10°C
A. elatius					
mg N g^{-1} day^{-1}	NH_4-N	2.29	2.43	2.36	4.86
	NO_3-N	1.02	0.92	1.37	3.37
	NO_3:NH_4-N	0.44	0.38	0.58	0.69
F. ovina					
mg N g^{-1} day^{-1}	NH_4-N	4.96	7.37	4.45	8.90
	NO_3-N	3.88	4.46	5.21	3.41
	NO_3:NH_4-N	0.78	0.61	1.17	0.38

Tucker 1972) and ryegrass (Clarkson & Warner 1979). This was not so in the case of *F. ovina*.

In *F. ovina*, the SAR for ammonium was higher at higher root temperatures and was significantly higher than that of *A. elatius* at all temperatures. The SAR for nitrate was not significantly affected by temperature but was significantly higher than that in *A. elatius* in all but the highest temperature treatment.

Translocation. In both *A. elatius* and *F. ovina* translocation to shoots of nitrogen taken up as ammonium showed a marked dependence on root temperature (Table 9.4). Translocation to shoots of nitrogen taken up as nitrate was dependent on shoot temperature in *A. elatius* but there was little response in *F. ovina*.

TABLE 9.4. Percentage of nitrogen taken up as ammonium or nitrate translocated to shoots. Mean values of four replicates.

		Treatments 1	2	3	4
	Root	5°	10°	10°	15°C
A. elatius	Shoot	5°	5°	10°	10°C
% Translocation	NH_4-N	18	36	35	56
	NO_3-N	36	31	47	42
F. ovina					
% Translocation	NH_4-N	26	55	48	60
	NO_3-N	29	43	44	44

Tissue concentration of ammonium and nitrate. The ammonium concentration in roots of *A. elatius* was markedly increased at lower root temperatures (Table 9.5). The ammonium concentration in shoots also increased at lower temperatures, but less markedly. The ammonium concentration in roots and shoots of *F. ovina* was apparently insensitive to temperature, and values obtained were lower than those for *A. elatius* in all treatments.

The nitrate concentration in roots of *A. elatius* showed a marked decrease

TABLE 9.5. Tissue concentration (mg N g dry wt.$^{-1}$) of ammonium and nitrate at Day 25. Mean values of replicated analysis of bulked samples.

		Treatments 1	2	3	4
	Root	5°	10°	10°	15°C
	Shoot	5°	5°	10°	10°C
A. elatius					
NO_3-N, mg g^{-1}	Root	3.2	4.2	3.3	6.9
	Shoot	2.1	1.6	1.8	2.7
F. ovina					
NO_3-N, mg g^{-1}	Root	3.0	3.7	3.3	3.5
	Shoot	0.3	0.9	0.5	1.0
A. elatius					
NH_4-N, mg g^{-1}	Root	4.8	3.2	3.4	1.6
	Shoot	1.9	2.0	1.7	1.2
F. ovina					
NH_4-N, mg g^{-1}	Root	1.4	1.2	1.4	1.5
	Shoot	0.8	0.9	1.0	0.9

at lower root temperatures, particularly compared with Treatment 4. A higher value for shoot nitrate concentration was recorded at the highest temperature (Treatment 4). The nitrate concentration in shoots of *F. ovina* was dependent on root temperature; that in the roots was also dependent on root temperature but the effect was much less marked. The nitrate concentration in roots and shoots of *F. ovina* was lower than in *A. elatius* in all treatments.

DISCUSSION

Arrhenatherum elatius absorbed relatively more ammonium than nitrate at low temperatures. The previous experiment illustrated the marked effect of low temperature on the uptake of nitrate when only nitrate was supplied. This leads to the suggestion that at low temperatures *A. elatius* was able to compensate in part for the reduced uptake of nitrate by maintaining a proportionately greater uptake of ammonium when both were supplied. In the field this would be of advantage to the plant when at low temperatures (autumn and winter) proportionally more of the available nitrogen is in the ammonium form. It should be noted, however, that lower temperatures resulted in increased concentrations of free ammonium in the roots and to some extent in the shoots because assimilation did not keep pace with uptake. Thus of the ammonium taken up at low temperatures a fraction was not assimilated and could have a deleterious effect on the plant.

Absorption rates were measured on plants in solutions containing 4 mM ammonium nitrate. It is likely that this concentration was saturating both the ammonium and nitrate uptake systems. Clarkson & Warner (1979) and Clarkson (1981, unpublished) have shown that in ryegrass the preference for ammonium relative to nitrate at 3°C may be much greater at non-saturating concentrations closer to the K_m of the uptake systems (e.g. 10–50 μM ammonium nitrate).

Predictably, *F. ovina* was able to maintain rates of absorption of both ammonium and nitrate in excess of those of *A. elatius* at lower temperatures. *F. ovina* was able to maintain assimilation of both ammonium and nitrate at low temperatures; values for tissue concentration of unreduced, unassimilated nitrate and ammonium being lower than in *A. elatius* in all treatments and less dependent on temperature. Higher concentrations of ammonium in *A. elatius* were considered to be due to a direct effect of low temperature upon the nitrogen assimilation pathway. Indirect effects via a reduction in the supply of carbon skeletons for incorporation of ammonium into amino acids are unlikely to limit nitrogen assimilation at lower temperatures since it is reported that a range of species maintain photosynthesis to a greater extent than growth (dry matter production) at low temperatures. This leads to increased concentrations of soluble carbohydrate (Brown & Blaser 1970; Regehr & Bazzaz 1976; Barlow, Boersma & Young 1976) and free amino-nitrogen (Chowdhury & Zubriski 1973; Patterson, Grunes & Lathwell 1972) in the tissues. Preliminary results from analysis of *A. elatius* for soluble amino-nitrogen confirm this; at low temperatures there was an increase in the concentration of amino acids in both roots and shoots.

Lower concentration of nitrate in roots of *A. elatius* at lower temperatures was attributed to a reduced flux into the root relative to nitrate reductase activity. Higher shoot concentrations of nitrate in *F. ovina* when shoot temperature was less than that of the root were due to a higher input from roots relative to shoot nitrate reductase activity.

Values quoted for percentage translocation refer only to the amount of ^{15}N which was absorbed in the ammonium or nitrate form and not to the form in which the nitrogen was translocated. Much of the ammonium entering the root must be assimilated to amide or amino acid for reasons of pH control in the cells (Raven & Smith 1976). The reduced translocation to shoots of nitrogen taken up as ammonium at low temperatures was due to one or a combination of the following:

1. Reduced capacity of the root to assimilate ammonium and a lack of transport of ammonium to shoots.

Root concentration of ammonium in *A. elatius* was increased at low temperatures indicating that the rate of assimilation was reduced relative to the rate of uptake. However, as previously mentioned, the concentration of free amino acids in the roots was higher at lower temperatures indicating a supply of assimilated nitrogen available for transport. Thus reduced capacity of the roots to assimilate ammonium at low temperatures was probably not directly limiting translocation to shoots of ^{15}N absorbed in the ammonium form.

2. The requirement for relatively greater quantities of assimilate in the root to permit the increased root growth relative to shoot growth which occurred at low temperatures.

In both species, particularly *A. elatius*, root temperature had a marked effect on root/shoot dry weight ratio and on the percentage translocation to shoots of ^{15}N absorbed as ammonium.

3. Reduced capacity to transport ^{15}N ammonium and more importantly the assimilated ^{15}N from roots to shoots at lower root temperatures.

Nitrate entering the roots may be transported to shoots via the xylem directly as nitrate, enter a vacuolar pool in the roots and thus become at least temporarily unavailable for assimilation and/or be reduced immediately to ammonium and thence to amide or amino nitrogen. In *A. elatius*, lower shoot temperature reduced the percentage translocation to shoots of nitrogen absorbed as nitrate. This may represent the following:

1. Reduced capacity to transport nitrate from roots to shoots.
2. Increased amounts of nitrate entering root vacuoles and thus not being immediately available for transport.
3. An increase in the proportion of nitrate entering the root which is reduced to amino acids in the roots and is then affected by lower temperatures in the same manner as described above.

Osborne & Whittington (1981) state that in *Agrostis* spp. the roots contribute only about 6–7% of the total nitrate reductase activity *in vivo*. This would suggest that nitrate reduction in roots is of only minor importance to the nitrogen economy of the plant. However, Deane-Drummond, Clarkson & Johnson (1980) have shown that lowered root temperatures in *Hordeum vulgare* L. may markedly increase the contribution of the root to total plant nitrate reductase activity especially when activity during periods of darkness is taken into account. More detailed studies on the possible significance of the effects of temperature on sites of nitrate reduction are presented elsewhere in this volume (Clarkson & Deane-Drummond, p. 211, this volume).

CONCLUSIONS

The aim of this paper has been to aid understanding of the survival of some native herbaceous species by explaining observed ecological and phenological differences in terms of quantifiable physiological responses.

The problem has been considered largely in relation to populations of two species of grass which grow adjacent to each other in a shallow rendzina soil on the south-facing slope of a dry ungrazed limestone dale in Derbyshire. Although total nutrient resources are high (>70% organic nitrogen), concentrations of available nutrients are low and are released slowly and in variable amounts. Nitrogen is of particular interest in this regard for reasons given earlier.

The juxtaposition of the two species in the field is also noteworthy. *Arrhenatherum elatius* is normally associated with deep, base-rich soils but is also capable of colonizing limestone screes (Grime & Lloyd 1973). *F. ovina* is more widespread and may also be found on dry eroded slopes. *A. elatius* is potentially tall, fast-growing, deep-rooting and susceptible to grazing pressure (Pfitzenmeyer 1962). *F. ovina* has a moderate growth rate with limited extension both above and below ground, but is tolerant of heavy grazing (Hubbard 1968).

Some of these attributes were confirmed experimentally, with *A. elatius* growing faster and taking up more nitrogen under favourable conditions of nitrogen supply and temperature than *F. ovina*. Under unfavourable conditions, as with a drop in temperature, the roles were reversed; *F. ovina* maintained its rate of growth and assimilation of nitrogen to a greater degree than *A. elatius*. It did so regardless of the form of nitrogen supplied, while in *A. elatius* the proportion of ammonium to nitrate taken up was increased at low temperatures. In *F. ovina* there was an increase in tissue nitrogen concentration at low temperature (see Fig. 9.5) with continuing growth which indicated a build-up of some form of nitrogen at low temperature. In *A. elatius* a decreased tissue nitrogen concentration at low temperature indicated a lack of storage. In both species, therefore, there was no evidence of temporal separation of growth from uptake but an indication that in *F. ovina* a build-up of nitrogen at low temperature occurred which might be utilized as temperatures rose (as in spring). This was not the case with *A. elatius*. These detailed laboratory findings were confirmed in the field as the plants grew through summer to winter conditions (see Fig. 9.3).

Under the infertile field conditions from which the species were taken, they both survive despite their differential growth potential. The occurrence of *A. elatius* with *F. ovina* together in the unmanaged limestone grassland may be due, at least partly, to the advantage which *F. ovina* is able to gain over *A. elatius* in terms of nitrogen-uptake and growth during winter conditions which severely inhibit the activity of *A. elatius*. This is to the advantage of *F. ovina*, particularly in the seedling phase, since both species are capable of germinating in the late summer. The capacity for winter growth could mean that *F. ovina* is better established than *A. elatius* by the following spring and that absolute growth might even exceed that of *A. elatius* despite *F. ovina*'s lower potential relative growth rate. Thus, the questions posed on p. 193 about the relationship of growth and uptake with the external factors of nitrogen supply and temperature regime have been answered in a way which indicates specific adaptation.

These results emphasize the need for comparative studies of plant responses and point to differences between potentially fast-growing species such as *A. elatius* and the slow-growing *F. ovina* which, when measured over a complete annual cycle, may produce very similar absolute growth. Similar results were obtained for the fast-growing *Holcus lanatus* and the slow-growing *Deschampsia flexuosa* (Rorison 1981).

In contrast, there is a danger of extrapolation from experiments which may be carried on for a year under controlled conditions which do not allow for a period of low temperature and short days. Thus the findings of Mahmoud & Grime (1976), that *A. elatius* can outgrow *F. ovina* even under conditions of very low nitrogen supply in a glasshouse for one year might well be reversed under natural conditions. The balance between two species may be shifted by other factors, such as summer drought (Grime & Curtis 1976), or a particularly mild or severe winter, by heavy grazing or the unlikely event of increased fertility. The importance of nitrogen relative to these other factors has still to be assessed. This and the effects of soil reaction are considered in detail elsewhere (Peterkin 1981).

ACKNOWLEDGEMENTS

We thank Dr P. L. Gupta, Mr M. G. Johnson and Mr W. Downs for analyses of soil and plant samples and the Natural Environment Research Council for a CASE award to one of us (JHP).

REFERENCES

Al-Mufti M.M., Sydes C.L., Furness S.B., Grime J.P. & Band S.R. (1977) A quantitative analysis of shoot phenology and dominance in herbaceous vegetation. *Journal of Ecology*, **65**, 759–91.

Barlow E.W.R., Boersma L. & Young J.L. (1976) Root temperature and soil water potential effects on growth and soluble carbohydrate concentration of corn seedlings. *Crop Science*, **16**, 59–62.

Berry J. & Björkmann O. (1980) Photosynthetic response and adaptation to temperature in higher plants. *Annual Review of Plant Physiology*, **31**, 491–543.

Bowen G.D. & Smith S.E. (1981) The effects of mycorrhizas on nitrogen uptake by plants. In *Terrestrial Nitrogen Cycles* (eds F. E. Clark & T. Rosswall), pp. 237–47. Ecological Bulletins 33. Swedish National Science Research Council, Stockholm.

Brown R.H. & Blaser R.E. (1970) Soil moisture and temperature effects on growth and soluble carbohydrates of Orchardgrass (*Dactylis glomerata*). *Crop Science*, **10**, 213–16.

Caldwell M.M. (1979) Root structure: the considerable cost of below ground function. In *Topics in Plant Population Biology* (eds O.T. Solbrig, S. Jain, G.B. Johnson & P.H. Raven), pp. 408–27. Macmillan Press, London.

Chowdhury I.R. & Zubriski J.C. (1973) Effects of temperature and nitrogen supply on four nitrogen fractions in Barley. *Agronomy Journal*, **65**, 529–32.

Clapham A.R., Tutin T.G. & Warburg E.F. (1962) *Flora of the British Isles*, 2nd edn. Cambridge University Press, Cambridge.

Clark S.C. (1974) Biological flora of the British Isles. *Catapodium rigidum* (L.) C.E. Hubbard. *Journal of Ecology*, **62**, 937–58.

Clarkson D.T. & Warner A.J. (1979) Relationships between root temperature and the transport of ammonium and nitrate ions by Italian and perennial ryegrass (*Lolium multiflorum* and *Lolium perenne*). *Plant Physiology*, **64**, 557–61.

Cooper A.J. (1973) Root temperature and plant growth. *Research Review No. 4*. Commonwealth Bureaux of Horticulture and Plantation Crops, Farnham Royal, Bucks.

Davy A.J. & Taylor K. (1974) Seasonal patterns in nitrogen availability in contrasting soils of the Chiltern Hills. *Journal of Ecology*, **62**, 793–807.

Deane-Drummond C.E., Clarkson D.T. & Johnson C.B. (1980) The effect of differential root and shoot temperature on the nitrate reductase activity, assayed *in vivo* and *in vitro* in roots of *Hordeum vulgare* (Barley). *Planta*, **148**, 455–61.

Ellenberg H. (1964) Stickstoff als. Standortsfaktor. *Berichte der Deutschen chemischen Gesellschaft*, **77**, 82–92.

Frota J.N.E. & Tucker T.C. (1972) Temperature influence on ammonium and nitrate absorption by lettuce. *Proceedings of the Soil Science Society of America*, **36**, 97–100.

Gates D.M. (1968) Transpiration and leaf temperature. *Annual Review of Plant Physiology*, **19**, 211–38.

Gigon A. & Rorison I.H. (1972) The response of some ecologically distinct species to nitrate- and to ammonium-nitrogen. *Journal of Ecology*, **60**, 92–102.

Grime J.P. & Curtis A.V. (1976) The interaction of drought and mineral nutrient stress in calcareous grassland. *Journal of Ecology*, **64,** 976–98.

Grime J.P. & Hunt R. (1975) Relative growth rate: its range and adaptive significance in a local flora. *Journal of Ecology*, **63**, 393–422.

Grime J.P. & Lloyd P.S. (1973) *An Ecological Atlas of Grassland Plants*. Edward Arnold, London.

Gupta P.L. & Rorison I.H. (1975) Seasonal differences in the availability of nutrients down a podzolic profile. *Journal of Ecology*, **63**, 521–34.

Havill D.C., Lee J.A. & Stewart G.R. (1974) Nitrate utilisation by species from acidic and calcareous soils. *New Phytologist*, **73**, 1221–31.

Haynes R.G. & Goh K.M. (1978) Ammonium and nitrate nutrition of plants. *Biological Reviews*, **58**, 465–510.

Hewitt E.J. (1966) *Sand and Water Culture Methods used in the Study of Plant Nutrition*. 2nd edn. Commonwealth Agricultural Bureaux, Farnham Royal, Bucks.

Hubbard C.E. (1968) *Grasses*. 2nd edn. Penguin Books, Harmondsworth, Middlesex.

Hunt R. & Parsons I.T. (1974) A computer program for deriving growth functions in plant growth analysis. *Journal of Applied Ecology*, **11,** 297–307.

Ingestad T. (1979) A definition of optimum nutrient requirements in birch seedlings. III. Influence of pH and temperature of the nutrient solution. *Physiologia Plantarum*, **46**, 31–5.

Lee J.A. & Stewart G.R. (1978) Ecological aspects of nitrogen assimilation. *Advances in Botanical Research*, **6**, 1–43.

McCown B.H. (1978) The interactions of organic nutrients, soil nitrogen and soil temperature and plant growth and survival in the Arctic environment. In *Vegetation and Production Ecology of an Alaskan Arctic Tundra* (ed. L.L. Tienszen), pp. 435–56. Ecological Studies No. 29. Springer-Verlag, Berlin.

McGrath S.P. (1979) Growth and distribution of *Holcus lanatus* L. populations with reference to nitrogen source and aluminium. Ph.D. thesis, University of Sheffield.

Mahmoud A. & Grime J.P. (1976) An analysis of competitive ability in three perennial grasses. *New Phytologist*, **77**, 431–5.

Moraghan J.T. & Porter O.M. (1975) Maize growth as affected by root temperature and form of nitrogen. *Plant and Soil*, **43**, 479–87.

Newbould P. (1969) The absorption of nutrients by plants from different zones in the soil. In *Ecological Aspects of the Mineral Nutrition of Plants* (ed. I.H. Rorison), pp. 177–90. Blackwell Scientific Publications, Oxford.

Osborne B.A. & Whittington W.J. (1981) Eco-physiological aspects of inter-specific and seasonal variation in nitrate utilization in the genus *Agrostis*. *New Phytologist*, **87**, 595–614.

Patterson R.P., Grunes D.L. & Lathwell D.J. (1972) Influence of root-zone temperature and P supply on total and inorganic P, free sugars, aconitate and soluble amino N in corn. *Crop Science*, **12**, 227–30.

Peterkin J.H. (1981) Plant growth and nitrogen nutrition in relation to temperature. Ph.D. thesis, University of Sheffield.

Pfitzenmeyer C.D.C. (1962) Biological flora of the British Isles *Arrhenatherum elatius* (L.) J. & C. Presl. *Journal of Ecology*, **50**, 235–45.

Ratcliffe D. (1961) Adaptation to habitat in a group of annual plants. *Journal of Ecology*, **49,** 187–203.

Raven J.A. & Smith F.A. (1976) Nitrogen assimilation and transport in vascular land plants in relation to intra-cellular pH regulation. *New Phytologist*, **76**, 415–31.

Read D.J., Koucheki H.K. & Hodgson J.G. (1976) Vesicular-arbuscular mycorrhizae in natural vegetation systems. I. The occurrence of infection. *New Phytologist*, **77**, 641–53.

Regehr D.L. & Bazzaz F.A. (1976) Low temperature photosynthesis in successional winter annuals. *Ecology*, **57**, 1297–303.

Rorison I.H. (1981) Plant growth in response to variations in temperature: field and laboratory studies. In *Plants and Their Atmospheric Environment* (eds J. Grace, E.D. Ford & P.G. Jarvis), pp. 313–32. Blackwell Scientific Publications, Oxford.

Rorison I.H. & Sutton F. (1976) Climate, topography and germination. In *Light as an Ecological Factor II* (eds G.C. Evans, R. Bainbridge & O. Rackham), pp. 361–83. Blackwell Scientific Publications, Oxford.

Runge M. (1974) Die Stickstoff-Mineralisation im boden eines Sauerhumus—Buchenwaldes II. Die nitratproduktion. *Oecologia Plantarum*, **9**, 219–30.

Smith S.E. (1980) Mycorrhizas in autotrophic higher plants. *Biological Reviews*, **55**, 475–510.

Sydes C.L. (1980) Some aspects of competition and co-existence in various types of herbaceous vegetation. Ph.D. thesis, University of Sheffield.

Van de Dijk S.J. & Troelstra S.R. (1980) Heterotrophic nitrification in a heath soil demonstrated by an *in-situ* method. *Plant and Soil* **57**, 11–21.

Welbank P.J. (1962) The effects of competition with *Agropyron repens* and of nitrogen and water supply on the N content of *Impatiens parviflora. Annals of Botany*, **26**, 361–73.

Went F.W. (1953) The effect of temperature on plant growth. *Annual Review of Plant Physiology*, **4**, 347–62.

Williams D.E. & Vlamis J. (1962) Differential cation and anion absorption as affected by climate. *Plant Physiology*, **37**, 198–202.

Williams J.T. (1969) Mineral nitrogen in British grassland soils. I. Seasonal patterns in simple models. *Oecologia Plantarum*, **4**, 307–20.

Williams R.F. (1946) The physiology of plant growth with special reference to the concept of net assimilation rate. *Annals of Botany*, **10**, 41–72.

10. THERMAL ADAPTATION OF NITRATE TRANSPORT AND ASSIMILATION IN ROOTS?

D. T. CLARKSON AND CELIA E. DEANE-DRUMMOND
ARC Letcombe Laboratory, Wantage OX12 9JT, and Plant Science Laboratories, University of Reading, Reading RG1 5AQ

SUMMARY

Evidence is presented that ion transport systems in roots which have become acclimatized to low temperatures may increase their capacity to absorb nutrients, including nitrate. The response seems to be part of the cold-hardening process and results from an expansion of the systems rather than from any change in their temperature sensitivity.

Increased capacity for nitrate absorption at low root temperature is matched by increased nitrate reductase activity in the root. This shifts the balance of nitrate assimilation in the whole plant towards the root system. There is, additionally, some indication that the properties of nitrate reductase itself may be modified by environmental temperature.

The extent to which these changes should be regarded as adaptive responses, and part of a mechanism for temperature compensation, is discussed in relation to changes in the distribution of resources between root and shoot. However they are regarded they are likely to assume the greatest significance when the shoot is warmer than the root as in the late winter and spring.

INTRODUCTION

It is a matter of common observation and commonsense that nearly all plants grow more slowly in cool conditions than in warm ones. The temperature coefficient for growth is usually large, especially when the limits to which a species is adapted are approached. Temperature has effects on cell division and cell expansion which are determined by interactions with biochemical events (e.g. nucleic acid metabolism, protein synthesis) and hormonal events (e.g. cytokinin, auxin metabolism). But cell proliferation and growth can also be influenced by the supply of essential substrates and building materials. Restriction of photosynthesis obviously influences the rate and pattern of growth and so does the supply of mineral nutrients. In soils that are not amended by fertilizers, the supply of nutrients frequently reduces plant growth and performance below the limits set by other environmental factors.

When physiologists consider the effect of temperature on plant development, with rare exceptions they treat the whole plant to a given temperature and compare results with another plant treated at a different one. In nature, however, shoot and root are rarely at the same temperature and so such experimental designs miss a very important aspect of the internal strains or stresses which can result from a temperature differential. In the late winter or spring in temperate regions it is not uncommon that the leaves of overwintering plants experience conditions which are conducive to growth, i.e. warm temperatures, high and unobstructed illumination and adequate water, while the roots of the plant exist, or are developing, in soil which is still quite cold at depths below about 5 cm. In such circumstances the shoots of some species grow very poorly (e.g. *Zea mays*) while others seem to be able to develop the leaf canopy vigorously. What conclusions should we draw about the influence of the cold soil on the performance of the plant? There are at least two ways in which a slow-growing root might restrict shoot development. It might deliver some hormonal message at a rate which indicates its slow growth (Atkin, Barton & Robinson 1973) or it might fail to provide an adequate absorbing system to support the shoot. An adequate absorption system is determined by root proliferation in the soil, particularly for slow-diffusing nutrients like phosphate, and by the ability of ion transport systems to cope with the demand placed on them by the shoot.

In this paper we shall describe a number of physiological changes in the nitrate transport and assimilation in barley roots which occur when they are grown at temperatures which are low relative to those of the shoots. We shall then consider how far one might be justified in regarding these changes as being of adaptive value. We reach the conclusion that plant factors in the absorption and assimilation of nitrogen are unlikely to limit shoot growth when roots are cool.

EXPERIMENTAL EVIDENCE AND ITS INTERPRETATION

A number of techniques have been used to assemble the evidence to be presented. We have not felt it appropriate to describe all of them since that has been done elsewhere. References are given as footnotes to the various tables and figures.

One matter which does require comment is the selection of fresh weight as the basis on which comparisons are made between plants grown at different temperatures. One consequence of growing plants for extended periods of time at low root temperature is an increase in the percentage dry matter of the plant. The increase is of such a magnitude that it can be accounted for only by an increase in cell wall thickness; soluble materials within the cells, although statistically and physiologically significant, contribute in only minor ways to the increased percentage of dry matter. The extent of root systems grown at different temperatures is more reliably inferred from their fresh, than from their dry, weights.

Effects of pretreatment temperature on ion transport

When intact barley plants have the solution around their roots cooled from 20°C to 8°C for two or more days, while their shoot remains at 20°C, their capacity for ion transport increases quite markedly. To study this we have used root systems excised from plants which have been pretreated at two temperatures and gathered the sap which exudes from the cut ends of the xylem. Table 10.1 shows that the flow of sap at any given temperature is much greater from roots which have been pretreated at 8°C for 72 hours than from those kept at 20°C prior to the experiment. The increased flow of sap is accounted for by marked increases in the flows of solutes from the root into the xylem. The table shows that the flow of both potassium (which is a major osmotic component of the sap) and phosphate (whose movement into the sap

TABLE 10.1. Effect of pretreatment temperature on flows of xylem sap and ion fluxes into the xylem of detached root systems of barley measured at various temperatures. The measurements were made on plants which were 16 days old. In the three day period prior to the experiment roots were kept in solutions at 20°C or 8°C while the shoots were at 20°C in both instances. For further analysis of results of this kind, see Clarkson (1976).

Experiment temperature (°C)	Sap flow* (mg h^{-1} g root fresh wt.$^{-1}$)		Ion fluxes* (μmol h^{-1} g root fresh wt.$^{-1}$) Potassium		Phosphate	
	20°C control	8°C pretreat	20°C control	8°C pretreat	20°C control	8°C pretreat
25	38	215	0.39	2.31	0.06	0.45
22.5	28	251	0.35	2.87	0.08	0.45
20	25	146	0.42	2.06	0.08	0.38
14	14	86	0.29	1.41	0.06	0.17
10	11	64	0.21	1.62	0.05	0.16
9	6	46	0.14	0.72	0.03	0.10
4	4	18	0.07	0.28	0.01	0.02

* Measured over the first four hours following excision.

depends on metabolic events in the sugar phosphate pool—see Loughman 1966) is markedly increased by pretreatment at 8°C. It is worth noting that the potassium and phosphate fluxes at 8°C in the cool acclimatized roots are as great, or greater than those from 20°C-grown roots at high temperatures.

Analysis of data of this kind from both barley and rye leads to the conclusion that cooling results in an expansion of the ion-transporting capacity rather than from any change in temperature sensitivity of the various processes involved (Clarkson 1976).

Effect of pretreatment temperature on the composition of the roots

When roots are cooled down to 10°C while the shoots are at 20°C there is a reduction in the relative size of the root system. Over the 72 hours used in the above experiment the shoot/root fresh weight ratio increased from approximately 1.2 to 1.6. Over longer periods of time at 10°C the value of the shoot/root ratio stabilizes at about 1.7 (Deane-Drummond, Clarkson & Johnson 1980). The reduction in the rate of root growth is accompanied by increases in the soluble carbohydrate, malate, and dry matter contents and oxygen consumption, but there is no change in the soluble protein content (Table 10.2). The increases in substrates may be due to reduced consumption in growth.

Effect of pretreatment temperature on the absorption and transport of labelled nitrogen from nitrate

When labelled ions are added to the solution bathing excised roots it is found that the influx of labelled nitrogen into the xylem is greatly increased by pretreatment at cool temperatures. This is illustrated in Table 10.3 by an experiment where $^{15}NO_3^-$ was present in the external solution bathing roots of intact plants as well as of those which had been excised. The data are tabulated in such a way that values from roots functioning at the temperature in which they had been grown, i.e. 20/20 and 10/10, are adjacent. It is evident that the shoots of plants in either of these treatments would have received about the same *amount* of labelled nitrogen. Although the rate of transport to the shoot by 10/10 roots was 40% greater than by 20/20 roots, the root system was 30% smaller relative to the shoot. The extra transport has compensated, therefore, for the smaller root system. The table also makes it clear that the difference caused by pretreatment temperature is more marked in the translocated fraction of $^{15}NO_3^-$ uptake than in accumulation by the root tissue itself. The degree of adaptation which had occurred after four days' acclimatization can be seen by comparing the 20/10 and 10/10 treatments.

TABLE 10.2. Composition of barley roots grown at 20°C or 10°C. Carbohydrate and malate were measured in 80% ethanol extracts of roots using methods of Fairbairn (1953) and Williamson & Corkley (1969) respectively. Water-soluble protein from root homogenates was determined by the method of Sedmak & Grossberg (1977). Oxygen consumption was measured by a manometric method.

Measurement	20°C control	10°C-grown
Soluble carbohydrate (mg g^{-1})	1.0	3.4
Soluble protein (mg g^{-1})	7.1	8.0*
Malate (μmol g^{-1})	0.01	2.1
Dry matter (%)	5.9	7.7†
Oxygen consumption (μl h^{-1} g^{-1}) at 20°C	215	379

* Roots grown for three days at 10°C.
† Roots grown for 13 days at 10°C.

TABLE 10.3. Comparison of the effects of pretreatment at 20°C and 10°C on the absorption and long-distance transport of labelled nitrogen from nitrate by intact plants and excised roots of barley. ^{15}N measured by mass spectrometry after Dumas combustion of freeze-dried tissue and exudate samples.

Temperature pretreatment/experiment	Uptake of labelled nitrogen from nitrate* (μmol h^{-1} g root fresh wt.$^{-1}$) Intact plants 20/10	20/20	10/10	10/20	Excised roots 20/10	20/20	10/10	10/20
Shoot	0.60	1.15	1.64	2.44	—	—	—	—
Exudate	—	—	—	—	0.27	0.50	0.78	1.34
Root	1.23	2.32	2.09	2.99	1.11	1.63	1.71	1.99
Total	1.83	3.47	3.73	5.43	1.38	2.13	2.49	3.33
% Translocated	33	33	44	45	20	23	31	40

* Measured for five hours during hours 5–10 of the photoperiod.

Effect of pretreatment temperature on nitrate assimilation in roots

Nitrate absorbed by the transport system may be either moved into the xylem unchanged or as amino acids after metabolic assimilation. Table 10.4 shows that, in roots grown at 20°C, 20–25% of the total nitrogen entering the xylem was amino acid, whereas in roots grown at 10°C 30–33% was reduced nitrogen. The fluxes of both nitrate and, particularly, amino acid were much

TABLE 10.4. Solute flows* into the xylem of excised barley roots. These measurements were made on plants which were nine days old; in the 42 hours prior to the experiment half of the plants had their roots treated at 10°C while the other remained at 20°C. A shoot temperature of 20°C was maintained in both treatments. The roots were treated with culture solution containing 10 mmol nitrate throughout pretreatment and experiment.

	Experiment temperature 20°C Pretreatment temperature			Experiment temperature 10°C Pretreatment temperature		
Solute	20°C control	10°C pretreated	10°:20° ratio	20°C control	10°C pretreated	10°:20° ratio
Nitrate	1.02	2.0	**1.96**	0.50	1.38	**2.76**
Amino acids†	0.21	0.90	**4.29**	0.16	0.69	**4.31**

* μmol h^{-1} g fresh wt.$^{-1}$.
† Glycine equivalents assayed by method of Wylie & Johnson (1962).

greater in roots grown at 10°C irrespective of the temperature at which the experiment was conducted. These results suggest that nitrate assimilation is greater in cooled roots and that the activity of the enzyme which is generally regarded as rate limiting, i.e. nitrate reductase, must also be greater. Measurements of nitrate reductase activity using assays both *in vitro* and *in vivo* have shown that this is so (Deane-Drummond, Clarkson & Johnson 1980). Fig. 10.1 shows that, at all times during a 24-hour cycle, nitrate reductase activity measured by the assay *in vivo* was greater in roots which had been grown at 10°C than in those grown at 20°C. The difference in activity was particularly marked during the dark period when values close to zero were found in 20°C-grown roots. It was shown (Deane-Drummond, Clarkson & Johnson 1980) that the greater nitrate reductase activity in cool-grown roots in the night and early part of the photoperiod was associated with increased respiratory substrate supply, particularly malate (see Table 10.2). These results indicate that, when roots are cool relative to leaves, the proportion of the incoming nitrate reduced in roots increases so that at high external nitrate concentration which must saturate the high affinity nitrate uptake system, 40–50% of absorbed nitrate is assimilated in the root. In 20°C-grown roots the proportion is much lower because there would appear to be negligible nitrate reductase activity throughout the dark period. These conclusions are in accordance with observations of Kirkman & Mifflin (1979) who showed that, at low rates of fertilizer nitrogen, soil-grown spring barley plants reduced more than 50% of incoming nitrate in the roots.

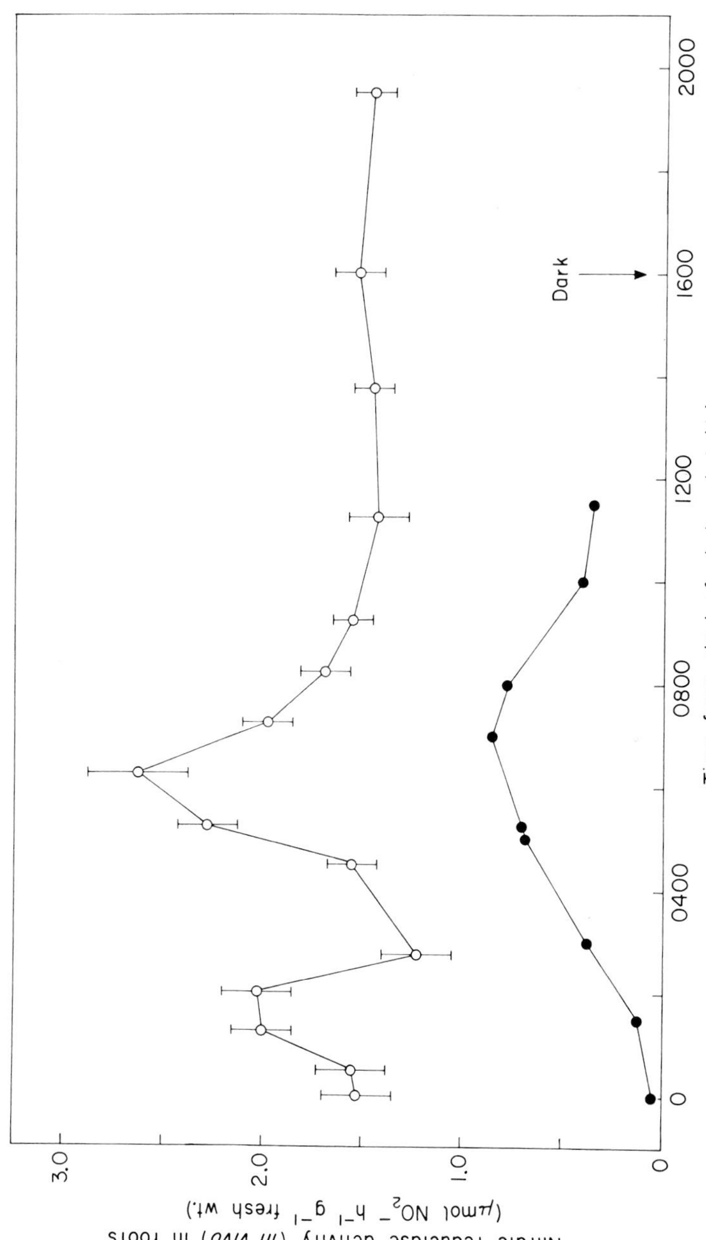

FIG. 10.1. Diurnal variation in nitrate reductase activity (*in vivo*) of barley roots. Roots grown for 13 days at 10°C (open circles) or 20°C (closed circles). Bar lines indicate standard errors of mean values of at least five determinations. Enzyme assay as described in Deane-Drummond, Clarkson & Johnson (1979).

Effect of growth temperature on properties of nitrate reductase

Nitrate reductase activity in roots is induced by the presence of nitrate in the solution bathing them. In roots grown at 10°C, nitrate reductase activity was much greater, particularly at lower nitrate concentrations, but maximal activity was induced by treating roots with 0.5 mmol nitrate; maximal activity in 20°C-grown roots required more than twice this concentration of nitrate. The nitrate reductase acivity of cooled roots is, therefore, more readily induced by nitrate. This may be a consequence of the enhanced rate of nitrate uptake into the cells.

The affinity of the enzyme for nitrate is measured by the Michaelis constant ($K_mNO_3^-$); it is determined by the concentration of nitrate which will induce half the maximum rate of nitrate reduction. When semi-purified nitrate reductase from roots grown at 20°C and 10°C was compared it was clear that the value of $K_mNO_3^-$ was influenced both by the temperature of the incubation medium and by the previous growth temperature. The K_m was maximal at the growth temperature and was distinctly smaller at 5°C below the growth temperature (Fig. 10.2). The response at higher temperatures than those at which growth had occurred was complicated by there being two apparent Michaelis constants (Deane-Drummond 1980; Eaglesham & Hewitt 1975). For convenience the mean value of these two $K_mNO_3^-$ is shown in the figure. Below the growth temperature there was only one $K_mNO_3^-$.

Thus there was the same pattern of behaviour in enzymes biosynthesized at different temperatures but it was centred around the growth temperature in each case. This suggests that there may be some difference in the structure of composition of the enzyme synthesized at different temperatures. There are no reports, as far as we know, that nitrate reductase isozymes or allozymes exist and much more stringent investigation would be necessary to establish whether the behaviour in Fig. 10.2 is due to such a cause. Changes in the relative amounts of malic dehydrogenase isozymes have been detected in *Typha latifolia* as its growth temperature is changed (Jones, Hancock & Liu 1979). The thermal properties of malic dehydrogenase have also been found to vary in *Lathyrus japonicus* populations adapted to different temperature environments (Simon 1979a & b).

GENERAL DISCUSSION

There can be little doubt that the evidence presented above indicates changes in the capacity for the transport of nitrate and other ions and for the reduction of nitrate in the root. It is evident that these processes can proceed as rapidly at a root temperature of 10°C as at 20°C once the plants have become

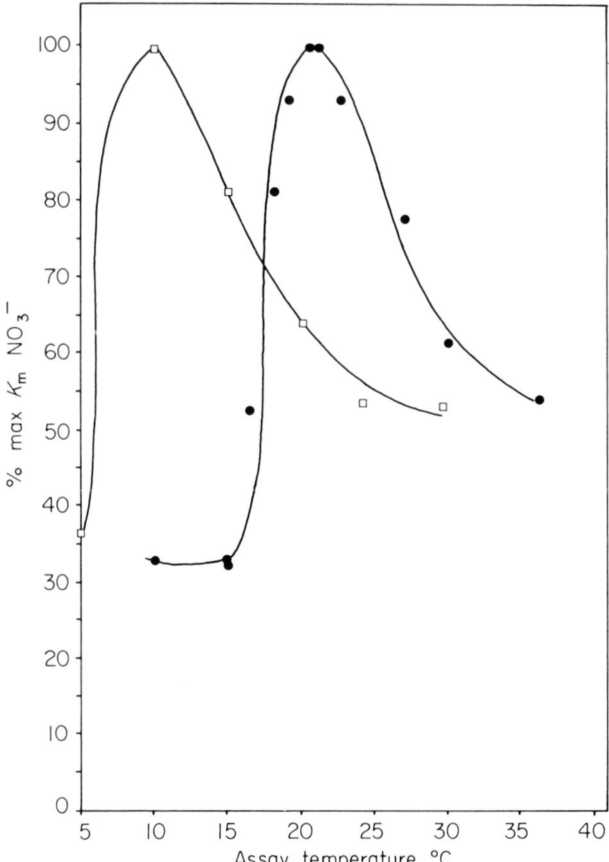

FIG. 10.2. Temperature dependence of the Michaelis constant for nitrate ($K_mNO_3^-$) for nitrate reductase partially purified from roots grown for two weeks at 20°C (●) and 10°C (□). Extracts of nitrate reductase from roots were prepared as described in Deane-Drummond, Clarkson & Johnson (1979) and then desalted on a Sephadex G100 column as described in Deane-Drummond (1980). The incubation mixture for the assay contained 10 μmol flavin adenine dinucleotide, 0.2 mg ml^{-1} NADH and various concentrations of potassium nitrate.

acclimatized. We now consider how these changes may be brought about and to what extent we might regard them as having adaptive significance.

Ecologists should be the first to point out that what we have called the 'controls' in these experiments should be, perhaps, more properly thought of as our variant. Root temperatures of 20°C are quite uncommon, at least throughout the entire root system, in the climate of Northern Europe. The distribution of resources between the root and shoot system when they are at the same relatively high temperature can hardly be regarded as the norm against which more realistic combinations of shoot and root temperature are compared. This question of which is control and which is treatment needs answering, not simply to satisfy pedantry but to indicate more clearly the types of experiments which need to be done. The relationship between root temperature and shoot growth can vary greatly with species and it is, therefore, difficult to make generalizations. With both C_3 and C_4 photosynthesizing members of the Gramineae, reductions in root temperature appreciably below that of the shoot result in a marked reduction in the relative size of the root system. In *Zea mays* the shoot/root dry weight ratio of plants growing rapidly at 20°C was 1:8, whereas in plants growing slowly with roots at 10–13°C the ratio increased to 6:1 (Clarkson & Gerloff 1981). In the present work a brief period with roots at 10°C caused a shift in the shoot/root fresh weight from 1.17 at 20°C to 1.57 at 10°C after three days. In unpublished work with *Lolium perenne* the shoot/root fresh weight ratio of plants grown for three weeks with their roots at 5°C and leaves at 20–24°C reached a steady state at about 4.5:1. Six days after the root temperature was increased to 17°C, a rapid increase in root growth doubled its size relative to the shoot and the ratio fell to 2:1. Precise quantitative information of this kind is hard to come by, both in the field and in the published literature; but the work of Davidson (1969) appears, on first reading, to contradict the above results. In his experiments, however, root temperatures were controlled whereas the shoots were exposed to daily variations in temperature between 2°C and 20°C; this is, therefore, quite a different state of affairs compared with the present work where a constant differential temperature was maintained between shoot and root. Davidson's results resemble those from circumstances when the whole plant is cooled down when it has been found that the root becomes relatively larger (Brouwer 1962; Boote 1977; Rorison *et al.*, p. 189, this volume).

It is also shown repeatedly that the abundance of respiratory substrate in roots increases as their temperature is reduced. In some ways one might regard this as a reduction in growth demand for photosynthate with a concomitant increase in the availability of substrates for other purposes. The marked increase in dry matter content brought about by cellulose deposition in thicker walls, the increases we report in ion transport activity and the increased activity of nitrate reductase all seem to be related, directly or indirectly to this changed pattern in which resources are utilized.

In addition to changes of the kind described above, which might be thought of as fine tuning of the internal economy of the root, there are biochemical modifications as the root becomes acclimatized to a new temperature. We have not discussed in this paper the role of fatty acid desaturation in membrane lipids in relation to cool temperature acclimatization (Clarkson, Hall & Roberts 1980) but this phenomenon is very widely reported and believed to maintain membrane fluidity at a relatively constant value over the temperature range to which a species is adapted. This really is an adaptation to temperature and it is not hard to see that it is equally important for rapid- as for slow-growing plants that they should maintain the integrity and function of the cell membranes over a wide range of temperature. What is perhaps less evident is why there should be any need to take steps to compensate ion transport systems for changing temperature. The relationship between the rate at which most nutrients, and particularly nitrogen, are absorbed and the rate at which plants grow is an intimate one. If growth is slowed down by low temperature, nitrogen absorption and metabolism are likely to slow down to a very similar extent (Ingestad 1979). If nitrogen absorption were less sensitive to low temperature (i.e. compensated) than growth, we might expect to find that plants contained higher nitrogen concentrations within them when they are grown at low tempertures than when they are grown at warmer ones, implying their uptake would be exceeding their growth requirement. Our own work and that of Ingestad suggests that this is only rarely the case. On the other hand, if low temperature makes the root system relatively smaller, the demand placed on each unit of root increases and thus there must be an equivalent increase in the rate of transport if the nutrient levels in the shoot are to be sustained. The fact that the rate of uptake, and particularly of long distance transport into the xylem, is increased for all of the major ions in barley (Clarkson, Shone & Wood 1974) and in birch (Ingestad 1979) lends support to the idea that what we have been observing is an adjustment of root activity to demand set by the relative shoot/root size.

There are, therefore, some parallels between the present results and those which are found when certain nutrients become deficient in the plant. The size or activity of the phosphate (Bowen 1970; Clarkson, Sanderson & Scattergood 1978) and potassium (Glass 1977; Pitman 1972) transport systems respond very readily to the increased demand created by nutrient stress. This identifies what is happening without explaining how it happens. Kinetic analyses in nutrient-deficient plants indicate unequivocally that the capacity, or V_{max}, of the transport system for the nutrient in short supply is increased by demand. We can most easily envisage this as an expansion of the numbers of 'porters' involved in the transport. An explanation of our results from temperature studies in such terms suggests that increased ion transport capacity may be due to a new balance being struck between synthesis and turnover of all of the ion 'porters' when temperature conditions are altered. A similar type of explanation has been given for the increased levels in phospholipids (Clarkson, Hall & Roberts 1980) and of nitrate reductase (Deane-Drummond, Clarkson & Johnson 1980) in cooled barley roots.

If plants in nature respond to cool temperatures around their roots in the same way that we have seen in laboratory-grown crop plants (there are those who might regard this as a fairly substantial 'if'), then it seems most unlikely that shoot development will be limited by a failure of roots to provide an adequate supply of nitrogen in reduced form or as nitrate, assuming that there is a source of nitrogen available for absorption by the roots. It is also rather unlikely that shoot development will be limited by the rate of photosynthesis since there have been repeated demonstrations that this process is *less* sensitive to temperature than growth. It therefore seems likely that, when roots are cool, shoot development is regulated by internal constraints which are possibly of hormonal origin.

Cool root temperature in our experiments shifted the emphasis of nitrate assimilation towards the root system, but Lee & Stewart (1978) observed appreciable shoot nitrate reductase activity in alpine tundra plants in the field. There is evident scope here for work on species from a wide range of environments and their response to combinations of shoot and root temperature. The changes described here can be related to the distribution of resources within the plant and do not necessarily imply any biochemical adaptation *per se*. The increased availability of malate in cooled roots provides an essential source of reducing power for nitrate reductase (Lee 1980). By contrast, one might speculatively attribute adaptive significance to changes in the substrate affinity of the nitrate reductase protein present in roots grown at different temperatures. At first sight it seems odd that the affinity between enzyme and substrate is lowest at the growth temperature (i.e. $K_mNO_3^-$ at a maximum value). It is not, however, hard to envisage a system in which a drop in temperature of 5°C below the growth temperature, which will tend to slow down the rate of chemical reactions involved in nitrate reduction, could be offset by an increase in affinity of the enzyme and substrate (decrease in $K_mNO_3^-$) which will tend to increase the rate of reaction. This simple balancing system would work only where the concentration of nitrate inside the cells was about the same, or lower than $K_mNO_3^-$, i.e. about 0.6 mM. For plants growing in natural soils free from added nitrate fertilizer this condition is most likely to be encountered. Preliminary results from measurements of nitrate reductase activity *in vitro* confirm that there is no change in the rate of nitrate reduction over a range of 5°C below the growth temperature (Deane-Drummond 1980) and therefore encourage us to believe that there may be adaptive significance in the temperature sensitivity of $K_mNO_3^-$ which would allow homeostasis in nitrate assimilation over a limited range of temperature (McNaughton 1972). Clearly this is a matter which merits more thorough investigation and should involve liaison between biochemists and ecologists so as to help us undertand one of the ways in which plants cope with fluctuating temperature.

ACKNOWLEDGEMENTS

We gratefully acknowledge the help and guidance given in this work by Dr C. B. Johnson and Dr R. B. Lee and the analyses of ^{15}N by mass spectrometry by Mr M. G. Johnson of Letcombe Laboratory. We also thank other members of the Chemistry and Electronics Department for inorganic analyses. C.D.-D. was supported by a CASE studentship awarded by the Science Research Council.

REFERENCES

Atkin R.M., Barton G.E. & Robinson D.K. (1973) Effect of root-growing temperature on growth substances in xylem exudate of *Zea mays*. *Journal of Experimental Botany*, **24**, 475–87.

Boote K.J. (1977) Root:shoot relationships. *The Soil and Crop Science Society of Florida Proceedings*, **36**, 15–23.

Bowen G.D. (1970) Early detection of phosphorus deficiency in plants. *Soil Science and Plant Analysis*, **1**, 293–8.

Brouwer R. (1962) Distribution of dry matter in the plant. *Netherlands Journal of Agricultural Science*, **10**, 361–76.

Clarkson D.T. (1976) The influence of temperature on the exudation of xylem sap from detached root systems of rye (*Secale cereale*) and barley (*Hordeum vulgare*). *Planta*, **132**, 297–304.

Clarkson D.T. & Gerloff G.J. (1980) Growth and nutrient absorption by roots of maize genotypes at low temperatures. In *Production and Utilization of the Maize Crop* (ed. E.S. Bunting), pp. 179–87. Packard Publishing, Chichester, Sussex.

Clarkson D.T., Hall K.C. & Roberts J.K.M. (1980) Phospholipid composition and fatty acid desaturation in roots of rye during acclimatization to low temperature. Positional analysis of fatty acids. *Planta*, **149**, 464–71.

Clarkson D.T., Sanderson J. & Scattergood C.B. (1978) Influence of phosphate stress on phosphate absorption and translocation by various parts of the root system of *Hordeum vulgare* L. (barley). *Planta*, **137**, 47–53.

Clarkson D.T., Shone M.G.T. & Wood A.V. (1974) The effect of pretreatment temperature on the exudation of xylem sap by detached barley root systems. *Planta*, **121**, 81–92.

Davidson R.L. (1969) Effect of root/leaf temperature differentials on root/shoot ratios in some pasture grasses and clover. *Annals of Botany*, **33**, 561–9.

Deane-Drummond C.E. (1980) Nitrate reduction in barley plants. Ph.D. thesis, University of Reading.

Deane-Drummond C.E., Clarkson D.T. & Johnson C.B. (1979) The effect of shoot removal and malate on the activity of nitrate reductase assayed *in vivo* in barley roots (*Hordeum vulgare* cv. Midas). *Plant Physiology*, **64**, 660–2.

Deane-Drummond C.E., Clarkson D.T. & Johnson C.B. (1980) The effect of differential root and shoot temperature on the nitrate reductase activity, assayed *in vivo* and *in vitro*, in roots of *Horduem vulgare* (barley). Relationship with diurnal changes in edongenous malate and sugar. *Planta*, **148**, 455–61.

Eaglesham A.R.J. & Hewitt E.J. (1975) Inhibition of nitrate reductase from spinach (*Spinacea oleracea*) leaf by adenine nucleotides. *Plant and Cell Physiology*, **16**, 1137–49.

Fairbairn N.J. (1953) A modified anthrone reagent. *Chemistry and Industry*, **72**, 86.

Glass A.D.M. (1977) Regulation of K^+ influx in barley roots: evidence for direct control by internal K^+. *Australian Journal of Plant Physiology*, **4**, 313–18.

Ingestad T. (1979) A definition of optimum nutrient requirements in birch seedlings. III. Influence of pH and temperature of the nutrient solution. *Physiologia Plantarum*, **46**, 31–5.

Jones J.C., Hancock J.F. & Liu E.H. (1979) Biochemical and morphological effects of temperature on *Typha latifolia* L. (Typhaceae) originating from different ends of a thermal gradient. I. Controlled environmental studies. *American Journal of Botany*, **66**, 902–6.

Kirkman M.A. & Miflin B.J. (1979) The nitrate content and amino acid composition of the xylem fluid of spring wheat throughout the growing season. *Journal of the Science of Food and Agriculture*, **30**, 653–60.

Lee J.A. & Stewart G.R. (1978) Ecological aspects of nitrogen assimilation. *Advances in Botanical Research*, **6**, 1–43.

Lee R.B. (1980) Sources of reductant for nitrate assimilation in non-photosynthetic tissue: a review. *Plant, Cell and Environment*, **3**, 65–90.

Loughman B.C. (1966) The mechanism of absorption and utilization of phosphate by barley plants in relation to subsequent transport to the shoot. *New Phytologist*, **65**, 388–97.

McNaughton S.J. (1972) Enzymic thermal adaptation: the evolution of homeostasis in plants. *American Naturalist*, **106**, 165–72.

Pitman M.G. (1972) Uptake and transport of ions in barley seedlings. III. Correlation between transport to the shoot and relative growth rate. *Australian Journal of Biological Science*, **25**, 905–19.

Sedmak J.J. & Grossberg S.E. (1977) A rapid, sensitive and versatile assay for protein using coumassie brilliant blue G250. *Analytical Biochemistry*, **79**, 544–52.

Simon J.-P. (1979a) Adaptation and acclimation of higher plants at the enzyme level: temperature-dependent substrate binding ability of NAD-malate dehydrogenase in four populations of *Lathyrus japonicus*Willd. (Leguminosae). *Plant Science Letters*, **14**, 113–20.

Simon J.-P. (1979b) Differences in thermal properties of NAD malate dehydrogenase in genotypes of *Lathyrus japonicus* Willd. (Leguminosae) from maritime and continental sites. *Plant, Cell and Environment*, **2**, 23–33.

Williamson J.R. & Corkley R.E. (1969) Assays of intermediates of the citric acid cycle and related compounds by fluorometric enzyme methods. In *Methods in Enzymology*, Vol. 13 (ed. J.M. Lowenstein), pp. 434–513. Academic Press, London & New York.

Wylie E.B. & Johnson M.J. (1962) Effect of penicillin on the cell wall of *E. coli*. *Biochemica & Biophysica Acta*, **59**, 450–7.

11. PATTERNS OF NITROGEN METABOLISM IN HIGHER PLANTS AND THEIR ECOLOGICAL SIGNIFICANCE

J. S. PATE

Botany Department, University of Western Australia

SUMMARY

Case histories are presented of the behaviour of different species or groups of allied species in relation to the form of nitrogen absorbed from the environment, the nature and siting of processes of nitrogen assimilation, transport and storage of nitrogen, and the partitioning of nitrogen alongside carbon assimilation and growth. Key events in these processes are considered, and examples provided in which accumulation of a specific solute or set of solutes dominates the nitrogen metabolism of a species. Formation of toxic solutes containing nitrogen is discussed in the context of predation by animals.

Cases considered in an ecological setting involve both '*r*-selected', short-lived species and '*K*-selected' perennials of herbaceous or woody character. Utilization of nitrate is considered in relation to root- or shoot-dominated patterns of nitrate reduction, utilization of ammonium from the soil in relation to the possible role of mycorrhizas, nitrogen fixation in relation to the alternative use of inorganic soil nitrogen. Parasitic angiosperms and carnivorous plants are selected to illustrate instances in which a conventional nutrition, based on inorganic nitrogen, is supplemented with, or supplanted by, a diet of organic nitrogen.

These different forms of nitrogen nutrition are discussed in terms of resource deployment and competitiveness of species.

INTRODUCTION

Extreme variability is displayed amongst species of higher plants in content of total nitrogen in dry matter, in distribution of this nitrogen between aqueous and insoluble fractions of different tissues and organs, and in the types and quantities of dissolved forms of nitrogen selected for transport, storage and defence against animals. Higher plants also exhibit equally great diversity in how they acquire and metabolize nitrogen, and gear the primary and secondary products of their metabolism to growth of shoot and root systems.

Unfortunately, we know very little about the nitrogen metabolism of all save a handful of intensively investigated species, even less of how species react in competition for nitrogen; but certain general principles have already emerged of ecological significance. This review will concentrate on these.

GENERAL FEATURES OF PLANT NITROGEN NUTRITION

This section attempts to summarize the basic biochemical and physiological steps in the assimilation of different forms of inorganic nitrogen, and in the synthesis of key solutes concerned with transport, storage and protein synthesis. Attention is directed to unusual secondary products of nitrogen and their possible functions are discussed.

Uptake of ammonium and nitrate, and the reduction of nitrate and nitrogen to ammonium

Ammonium and nitrate are the common forms of nitrogen available to higher plants, and present evidence suggests that they are absorbed by roots by means of inducible carriers, probably located in the plasma membranes of epidermal and cortical tissues of the root (Huffaker & Rains 1978; Jackson 1978). Ammonia, a highly toxic metabolite, is immediately converted to organic solutes of nitrogen, a proportion of which may be stored in the root, the remainder exported to the shoot in the xylem (Ivankov & Ingversen 1971; Pate 1973). The situation for nitrate is more complex since this ion may transfer directly to sites of storage in the root, move to the shoot in the xylem, or be reduced by respiration-linked nitrate reductase systems in the root. This last activity leads to synthesis of organic solutes of nitrogen: these in turn may be stored *in situ* or exported (Schrader 1978; Pate 1980).

The extent to which nitrate is reduced or stored in the root, and the proportions of nitrate and products of its reduction which are exported to plant tops, are highly variable features, depending on species and, within species, on plant age and environmental factors, especially the previous and current availability of nitrate to the roots. Assimilation of nitrate by roots also changes diurnally in accordance with transpirationally induced fluctuations in rates of flow of nitrate and water through apoplast and symplast of the root (Clarkson 1974; Shaner & Boyer 1976; Schrader 1981). Another important factor determining a root's assimilatory capacity is the availability of translocated sugar from the shoot, since, when metabolized, this provides reductant for converting nitrate to ammonium and the reductant, ATP, and carbon skeletons needed for ammonium assimilation into organic solutes of nitrogen (Lee 1980; Pate & Atkins, 1983).

Nitrate escaping to the shoot in unreduced form may be abstracted from the ascending xylem stream by stem, petioles and reproductive organs, where it may be stored as such or subjected to reduction. If passing to leaves, nitrate is usually assimilated immediately by photosynthetically assisted nitrate-reducing systems.

The nitrate reductase of leaves is a soluble enzyme of chloroplasts and uses NADH (or in certain cases NADPH) as reductant for producing nitrite (Beevers & Hageman 1969; Jolly, Campbell & Tolbert 1976). Nitrite reductase is located in chloroplasts and its activity is mediated by reduced ferredoxin formed in photosynthesis (Miflin 1974). Synthesis of amino compounds from the ammonium formed during nitrate reduction is also stimulated by photosynthesis (Canvin & Atkins 1974; Bauer, Urquhart & Joy 1977); this finding is possibly related to the complex diurnal fluctuations in enzymes of primary metabolism of nitrogen observed in leaves (Duke *et al.* 1978). Also, since nitrate reductase is inducible by nitrate, its activity rises and falls with changes in supply of nitrate to the leaf through the xylem (Wallace & Pate 1965; Nicholas, Harper & Hageman 1976).

The nitrate reductase systems of roots and other non-photosynthetic tissues have been inadequately studied, but the enzyme is believed to be cytoplasmic and to have its activity governed by the flux of nitrate to the cytoplasm of the tissue (Shaner & Boyer 1976). Regulation of nitrate reductase, additional to its induction by nitrate, has been variously suggested to be mediated by specific proteases (Wallace 1974), by end product repression (Radin 1975), or by some as yet unspecified process which controls the supply of nitrate to sites of nitrate reductase. This last suggestion embraces the concept of a 'nitrate operon' (Butz & Jackson 1977), responsible for coordinating uptake, transport and assimilatory functions relating to nitrate. According to Breteler & Hanisch-Tencate (1980) active and passive components of nitrate uptake exist in roots, but only the active component is involved in regulating nitrate reductase activity.

Utilization of nitrogen gas by higher plants is restricted to species forming loosely organized rhizosphere associations with nitrogen-fixing bacteria (e.g. *Azospirillum*), or to species forming more easily recognizable symbioses with an actinomycete, cyanobacterium or bacterium (e.g. *Rhizobium*). In the latter cases the microsymbiont is housed in a specialized plant structure such as a coralloid root, root nodule, or cavity in the lamina or petiole of a leaf; and the nitrogen-fixing enzyme (nitrogenase) of the microbial partner is provided with reductant and ATP by metabolism of carbohydrate-type materials provided from the host plant.

The first stable product of nitrogenase activity in nitrogen-fixing organisms is ammonium (Bergersen 1971; Stewart 1978) and in most symbiotic associations the cytoplasm of host tissue surrounding the endophyte is responsible for assimilating ammonium into organic solutes of nitrogen (see other articles in this symposium).

Nitrogenase functions optimally under near anaerobic conditions, and in certain symbiotic associations (e.g. the legume–*Rhizobium* association) the microsymbiont occupies cellular compartments where access to oxygen is limited and controlled. In associative rhizosphere fixation, where the nitrogen-fixing bacteria inhabit root surfaces or gel-filled spaces in outer regions of the root, nitrogenase is likely to be subject to wide variations in Po_2, especially due to changes in soil water content. The association is then likely to function inefficiently in terms of the amount of carbon source consumed per unit of nitrogen fixed, particularly if the bacterium protects its nitrogenase by a 'sparging-type' of respiration.

The nitrogenase systems of many, perhaps all, nitrogen-fixing organisms can reduce protons to hydrogen gas as well as forming ammonium from nitrogen, and some nitrogen fixers have the capacity to reassimilate part or all of the evolved hydrogen by means of unidirectional uptake hydrogenase systems. The relevance of protons and hydrogen metabolism to the energy demands of nitrogen fixation has been recently evaluated (Phillips 1980; Schubert & Ryle 1980; Pate, Atkins & Rainbird 1981) and is considered by Sprent (see p. 35, this volume).

Ammonia assimilation and subsequent transformations involving amino nitrogen

The demonstration by Lea & Miflin (1974) of a ferredoxin-dependent glutamate synthase (GOGAT) in chloroplasts of *Pisum* has led to a series of studies concluding that ammonium assimilation by photosynthesizing tissues proceeds by coupled activity of GOGAT and glutamine synthetase (GS) (Miflin & Lea 1977), not, as suggested in earlier studies, by reductive amination of 2-oxoglutarate by glutamate dehydrogenase (GDH). In this GS-GOGAT system, the GS engages in primary assimilation of ammonium to glutamine; the GOGAT generates glutamate from the glutamine and 2-oxoglutarate, while glutamate has the dual role of initial recipient of ammonium and end product of the assimilation sequence.

Assimilation of ammonium in non-photosynthetic tissue is less well studied, but root tissue may contain NADH-specific GDH systems, inducible by ammonium (Weissman 1972; Joy 1973), as well as GS-GOGAT systems (Miflin & Lea 1977; Probyn & Lewis 1979). Similarly, both GDH and

GS-GOGAT are recorded for nitrogen-fixing systems (Stewart *et al.* 1979; Boland, Fordyce & Greenwood 1978) and developing fruits and seeds (Miflin & Lea 1977). In these organs, as in leaves, GS-GOGAT is the favoured agent of assimilation, in view of its low K_m for ammonium compared with most GDH systems; but placing reliance on assays *in vitro* of substrate affinity may be unwise. GDH might well function as a 'back-up' system, operating when ammonium production rises transiently, or when GS-GOGAT becomes inoperative. It might be the major route of assimilation under conditions of ammonium accumulation, for example in the endosperm of developing embryos (Atkins, Pate & Sharkey 1975) or during leaf senescence. The GS-GOGAT pathway and GDH are also apparently the normal means of assimilation of ammonium in roots and shoots of a range of grassland species under situations where nitrate and ammonium serve as major sources of nitrogen (Taylor & Havill 1981).

In view of its central position in many metabolic pathways, glutamate has long been considered as the major amino donor for amino acid synthesis, whether by transamination or other transformations. Glutamine may behave similarly, being especially implicated in synthesis of arginine and asparagine (Streeter 1977; Miflin & Lea 1977; Reynolds & Farnden 1979). It may also be involved in purine biosynthesis, and thus be involved in nucleic acid synthesis or formation of ureides (Pate & Atkins 1983).

As well as functioning in connection with the assimilation of inorganic nitrogen, the GS, GDH, and GS-GOGAT systems of plants may also serve as agents of cycling of nitrogen in essentially non-assimilatory situations, notably the conservation of nitrogen in senescing organs, the flow of nitrogen accompanying photorespiration (Keys *et al.* 1978), and the utilization of amides and ureides as nitrogen sources in shoot and root apices and developing seeds (Miflin & Lea 1977; Pate 1980).

Storage of nitrogen

Plant dry matter contains a fairly constant proportion of carbon, but a very variable amount of nitrogen, so that its carbon/nitrogen ratio can vary from as little as 5:1 to up to 150:1 (Pate & Layzell 1981). Lowest nitrogen contents (0.3% or less) are found in the dry matter of old stems or woody tissues, carrying low fractional volumes of living cells and high amounts of cellulose and lignin; roots are also low in nitrogen (1–2% in dry matter); leaves and young parts of shoot are of much higher nitrogen content (4–5%); and highest values of nitrogen (7–8%) are recorded for storage parenchyma of seeds specializing in protein storage (Pate & Layzell 1981). Across this broad range runs great variability in ratios of soluble to insoluble nitrogen.

Storage may be defined simply as the accumulation of nitrogen in amounts manifestly exceeding immediate demands for growth and maintenance of tissues; but it is by no means easy to assess the magnitude of storage or its significance nutritionally to later stages of growth. Moreover, the situation within the whole plant is complicated by interdependencies of plant parts for nitrogen, ontogenetic changes in organ capacities for assimilation, storage and cycling of nitrogen, and difficulty in measuring which fractions of an organ's complement of nitrogen, as opposed to resident pools of nitrogen, are currently devoted to transport. Accordingly, it is virtually impossible to measure precisely the basic needs of a plant organ for nitrogen.

Despite what is said above, evidence of storage is not hard to find in higher plants. Firstly, microscopically recognizable stores of protein may be encountered. Secondly, pools of unusual solutes of nitrogen may be present and found to be in a state of low turnover, and therefore likely to be only distantly related to the day to day metabolism of the tissue in which they currently reside. Thirdly, in comparison with the low pool sizes of free amino compounds in microorganisms and non-vacuolate meristematic tissue, the mature tissues of many plant organs contain high levels of soluble nitrogen, often in the form of nitrogen-rich compounds such as amides, ureides, arginine, citrulline—all potential sources of nitrogen for growth, but clearly not being utilized by the tissue at the time of examination.

If stores of soluble nitrogen are not utilized in the life of an organ or tissue, the nitrogen may be mobilized to other organs and may even cycle through several generations of plant parts before reaching a region equipped enzymatically to use the relevant solutes. When this happens the compounds used for storage are usually the same as, or closely related to, those used for transport of nitrogen—a common situation in higher plants (Pate 1980) and well demonstrated in white lupin (*Lupinus albus* L.) in respect of the amides, glutamine and asparagine (Pate *et al.* 1981).

Plants may accumulate a range of nitrogen-containing compounds which apparently function in aiding survival of species in inhospitable environments; for example, the accumulation of certain solutes under salt or water stress, and the synthesis of compounds conferring resistance against attack by grazing animals or insects. In each case significant fractions of the plant's nitrogen may be associated with the protective compounds in question.

Accumulation of proline and, in certain species, of glycine-betaine and other methylated quaternary ammonium compounds, is now well documented for plants under osmotic stress, particularly amongst species of halophytes (Stewart & Lee 1974; Flowers, Troke & Yeo 1977; Cavalieri & Huang 1979; Storey & Wyn-Jones 1979; Stewart *et al.* 1979). These solutes, and other non-nitrogenous solutes (e.g. sugars) which may also accumulate, are generally believed to function as metabolically compatible counter-osmotica to salt absorbed by the tissue (Stewart *et al.* 1979; Greenway & Munns 1980; Winter, Osmond & Pate 1981). To be effective in this regard the organic solutes in question would have to accumulate in cytoplasm, thereby countering salt accumulation in vacuoles and protecting enzyme systems from the toxic effects of salt.

Nitrogenous compounds are prominent amongst the chemical defence agents employed by plants. They include outright poisons, often with sensory warnings in terms of bitterness of taste (e.g. alkaloids) or subsequent discomfort (e.g. cyanogens, seleno-amino compounds, neurotoxins); or toxic materials of less immediate effect (e.g. anti-nutritional non-protein amino acids, haemolytic or hypoglycaemic principles, depilatory agents, and a range of nitrogenous substances suspected to have carcinogenic properties (Bell 1980)). Often one part of the plant, for example the leaves or the embryos of the seed, may be highly poisonous, whereas other parts, say the fleshy coat of an animal-dispersed seed, are attractive and palatable.

Many of the above toxic substances are likely to have co-evolved with the development of detoxifying mechanisms on the part of specific predators (Main 1981): this may explain why such high concentrations of certain toxins are present, and why only a few predators turn out to be resistant to their effect. Examples are afforded by the quite massive accumulation of L-canavanine and certain other nitrogenous substances in seeds of legumes, and the unique resistance to these compounds displayed by certain seed-eating bruchid beetles (Rosenthal, Dehlman & Janzen 1976); and the lack of harmful effects of leaves of *Acacia* on insects adapted to eating these leaves, versus the highly lethal effect of *Acacia* leaves to insects not normally using this plant for food. The active principles in *Acacia* are believed to be homoarginine, pipecolic acid and 4-hydroxy-pipecolic acid (Bell 1980).

Evolution of autotoxicity in plants is a one-way process, reinforced and maintained by predatory agents. Disastrous effects are evident once a species loses its ability to produce a toxin. This is well documented for the naturally occurring acyanogenic lines of *Lotus corniculatus* L. and *Trifolium repens* L. (Foulds & Young 1977). It can also be observed experimentally by raising plants of *Astragalus* in selenium-deficient conditions, whereupon, being unable to synthesize toxic seleno-amino compounds, plants become readily susceptible to attack by a range of insects (J.S. Pate, unpublished). Amongst cultivated plants the sad state of many grain legumes in terms of yield losses through disease and predation by animals may be directly attributed to the removal by selective breeding of a range of toxic and antinutritional factors.

The ideal situation, of course, would be for a substance to serve first as agent of survival, then as a source of nitrogen for plant growth. There is some evidence of this. Proline, for instance, is metabolized once water or salt stress is relieved; alkaloids may be utilized at the end of the growth cycle of an annual plant as sources of nitrogen for filling of seeds; and certain toxic non-protein amino compounds achieving high concentration in seeds (e.g. canavanine, 3,4-hydroxyphenylalanine (Bell 1980)) are likely to serve as nitrogen sources once germination has occurred and danger from seed-eating beetles has passed.

PATTERNS OF UPTAKE, TRANSPORT AND UTILIZATION OF NITROGEN IN SINGLE SPECIES OR GROUPS OF SPECIES WITH SIMILAR BEHAVIOUR

Most studies of this nature have concerned cultivated annual plant species grown under defined nitrogen regimes in glasshouse or laboratory. Virtually all have dealt with the utilization of nitrate. The examples given below focus, wherever possible, on the siting and relative activity of assimilatory processes, types of solutes used for transport, the storage potential of plant organs for nitrogen, and the nutritional dependencies of plant parts during growth and development. In certain instances uptake and partitioning of nitrogen are viewed alongside the flow of photosynthetically fixed carbon within the plant. A separate section is devoted to grain legumes, in view of the author's familiarity with these species.

Annual species with a shoot-dominated pattern of nitrate assimilation

Species in this category are readily recognized by a low or undetectable level of nitrate reductase activity in roots. Their root xylem sap regularly contains upwards of 95% of its total nitrogen as nitrate, provided nitrate is freely available in the rooting medium. When luxury uptake of nitrate has occurred, root, stem and petioles build up large reserves of nitrate (Schrader 1978; Pate 1980). Pools of soluble organic nitrogen are usually very small; in *Xanthium*, for example, they amount to only a few percent of the total soluble nitrogen (Wallace & Pate 1967).

Assays for nitrate reductase indicate that the bulk of nitrate assimilation takes place in leaves (Wallace & Pate 1967 (for *Xanthium*); Radin, Sell & Jordan 1975; Radin 1977 (for *Gossypium*); Olday, Barker & Maynard 1976 (for *Cucumis*)). By analogy with phloem sap analyses for other species (Pate 1980), and information on phloem exudates of *Ricinus*, a species-reducing nitrate strongly in its tops (Kirkby & Armstrong 1980), phloem translocation from leaves apparently involves organic forms of nitrogen, not nitrate. Leaves are thus likely to be a major source of reduced nitrogen for growth of both shoot and root, although some nitrate may reach fruits, young leaves and shoot apices through the xylem, in accordance with transpirational loss. This may explain the nitrate reductase activity exhibited by such parts (Wallace & Pate 1967 (*Xanthium*)). Also, if the supply of nitrate to the root is withdrawn, nitrate previously stored in stems becomes available as a nitrogen source for growth, although the manner of its utilization is not known (Wallace & Pate 1967 (*Xanthium*)).

The nitrogen nutrition of the roots of species reducing nitrate in their shoots is still a matter of debate. Radin (1977, 1978), studying cotton (*Gossypium*), and using an assay *in vivo* for nitrate reductase, concluded that the low level of nitrate reductase expressed by the root would be sufficient to satisfy the roots' relatively modest demand for reduced nitrogen. In contrast, Wallace & Pate (1967), studying cocklebur (*Xanthium*), were unable to demonstrate nitrate reductase activity in roots by assay *in vitro*, and suggested that the root must subsist on amino-nitrogen translocated from the shoot in the phloem. Unfortunately, phloem exudates have not been obtained from these species, so it is not possible to determine how much reduced nitrogen, if any, is delivered to roots with sugar and other solutes in the translocation stream. Again, by analogy with data on phloem sap composition of *Ricinus* (Hall & Baker 1972), levels of reduced nitrogen in translocate are likely to be significant.

According to Dijkshoorn (1971) and Ben-Zioni, Vaadia & Lips (1971) plants reducing nitrate mainly in their shoots are capable of generating a pattern of ion flow between root and shoot which might regulate nitrate uptake from the medium. The argument reads as follows:

1. The principal charge-determining ions moving from root to shoot in xylem are potassium and nitrate.
2. Reduction of nitrate in leaves results in the production of an equivalent amount of organic acid anions ($RCOO^-$).
3. The principal charge-determining ions moving from shoot to root in phloem are potassium and $RCOO^-$.
4. Metabolism of this $RCOO^-$ in the root results in production of bicarbonate ions. Translocated potassium cycles back to the shoot via the xylem.
5. Extrusion of bicarbonate to the rooting medium results in uptake of an equivalent amount of nitrate by the root system.

These suggestions are supported by observations of increased pH and excess anion uptake in water cultures of plants grown with nitrate as sole source of nitrogen; they are compatible with the principles of internal regulation of pH in plant tissues as defined by Raven & Smith (1976); and they receive direct experimental confirmation in studies on the composition of xylem and phloem exudates and the cation-anion balance of castor bean (*Ricinus*) by Kirkby & Armstrong (1980). These workers found that efflux of bicarbonate ions accounted for 56–63% of the anion charge in water cultures of *Ricinus* at a high (10 m-equiv. l^{-1}) level of nitrate where most nitrate was reduced in shoots, but only 23% of the anion charge at a low (1 m-equiv. l^{-1}) nitrate level, where a larger fraction of the absorbed nitrate was likely to have been assimilated by the root. As the hypothesis stated above would suggest, xylem ions were principally potassium and nitrate, those of phloem principally potassium and $RCOO^-$.

There is evidence that charge regulation in plants reducing nitrate in their shoots does not necessarily involve a massive translocation of organic acid anions to roots. Many species accumulate organic acids in shoots, in amounts more or less matching shoot activity in assimilating nitrate. Tomato (*Lycopersicon*) is a case in point—a species which, unlike *Ricinus*, carries much higher levels of organic acids in its tops than in its roots when feeding on nitrate (Armstrong & Kirkby 1979).

Species showing marked ability to reduce nitrate in their roots

These species characteristically show high levels of nitrate reductase activity in root tissue and produce a xylem sap in which most or all of the nitrogen is in organic form, as opposed to nitrate (Bollard 1960; Pate 1973). Organic solutes of nitrogen also predominate in storage pools of nitrogen in the shoot, although some nitrate may accumulate in stem and petioles if nitrate is available in luxury amounts and has escaped the nitrate reductase systems of the root to reach the shoot in the transpiration stream. Leaf tissue rarely accumulates nitrate nitrogen, even under very high levels of applied nitrate thus suggesting that the photosynthetically linked nitrate reductase can easily cope with nitrate spilling over from root and stem to leaves.

Annual cultivated species falling into this category are certain cereals (e.g. *Hordeum*, *Triticum*, *Zea*) and radish (*Raphanus*), all of which store and transport glutamine as the principal solute of nitrogen. Examples of woody species are apple (*Pyrus*) and peach (*Prunus*), both of which store soluble nitrogen largely as arginine and asparagine (Bollard 1960; Taylor & May 1967; Tromp & Oova 1976, 1979). In apple, nitrate is detected in xylem (tracheal) sap only after heavy applications of nitrogen fertilizer: a treatment causing elevated levels of organic forms of soluble nitrogen in root and trunk (Tromp & Oova 1979), and induction of nitrate reductase in leaves (Beevers & Hageman 1969). With lower availability of nitrate, little or no nitrate escapes to the shoot (Bollard 1960), indicating that the nitrate reductase system present in fine feeding roots (Grasmanis & Nicholas 1967) is a major site of assimilation.

Stem tissue of apple can abstract a range of amino compounds from the xylem stream, as shown in a series of experiments involving application of ^{14}C-labelled amino compounds to shoot segments of the species (Hill-Cottingham & Lloyd-Jones 1968, 1973). This may be the mechanism whereby the trunk replenishes storage pools of nitrogen during the growing season. These stem reserves are mobilized to buds and young foliage in spring, whereupon xylem sap can become greatly enriched with nitrogen (Bollard 1960). A similar seasonal cycle of storage and release of nitrogen may be common amongst deciduous species of trees (Taylor 1967; Sauter, Iten & Zimmermann 1973; Ziegler 1975).

Annual grain legumes feeding on nitrogen, nitrate, or nitrogen and nitrate

Nodulated plants grown without combined forms of nitrogen assimilate the ammonium produced in nitrogen fixation into a range of solutes of nitrogen (Pate & Atkins 1983), and the bulk of these pass in concentrated form from nodule to host through the xylem (Pate, Gunning & Briarty 1969; Wong & Evans 1971; Streeter 1979). Some tropical species of grain legumes (e.g. of the genera *Vigna*, *Phaseolus*, *Psophocarpus*, *Cyamopsis*, *Cajanus*, *Glycine*, *Macrotyloma*) specialize in forming ureides as major products of nitrogen fixation and may also store these compounds in stem and petioles during vegetative growth (Herridge *et al.* 1978; McClure & Israel 1979; Streeter 1979; Pate *et al.* 1980). These species mostly possess nodules with a determinate pattern of growth and an open vascular network between nodule and root (Sprent 1980). Other grain legumes (e.g. *Lupinus*, *Pisum*, *Lathyrus*, *Vicia*, *Lens*), largely of temperate origin, and with nodules of indeterminate growth and closed vascular network, tend to form asparagine and glutamine as major products of fixation and to store these and related amino compounds in their stem and leaves (Wallace & Pate 1965; Pate *et al.* 1979; Sprent 1980, Pate & Atkins 1983).

In *Pisum* and *Lupinus* phloem exudates have been collected and found to carry 70% of their nitrogen as asparagine and glutamine, just as in xylem transport of the same species (Lewis & Pate 1973; Pate *et al.* 1979). This phloem supply of amide-nitrogen is of special significance to nutrition of shoot apices and developing fruits, in view of the poor ability of such organs to attract xylem-borne solutes through transpirational loss of water (Pate, Sharkey & Atkins 1977; Layzell *et al.* 1981).

It remains to be seen whether the ureides allantoin and allantoic acid fulfil a similar role in phloem translocation of nitrogen-fixing legumes in which these solutes feature prominently in xylem transport. According to data provided by Sprent (1980), there may be problems of solubility in the use of ureides for transport and storage; and, as suggested by Pate & Atkins (1983) catabolism of these compounds in the shoot, following their receipt from nodules via the xylem, may offer a less direct route for nitrogen to protein synthesis than in the case of amides or amino acids. Nevertheless, more economical usage of translocated carbon has been observed in nodules of a ureide-producing than in an amide-producing grain legume (Pate, Atkins & Rainbird 1981), and such an advantage would have to be weighed against some of the disadvantages of using ureides as suggested above.

Amide-producing legumes export virtually identical sets of organic solutes from their roots when utilizing ammonium or nitrate, as when relying on nitrogen (Pate & Wallace 1964; Atkins, Pate & Layzell 1979). Accordingly, apart from the possible presence in xylem of nitrate as evidence of access to soil nitrogen, it is not possible to determine from xylem sap analysis how heavily a nodulated plant is relying on soil or fertilizer nitrogen. In marked contrast, legumes forming ureides when fixing nitrogen tend to produce more amides and amino acids, but very much less ureide when utilizing ammonium and nitrate (McClure & Israel 1979; Thomas, Feller & Erismann 1979). For example, in cowpea (*Vigna unguiculata* Walp.) nodulated plants exposed to a range of levels of nitrate were shown to vary widely in composition of xylem sap, depending on the proportions of their nitrogen derived from nitrate (Pate *et al.* 1980). With increased dependence on nitrogen as opposed to nitrate, the proportion of xylem sap nitrogen as ureide increased, as did the ratio of glutamine to asparagine in xylem sap whereas, with increasing dependence on nitrate, there was a progressive increase in the proportion of xylem sap nitrogen as nitrate, and increases in the ratio of asparagine + nitrate to glutamine + ureide. These compositional differences between nitrogen- and nitrate-dependent plants proved to be true of a wide variety of ureide-producing grain legumes, thus encouraging the use of xylem sap analysis as a field assay of crop reliance on symbiotically fixed nitrogen as opposed to soil- or fertilizer-nitrogen (Pate *et al.* 1980).

Annual grain legumes vary appreciably in the extent to which their roots assimilate nitrate. The picture so far is that amide-producing species of temperate origin possess highly active, root-based nitrate reductases, e.g. *Pisum*, *Vicia* and *Lupinus* (Pate 1973; Atkins, Pate & Layzell 1979), while ureide-forming tropical species reduce the bulk of absorbed nitrate in their shoots provided, of course, that nitrate is freely available from the rooting medium (Martin 1976; McClure & Israel 1979; Thomas, Feller & Erismann 1979; Atkins *et al.* 1980). If this is generally true for legumes, a study of its implications in terms of the ecology of the ancestral species would be of great interest.

Recent studies on white lupin (*Lupinus albus* L.) have attempted to show how carbon- and nitrogen-containing solutes are partitioned and utilized during growth of nodulated plants relying on nitrogen gas as the source of nitrogen. To this end, empirically based models of uptake, flow and utilization of carbon and nitrogen have been constructed for the species over various intervals of growth (Pate, Layzell & McNeil 1979; Layzell *et al.* 1981). Construction of the models requires measurements of increments of carbon and nitrogen in dry matter of plant parts, ^{14}C-feeding studies of translocate distribution from leaves, assessments of gains and losses of carbon as carbon dioxide from plant parts, and determinations of weight ratios of carbon to nitrogen in xylem and phloem streams serving various regions of the plant. The exercise consists of allocating the plant's net intake of carbon and nitrogen to specific organs through xylem and phloem, on the assumptions that transport involves mass flow, and that carbon and nitrogen move through phloem and xylem in relative amounts as suggested from carbon/nitrogen ratios of xylem and phloem exudates. The model thus predicts the mixtures of xylem and phloem streams which meet precisely the recorded amounts of carbon and nitrogen produced or consumed by each part of the plant during the period of study.

A principal outcome of the modelling exercises has been to indicate how nodulated root and apical growing parts of the shoot of white lupin draw upon the carbon and nitrogen available from mature parts of the shoot system. The situation for the period 51–58 days after sowing is shown in Fig. 11.1, using data provided by Layzell *et al.* (1981) and Pate & Layzell (1981). Apical shoot parts, including the inflorescence and upper axillary shoots at this stage of development, have a relatively small requirement for carbon but a relatively large one for nitrogen. This reflects the low carbon/nitrogen ratio (12–15:1) of their dry matter, and an ability, through possession of chlorophyll, to maintain themselves at, or slightly above, compensation point for carbon dioxide during daytime. These apical parts derive virtually all of their carbon as sugar-rich translocate (carbon/nitrogen ratio of 60–80:1) from the uppermost strata of leaves; while their main supply of nitrogen comes from other sources, namely direct attraction of xylem fluid (carbon/nitrogen ratio of 2:1), and a process of 'xylem to phloem transfer' in upper regions of the stem, by which amides and other nitrogen solutes of xylem enrich the translocate stream to the apex with nitrogen (Fig. 11.1). Nitrogen attracted to the shoot apex through xylem is pictured as consisting of two components, one representing the amount of nitrogen which would be delivered should xylem fluid reach the apex at the same nitrogen concentration at which it left

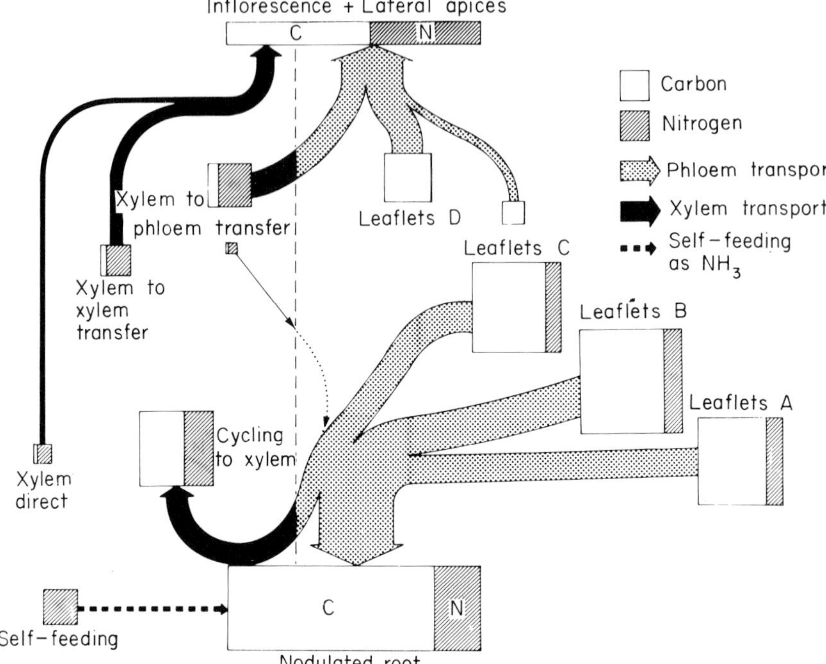

FIG. 11.1. Diagrammatic representation of the source agencies of the mature shoot which supply carbon and nitrogen to the young apical region of the shoot (inflorescence + lateral apices) and the nodulated root. Relative amounts of carbon and nitrogen supplied from each source to the two sink regions are indicated using flow lines of varying thickness, and rectangles (sinks) or squares (sources) of varying area. An area scale is used in which 10 units by weight of carbon are equivalent in area to one unit by weight of nitrogen.

Leaflets A–D refer to four strata of leaves on the main stem. Each stratum consists of four leaves, the lowest three strata supplying the root the upper stratum (D) the inflorescence and shoot apex. The data apply to the stage 51–58 days after sowing. The item marked 'self-feeding' refers to the capacity of the root nodules to incorporate fixed nitrogen directly into their dry matter (from Pate & Layzell 1981).

the root (item marked 'xylem direct', Fig. 11.1), the other consisting of the extra nitrogen reaching the apex due to xylem to xylem transfer in the stem. In the latter process nitrogen committed through leaf traces to lower leaves is envisaged as being abstracted from these traces at nodes of the stem, and then passing to xylem streams moving on up the shoot (item marked 'xylem to xylem transfer', Fig. 11.1). The net effect of these vascular exchanges in the stem is that apically sited organs acquire substantially more nitrogen than they would be able to attract were they to rely on phloem translocate at the nitrogen content at which it left upper leaves, and on xylem fluid at the concentration of nitrogen at which it was exported from the root. Transfer cells, which are believed to mediate xylem to xylem and xylem to phloem exchanges in stems, are prevalent in grain legumes (Kuo *et al.* 1980).

The model illustrated in Fig. 11.1 depicts the nodulated root of *L. albus* as requiring much more carbon relative to nitrogen (45:1) than in the case of the apical parts of the shoot (12–15:1). This stems from the massive loss of respired carbon dioxide from the root and the much lower carbon/nitrogen ratio in dry matter of root than of shoot (Pate & Layzell 1981). The root monopolizes the translocate of the three lower strata of leaves to meet its large requirement for carbon. In so doing it gains more nitrogen from the shoot than is required for root growth; and, since nodules and possibly proximal parts of the root can derive part of their growth requirement for nitrogen by direct incorporation of recently fixed nitrogen (see Layzell *et al.* 1979 and item marked 'self-feeding', Fig. 11.1), the nodulated root, at this stage of growth, is clearly in positive balance in respect of nitrogen. It is suggested that this surplus nitrogen returns to the shoot in the xylem (item marked 'cycling to xylem', Fig. 11.1), thereby effecting further differential enrichment of plant tops with nitrogen.

It is not known how widely mechanisms similar to those of Fig. 11.1 apply amongst higher plants. However, xylem to phloem transfer appears to be prevalent in both woody and herbaceous species (Pate 1980), suggesting that this process, at least, might be a common mechanism for partitioning nitrogen preferentially to growing parts of the shoot. Nevertheless, exchange systems for nitrogen in stems might be necessary to much lesser extent in species generating reduced nitrogen largely by nitrate reduction in leaves. Also, as suggested by Pate & Layzell (1981), the potential exists in species possessing internal and external phloem for carbon and nitrogen to be partitioned to roots and shoots using streams of widely differing carbon/nitrogen ratio; the high requirement of the root for carbon and that of the shoot for nitrogen may thereby possibly be satisfied without recourse to xylem to xylem or xylem to phloem exchanges in the stem.

PATTERNS OF NITROGEN METABOLISM IN RELATION TO LIFE-FORM AND ECOLOGICAL CONDITIONS

Virtually every ecosystem displays a range of contrasting life-forms and nutritional strategies, whether one recognizes these in terms of morphologically or physiologically based adaptations. In the sections which follow, some of these classes of plants will be examined in an attempt to discover whether consistent patterns of uptake, assimilation and storage of nitrogen occur and, if such are evident, how they might relate to the growth cycle and reproductive features of the species under consideration.

Annual or short-lived perennial species

These plants usually fall under the heading of '*r*-selected' species (MacArthur & Wilson 1967; Stearns 1976), characterized by fast growth rate, short life cycle with high reproductive achievement, and ability to exploit environments which are transiently non-limiting in water and nutrients. These 'opportunist' species are common in the ephemeral flora developing in arid zones after occasional rain; in disturbed areas of forest and open ground, say after flood, landslide or effects of storm; and as pioneer species after fire. Many are now important weeds of agriculture. In the case of *r*-selected 'fire weeds' the species are established only after fire and complete their growth cycle before perennial species have regenerated fully from fire-resistant stocks. Fire weeds can exhibit rates of growth up to 15–20 times those of sympatric seedlings of perennial climax species, even when the latter (but not the fire weeds) are legumes and capable of fixing nitrogen (J.S. Pate, unpublished data for sand plain ecosystems in S.W. Australia).

Were nitrogen to be available at a constant, near-limiting rate throughout the cycle of growth of an *r*-selected species, decided advantage might be obtained in terms of competitive ability if the species invested all of its newly absorbed nitrogen immediately into new leaf and root tissue, thereby maximizing further gain of dry matter. On this argument, a species recoursing to storage of nitrogen, say in fleshy roots or stems, would be severely disadvantaged. However, in many situations in which *r*-selected species abound, the very environmental stimulus (e.g. onset of rain, fire, or other disturbance) prompting establishment of the species also fosters release of a pool of soil nitrogen, albeit of small size, yet unlikely to be limiting for at least the early period of growth of the ephemeral species. In this situation a species engaging in temporary storage of nitrogen might compete on more equal terms, since the store of nitrogen would enable it to extend its growth cycle beyond the time when soil nitrogen had been depleted, and long after non-storing species had senesced due to nitrogen deficiency and the demands of nitrogen in reproduction.

To explore the above suggestions experimentally, the *r*-selected ephemeral flora of mulga (*Acacia aneura*) scrub was recently examined following good winter rainfall at a site near Meekatharra, West Australia (J.S. Pate, unpublished data). The soil was semi-saline and quite rich in nitrate. A wide range of species and families was represented amongst the ephemerals, including Asteraceae, Apiaceae, Poaceae, Portulacaceae, Solanaceae, Boraginaceae, Lobeliaceae, Campanulaceae, Chenopodiaceae, Polygonaceae, Zygophyllaceae, Amaranthaceae and Goodeniaceae, and the species encompassed a varied range of morphologies and reproductive behaviour. When examined just prior to, or during, flowering some of the species proved strongly positive for stored nitrate but only weakly positive for ninhydrin-positive soluble nitrogen thereby suggesting weak nitrate reductase activity in roots and good storage capacity for unreduced nitrate nitrogen. Many of these species had fleshy stems and leaves and tended to be late-flowering. A second group of species proved negative in their shoots for nitrate yet were high in organic soluble nitrogen, indicating reduction of nitrate immediately upon absorption from the soil and a capacity to form temporary reserves of reduced nitrogen. A range of fleshy and non-fleshy species was represented in this, the commonest category. The third category consisted of species giving only weak reactions for nitrate and soluble amino-nitrogen, implying a direct coupling of nitrogen uptake to growth. Most species in this group were small and early-flowering, especially members of the Asteraceae.

Unfortunately, it has not yet proved possible to examine the complementary ephemeral flora of the Meekatharra study region developing after occasional summer rains, but the data so far suggest that species differing greatly in pattern of nitrogen metabolism may coexist in tightly knit communities, presumably sharing a common pool of nitrogen. One pattern of metabolism may carry subtle advantages over others in one set of conditions; and, should this be reversed in other seasons or sets of conditions, one might discover at least part of the reason why dominance patterns amongst these ephemeral populations change so noticeably from site to site and season to season. Of course, other environmental influences, such as intensity, timing and sequence of fall of rains, might be equally important in determining the mix and relative success of species, and these would have to be considered alongside behaviour in relation to nitrogen metabolism, before a full analysis of the formative influences within the ecosystem could be obtained. Much further study of this type of situation is obviously required.

Herbaceous perennials

Most plants in this category come under the heading of '*K*-selected' species, namely slow-growing elements of climax or near-climax vegetation, with prolonged seedling and vegetative stages and an ability to utilize environmental resources efficiently in terms of conservation of nutrients and water during growth and reproduction (Stearns 1976).

Species of this kind often exhibit perennation by means of rhizomes, stem and root tubers, tap roots, corms and bulbs; and, as in the case of the *r*-selected species mentioned earlier, examples may be found of species which store soluble nitrogen principally as nitrate (e.g. *Menyanthes*, *Borago*, perennial Chenopodiaceae (McKee 1962; Pate 1980)), as ureides (e.g. *Symphytum* (Reinbothe & Mothes 1962)), or as nitrogen-rich amino compounds such as amides, non-protein amino acids, arginine and citrulline. Table 11.1 gives examples involving this last group of compounds as encountered in the aestivating storage organs of a series of West Australian geophytes. Amongst these geophytes, water content of dormant storage organs ranged from 2–98% of fresh weight; total nitrogen from 2–36 mg N g dry wt.$^{-1}$; protein content (ethanol-insoluble nitrogen $\times$ 6.25) from 0.2–18% of dry weight; percentage of total nitrogen in ethanol-soluble form from 5–95% (Pate & Dixon 1981; Pate & Dixon 1982). With this bewildering array of storage propensities for water, and soluble and insoluble nitrogen, it becomes most interesting to discover how combinations of specific storage characteristics might equip species for survival in particular climatic and edaphic conditions.

Regardless of the form in which nitrogen may be stored, the reserves of fleshy underground storage organs are of obvious advantage to species inhabiting regions of seasonal drought and low availability of water and soil nutrients. For example, in the mediterranean climate of the sand plains and coastal forests of S.W. Australia, geophytes often represent the most successful elements of the herbaceous winter ground flora. They appear to outcompete annual species in many situations by employing their substantial aestivating resources of nitrogen, water and other nutrients to establish new growth rapidly after, or even slightly in anticipation of, winter rains, and by carrying over in their storage organs from season to season supplies of nitrogen, phosphorus and other minerals, abstracted from the soil over many seasons of growth. The efficiency of carry over of minerals from one season's vegetative growth to a new set of storage organs may be extremely high, with values of 80–95% being recorded for the retrieval of nitrogen, phosphorus and potassium from leaf stem and root to developing corms or tubers of a variety of species (Pate & Dixon 1982).

Parasitic angiosperms

Most ecosystems possess one or more parasitic angiosperms. Their worldwide taxonomic diversity is matched by equally great variation in life-form and nutritional adaptation of species, with examples being known of minute epidermal holoparasites (e.g. *Pilostyles*, *Arceuthobium*), hemiparasitic epiphytes (*Viscum*, *Loranthus*), scrambling aerial holoparasites (*Cuscuta*, *Cassytha*), subterranean root holoparasites (*Lathraea*, *Striga*, *Orobanche*), and

TABLE 11.1. Storage of amino compounds in aestivating storage organs of a range of Western Australian plants[1] (data from Pate & Dixon, in press).

Category 1 Arginine as principal compound (27 species)

(a) Arginine only as major compound: *Tacca maculata; Stylidium carnosum; Platysace deflexa; P. maxwellii; Corybas dilatatus; Elythranthera brunonis; Ophioglossum* sp; *Phylloglossum drummondii; Schoenolaena juncea; Isoetes muelleri; Utricularia menziesii; Arthropodium capillipes; Poa drummondiana; Clematicissus angustissima; Villarsia albiflora; Drosera erythrorhiza; D. gigantea; D. macrantha; D. stolonifera.*

(b) Arginine + Serine[2] *Clitoria australis*

(c) Arginine + Asparagine *Stylidium petiolare*

(d) Arginine + Asparagine + Glutamic Acid *Philydrella pygmaea*

(e) Arginine + γ-methylene glutamic acid + γ-methylene glutamine *Burchardia umbellata, B. multiflora*

(f) Arginine + γ-Amino Butyric Acid + Alanine *Stirlingia latifolia*

(g) Arginine + Glutamine *Thysanotus speckii, Thysanotus arbuscula*

Category 2 Asparagine as principal compound (16 species)

(a) Asparagine only as major compound *Platysace cirrosa, Trichocline spathulata, Ipomoea gracilis, Abelmoschus moschatus, Tribonanthes variabilis, Commelina lanceolata, Chamaescilla spiralis, Brachychiton gregorii*

(b) Asparagine + Arginine[2] *Pterostylis recurva, Spiculaea ciliata, Thelymitra fuscolutea, Chamaescilla corymbosa, Triglochin procera, Lyperanthus nigricans, Haemodorum spicatum*

(c) Asparagine + Aspartic Acid *Tricoryne* sp.

Category 3 Glutamine as principal compound (3 species)

(a) Glutamine only as major compound *Stylidium caricifolium, Pentapeltis peltigera*

(b) Glutamine + Asparagine[2] *Clematis pubescens*

Category 4 Citrulline as principal compound (3 species)

(a) Citrulline only as major compound *Thysanotus patersonii*

(b) Citrulline + Glutamine[2] *Caesia* sp.

(c) Citrulline + Arginine *Macrozamia riedlei*

Category 5 Variety of compounds at low concentration (8 species)

Calothamnus tuberosus, Colocasia antiquorum, Hypoxis occidentalis, Dioscorea hastifolia, Gastrodia sesamoides, Juncus holoschoenus, Xanthorrhoea preissii, Ipomoea longiflora

[1] All organs collected from native habitat at commencement of dormancy.

[2] A compound or compounds additional to a principal compound are included in a subcategory if together or singly they account for at least 30% of the soluble nitrogen.

root hemiparasites ranging from small-sized herbs (*Euphrasia, Melampyrum, Rhinanthus*) to sizeable trees and shrubs (*Santalum, Nuytsia*).

Generally, hemiparasites establish haustorial connections principally with the xylem of their hosts, and are therefore likely to benefit mainly from the minerals and organic nitrogenous solutes in the transpiration stream of their hosts (Govier, Brown & Pate 1968; Kuijt 1969). Holoparasites tap phloem (and sometimes also xylem), deriving carbohydrate and other organic solutes from the host translocation stream (Kuijt 1969).

Strong compositional resemblances in storage pools of nitrogen have been recorded between parasitic angiosperms and their hosts (Greenham & Leonard 1965; Govier, Nelson & Pate 1967), though ability to metabolize a range of amino compounds is of obvious advantage to promiscuous species capable of parasitizing many different hosts. For instance, red bartsia (*Odontites verna* (Bell.) Dum) utilizes glutamine, provided as main nitrogen solute from *Stellaria* and *Hordeum*, as easily as it does asparagine, the principal xylem solute which it derives from *Trifolium* (Govier, Nelson & Pate 1967; Govier, Brown & Pate 1968). Similarly, the hemiparasitic epiphyte *Tapinanthus bangwensis* Danser can flourish on host species of very different nitrogen content. Thus, total nitrogen content of the hemiparasite was shown to range from 18–33 mg N g dry wt.$^{-1}$, highest values for nitrogen being found when parasitizing hosts richest in nitrogen (Stewart & Orebamjo 1980). *Tapinanthus* showed induction of nitrate reductase when its shoots were infiltrated with nitrate, but nitrate reductase levels of it and its hosts were low when examined in natural habitats (Stewart & Orebamjo 1980). Thus, nitrate may not feature prominently in the normal nutrition of the species; the same conclusions have been drawn for the holoparasite *Cuscuta* by Srivastava & Chauhen (1977).

Herbaceous root hemiparasites (e.g. *Rhinanthus*, *Odontites*, *Orthocarpus*) can be grown without a host if provided with nitrate as a source of nitrogen, but their growth and reproductive output are greatly improved if a suitable host is parasitized. The nature of the stimulus to growth is unclear; organic solutes of nitrogen, or hormone-type substances abstracted from the xylem of the host might be responsible, though some more subtle form of exchange of stimulatory materials at the haustorial junction is equally possible. These suggestions are discussed more fully by Klaren & Janssen (1978), and Heineman & Atsatt (1978). In any event, once haustorial connections are established, a holoparasite or hemiparasite may drain its host most effectively of nitrogen and other solutes. This is well known to agronomists for troublesome parasitic weeds such as *Orobanche*, *Striga* and *Cuscuta* which, when present in large numbers, may cause failure or much reduced yield of the herbaceous crop species which they are exploiting.

Carnivorous plants

These plants typically occur in nutrient-poor environments or environmental niches, where they gain advantage over species with a conventional nitrogen nutrition by catching and digesting various forms of animal life, especially insects. The species studied so far in axenic culture turn out to be facultative heterotrophs, able to complete their growth cycle solely using inorganic nitrogen absorbed by their roots, but benefiting measurably in terms of growth and prolificity of reproduction when having access to suitable animals (Harder 1964; Small *et al.* 1977; Simola 1978).

Heslop-Harrison (1978) has recently summarized the different attributes of the principal genera and families of carnivorous plants in relation to types of trapping mechanisms and means of digestion of captured animals. She points to the basic difference between the digestion process of *Nepenthaceae* and *Lentibulariaceae*, occurring in a liquid-filled pitcher or bladder, and the essentially 'dry' digestion process on the leaf surfaces shown in other members (e.g. *Droseraceae*, *Byblidaceae*). Deployment of glandular tissues for purposes such as the production of mucilage, the secretion of digestive enzymes, and the absorption of digestion products also varies between genera. For example, in *Drosera* spp. the long, glandular tentacles on the leaves probably serve all of the above purposes, while in *Pinguicula* large stalked glands secrete mucilage while the small, semi-recessed glands scattered on leaf surfaces between the stalked glands produce digestive enzymes (Heslop-Harrison 1978).

Studies of the complements of digestive enzymes of carnivorous plants have shown proteases to be principal, regular components, implying benefit to the plant in terms of nitrogen from protein of animals carcasses. Acid phosphatases and esterases are often encountered, and lipases, ribonucleases, peroxidases and amylases have been recorded in certain instances (Heslop-Harrison 1978). This indicates that macromolecules other than protein might be utilized.

Release of sulphur, phosphorus and trace elements during digestion of insects has been suggested to be of benefit to certain Australian species of *Drosera*, especially in habitats deficient in these elements (Chandler & Anderson 1976; Pate & Dixon 1978). The West Australian tuberous species *Drosera erythrorhiza* Lindl has been shown to acquire ^{14}C and ^{15}N when its leaf rosettes are fed with *Drosophila* flies tagged with these isotopes, absorbing ^{15}N with 76% efficiency and, by the end of the growing season, mobilizing 70% of this absorbed ^{15}N to the new season's tubers (Dixon, Pate & Bailey 1980).

Observations on *D. erythrorhiza* in natural habitats showed that clones of the species caught a diverse range of arthropods during a growing season, with springtails (Collembola) most important in terms of numbers and amounts of nitrogen provided. It was estimated that animal prey captured by leaf rosettes provided from 11–17% of the net intake of nitrogen made seasonally by clones of the species. Indications were also obtained that the level of arginine in dormant tubers of a clone might be positively related to success in capture of animal prey during the previous growing season (Dixon, Pate & Bailey 1980; K.W. Dixon & J.S. Pate, unpublished data).

Trees and shrubs

The principal non-living reserve of nitrogen in most forest ecosystems comprises organic matter associated with litter decomposition in the surface layers of the soil, and since the acidity and high tannin content of these layers may severely restrict nitrification, the principal form of nitrogen provided to the feeding roots of plants may be ammonium, not nitrate (Rice & Pancholy 1972; Jordan, Todd & Escalante 1979; see also Lee, Harmer & Ignaciuk, p. 108, this volume).

Mycorrhizal fungi abound in heath and forest plant communities, not only in acid soils rich in organic matter (Mosse 1973; Tinker 1975; Bowen 1981), and, although it is widely conceded that woody plants gain minerals such as phosphorus, sulphur and zinc through association with such fungi (Sanders & Tinker 1973; Abbott & Robson 1977; Cox *et al.* 1980), possible benefit from mycorrhizal associations in terms of nitrogen nutrition has been less extensively investigated (see Alexander, p. 69, this volume). The most obvious forms of benefit from the fungal partner would be by making nitrogen available to its host by the digestion of soil organic matter, or by overcoming localized depletion of ammonium in the rhizosphere due to adsorption of this cation on colloids or other soil fractions.

Species inhabiting waterlogged acid peat soils also receive their nitrogen in forms other than nitrate, and probably largely by associating with mycorrhizal fungi. Indeed, such species may lack the ability to absorb and utilize nitrate, as indicated in studies on *Vaccinium* by Greidanus *et al.* (1972). The same probably applies to other members of the Ericaceae (Lee & Stewart 1978).

Most forest ecosystems carry woody species which fix nitrogen. Some of these are clearly of '*r*-selected' character, injecting nitrogen capital into the ecosystem over a short growth cycle after fire (e.g. the situation recorded for *Acacia pulchella* by Monk, Loneragan & Pate (1981) for forest ecosystems of S.W. Australia); others are '*K*-selected', making regular annual returns of nitrogen during a long growth cycle as climax members of an ecosystems (e.g. the situation for *Macrozamia riedlei* recorded for S.W. Australia by Halliday & Pate (1976)). The ecological significance of these and other nitrogen-fixing members of an ecosystem has been widely studied using the acetylene reduction assay and techniques based on ^{115}N (see Rennie 1979; W.D.P. Stewart, p. 1, this volume). Xylem sap analysis also offers promise as a means of assessing reliance on nitrogen in woody species forming ureides as products of fixation, using similar criteria to those mentioned above for annual grain legumes.

Analyses of tracheal (xylem) sap of woody species harvested from natural ecosystems often show absence of nitrate but presence of high levels of reduced nitrogen (Bollard 1960; Pate 1971). The organic solutes of nitrogen encountered are varied, but species-specific; for example, citrulline in *Alnus*, ureides in *Acer*, glutamine in *Pinus*, and arginine and asparagine in many species (see reviews of Ziegler 1975; Pate 1980). This situation probably reflects low availability of nitrate and the existence in roots and their mycorrhizal partnerships of nitrate reductase systems more than adequate to assimilate any nitrate absorbed from the soil.

There are, however, forest ecosystems in which nitrate may be regularly available in reasonably high amounts, or in which nitrate arises seasonally, or after events such as fire, clearing, or select felling of trees. To benefit from such regular or transient supplies of nitrate, trees and shrubs may have occasionally to deploy their above-ground parts for storing or reducing nitrate nitrogen. The ability to do this has been recently demonstrated for woody species of certain West African ecosystems, the species studied all showing massive induction of nitrate reductase in their above-ground parts following application of nitrate (see Stewart & Orebamjo, p. 167).

NITROGEN METABOLISM AND RESOURCE DEPLOYMENT OF PLANTS

The case histories presented in this chapter show great differences between plants in the manner in which they acquire, assimilate, store and utilize nitrogen. Each pattern of metabolism is patently successful in the context of species life-form and ecological circumstances, yet one can only guess at how specific combinations of metabolic and physiological characteristics equip plants to interact competitively or synergistically with other vegetational elements of an ecosystem. Clues as to what might be of adaptive significance are few and highly tentative, and cannot be properly evaluated without much more detailed information on the relative cost benefits or disadvantages of using different patterns of nitrogen metabolism in relation to the overall energy budget of the plant. Some of the problems encountered in an analysis of this kind are to be found in the paragraphs which follow.

Firstly, there is the basic question of the relative costs of assimilating different forms of nitrogen into organic solutes of nitrogen. It is here that plants are likely to differ most significantly. Assuming that these processes are implemented by complete oxidative phosphorylation of glucose at a $P/2\bar{e}$ ratio of 3 (6.3 mol ATP/mol CO_2), the theoretical maximum cost of nitrate reduction is 12 mol ATP equivalents/mol ammonium produced, that of nitrogen fixation within the range of 12.5–26.5 mol ATP/mol ammonium; the latter depending on the H_2/H^+ metabolism of the nitrogen-fixing organism

(Phillips 1980; Pate, Atkins & Rainbird 1981). Costs of assimilating ammonium into organic solutes depend, of course, on the solutes which are formed, but according to a recent analysis (Pate, Atkins & Rainbird 1981) these vary little between, say, amide- and ureide-producing species, and probably lie within the range 2.5–3.0 mol ATP (equivalents)/mol ammonium assimilated, i.e. a much lower relative cost than that of nitrogen or nitrate reduction to the level of ammonia.

One concludes from the above that, on strictly theoretical grounds, and considering only the biochemical reactions involved, nitrogen fixation is the most costly form of assimilation, respiration-linked nitrate reduction the next most costly, direct uptake of ammonium much less costly than either nitrogen or nitrate assimilation, and direct acquisition of already synthesized amino compounds (e.g. in parasites or carnivorous plants) the least costly of all. But other obvious 'costs' are involved in nitrogen assimilation. In nitrogen-fixing plant associations one must add the cost of formation and maintenance of symbiotic organs; in some ammonium assimilating plants, the maintenance of mycorrhizal associations; in parasitic plants, the cost of developing haustoria and of creating physiological conditions within the parasite promoting abstraction of solutes from the host; and in carnivorous plants, the cost of development and maintenance of systems for trapping and digesting insects. Placing values on these and other items in relative or absolute terms is just not possible on the information currently available.

Secondly, when considering plants subsisting on nitrate there is the problem of evaluating the use of above-ground or below-ground organs as sites of assimilation, and the related question of the disadvantages or advantages which might accrue from storing nitrogen in soluble form as unreduced nitrate or as already reduced forms of nitrogen. As has already been stated, reduction principally in roots is usually coupled with storage of organic nitrogen, while reduction of nitrate in shoots is usually associated with nitrogen being stored mainly as nitrate. Shoot-reducing species have the obvious cost advantage of utilizing reductant and ATP produced in photosynthesis for assimilating nitrate, although the advantage would be negated were nitrate reduction to have to compete for reductant with the photoassimilation of carbon dioxide (Canvin & Atkins 1974; Schrader 1981). Nitrate reduction in leaves is thus likely to be of maximum benefit to species at non-saturating light intensities, of least benefit, say, to shade plants. Despite its obvious cost in terms of respiration of carbohydrate, reduction of nitrate in roots may be advantageous under certain conditions. For example, in a deciduous tree in winter, deployment of fine feeding roots for absorption and reduction of nitrate in preparation for the peak demand of nitrogen for renewal of shoot growth in spring may be the strategy giving most effective use of overwintering resources of carbohydrate. The same may apply to herbaceous plants under conditions where shoot growth is limited by temperature but where roots at higher temperatures might still be employed for reducing nitrate. Moreover, reduction of nitrate in roots, with the possibility of direct extrusion of hydroxyl bicarbonate ions to the medium by the very tissues which are assimilating nitrate may carry significantly less cost in terms of internal pH regulation than in the case of nitrate reduction in leaves, where organic acids may have to be synthesized to maintain charge balance. One may also view the storage of nitrogen in the form of unreduced nitrate as a means of rapidly establishing an input of nitrogen from the environment at minimum cost in terms of consumption of carbohydrate. Nitrate storing species are commonly encountered in arid regions, where uptake of nitrate may be limited to relatively short periods after rain—the very situation in which rapid, low-cost uptake and storage may be crucial to survival in competition with other species. Once drought reconvenes, the store of nitrate can be slowly reduced as nitrogen source for growth.

Finally, there is the open-ended question of costing the processes in plants whereby resources of nitrogen are utilized for protective purposes rather than for growth, as has already been mentioned in relation to osmotic regulation in salt stress and defence against attack by animals. There may be other instances in this category, for example, in relation to the use of nitrogenous compounds as allelopathic substances, as antifungal agents, and even in relation to defence against viruses. In all these instances very little, if anything, is known of ultimate costs in terms of plant growth and productivity, and of the balance between these entities and the competitive benefits to be gained through resistance to predation or the development of substances harmful to surrounding vegetation. Here is to be found an area of intense ecological interest, long in need of further detailed investigation.

REFERENCES

Abbott L.K. & Robson A.D. (1977) The distribution and abundance of vesicular-arbuscular endophytes in some Western Australian soils. *Australian Journal of Botany*, **25**, 515–22.

Armstrong M.J. & Kirkby E.A. (1979) Estimation of potassium recirculation in tomato plants by comparison of the rates of potassium and calcium accumulation in the tops with their fluxes in the xylem stream. *Plant Physiology*, **63**, 1143–8.

Atkins C.A., Pate J.S., Griffiths F.J. & White S.T. (1980) Economy of carbon and nitrogen in nodulated and non-nodulated (NO_3-grown) cowpea (*Vigna unguiculata* (L.) Walp.). *Plant Physiology*, **66**, 978–83.

Atkins C.A., Pate J.S. & Layzell D.B. (1979) Assimilation and transport of nitrogen in non-nodulated (NO_3-grown) *Lupinus albus* L. *Plant Physiology*, **56**, 807–12.

Atkins C.A., Pate J.S. & Sharkey P.J. (1975) Asparagine metabolism key to nitrogen nutrition of developing legume seeds. *Plant Physiology*, **56**, 807–12.

Bauer A., Urquhart A.A. & Joy K.W. (1977) Amino acid metabolism of pea leaves. Diurnal changes and amino acid synthesis from ^{15}N-nitrate. *Plant Physiology*, **59**, 915–19.

Beevers L. and Hageman R.H. (1969) Nitrate reduction in higher plants. *Annual Review of Plant Physiology*, **20**, 495–522.

Bell E.A. (1980) The non-protein amino acids of higher plants. *Endeavour*, **4**, 102–7.

Ben-Zioni A., Vaadia Y. & Lips S.H. (1971) Nitrate uptake by roots as regulated by nitrate reduction products in the shoot. *Physiologia Plantarum*, **24**, 288–90.

Bergersen F.J. (1971) Biochemistry of symbiotic nitrogen fixation in legumes. *Annual Review of Plant Physiology*, **22**, 121–40.

Boland M.J., Fordyce A.M. & Greenwood R.M. (1978) Enzymes of nitrogen metabolism in legume nodules: a comparative study. *Australian Journal of Plant Physiology*, **5**, 553–61.

Bollard E.G. (1960) Transport in the xylem. *Annual Review of Plant Physiology*, **11**, 141–66.

Bowen G.D. (1981) Coping with low nutrients. In *Biology of Australian Plants* (eds J.S. Pate & A.J. McComb), pp. 33–64. University of Western Australia Press, Perth.

Breteler H. & Hanish-Tencate C.H. (1980) Fate of nitrate during initial nitrate utilization by nitrogen-depleted dwarf bean. *Physiologia Plantarum*, **48**, 292–6.

Butz R.G. & Jackson W.A. (1977) A mechanism for nitrate transport and reduction. *Phytochemistry*, **16**, 409–17.

Canvin D.T. & Atkins C.A. (1974) Nitrate, nitrite, and ammonia assimilation by leaves: effect of light, carbon dioxide and oxygen. *Planta*, **116**, 207–24.

Cavalieri A.J. & Huang A.H.C. (1979) Evaluation of proline accumulation in the adaptation of diverse species of marsh halophytes to the saline environment. *American Journal of Botany*, **66**, 307–12.

Chandler G.E. & Anderson J.W. (1976) Studies on the nutrition and growth of *Drosera* species with reference to the carnivorous habit. *New Phytologist*, **76**, 129–41.

Clarkson D.T. (1974) *Ion Transport and Cell Structure in Plants.* McGraw-Hill, London.

Cox G., Moran K.J., Sanders F., Nockolds C. & Tinker P.B. (1980) Translocation and transfer of nutrients in vesicular-arbuscular mycorrhizas III. Polyphosphate granules and phosphorus translocation. *New Phytologist*, **84**, 649–59.

Dijkshoorn W. (1971) Partition of ionic constituents between organs. In *Recent Advances in Plant Nutrition* (ed. R.M. Samish), pp. 447–76. Gordon and Breach Science Publishers, New York.

Dixon K.W., Pate J.S. & Bailey W. (1980) Nitrogen nutrition of tuberous sundew *Drosera erythrorhiza* Lindl. with special reference to catch of arthropod fauna by its glandular leaves. *Australian Journal of Botany*, **28**, 283–97.

Duke S.H., Friedrich J.W., Schrader L.E. & Koukkari W.I. (1978) Oscillations in the activities of enzymes of nitrate reduction and ammonia assimilation in *Glycine max.* and *Zea mays. Physiologia Plantarum*, **42**, 269–76.

Flowers T.J., Troke P.F. & Yeo A.R. (1977) The mechanism of salt tolerance in halophytes. *Annual Review of Plant Physiology*, **28**, 89–121.

Foulds W. & Young L. (1977) Effect of frosting, moisture stress and potassium cyanide on the metabolism of cyanogenic, and acyanogenic phenotypes of *Lotus corniculatus* L. and *Trifolium repens* L. *Heredity*, **38**, 19–24.

Govier R.N., Brown J.G.S. & Pate J.S. (1968) Hemiparasitic nutrition in angiosperms II. Root haustoria and leaf glands of *Odontites verna* (Bell.) Dum. and their relevance to the abstraction of solutes from the host. *New Phytologist*, **67**, 963–72.

Govier R.N., Nelson M.D. & Pate J.S. (1967) Hemiparasitic nutrition in Angiosperms I. The transfer of organic compounds from host to *Odontites verna* (Bell.) Dum. (Scrophulariaceae). *New Phytologist*, **66**, 285–97.

Grasmanis V.O. & Nicholas D.J.D. (1967) A nitrate reductase from apple roots. *Phytochemistry*, **6**, 217–18.

Greenham C.G. & Leonard O.A. (1965) The amino acids of some mistletoes and their hosts. *American Journal of Botany*, **52**, 41–7.

Greenway H. & Munns R.A. (1980) Mechanisms of salt tolerance in non-halophytes. *Annual Review of Plant Physiology*, **31**, 149–90.

Greidanus T., Paterson L.A., Schraeder L.E. & Dana M.N. (1972) Essentiality of ammonium for cranberry nutrition. *Journal of the American Society of Horticultural Science*, **97**, 272–7.

Hall S.M. & Baker D.A. (1972) The chemical composition of *Ricinus* phloem exudate. *Planta*, **106**, 131–40.

Halliday J. & Pate J.S. (1976) Symbiotic nitrogen fixation by coralloid roots of the cycad, *Macrozamia riedlei*: Physiological characteristics and ecological significance. *Australian Journal of Plant Physiology*, **3**, 349–58.

Harder R. (1964) Vegetative development and flower formation in axenic cultures of the insectivorous plant *Drosera pygmaea*. *Planta*, **63**, 316–25.

Heineman R.T. & Atsatt P.R. (1978) Nutritional studies on hemiparasitic *Orthocarpus* (Scrophulariaceae). *Journal of Experimental Botany*, **29**, 789–9.

Herridge D.F., Atkins C.A., Pate J.S. & Rainbird R.M. (1978) Allantoin and allantoic acid in the N economy of the cowpea (*Vigna unguiculata* (L.) Walp). *Plant Physiology*, **62**, 495–8.

Heslop-Harrison Y. (1978) Carnivorous plants. *Scientific American*, **238**, 104–15.

Hill-Cottingham D.G. & Lloyd-Jones C.P. (1968) Relative mobility of some organic nitrogenous compounds in the xylem of apple shoots. *Nature*, **220**, 389–90.

Hill-Cottingham D.G. & Lloyd-Jones C.P. (1973) Seasonal variations in absorption and metabolism of carbon-14-labelled arginine in intact apple stem tissue. *Physiologia Plantarum*, **29**, 35–44.

Huffaker R.C. & Rains D.W. (1978) Factors influencing nitrogen acquisition by plants: assimilation and fate of reduced nitrogen. In *Nitrogen in the Environment 2: Soil–Plant–Nitrogen Relationships*, (eds D.R. Nielsen & J.G. MacDonald) pp. 1–43. Academic Press, London & New York.

Ivankov S. & Ingversen J. (1971) Investigations on the assimilation of nitrogen by maize roots and the transport of some major nitrogen compounds by xylem sap. III. Transport of nitrogen compounds by xylem sap. *Plant Physiology*, **24**, 355–62.

Jackson W.A. (1978) Critique—of factors influencing nitrate acquisition of plants: assimilation and fate of reduced nitrogen. In *Nitrogen in the Environment 2: Soil–Plant–Nitrogen Relationships*, (eds D.R. Nielsen & J.G. MacDonald), pp. 45–88. Academic Press, London & New York.

Jolly S.O., Campbell W. & Tolbert N.E. (1976) NADPH- and NADH-nitrate reductases from soybean leaves. *Archives of Biochemistry and Biophysics*, **174**, 431–9.

Jordan C.F., Todd R.L. & Escalante G. (1979) Nitrogen conservation in a tropical rain forest. *Oecologia (Berlin)*, **39**, 123–8.

Joy K.W. (1973) Control of glutamate dehydrogenase from *Pisum sativum* roots. *Phytochemistry*, **12**, 1031–40.

Keys A.J., Bird I.F., Cornelius M.J., Lea P.J., Wallsgrove R.M. & Miflin B.J. (1978) Photorespiratory nitrogen cycle. *Nature*, **273**, 741–3.

Kirkby E.A. & Armstrong M.J. (1980) Nitrate uptake by roots as regulated by nitrate assimilation in the shoot of castor oil plants. *Plant Physiology*, **65**, 286–90.

Klaren C.H. & Janssen G. (1978) Physiological changes in the hemiparasite *Rhinanthus serotinus* before and after attachment. *Physiologia Plantarum*, **42**, 151–5.

Kuijt J. (1969) *The Biology of Parasitic Flowering Plants*. University of California Press, Berkeley.

Kuo J., Pate J.S., Rainbird R.M. & Atkins C.A. (1980) Internodes of grain legumes—new location for xylem parenchyma transfer cells. *Protoplasma*, **104**, 181–5.

Layzell D.B., Pate J.S., Atkins C.A. & Canvin D.T. (1981) Partitioning of carbon and nitrogen and the nutrition of root and shoot apex in a nodulated legume. *Plant Physiology*, **67**, 30–6.

Layzell D.B., Rainbird R.M., Atkins C.A. & Pate J.S. (1979) Economy of photosynthate use in N-fixing legume nodules. Observations on two contrasting symbioses. *Plant Physiology*, **64**, 888–91.

Lea P.J. & Miflin B.H. (1974) Alternative route for nitrogen assimilation in higher plants. *Nature*, **251**, 614–16.

Lee J.A. & Stewart G.R. (1978) Ecological aspects of nitrogen assimilation. *Advances in Botanical Research*, **6**, 1–43.

Lee R.B. (1980) Sources of reductant for nitrate assimilation in non-photosynthetic tissue: a review. *Plant, Cell and Environment*, **3**, 65–90.

Lewis O.A.M. & Pate J.S. (1973) The significance of transpirationally derived nitrogen in protein synthesis in fruiting plants of pea (*Pisum sativum* L.). *Journal of Experimental Botany*, **24**, 596–606.

MacArthur R.H. & Wilson E.O. (1967) *The Theory of Island Biography*. Princetown University Press, Princetown, New Jersey.

McClure P.R. & Israel D.W. (1979) Transport of nitrogen in the xylem of soybean plants. *Plant Physiology*, **64**, 411–16.

McKee H.S. (1962) Nitrogen Metabolism in Plants. Clarendon Press, Oxford.

Main A.R. (1981) Plants as animal food. In *Biology of Australian Plants* (eds J.S. Pate & A.J. McComb), pp. 342–61. University of Western Australia Press, Perth.

Martin P. (1976) Verteilung von Stickstoff aus spross und Wurzel bei jungen bohnenpflanzen nach der Aufnahme von NO_3^- and NH_4^+. *Zeitschrift für Pflanzen Bodenkunde*, **2**, 181–93.

Miflin B.J. (1974) The location of nitrite reductase and other enzymes related to amino acid biosynthesis in the plastids of roots and leaves. *Plant Physiology*, **54**, 550–5.

Miflin B.J. & Lea P.J. (1977) Amino acid metabolism. *Annual Review of Plant Physiology*, **28**, 299–329.

Monk D., Loneragan W.A. & Pate J.S. (1981) Biology of *Acacia pulchella* with special reference to symbiotic nitrogen fixation. *Australian Journal of Botany*, **29**, 579–92.

Mosse B. (1973) Advances in the study of vesicular-arbuscular mycorrhiza. *Annual Review of Phytopathology*, **11**, 171–96.

Nicholas J.C., Harper J.F. & Hageman R.H. (1976) Nitrate reductase activity in soybeans (*Glycine max* (L.) Merr.) II. Energy limitations. *Plant Physiology*, **58**, 736–9.

Olday F.S., Barker A.V. & Maynard D.N. (1976) A physiological basis for different patterns of nitrate accumulation in cucumber and pea. *Journal of the American Society of Horticultural Science*, **101**, 219–21.

Pate J.S. (1971) Movement of nitrogenous solutes in plants. In *Nitrogen-15 in Soil-Plant Studies*, pp. 165–87. International Atomic Energy Agency, Vienna.

Pate J.S. (1973) Uptake, assimilation and transport of nitrogen compounds by plants. *Soil Biology and Biochemistry*, **5**, 109–19.

Pate J.S. (1980) Transport and partitioning of nitrogenous solutes. *Annual Review of Plant Physiology*, **31**, 313–40.

Pate J.S. & Atkins C.A. (1983) Nitrogen, uptake, transport and utilization. In *Ecology of Nitrogen Fixation*, Vol. 3, *Legumes* (ed. W.J. Broughton), pp. 245–97. Oxford University Press, Oxford.

Pate J.S., Atkins C.A., Hamel K., McNeil D.L. & Layzell D.B. (1979) Transport of organic solutes in phloem and xylem of a nodulated legume. *Plant Physiology*, **63**, 1082–8.

Pate J.S., Atkins C.A., Herridge D.F. & Layzell D.B. (1981) Synthesis, storage and utilization of amino compounds in white lupus (*Lupinus albus*). *Plant Physiology*, **67**, 37–42.

Pate J.S., Atkins C.A. & Rainbird R.M. (1981) Theoretical and experimental costing of nitrogen fixation and related processes in nodules of legumes. *Proceedings of 4th International Congress on N_2 Fixation*, pp. 105–16. Canberra, Australia.

Pate J.S., Atkins C.A., White S.T., Rainbird R.M. & Woo K.C. (1980) Nitrogen nutrition and xylem transport of nitrogen in ureide-producing grain legumes. *Plant Physiology*, **65**, 961–5.

Pate J.S. & Dixon K.W. (1978) Mineral nutrition of *Drosera erythrorhiza* Lindl. with special reference to its tuberous habit. *Australian Journal of Botany*, **26**, 455–64.

Pate J.S. & Dixon K.W. (1981) Plants with fleshy underground storage organs. In *Biology of Australian Plants* (eds J.S. Pate & A.J. McComb). University of Western Australia Press, Perth.

Pate J.S. & Dixon K.W. (in press) *Bulbous, Cormous and Tuberous Plants of West Australia.* University of Western Australia Press, Perth.

Pate J.S., Gunning B.E.S. & Briarty L. (1969) Ultrastructure and functioning of the transport system of the leguminous root nodule. *Planta*, **85**, 11–34.

Pate J.S. & Layzell D.B. (1981) Carbon and nitrogen partitioning in the whole plant—a thesis based on empirical modelling. In *The Physiology and Biochemistry of Plant Productivity* (ed. J.D. Bewley), pp. 94–135. Martinus Nijhoff, Netherlands.

Pate J.S., Layzell D.B. & McNeil D.L. (1979) Modelling the transport and utilization of C and N in a nodulated legume. *Plant Physiology*, **63**, 730–8.

Pate J.S., Sharkey P.J. & Atkins C.A. (1977) Nutrition of a developing fruit. Functional economy in terms of carbon, nitrogen and water. *Plant Physiology*, **59**, 506–10.

Pate J.S. & Wallace W. (1964) Movement of assimilated nitrogen from the root system of the field pea (*Pisum arverse* L.). *Annals of Botany*, **28**, 80–99.

Phillips D.A. (1980) Efficiency of symbiotic nitrogen fixation in legumes. *Annual Review of Plant Physiology*, **31**, 29–49.

Probyn T.A. & Lewis O.A.M. (1979) The route of nitrate-nitrogen assimilation in the root of *Datura stramonium* L. *Journal of Experimental Botany*, **30**, 299–305.

Radin J.W. (1975) Differential regulation of nitrate reductase induction in roots and shoots of cotton plants. *Plant Physiology*, **55**, 178–82.

Radin J.W. (1977) Contribution of the root system to nitrate assimilation in whole cotton plants. *Australian Journal of Plant Physiology*, **4**, 811–19.

Radin J.W. (1978) A physiological basis for the division of nitrate assimilation between roots and leaves. *Plant Science Letters*, **13**, 2–25.

Radin J.W., Sell C.R. & Jordan W.R. (1975) Physiological significance of the in-vivo assay for nitrate reductase in cotton seedlings. *Crop Science*, **15**, 710–13.

Raven J.A. & Smith F.A. (1976) Nitrogen assimilation and transport in vascular plants in relation to intracellular pH regulation. *New Phytologist*, **76**, 415–31.

Reinbothe H. & Mothes K. (1962) Urea, ureides, and guanidines in plants. *Annual Review of Plant Physiology*, **13**, 129–50.

Rennie R.J. (1979) Comparison of ^{15}N-aided methods for determining symbiotic dinitrogen fixation. *Review of Ecology and Biology of Soil*, **16**, 455–63.

Reynolds P.H.S. & Farnden K.J.F. (1979) The involvement of aspartate aminotransferases in ammonium assimilation in lupin nodules. *Phytochemistry*, **18**, 1625–30.

Rice E.L. & Pancholy S.K. (1972) Inhibition of nitrification by climax ecosystems. *American Journal of Botany*, **59**, 1033–40.

Rosenthal G.A., Dehlman D.L. & Janzen D.H. (1976) A novel means for dealing with L-canavanine, a toxic metabolite. *Science*, **192**, 256–8.

Sanders F.E. & Tinker P.B. (1973) Phosphate flow into mycorrhizal roots. *Pesticide Science*, **4**, 385–95.

Sauter J.J., Iten W. & Zimmermann M.H. (1973) Studies on the release of sugar into the vessels of sugar maple (*Acer saccharum*). *Canadian Journal of Botany*, **51**, 1–8.

Schrader L.E. (1978) Uptake, accumulation, assimilation and transport of nitrogen in higher plants. In *Nitrogen in the Environment, 2. Soil–Plant–Nitrogen Relationships* (eds D.R. Nielsen & J.G. MacDonald), pp. 101–41. Academic Press, London & New York.

Schrader L.E. (1981) Nitrate utilization by plants. In *The Physiology and Biochemistry of Plant Productivity* (ed. J.D. Bewley). Martinus Nijhoff, Netherlands.

Schubert K.R. & Ryle G.J.A. (1980) The energy requirements for N_2 fixation in nodulated legumes. In *Advances in Legume Science* (eds R.J. Summerfield & A.H. Bunting), pp. 85–96. Royal Botanical Gardens, Kew, Surrey.

Shaner D.L. & Boyer J.S. (1976) Nitrate reductase activity in maize (*Zea mays* L.) leaves. I. Regulation by nitrate flux. *Plant Physiology*, **58**, 499–504.

Simola L.K. (1978) The effect of several amino acids and some inorganic nitrogen sources on the growth of *Drosera rotundifolia* in long- and short-day conditions. *Zeitschrift fur Pflanzenphysiologie*, **90**, 61–8.

Small J.G.C., Onraët A., Grierson D.S. & Reynolds G. (1977) Studies on insect-free growth, development and nitrate-assimilating enzymes of *Drosera aliciae* Hamet. *New Phytologist*, **79**, 127–33.

Sprent J.I. (1980) Root nodule anatomy, type of export product and evolutionary origin in some Leguminosae. *Plant, Cell and Environment*, **3**, 35–43.

Srivastava H.S. and Chauhan J.S. (1977) Seed germination, seedling growth and nitrogen and pigment concentration in dodder as affected by inorganic nitrogen. *Zeitschrift für Pflanzenphysiologie*, **84**, 391–8.

Stearns S.C. (1976) Life history tactics: a review of ideas. *Quarterly Review of Biology*, **51**, 3–47.

Stewart G.R., Larher F., Ahmad I. & Lee J.A. (1979) Nitrogen metabolism and salt tolerance in higher plant haplophytes. In *Ecological Processes in Coastal Environments* (eds R.L. Jeffries & A.J. Davy), pp. 211–27. Blackwell Scientific Publications, Oxford.

Stewart G.R. & Lee J.A. (1974) The role of proline accumulation in halophytes. *Planta*, **120**, 279–89.

Stewart G.R. & Orebamjo T.O. (1980) Nitrogen status and nitrate reductase activity of the parasitic angiosperm *Tapinanthus bangwensis* (Engl and K. Krause) Danser growing on different hosts. *Annals of Botany*, **45**, 587–9.

Stewart W.D.P. (1978) Nitrogen-fixing cyanobacteria and their associations with eukaryotic plants. *Endeavour*, **2**, 170–9.

Storey R. & Wyn-Jones R.G. (1979) Responses of *Atriplex spongiosa* & *Suaeda monoica* to salinity. *Plant Physiology*, **63**, 156–62.

Streeter J.G. (1977) Asparaginase and asparagine transaminase in soybean leaves and root nodules. *Plant Physiology*, **60**, 235–9.

Streeter J.G. (1979) Allantoin and allantoic acid in tissues and stem exudate from field-grown soybean plants. *Plant Physiology*, **63**, 478–80.

Taylor A.A. & Havill D.C. (1981) The effect of inorganic nitrogen on the major enzymes of ammonium assimilation in grassland plants. *New Phytologist*, **87**, 53–62.

Taylor B.K. (1967) The nitrogen nutrition of the peach tree. I. Seasonal changes in nitrogenous constituents in mature trees. *Australian Journal of Biological Sciences*, **20**, 379–87.

Taylor B.K. & May L.H. (1967) The nitrogen nutrition of the peach tree. II. Storage and mobilization of nitrogen in young trees. *Australian Journal of Biological Sciences*, **20**, 389–411.

Thomas R.J., Feller U. & Erismann K.H. (1979) The effect of different inorganic nitrogen sources and plant age on the composition of bleeding sap on *Phaseolous vulgaris* L. *New Phytologist*, **82**, 657–70.

Tinker P.B. (1975) Effects of vesicular-arbuscular mycorrhizas on higher plants. *Symposia of the Society for Experimental Biology*, **29**, 325–49.

Tromp J. & Oova J.C. (1976) Effect of time of nitrogen application on amino nitrogen composition of roots and xylem sap of apple. *Physiologia Plantarum*, **37**, 29–34.

Tromp J. & Oova J.C. (1979) Uptake and distribution of nitrogen in young apple trees after application of nitrate or ammonium with special reference to asparagine and arginine. *Physiologia Plantarum*, **45**, 23–8.

Wallace W. (1974) Purification and properties of a nitrate reductase inactivating enzyme. *Biochimica Biophysica Acta*, **341**, 265–76.

Wallace W. & Pate J.S. (1965) Nitrate reductase in the field pea (*Pisum arvense* L.). *Annals of Botany*, **29**, 655–71.

Wallace W. & Pate J.S. (1967) Nitrate assimilation in higher plants with special reference to the cocklebur (*Xanthium pennsylvanicum* Wallr.). *Annals of Botany*, **31**, 213–28.

Weissman G.S. (1972) Influence of ammonium and nitrate nutrition on enzymatic activity in soybean and sunflower. *Plant Physiology*, **49**, 138–41.

Winter K., Osmond C.B. & Pate J.S. (1981) Coping with salinity. In *Biology of Australian Plants* (eds J.S. Pate & A.J. McComb), pp. 88–113. University of Western Australia Press, Perth.

Wong P.P. & Evans H.J. (1971) Poly-β-hydroxybutyrate utilization by soybean (*Glycine max* Merr.) nodules and assessment of its role in maintenance of nitrogenase activity. *Plant Physiology*, **47**, 750–5.

Ziegler H. (1975) Nature of transported substances. In *Transport in Plants. I. Phloem Transport* (eds H.M. Zimmermann & J.A. Milburn). Encyclopedia of Plant Physiology, New Series 1. Springer-Verlag, Berlin.

12. THE ROLE OF NITROGEN IN THE ECOLOGY OF GRASSLAND AUCHENORRYNCHA

R.A. PRESTIDGE* AND S. McNEILL

Department of Pure and Applied Biology, Imperial College of Science and Technology, Silwood Park, Ascot, Berkshire

The growth and fecundity of herbivorous insects is to some extent determined by the quality and quantity of the available nitrogen in the host plant (Feeny 1976; Wint 1979). Hill (1976) investigated the importance of the available nitrogen in *Holcus mollis* L., to a complex of Auchenorrhyncha (leafhoppers, planthoppers and froghoppers) and was able to show that the insects were closely cued into the nutrient flushing of the host. Subsequently, McNeill & Southwood (1978) and Lawton & McNeill (1979) have shown that the variation in the structure of the leafhopper community within and between seasons can be better explained by considering the mode of feeding of the insects, phloem feeders being closely linked to the plant soluble nitrogen levels in the stems and mesophyll feeders to the nitrogen levels in the leaves.

Evidence has now accumulated to suggest that subtle changes in amino acid composition may directly influence phytophagous insects irrespective of any changes in the levels of total nitrogen or total amino acids. For example, van Emden (1973) found that growth rates and fecundity of *Myzus persicae* and *Brevicoryne brassicae* were positively correlated with the levels of amide (glutamine + asparagine) in the leaves of their host plants. Other amino acids positively correlated with aphid performance included threonine and glutamic acid for *Brevicoryne* and methionine and leucine for *Myzus*. He also found that a number of amino acids were negatively correlated with aphid performance, e.g. phenylalanine and proline for *Brevicoryne* and γ-aminobutyric, tyrosine and proline for *Myzus*. Sogawa (1971a) and Cheng & Pathak (1972) linked fecundity levels in the rice leafhoppers *Nilaparvata lugens* and *Nephotettix virescens* to the total amino acid levels in the plant. The main amino acid implicated (Sogawa 1971b; Cagampang, Pathak & Juliano 1974) was asparagine, one of the constituents of the amide fraction. Harrewijn (1978) linked low population development in *M. persicae* to a reduced amide concentration and noted that low methionine content increased aphid

* Present address: Ruakura Agricultural Research Centre, Private Bag, Hamilton, New Zealand.

restlessness. Low levels of pre-alates were recorded on plants with a high proline content and a low amount of tyrosine, the latter known to stimulate wing development in this insect (Harrewijn 1976). Parry (1974) suggested that population build-up and rapid decline of *Elatobium abietinum* (Walker) on sitka spruce needles was related to a high level of essential amino acids (when temperature was not limiting) followed by a rapid decline, a view recently modified by Carter & Cole (1977), who suggested that the trigger to alate production may be subtle changes in amino acid ratios during the spring flush.

Hill (1976) noted a 34-fold difference in total essential amino acid levels in *H. mollis* leaves between samples taken only 28 days apart. The life histories of five species of Auchenorryncha were closely linked to the periods of high amino acid concentration in this plant. McNeill & Southwood (1978) suggest that very low methionine levels in *Holcus* in June may be responsible for alate production in *Holcaphis holci.*

Ten amino acids are generally regarded as being essential in insect diets: arginine, histidine, isoleucine, leucine, lysine, methionine, phenylalanine, threonine, tryptophane, and valine (House 1972). Retnakaren & Beck (1968), using the deletion technique with artificial diets, concluded that in *Acyrthosiphon pisum* (Harris) the absence of any of the ten essential acids plus cysteine resulted in an up to 75% weight reduction in the first generation nymphs and an impairment of reproduction. Dadd & Krieger (1968) concluded that only methionine, histidine, isoleucine and possibly lysine were essential for the growth of *Myzus*, the remaining essential amino acids being provided by the gut symbionts. Mittler (1970) found that the omission of methionine and histidine from a synthetic diet reduced ingestion by *Myzus* by 50% and 33% respectively. Artificial diets have, therefore, been very important in elucidating some of the finer points in aphid nutrition but, as van Emden & Bashford (1969) and van Emden (1973) have pointed out, it is impossible to alter the amino acid balance without upsetting other related factors in the diet such as pH.

This paper presents results from both field and laboratory studies showing how the grassland Auchenorrhyncha are affected by seasonal and phenological nutrient changes in their host plants. It is divided into four main sections. In the first two sections amino acids present in grasses and the abundance of Auchenorrhyncha associated with them are discussed. In the third section the influence of plant architecture on the structure of the leafhopper community is examined and an attempt is made to look at its interaction with food quality. The final section examines the main ways in which the insects exploit the quality of the food available to them and shows the importance of mobility to the leafhopper complex.

AMINO ACIDS IN GRASSES

Natural patterns

The seasonal changes in soluble nitrogen in grasses follow a distinct pattern (McNeill & Southwood 1978). In *Holcus lanatus* L. and *Holcus mollis* levels rapidly increase for a short period in spring as metabolites are moved to the growing points. They also rise in the autumn when metabolites are removed from the senescing leaves and also into the new hulms being formed at that time of year (sometimes these are separated in time giving a bimodal peak). There is often a less distinct flow in midsummer associated with the developing and filling of seed.

Most of the translocated nitrogen is in the form of soluble amino acids (Beevers 1976) so that the seasonal patterns of amino acid concentration follow the same seasonal pattern (Fig. 12.1). Levels are highest during the spring and autumn growth periods.

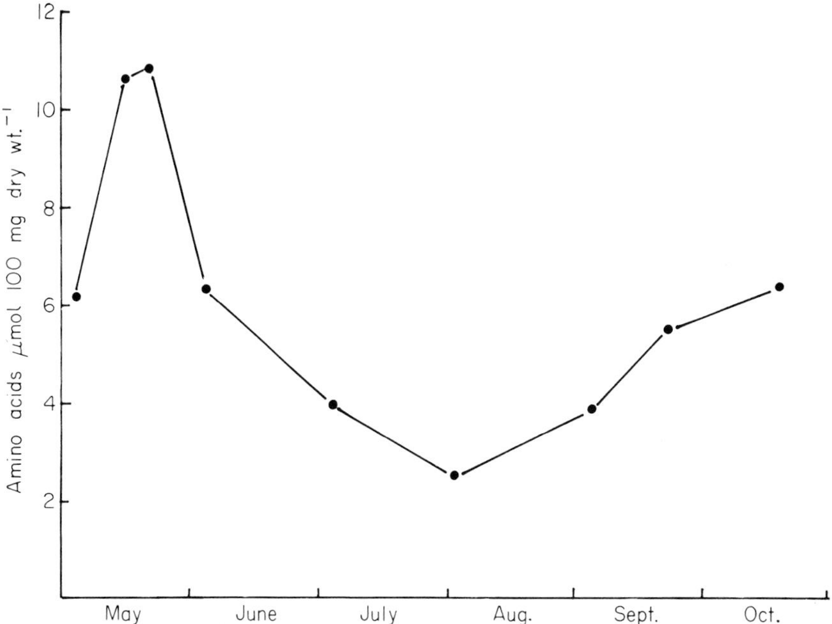

FIG. 12.1. Levels of leaf-soluble amino acids in *Holcus lanatus* in 1979.

The levels of individual amino acids in grass tissues follow three main patterns through the season; these three patterns are represented in Fig. 12.2a & b by amide, alanine and valine. The amino acids following a similar pattern

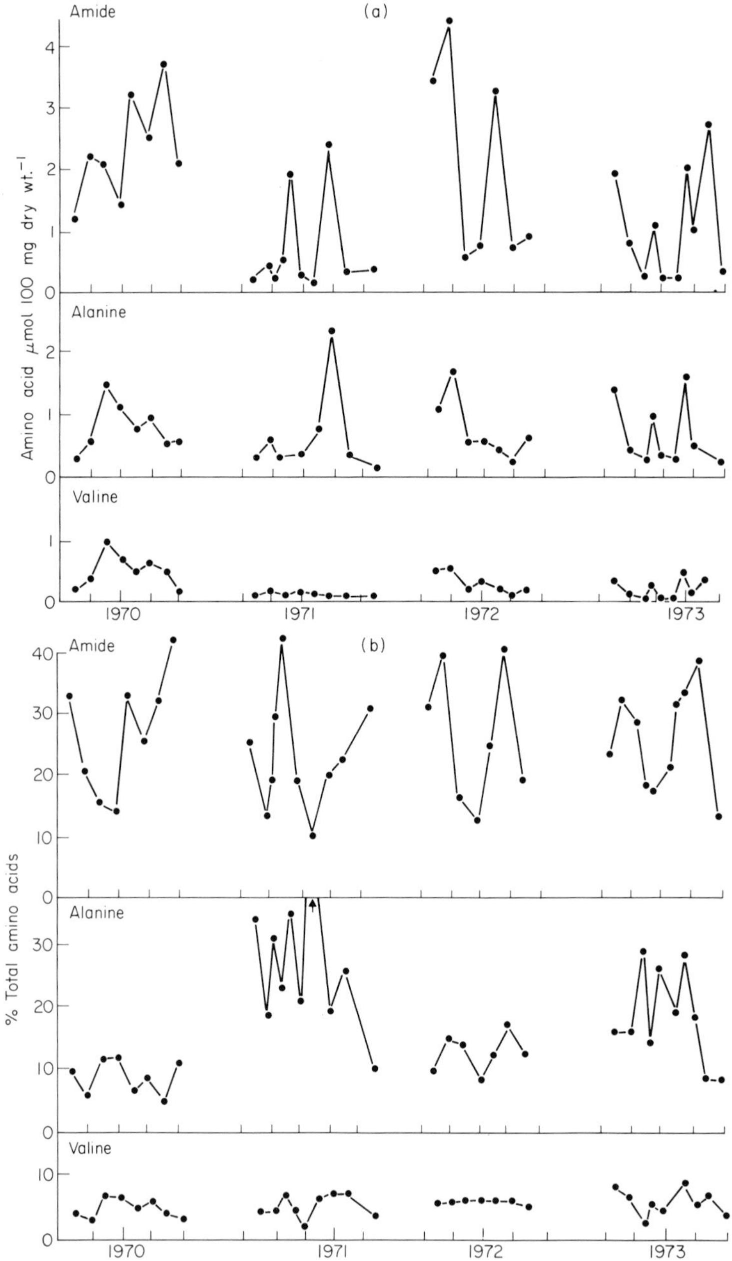

FIG. 12.2. (a) Concentrations of amides, alanine, and valine (mmol 100 mg dry wt.$^{-1}$); (b) the corresponding proportions of the total amino acid concentration in stems of *Holcus lanatus* 1970–1973.

to that of amide are listed in Group 1 of Table 12.1. This group has a sharp and well-defined peak in both the spring and the autumn with a low level throughout the summer. The height of the peaks and the general level of the summer lows are variable from season to season (see Fig. 12.2a), but, as proportions of the total amino acids in the samples, the members of this group are relatively constant between seasons (e.g. amide forms about 40% at the peak but only 10% in the troughs).

TABLE 12.1. Grouping of amino acids according to the seasonal patterns of abundance.

Group 1	Group 2	Group 3
Amide	Alanine	Arginine
Aspartic acid	γ-Amino butyric acid	β-Phenylalanine
Glutamic acid	Glycine	Isoleucine
Proline	L-dopa	Leucine
Threonine	Histidine	Lysine
		Methionine
		Tyrosine
		Valine

Amino acid quality index $= \Sigma$ (Group 1)/Σ (Group 2).

In Table 12.1 Group 2 amino acids (e.g. alanine) have a rather different seasonal and interseasonal pattern. They also tend to have spring and autumn peaks, although one or other of these peaks is usually very much weaker in any particular year. The summer low in this group tends to be at a higher level than for Group 1. The absolute concentrations in this group tend to be similar from season to season with the peak concentration, where it occurs, reaching between 1.5 and 2.5 μmol 100 mg dry wt. grass^{-1} and the troughs around 0.15 μmol 100 mg^{-1}. This means that, as a proportion of the total amino acids, this group is very variable from year to year, the peaks for alanine varying between 15% and 50% of the total and the troughs between 5% and 20%.

The third group all follow a pattern similar to valine, a relatively flat distribution, the level of which is variable between years but forms a constant percentage of the total—about 6 to 8% for valine.

The biochemical basis of these patterns

The current view of amino acid biosynthesis suggests that there are families of amino acids each of which arises from a single common precursor or head compound. These head compounds are the same in all orders of plants, although the emphasis may vary from order to order (Fowden 1967). Fig 12.3 shows the most important of these families of acids, those associated with glutamate and aspartate (Oaks & Bidwell 1970). In *Holcus* it is the Group 1 amino acids, comprising the amide forms, glutamine + asparagine, and their immediate derivatives, which show the most rapid rise and fall in concentration in the plant. Apparently, amide is being transported around the plant and converted into other amino compounds and proteins at the active sites of growth and development (Oaks & Bidwell 1970; Miflin & Lea 1977).

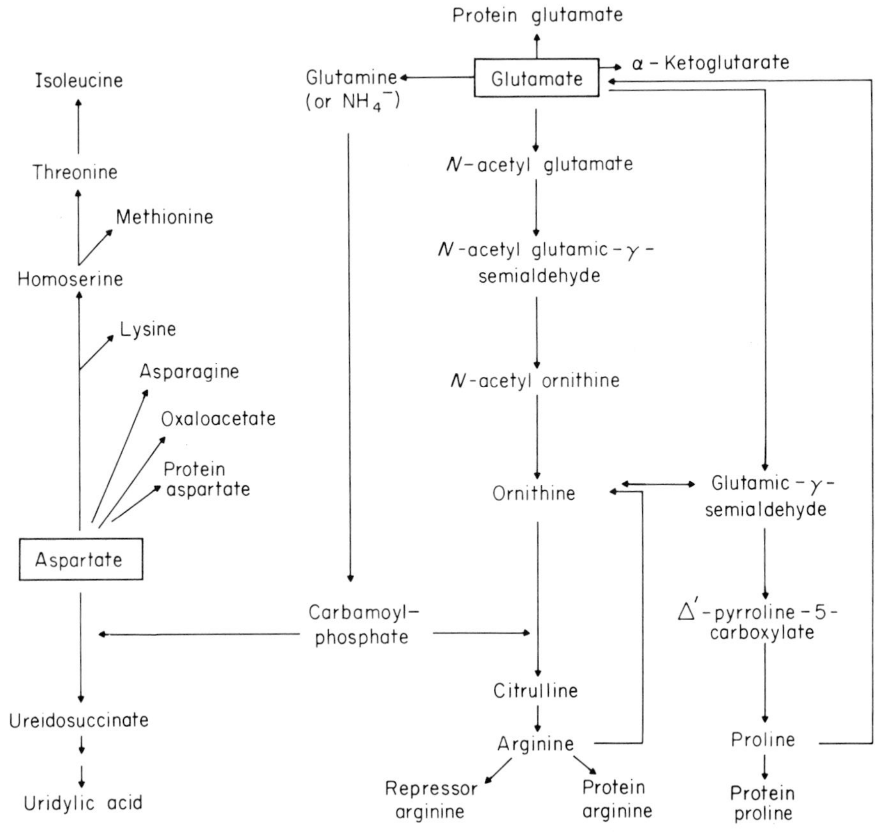

FIG. 12.3. Biosynthesis of amino acids from aspartate and glutamate (after Oaks & Bidwell 1970).

Insect performance has been shown to be positively correlated with most of the members of this group of acids, most of them indeed being essential for many insects (van Emden 1973; House 1972).

Group 3 amino acids are those belonging to the same families as the Group 1 acids but are further removed from the head compounds (see Fig. 12.3 and Oaks & Bidwell 1970). Their levels in the plant, therefore, are more a measure of the rate of conversion of Group 1 compounds into proteins, etc. and are a general measure of the level of protein metabolism in the plant. Several of them are essential components of insect diets, but as they are relatively constant in the plant they are of lesser interest to insect population studies.

Group 2 amino acids, on the other hand, include those acids known to be toxic, e.g. γ-aminobutyric acid and L-dopa (Bell & Janzen 1971, Rehr *et al.* 1973), and those that do not seem to have any significant effect on insect performance, e.g. alanine (House 1965). Most of this group of amino acids does not belong to the major protein-forming families of acids, e.g. glycine (Larsen 1977), and several may be regarded as possible plant defence compounds (Rehr *et al.* 1973; Rhoades & Cates 1976).

Experimental manipulation of natural patterns

Fertilizers have provided a useful tool with which to manipulate the levels of amino acids in grasses. The effects of fertilizers are relatively short-lived and although the levels of nitrogen and amino acids remain higher on fertilized than unfertilized plots, the effect of a spring fertilizer application is to increase the levels of these substances during the growth periods. The seasonal fluctuations, related to the times of nutrient mobilization within the plant, are still maintained.

Nitrogen fertilizer treatment has the effect of boosting the absolute levels of all the individual amino acids, though Group 1 acids were clearly raised more than others in both absolute and relative terms. Group 2 acids decreased in relative terms, and Group 3 compounds remained as an unchanged proportion of the amino acid pool.

THE AUCHENORRYNCHA COMPLEX

Year to year variation in natural grassland

The performance of some homopteran populations has been correlated with the levels of groups of amino acids in the food (McNeill & Southwood 1978) while other authors have been able to show a strong relationship between the levels of individual amino acids and insect growth and development (van Emden 1973, 1978; Auclair 1963; Carter & Cole 1977). It appears to us that both are likely to be of importance.

Data gathered during a long-term study of *Holcus* grassland by one of the authors (S. McNeill) suggests that the interseasonal variation of amino acids is important in determining the overall abundance of insect herbivores on this grass. Correlating the abundance of the five most important herbivores in each season (Table 12.2) with the peak concentrations of Group 1 amino acids (Fig. 12.4a), we can see that there is a reasonably strong relationship. There is no relationship, however, between insect abundance and the concentration of

Group 2 acids (Fig. 12.4b). This lack of a relationship is probably due to the fact that the levels were very similar from year to year and that therefore the expected negative relationship due to their adverse effects was not evident.

TABLE 12.2. The abundance of the dominant five species of insect herbivores on *Holcus mollis* 1970–1977. All figures are the maximum number m^{-2} in routine samples.

1970	*Aptinothrips* 2851, *Adarrus* 976, mite (unident) 963, *Holcaphis* 554, *Tytthus* 190.
1971	*Aptinothrips* 325, *Adarrus* 205, *Crypthaphis* 122, *Macustus* 93, *Tytthus* 89.
1972	*Holcaphis* 1849, *Adarrus* 803, *Zyginidia* 349, *Dicranotropis* 229, *Aptinothips* 205.
1973	*Dicranotropis* 1673, *Holcaphis* 1371, *Adarrus* 818, *Aptinothrips* 808, *Zyginidia* 304.
1974	*Dicranotropis* 1950, *Aptinothrips* 963, *Holcaphis* 742, *Diplocoelenus* 495, *Adarrus* 441.
1975	*Adarrus* 735, *Zyginidia* 537, *Elymana* 205, *Recilia* 191, *Diplocoelenus* 127.
1976	*Diplocoelenus* 558, *Psammotettix* 420, *Adarrus* 286, *Elymana* 129, *Zyginidia* 107.
1977	*Psammotettix* 244, *Diplocoelenus* 103, *Aphrodes* 102, *Elymana* 26, *Zyginidia* 14.

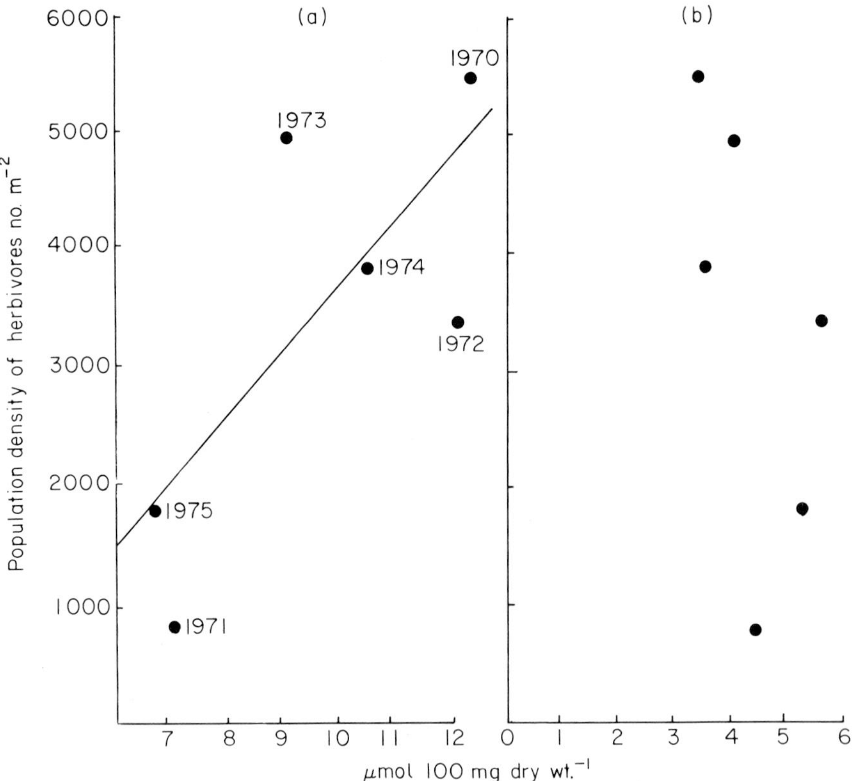

FIG. 12.4. The relationship between amino acid levels and the abundance of the five most numerous herbivores 1970–1975: (a) concentration of Group 1 amino acids, (b) concentration of Group 2 amino acids.

This overall pattern, however, means that in years when the total levels of amino acids are low, the balance between the beneficial Group 1 acids and the potentially harmful Group 2 acids is altered so that the Group 2 acids are relatively much more important. In such years they form almost half of the total, whereas in high nitrogen years they form only about 25%. If we now look at the relationship between insect herbivore abundance and the ratio of Group 1 to Group 2 acids as a measure of amino acid quality (Fig. 12.4c), we get a markedly better correlation than using Group 1 alone.

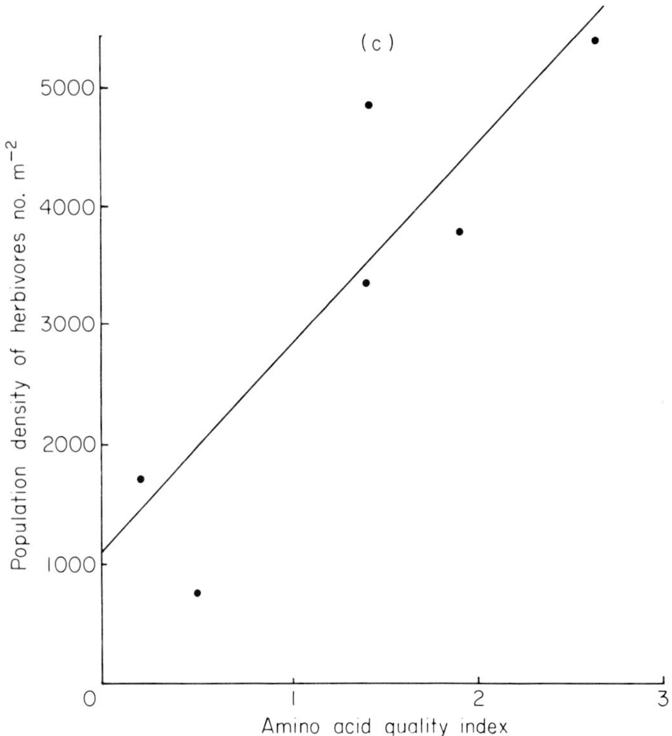

FIG. 12.4. (c) The relationship between the abundance of the five most numerous herbivores 1970–1975 and the amino acid quality index. (For details, see foot of Table 12.1.)

Experimental manipulation

To test the hypothesis suggested by this correlation, namely that the insect performance is markedly affected by the ratio of Group 1 to Group 2 acids, the interseasonal variation due to weather and other such factors needs to be removed from the system. We have done this by the use of fertilizers to generate plots with varying degrees of total available nitrogen in the same season; the response in terms of amino acid quality is similar to that between years. Thus we were able in 1978 and 1979 to investigate the response of the Auchenorrhyncha complex, the dominant guild in this grassland, to variations in availability and balance of amino acids in their host. The summarized results are shown in Figs 12.5 & 12.6.

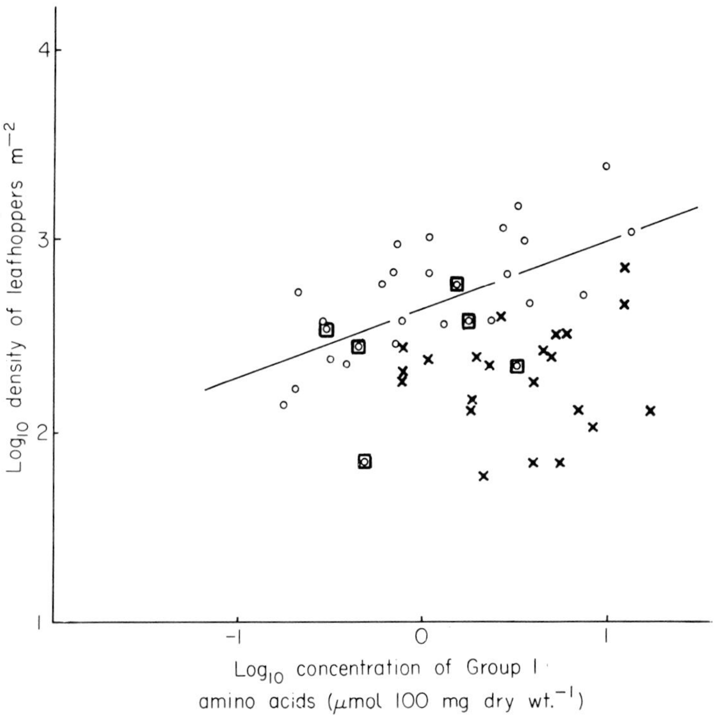

FIG. 12.5. The relationship between the concentration of Group 1 amino acids and leafhopper density. × data previous to 1st July, no significant relationship; ○ data after 1st July, regression significant $P<0.001$; ⊡ data for early July.

Fig. 12.5 shows a graph of the abundance of leafhoppers associated with the levels of Group 1 amino acids. Interpretation of this relationship requires a knowledge of the insect's phenology. The data fall into two distinct phases. There is no significant relationship early in the season when temperature restricts movement of the adults (Waloff 1980) and before nymphs are picked up in the D-vac samples. After the beginning of July, when temperatures are no longer limiting and the movement of the adult insects and this season's nymphs become samplable, there is a highly significant relationship ($P<0.001$). The relationship between the index of amino acid quality and leafhopper species' below) suggest that both food quality and available living a strong relationship ($P<0.001$) later. The scatter around both of these relationships is high, the extremes being contributed by samples taken in early July (shown ◙ in Fig. 12.6) before the relationship is firmly established.

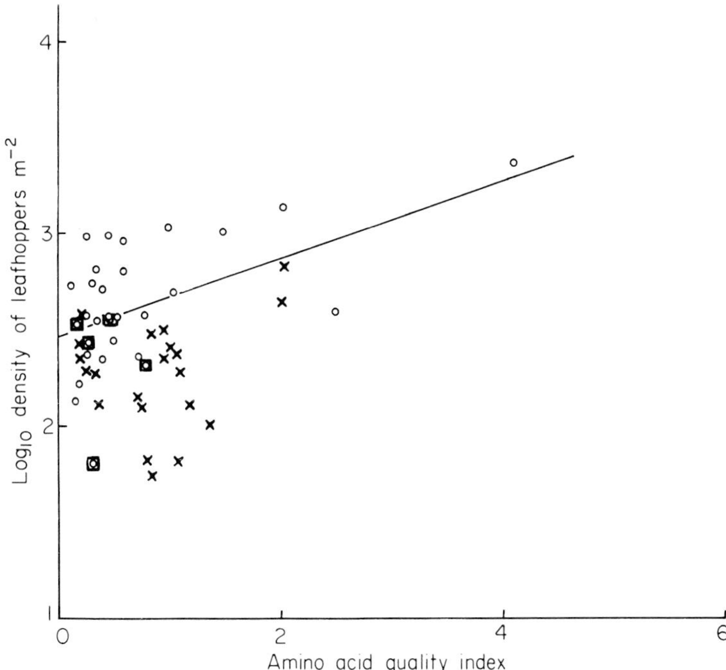

FIG. 12.6. The relationship between amino acid quality index and leafhopper density. × data previous to 1st July, no significant relationship; ○ data after 1st July, regression significant $P<0.001$; ◙ samples taken in early July.

This scatter may be due, at least in part, to variations in available living space as well as field sampling errors. The literature (Lawton 1978) and data from other experiments (discussed under the heading 'Plant architecture and leafhopper species' below) suggest that both food quality and available living space are very important factors determining both the abundance and the diversity of insect herbivores.

THE INFLUENCE OF PLANT ARCHITECTURE

Plant architecture and leafhopper abundance

There is a sizeable literature on the effects of plant architecture on the numbers of species of herbivores exploiting a host plant but much less on the effects of architecture on the abundance of these organisms. That there is a relationship between architecture and the abundance of leafhoppers in grassland has been shown, or at least implied, by Andrzejewska (1965), Morris (1971, 1973, 1974, 1978) and Waloff (1980).

It is obvious that plant architecture is a measure of the potential number of feeding sites and hence living space (Southwood 1977). If food is therefore in any way important in the ecology of herbivores, the number of feeding sites and their quality should be a major factor determining their numbers. The previous sections suggest that quality is certainly important for grassland leafhoppers and the present section will look at architecture.

The measurement of plant architecture often requires much time-consuming gathering of point quadrat data; grassland structure is, however, essentially simple, with linear leaves and stems as the main components. It was therefore decided that an adequate measure of architecture in this case would be: architecture index = biomass (kg m^{-2}) × mean pasture ht. (cm). This would provide a measure of living space as the number of sites would increase with the biomass and the degree of vertical packing (Andrzejewska 1965) would be affected by the height.

Fig. 12.7 shows a very good relationship between the architecture index on the fertilizer plots and the density of leafhoppers on those same plots throughout 1979 ($P < 0.001$) showing that this could be an equally good explanation for the variation in leafhopper density on the plots of different fertilizer levels. It is, however, much more likely that both factors are important in determining population levels as increases in quality will make

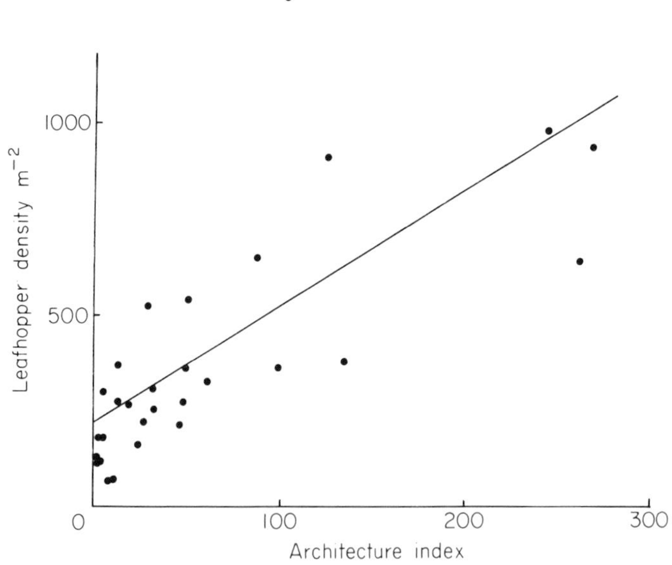

FIG. 12.7. Leafhopper density in relation to plant architecture, regression significant $P < 0.001$.

more potential sites available for utilization, resulting in an increase in species or numbers, or both (Lawton 1978). Data from another experiment where a number of sown single-species grass swards were fertilized suggest that this is the case (see under heading below). Both numbers and species increased immediately the nitrogen appeared in the plants—but before the structure altered—and a further increase occurred when the pulse of consequent growth of the grass occurred three weeks later.

In an attempt to combine the influence of quality and architecture on the abundance of leafhoppers on the experimental plots, the residuals, both before and after the begining of July, around the regression of abundance on amino acid quality (see Fig. 12.6) were calculated and these in turn plotted against the architecture index (Fig. 12.8). A very good relationship was found ($P < 0.001$) with a small scatter, considering that four independent sets of field data were involved (leafhopper density, amino acid levels, biomass, and pasture height). It thus appears that both food quality as indicated by amino acid ratios and living space as measured by the architecture index are important in the population dynamics of the grassland leafhoppers in this study.

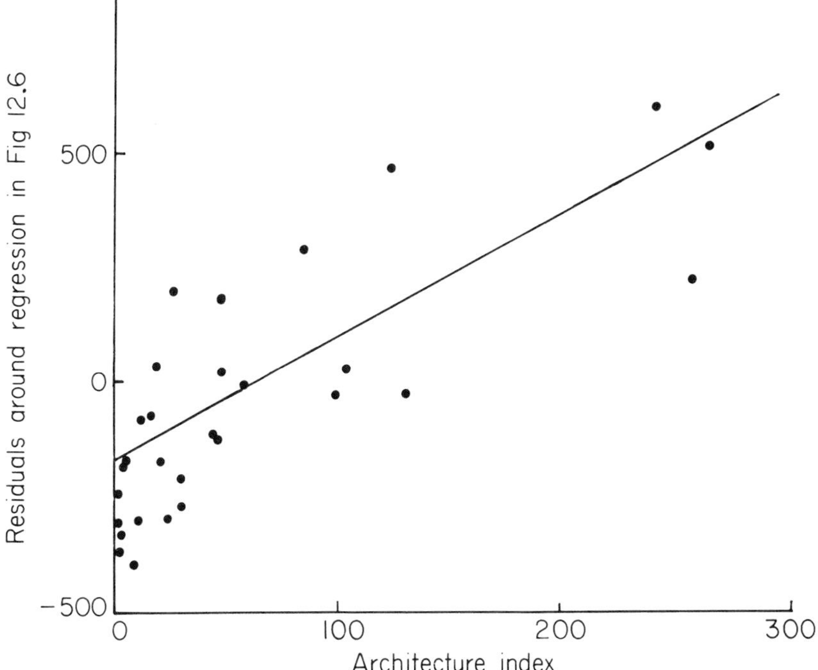

FIG. 12.8. The relationship between the residuals about the regression line in Fig. 12.6 and plant architecture, regression significant $P < 0.001$.

Plant architecture and leafhopper species

Many workers have shown that habitat complexity affects the diversity of leafhopper species (Andrzejewska 1965; Whittaker 1969; Murdoch, Evans & Peterson 1972; McClure & Price 1975, 1976; Denno 1977; Tallamy & Denno 1979). In general, all these studies show that grasses that are tall or have a complex structure support a larger and more diverse sap-feeding fauna than grasses which are shorter or have a simpler structure.

This concept was examined by Lawton (1976) who showed that in bracken (*Pteridium aquilinum* (L.) Kuhn) the number of herbivore species divided by the log of the living space remained more or less constant throughout the greater part of the growing season. Fig. 12.9a shows that the same is true for the leafhopper fauna in our experimental area; after an initial rapid rise the number of species per unit architecture remains at about four species per unit

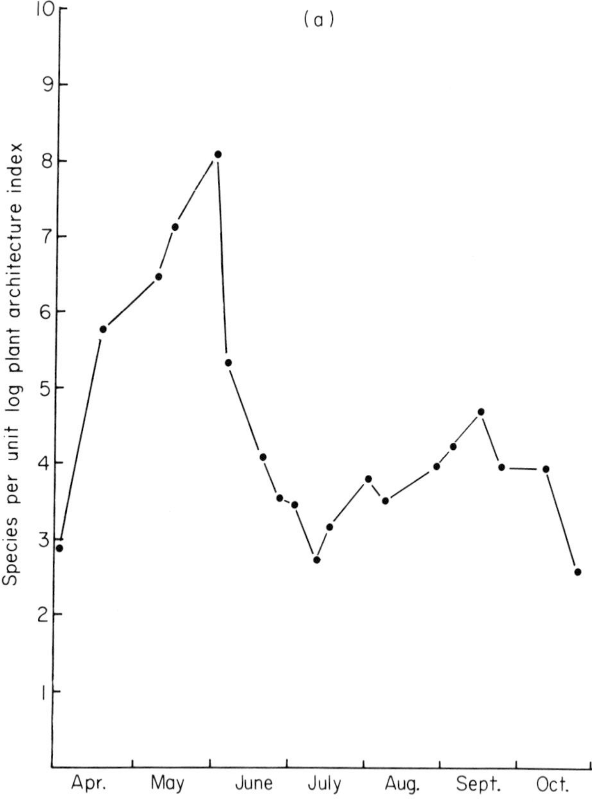

FIG. 12.9. (a) Leafhopper species per log unit of plant architecture index in 1979.

architecture. Fig. 12.9b shows the number of leafhopper species per unit amino acid quality index throughout the season, and this also remains level for much of the season (about three species per unit of amino acid quality index), the plateau being maintained from earlier in the season but dropping away earlier in the autumn than does the architectural relationship. There is no real difference in the 'stability' of these two relationships; both would appear to be as good an explanation of the species diversity throughout the season, just as in the discussion of abundance in the previous section. However, once again, as Lawton (1978) has pointed out, both are likely to be involved, with quality determining what proportion of the potentially available niches is exploitable at any time.

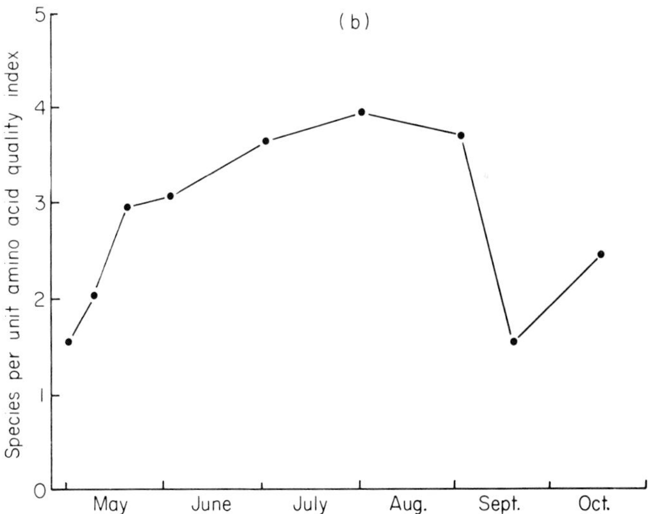

FIG. 12.9. (b) Leafhopper species per log unit of amino acid quality index in 1979.

That both quality and architecture are important is shown in Fig. 12.10. In this experiment a number of single-species swards were all fertilized on the same day and the nitrogen flows, plant growth and insect numbers were monitored before and after the fertilizer application. Very quickly after the application of the fertilizer the nitrogen was evident in the plants and was almost immediately picked up by the leafhoppers which quickly increased in both species and numbers; the flow of nitrogen, however, quickly peaked and fell back followed by the number of leafhopper species. About three weeks after the rise in nitrogen levels a steep increase in the growth of the grass and hence of architecture occurred, immediately followed by a further rise in

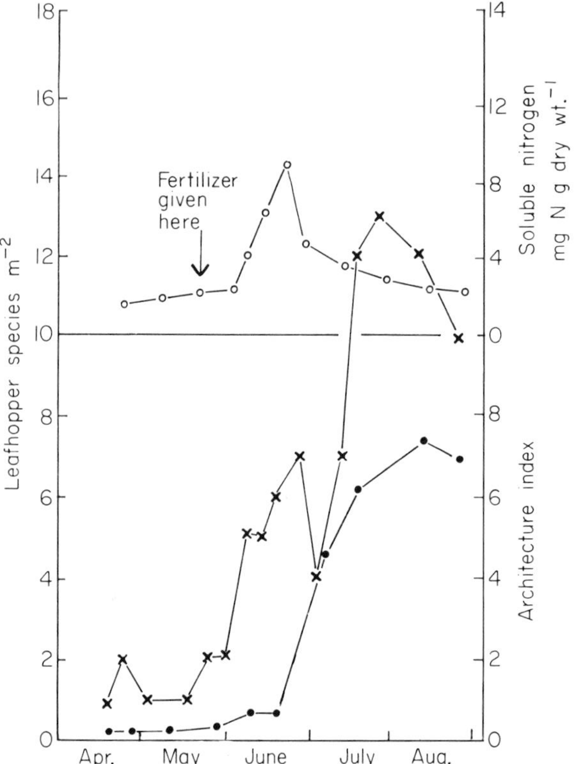

FIG. 12.10. Results of one of a number of trials in 1979 showing typical changes in (×) leafhopper species per m^2, (●) plant architecture, and (○) stem-soluble nitrogen in mg N g dry $wt.^{-1}$ before and after fertilizer application.

leafhopper species. This shows that the leafhoppers first responded to an increase in the available sites through increased quality without any real change in structure, and then to an increase in structure although the nitrogen levels were little raised from the previous levels.

NITROGEN, LEAFHOPPER REPRODUCTION AND MOBILITY

Reproduction and the utilization of food

In the 'evolutionary arms race' insects have evolved mechanisms to increase the availability to them of nitrogen in their food (McNeill & Southwood 1978). In grassland ecosystems where secondary chemicals are absent (Gibbs 1974) or in very low concentrations (Harborne & Williams 1976) close synchronization with the host-plant nitrogen level is one way to ensure a predictable high level in the food, when the nitrogen demand of the insect is at its highest. Hill (1976) determined nitrogen budgets for a number of leafhoppers in an attempt to determine periods of high nitrogen demand in the life cycle. He concluded that grass nitrogen levels were normally adequate to ensure nymphal and adult survival at any time of the year, but that the period of egg maturation created a very high nitrogen demand and this was the most likely period of nitrogen stress.

In support of this view, results of both our laboratory and field studies show that reproduction is more affected by available nitrogen than adult longevity or developmental period. For example, the multiplication rate of *Adarrus ocellaris* in the field, calculated as the number of nymphs per female, is greatest on experimental plots with high leaf nitrogen levels. Laboratory oviposition trials supported these studies; in several species, of which examples are given in Table 12.3, the rate of egg production is dependent on the levels of available nitrogen in the food. In most cases, however, egg

TABLE 12.3. Eggs female^{-1} day^{-1} for four species of leafhopper feeding on *Holcus mollis* fertilized at different rates with ammonium nitrate. All figures are the mean of results for females.

	N kg ha^{-1}					
Species	1200	900	600	300	100	0
Dicranotropis hamata	2.46	4.29	7.96	6.22	6.41	5.05
Elymana sulphurella	1.09	1.96	3.57	4.23	6.97	1.45
Eucelis incisus	—	—	3.2	3.56	1.14	1.13
Zyginidia scutellaris	4.05	2.17	1.93	1.67	1.27	1.42

production was not maximal at the highest level of available nitrogen. Adult longevity was unaffected by nitrogen level and hence total reproductive output paralleled the rate of egg production. This pattern of maximal utilization of nitrogen at intermediate levels of available nitrogen in the food was also seen in the nitrogen budgets of the reproducing females. The nitrogen utilization efficiencies, defined as in Waldbauer (1968), were calculated as a measure of the physiological performance of the leafhoppers. The results (Fig. 12.11) closely paralleled those of the oviposition experiments; in three of the species studied, both oviposition rate and nitrogen utilization efficiency peaked sharply over a very narrow range of available nitrogen levels. This was true for both phloem feeders such as *Elymana sulphurella* and mesophyll feeders such as *Zyginidia scutellaris*. In these species the relationship between the rate of egg production and nitrogen utilization efficiency was significant despite the very small amount of data available. In the case of *Dicranotropis*

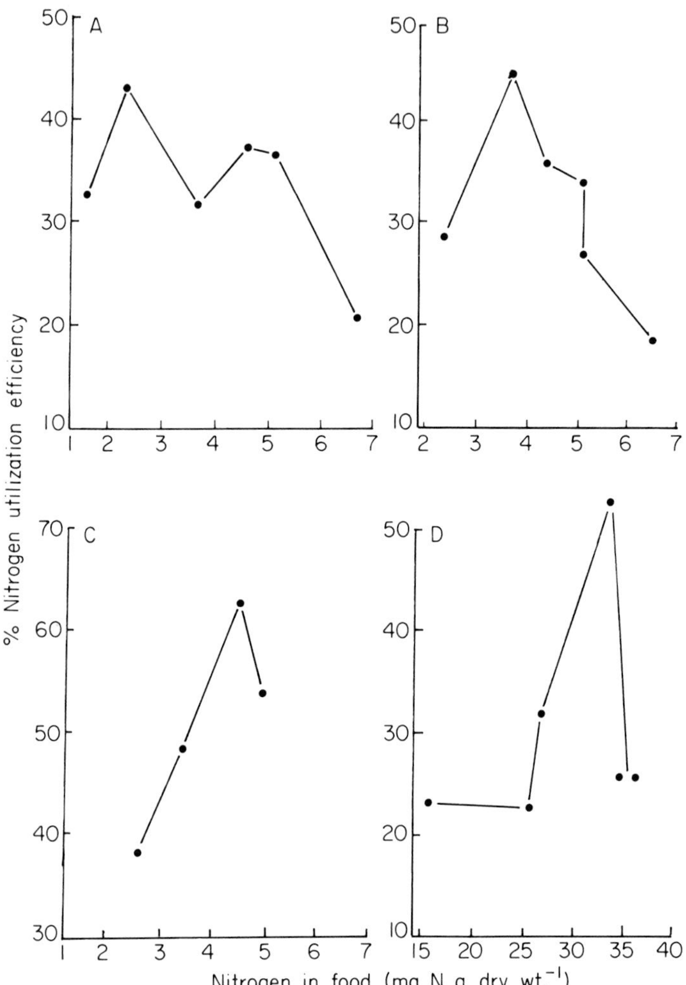

FIG. 12.11. The relationship between nitrogen utilization efficiency (NUE) and available nitrogen in food for reproducing females of *Dicranotropis hamata* (A), *Ellymana sulphurella* (B), *Eucelis incisus* (C) (phloem feeders) and *Zyginidi scutellaris* (D) (mesophyll feeder). $\text{NUE} = \frac{\text{nitrogen used}}{\text{nitrogen intake}} \times 100\%$.

hamata, however, the relationship was not nearly so good and this species is nowhere as sensitive to the nitrogen levels in the host (Fig. 12.12).

These results suggest that there may be two types of strategy within the leafhoppers to key their reproduction to adequate nitrogen levels in the host plant:

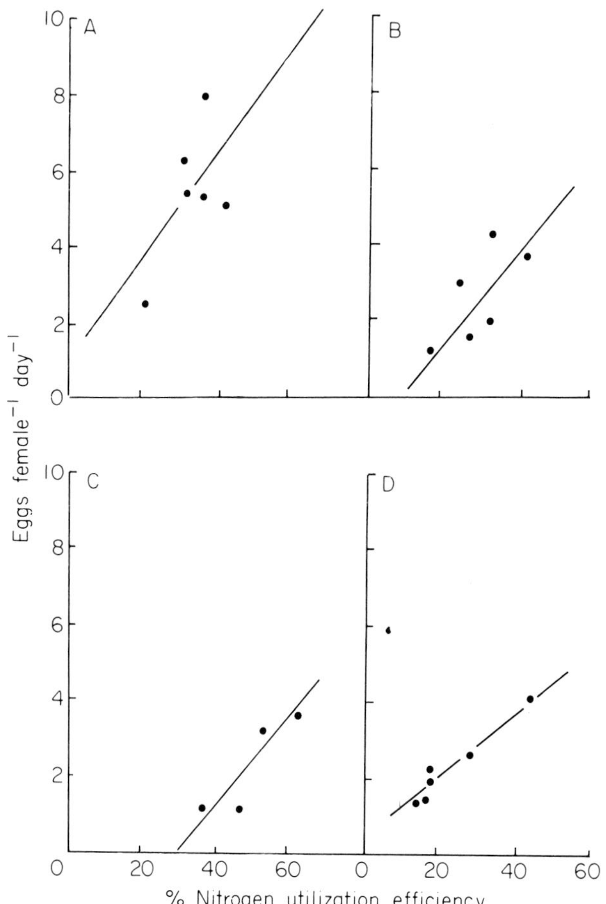

FIG. 12.12. The relationship between egg-laying rate and nitrogen utilization efficiency for *Dicranotropis hamata* (A) (NS), *Ellymana sulphurella* (B) ($P < 0.05$), *Eucelis incisus* (C) ($P < 0.05$), and *Zyginidia scutellaris* (D) ($P < 0.001$).

1. Highly mobile species which are relatively polyphagous in terms of host species but which are highly specific in their nitrogen requirements (van Emden 1978).
2. Less mobile species which are more tolerant of nitrogen levels but are closely tied to the phenology of their relatively specific hosts.

Food, nitrogen and mobility

Waloff (1980) has analysed the life history strategies of the grassland Auchenorryncha in terms of *r*- and *K*-strategies on the basis of their mobility, fecundity and host specificity. Her analysis included many of the species in this study and suggests that the two types of strategy outlined above correspond to her *r*- and *K*-strategists respectively, and that the majority of species tend toward the former strategy.

In order to look at the host specificity and mobility of the species in the current study, data from an experimental site on which plots of six of the commoner grasses had been sown were analysed. The total number of leafhoppers of each species caught in routine samples from each of these areas was expressed as a percentage of the total number of that species caught over the whole series of plots; the deviation of this from a uniform distribution of catches with respect to grass species was assessed as a χ^2. This was divided by the maximum possible χ^2 of 500 (for six species of host) to give an index between 0 for a complete generalist feeder and 1 for a complete specialist. The results are shown in Table 12.4, which show that there were only two host-specific species out of the ten species caught in usable numbers; the others were all near the polyphagous end of the scale, being almost equally abundant on all of the host species examined.

TABLE 12.4. Host specificity index for ten common leafhopper species on single-species swards. All figures in the body of the table are % total for each species. For derivation of specificity index, see text.

Leafhopper species	Grass species						Specificity index
	Lolium perrene	*Holcus lanatus*	*Festuca pratensis*	*Poa pratensis*	*Dactylis glomerata*	*Agrostis tenuis*	
Stenocranus minutus	0.21	0.84	0	0.63	98.33	0	0.96
Adarrus ocellaris	1.95	89.61	2.28	5.19	0.65	0.32	0.77
Macrosteles laevis	5.34	57.43	2.67	5.34	7.96	21.25	0.27
Artheldeus pascuellus	31.86	0.52	44.03	18.66	2.91	2.03	0.20
Ribautodelphax angulosa	10.17	2.28	6.82	12.42	21.48	46.84	0.16
Psammotettix confinis	23.89	3.01	10.47	16.43	1.51	44.70	0.16
Eucellis incisus	28.48	9.49	21.48	23.35	4.44	12.75	0.05
Zyginidia scutellaris	11.09	7.28	18.83	20.80	26.01	16.00	0.03
Javesella pellucida	13.19	14.07	23.80	18.33	16.50	14.12	0.01
Eucelis lineolatus	14.90	12.76	23.38	17.09	16.95	14.94	0.01

The association between leafhoppers and the nitrogen levels in the host was examined by plotting the average numbers of the mobile adults caught against the nitrogen levels of the host grasses at that time. Once again, two distinct patterns emerge: the more host-specific species such as *A. ocellaris* and *D. hamata* have a much flatter curve than polyphagous species such as *Zyginidia scutellaris* (Fig. 12.13).

In general, putting together the results of this and the previous sections, the majority of leafhoppers in our study showed a pattern of exploitation of food resources like that of *Z. scutellaris* (i.e. they are found on a range of hosts but tend to accumulate on those within a relatively narrow band of available food nitrogen levels). Their reproduction is also very dependent on the correct level of nitrogen in the food during maturation; their nitrogen utilization efficiencies are closely tied to egg production. These are Waloff's *r*-strategists.

A small number of species, often with the majority of the population with

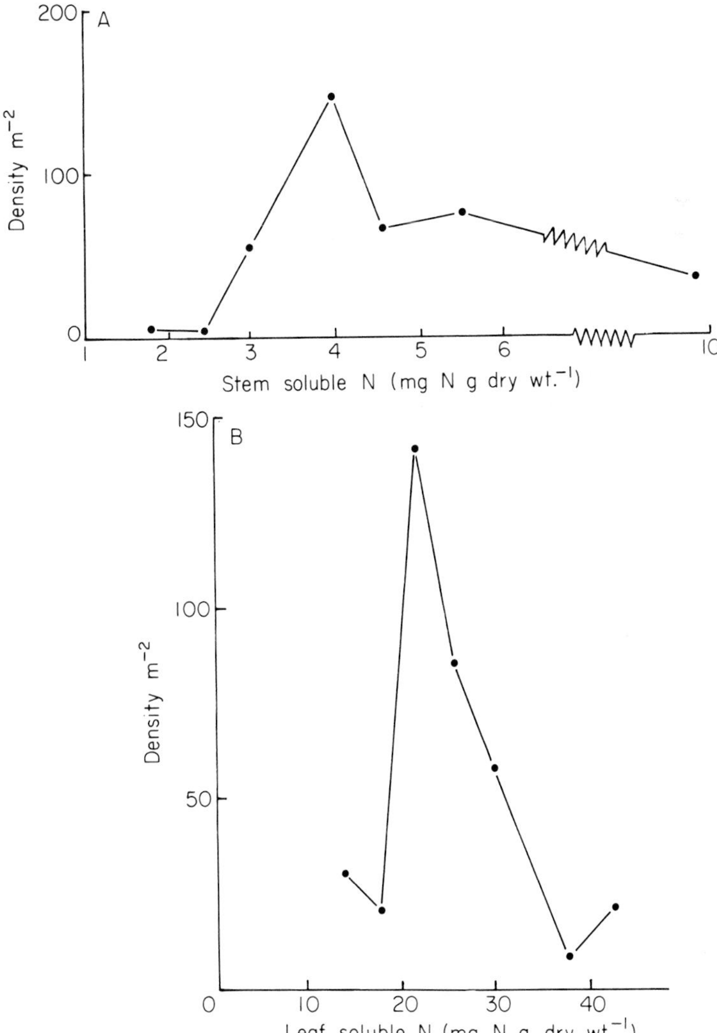

FIG. 12.13. Relationship between available levels of nitrogen in the food and the density of adult leafhoppers *Adarrus ocellaris* (A) and *Zyginidia scutellaris* (B) on plots in 1979.

reduced wings, and hence with low dispersal abilities, showed a pattern of exploitation of their food resources like that of *D. hamata* and *A. ocellaris*, i.e. they are largely confined to one host; they also show a much less strong tendency to accumulate on grasses with a particular nitrogen level, and have their reproduction and nitrogen utilization efficiencies less closely tied to the levels of available nitrogen in the food. These are Waloff's *K*-strategists.

The mechanisms used by the nitrogen specialist leafhoppers to find and track the grass nutrient levels at which they perform best can only be speculation at this stage, but Saxena & Saxena (1974) have shown that a variety of visual and short-range olfactory stimuli is important in the host-finding process.

CONCLUSIONS

Grasses, although having simple secondary chemistry, are highly variable both within and between seasons in the quantity and quality of amino acids available in their sap for exploitation by leafhoppers. These levels can easily be manipulated for experimental purposes by the use of nitrogen fertilizers.

The leafhopper complex shows a variety of responses to these changes in food quality but the overall pattern is for most species to be polyphagous in terms of host species but specialists in terms of nitrogen requirements; this in turn means that they are highly mobile, moving from host to host within and between seasons, tracking the most suitable hosts. A few much less mobile species are host specialists and nitrogen generalists, and as a result are much less tightly dependent for successful reproduction on available nitrogen levels. These species, however, also tend to keep their environment stable to them with respect to food quality by being closely synchronized in their life cycle to the phenology of their host.

In general, these strategies lead to an increase in both numbers and species with increasing levels of nitrogen in the host plants. This response is, however, linked also to changes in the architecture of the plant, as more 'living space' is made available by plant growth or by increasing quality, thus making more feeding sites exploitable.

The forces which govern the size and complexity of the leafhopper community on grasses are complex and many of the interactions are subtle, requiring careful field and laboratory experimentation to separate their effects. We have discussed only some of the more important factors associated with the host plant, but predators and parasites are probably almost as important in their effects; these are discussed more fully in Waloff's excellent review (Waloff 1980).

ACKNOWLEDGEMENTS

We wish to thank the many colleagues whom we have bored over the years on the subject of nitrogen and grassland insects—many of their criticisms and comments were useful. In particular, we should like to thank Drs N. Waloff, V. Brown, J. Lawton and J. Whittaker.

The work reported in this paper was supported by a grant from the Royal Society to S. McNeill, and an NRAC Doctoral Fellowship to R. A. Prestidge.

REFERENCES

Andrzejewska L. (1965) Stratification and its dynamics in meadow communities of Auchenorryncha (Homoptera). *Ekologia Polska Series A*, **13**, 685–715.

Auclair J.L. (1963) Aphid feeding and nutrition. *Annual Review of Entomology*, **8**, 439–90.

Beevers L. (1976) *Nitrogen Metabolism in Plants*. Edward Arnold, London.

Bell E.A. & Janzen D.H. (1971) Medical and ecological considerations of L-Dopa and 5-HTP in seeds. *Nature*, **229**, 136–7.

Carter C.I. & Cole J. (1977) Flight regulation in the green spruce aphid (*Elatobium abietinum*). *Annals of Applied Biology*, **86**, 137–51.

Cagampang G.B., Pathak M.D. & Juliano B.O. (1974) Metabolic changes in the rice plant during infestation by the brown planthopper, *Nilaparvata lugens*. *Applied Entomology & Zoology*, **9**, 174–84.

Cheng C.H. & Pathak M.D. (1972) Resistance to *Nephotettix virescens* in rice varieties. *Journal of Economic Entomology*, **65**, 1148–53.

Dadd R.H. & Krieger D.L. (1968) Dietary amino-acid requirements of the aphid *Myzus persicae* (Sulz). *Journal of Insect Physiology*, **14**, 741–64.

Denno R.F. (1977) Comparison of the assemblages of sap-feeding insects (Homoptera-Hemiptera) inhabiting two structurally different salt marsh grasses in the genus *Spartina*. *Environmental Entomology*, **6**, 259–72.

Feeny P.P. (1976) Plant apparency and chemical defence. *Recent Advances in Phytochemistry*, **10**, 1–40.

Fowden L. (1967) Aspects of amino-acid metabolism in plants. *Annual Review of Plant Physiology*, **18**, 85–106.

Gibbs R.D. (1974) *The Chemotaxonomy of Flowering Plants*. McGill-Queens University Press, London, Ontario.

Harborne J.B. & Williams C.A. (1976) Flavonoid patterns in the leaves of Gramineae. *Biochemical Systematics & Ecology*, **4**, 267–80.

Harrewijn P. (1976) Host plant factors regulating wing polymorphism in *Myzus persicae*. *Symposium Biologica Hungarica*, **16**, 79–83.

Harrewijn P. (1978) The role of plant substances in polymorphism of the aphid *Myzus persicae*. *Entomologia Experimentalis et Applicata*, **24**, 198–214.

Hill M.G. (1976) *The population and feeding ecology of five species of leafhopper (Homoptera) on* Holcus mollis. Ph.D. thesis, University of London.

House H.L. (1965) Insect nutrition. In *Physiology of the Insecta* (ed. M.E. Rockstein), Vol. 2, pp. 769–858. Academic Press, London & New York.

House H.L. (1972) Insect nutrition. In *Biology of Nutrition* (ed. R.N.T-W-Fiennes), *International Encyclopedia of Food and Nutrition*, Vol. 18, Section IV, pp. 513–73.

Larsen I. (1977) *Lysets og nitrogenilførslens indflydelse på aminosyrestofskiftet og stofproduktionen i italiensk rajgraes* (Lolium multiflorum) *studeret i karforsøg*. Ph.D. thesis, Royal Veterinary and Agricultural College, Copenhagen.

Lawton J.H. (1978) Hostplant influences on insect diversity: the effects of space and time. In *Diversity of Insect Faunas* (eds L.A. Mound & N. Waloff), pp. 105–25. Blackwell Scientific Publications, Oxford.

Lawton J.H. & McNeill S. (1979) Between the devil and the deep blue sea: on the problem of being a herbivore. In *Population Dynamics* (eds R.M. Anderson, B.D. Turner & L.R. Taylor), pp. 223–44. Blackwell Scientific Publications, Oxford.

McClure M.S. & Price P.W. (1975) Competition and coexistence among *Erythroneura* leafhoppers (Homoptera: Cicadellidae) on American Sycamore. *Ecology*, **56,** 1388–97.

McClure M.S. & Price P.W. (1976) Ecotype characteristics of co-existing *Erythroneura* leafhoppers (Homoptera: Cicadellidae) on Sycamore. *Ecology*, **57,** 928–40.

McNeill S. & Southwood T.R.E. (1978) The role of nitrogen in the development of insect/plant relationships. In *Biochemical Aspects of Plant and Animal Coevolution* (ed. J.B. Harborne), pp. 77–98. Academic Press, London & New York.

Miflin B.J. & Lea P.J. (1977) Amino-acid metabolism. *Annual Review of Plant Physiology*, **28,** 299–329.

Mittler T.E. (1970) Effects of dietary amino acids on the feeding rate of the aphid *Myzus persicae*. *Entomologia Experimentalis et Applicata*, **13,** 432–7.

Morris M.G. (1971) Differences between the invertebrate faunas of grazed and ungrazed chalk grassland. *Journal of Applied Ecology*, **8,** 37–52.

Morris M.G. (1973) The effect of seasonal grazing on the Heteroptera and Auchenorryncha (Hemiptera) on chalk grassland. *Journal of Applied Ecology*, **10,** 761–80.

Morris M.G. (1974) Auchenorryncha (Hemiptera) of the Burren with special reference to species associations of the grasslands. *Proceedings of the Royal Irish Academy*, **74,** 7–30.

Morris M.G. (1978) The effects of cutting on grassland Hemiptera: a preliminary report. *The Scientific Proceedings of the Royal Dublin Society, Series A*, **6**, 167–77.

Murdoch W.W., Evans F.C. & Peterson C.H. (1972) Diversity and pattern in plants and insects. *Ecology*, **53**, 819–28.

Oaks A. & Bidwell R.G.S. (1970) Compartmentation of intermediary metabolism. *Annual Review of Plant Physiology*, **21,** 43–66.

Parry W.H. (1974) The effects of nitrogen levels in sitka spruce needles on *Elatobium abietinum* (Walker) populations in north-east Scotland. *Oecologia*, **15,** 304–20.

Rehr S.S., Janzen D.H. & Feeny P.P. (1973) L-Dopa in legume seeds. A chemical barrier to insect attack. *Science*, **181**, 81–2.

Retnakaren A. & Beck S.D. (1968) Amino-acid requirements and sulphur amino-acid metabolism in the pea aphid, *Acyrthosiphon pisum* (Harris). *Comparative Biochemistry and Physiology*, **24**, 611–19.

Rhoades D.F. & Cates R.G. (1976) Towards a general theory of plant antiherbivore chemistry. *Recent Advances in Phytochemistry*, **10**, 168–213.

Saxena K.N. & Saxena R.C. (1974) Patterns of relationships between certain leafhoppers and plants. II. Role of sensory stimuli in orientation and feeding. *Entomologia Experimentalis et Applicata*, **17**, 493–503.

Sogawa K. (1971a) Feeding behaviour of the brown leafhopper and varietal resistance of rice to this insect. In *Symposium on Rice Insects. Proceedings of the Symposium on Tropical Agriculture Researches.* Serial No. 5, pp. 195–200, TARC, Tokyo.

Sogawa K. (1971b) Preliminary assay of antifeeding chemicals for the brown leafhopper, *Nilaparvata lugens* (Delphacidae). *Applied Entomology and Zoology*, **6**, 215–18.

Southwood T.R.E. (1977) Habitat, the templet for ecological strategies? *Journal of Animal Ecology*, **46**, 337–65.

Tallamy D.W. & Denno R.F. (1979) Responses of sap-feeding insects (Homoptera-Hemiptera) to simplification of hostplant structure. *Environmental Entomology*, **8**, 1021–8.

van Emden H.F. (1973) Aphid-host plant relationships: some recent studies. *Entomological Society of New Zealand Bulletin*, **2,** 25–43.

van Emden H.F. (1978) Insects and secondary plant substances—an alternative viewpoint with special reference to aphids. In *Biochemical Aspects of Plant and Animal Coevolution* (ed. J.B. Harborne), pp. 309–23. Academic Press, London & New York.

van Emden H.F. & Bashford M.A. (1969) A comparison of the reproduction of *Brevicoryne brassicae* and *Myzus persicae* in relation to soluble nitrogen concentration and leaf age in the brussels sprout plant. *Entomologia Experimentalis et Applicata*, **12,** 351–64.

Waldbauer G.P. (1968) The consumption and utilisation of food by insects. *Recent Advances in Insect Physiology*, **5,** 229–88.

Waloff N. (1980) Studies on grassland leafhoppers (Auchenorryncha, Homoptera) and their natural enemies. *Advances in Ecological Research*, **11,** 81–215.

Whittaker J.B. (1969) Quantitative and habitat studies of the froghoppers and leafhoppers (Homoptera, Auchenorryncha) of Wytham Woods, Berkshire. *Entomologist's Monthly Magazine*, **105**, 27–37.

Wint A.W. (1979) The ecology of the Wintermoth (*Operophthera brumata*) on five tree host species. D. Phil. thesis, University of Oxford.

13. NITROGEN IN A CROP–PEST INTERACTION; CEREAL APHIDS 1979

R. KOWALSKI AND P.E. VISSER
J.A. Pye Research Centre, Haughley Green, Suffolk IP14 3RS

SUMMARY

The interaction of cereal aphids with winter wheat was chosen for a study of the role of alternative farming systems in influencing pest incidence. During the summer of 1979 cereal aphid populations and other insect species were monitored in two crops of var. Maris Freeman, at Haughley in Suffolk. Leaf samples were taken at weekly intervals and analysed for free amino acid content by gas liquid chromatography. The conventional crop (treated with agrochemicals) developed a larger infestation of the aphid *Metopolophium dirhodum* (Walk.) than its organic counterpart. This crop also had higher levels of free protein amino acids in its leaves during June which were believed to have resulted from a nitrogen top dressing of the crop in early April. However, the difference in the aphid infestations between the crops was attributed to the aphids' response to the relative proportions of certain non-protein to protein amino acids in the leaves at the time of aphid settling in the crops. Natural enemies occurred in higher numbers on the conventional crop, reflecting their responsive role in pest incursions. The understanding of plant nutrition and the part it plays in plant resistance to insect attack is emphasized as a fundamental requirement for the development of integrated control programmes.

INTRODUCTION

Practitioners of organic farming consistently claim a virtual absence of pest problems which they attribute to natural enemy abundance within such organically managed crops (Oelhaf 1978). This view would tend to be supported by the studies of the Game Conservancy Council in West Sussex (Potts & Vickerman 1975) which showed that the levels of weeds in cereal crops, associated with more traditional farming methods, had a marked effect upon the arthropod fauna, especially predators and parasitoids. However, Lawton & McNeill (1979) have emphasized the importance of the reproductive rate of phytophagous insects in determining the ability of natural enemies to contain their populations. Indeed, Dempster & Pollard (1981) argued that the major component of insect population dynamics is the interaction between the insect and its food source. Therefore, in order to examine the claims of organic farmers thoroughly, it is necessary to examine not only the natural enemy abundance within the crop but also to look at the interaction between the crop plant and its associated phytophagous insect fauna.

For nearly 40 years, three alternative farming systems have been maintained on a 100 ha site at Haughley in Suffolk in order to compare the economic, ecological and nutritional viability of the three alternatives (Balfour 1975). Those alternatives are as follows:

1. An intensive arable farm with a limited rotation of mostly white straw crops, totally dependent upon agrochemical inputs.
2. A conventional ley-arable farm utilizing both farm wastes and agrochemicals on an eight-year rotation of four years of grass to four years of arable crops.
3. An organic ley-arable farm where no agrochemicals are used at all and which has the same rotation as its conventional counterpart.

This site provided an ideal setting for the examination of organic crops in comparison with those receiving chemical applications.

The crop and pest system chosen for study was that of winter wheat (*Triticum aestivum* L.) and cereal aphids. The widely held view that cereal aphids are becoming pests of increasing importance and that this increase derives from changes in cereal management (Daniels 1957; Adams & Drew 1965; Potts 1970; Baranyovits 1973; Kolbe 1973; Rautapää 1977) made them an obvious choice. Furthermore, both natural enemies and host plant interactions have been implicated in determining aphid population levels (Daniels 1957; van Emden 1966, 1972; van Emden & Bashford 1971; Potts & Vickerman 1975).

Reviews by Rodriguez (1960), Vickerman & Wratten (1979) and Carter *et al.* (1980) have all emphasized the desirability of studying the physiology of the host and the host–pest relationship in order to produce practical advice on crop management and to develop an enhanced understanding of plant resistance. The suggestion by McNeill & Southwood (1978) that plants of the family Gramineae rely upon temporarily limited nitrogen flushes to avoid herbivore damage and that plant nitrogen is a determining factor in insect fecundity, indicated that it was the crop nitrogen history which was most likely to be implicated in the interaction between farming practice and aphid numbers, outside the impact of natural enemies. Furthermore, they indicated that the nitrogen component most significantly correlated with aphid performance was likely to be the plant's free amino acid pools.

Therefore, in the summer of 1979 a study was made of two winter wheat fields and their arthropod fauna as set out below.

METHODS

Two fields, one from the organic farm and one from the conventional farm were selected for the study. Both fields were at the same point in the rotation, were drilled on the same day (13th October 1978) and were of approximately the same size (1.2 ha). The conventional crop received a seed-bed dressing of 250 kg ha^{-1} 9:25:25 NPK fertilizer on 11th October and the seed was treated with an organomercury fungicide dressing. A top dressing of 34.5% nitrogen applied at 187.5 kg ha^{-1} (i.e. 64.5 kg N ha^{-1}) was conducted on 11th April 1979. A broad-leaved weed herbicide, being a mixture of Dicamba, Mecoprop and MCPA, was applied at 5 l ha^{-1} on 8th May and a wild oat (*Avenc fatua* L.) herbicide, Difenzoquat, at 1.59 kg ha^{-1} on 15th May.

Insect samples were taken from the central half hectare of both fields from 18th April until 6th August. Wherever possible the fields were sampled twice a week. The method of sampling was to take sweeps, using a Univac suction sampler (Arnold, Needham & Stevenson 1973), of one minute's duration at the height of the topmost leaf, whilst walking slowly through the crop. Twenty samples were taken from each field on each sampling occasion, which always commenced at 0900 hours. The two fields alternated between sampling events as the site for taking samples first. Sampling never took longer than 1½ hours to complete. The samples were deep frozen and then sorted into alcohol for identification and counting later.

Whole plants, including surface rooting systems, were removed to the laboratory at 0800 hours at weekly intervals from 23rd April until 6th August from the central half hectare of each field, and the growth stage of the plants noted. The topmost leaves were excised and frozen immediately. After the full emergence of the flag leaf only this leaf was excised. All insect material was removed before processing the leaves. First samples also alternated between the two fields with each sampling event. The samples were freeze-dried, ground in a Moulinex blender and stored in plastic bags at room temperature prior to extraction and analysis.

Quadrat counts of the number of tillers (m^2) were taken on 28th June from the central half hectares of both fields. On 15th August random samples of ears were collected from both fields and the number of grains per ear counted. After, harvest samples of the grain were collected from the combine as representative of both fields and used to determine 1000 grain weights.

AMINO ACID ANALYSIS

Approximately 1 g dry wt. of ground leaf was suspended in 50 ml 0.01 N HCl and stirred for two hours at room temperature. The supernatant was then decanted, the cake of leaf debris being allowed to drain for a few minutes, and then centrifuged at 4000 r/min for ten minutes. This was followed by a further decantation and 0.3 ml 1.2 N trichloroacetic acid added and the volume measured. The mixture was centrifuged at 4000 r/min for a further ten minutes and an aliquot was taken to which was added a measured amount of cyclo-leucine, in 0.1 N HCl, as an internal standard. This solution was then subjected to cation exchange chromatography (Boila 1977) to remove contaminants from the sample. Ten 0.5 ml aliquots of the eluate were taken and stored at $-20°$ before derivatization for gas liquid chromatography.

Amino acids were analysed as the n-propyl N-O-heptafluorobutyryl esters using the method of March (1975) on a Pye Unicam Series 204 gas chromatograph. Peak areas were integrated by a DP88 computing integrator.

The use of the gas chromatography method has certain disadvantages over the conventional ion-exchange analyses. The major components of the nitrogen flushes, glutamine and asparagine (McNeill & Southwood 1978) are converted during derivatization to their respective acids and are measured in conjunction with the free glutamic and aspartic acids, respectively. Also, the process of derivatization is complex and more time-consuming than the preparations for ion exchange. Nevertheless, the analysis time on the machine, which is the usual constraint in any procedure, is much shorter than that for the conventional analysers.

The use of ten aliquots for separate analysis provided standard errors for the mean values of the acids detected. However, this statistic is purely a reflection of the degree of variation in the process of derivatization and analysis itself. It gives no indication of the degree of variation between plants nor does it cover any changes in the material during storage (Perkins 1961).

RESULTS

The use of a Univac suction sampler for taking insect samples had many drawbacks which are worthy of mention. The sample counts cannot be related to an area of crop or be presented as the number of insects per tiller, neither can the number of individuals of different species be related to each other as different species are subject to variation in susceptibility to capture (Henderson & Whittaker 1977). The use of a suction net or sweeping as a sampling technique (and this method is a cross between the two) has been criticized by Mayse, Kogan & Price (1978) and Byerly *et al.* (1978), both groups favouring a direct counting procedure or whole plant sample. However, since our intention was to compare two crops of broadly similar phenology the method was appropriate to our resources.

A further problem was presented when aphid populations became high in June and July in that the suction sampler often did much damage to the insect specimens. Furthermore, after storage in alcohol it became impossible to distinguish the identity of aphid nymphs with any certainty. However, all recognizable specimens of other species were removed and what remained were assumed to be *Metopolophium dirhodum* (Walk.) which, considering the proportions of the outbreak of this species in 1979, seemed to be a reasonable assumption.

The population development of the aphid *M. dirhodum* as reflected in the numbers per sample is presented in Fig. 13.1 for the two treatments. Fig. 13.2 presents similar information for the other cereal aphid, *Sitobion avenae* (F.), which occurred in very low, but nevertheless countable, numbers in July and

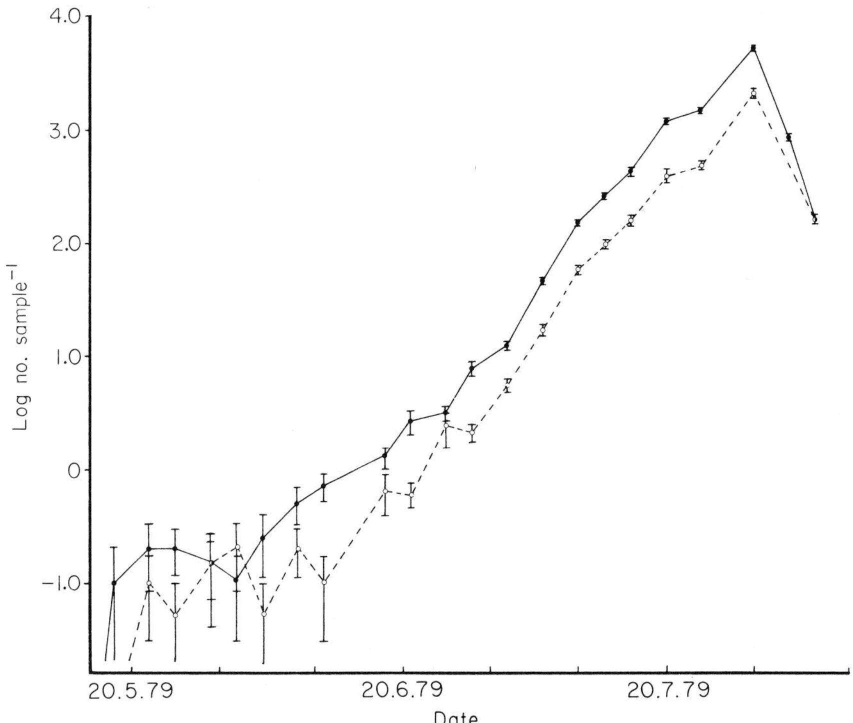

FIG. 13.1. *Metopolophium dirhodum* population development on ●——● conventional and ○ – – – – ○ organic winter wheat. Bars represent ± one standard error.

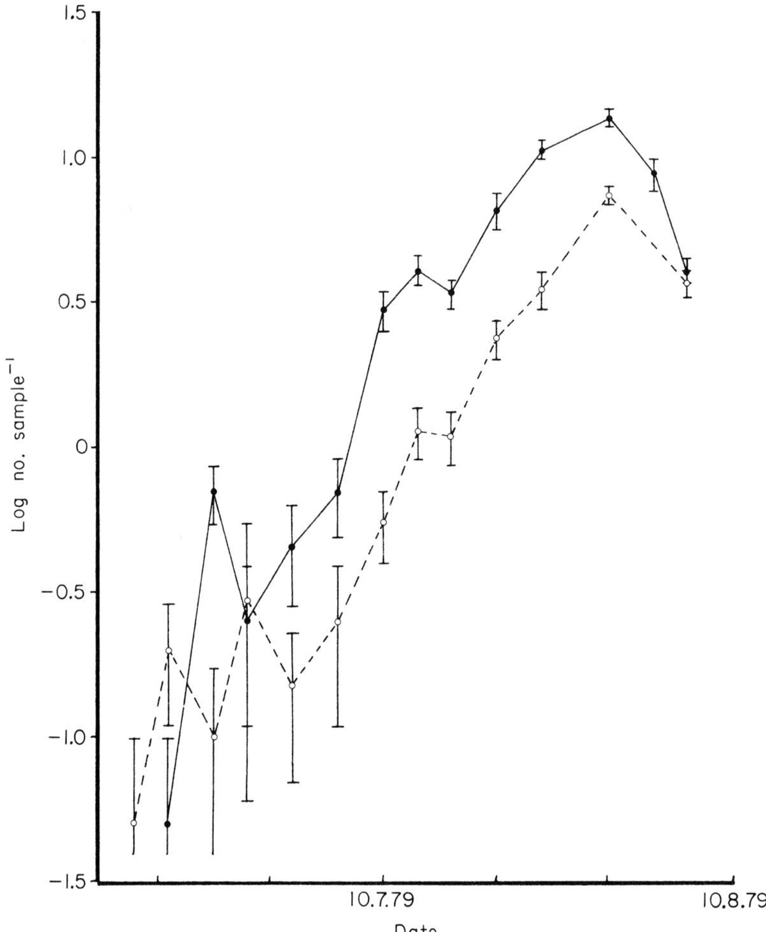

FIG. 13.2. *Sitobion avenae* population development on ●——● conventional and ○ - - - - ○ organic winter wheat. Bars represent ± one standard error.

August. Only a very small number of individuals of other aphid species were captured and these were of no value to the study. The use of the suction sampler was further mitigated by its ability to trace the development of the aphid populations when populations were too low for direct observations to be practical.

Large numbers of many other insect species were captured during the season, including parasitoids and polyphagous predator groups. In line with the observations of Potts (1970) the weed-dwelling Coleoptera were more numerous on the organic field indicating a prevalence of weeds in that crop. A full assessment of the differences in arthropod fauna will be made elsewhere, but indices of diversity (Williams 1947) were not significantly different between the crops, unlike the studies in West Sussex (Potts & Vickerman 1975) and natural enemies were more prevalent in the conventional crop, reflecting their dependence on and response to phytophagous insect numbers (Altieri & Whitcomb 1979).

Vickerman & Wratten (1979) stated that studies in West Sussex had shown a direct relationship between suction sample captures and aphid numbers directly observed on the crop. Thus the information presented in Figs 13.1 and 13.2 is a true reflection of population development for the respective crops. Both these figures demonstrate that the rates at which the populations of aphids grew were identical for both crops and thus predator pressure in both fields was the same, indicating that observed differences in the populations were not a consequence of different natural enemy abundance.

The changes in the total pool of free protein amino acids in the leaves of the two crops are summarized in Fig. 13.3. Individual protein amino acids by and large followed the pattern of the total pool. The initial difference in this pool between the treatments is almost certainly due to the inorganic nitrogen application made to the conventional crop on 11th April and is in line with observations of the responses of perennial grasses to fertilizer applications made by Prestidge (1980). Whilst the general trends in the changes of the pool of free protein amino acids are sustained, the fluctuations from week to week within a single treatment are likely to reflect changing environmental conditions affecting the diurnal levels of the free amino acids (Larsen 1977). Free protein amino acids occurred at high levels during June in the leaves of the conventional wheat crop. However, these differences did not occur in a way which paralleled the development of the respective *M. dirhodum* populations so that although higher free protein amino acids in the leaves were positively associated with higher aphid numbers a direct relationship was not to be seen.

Non-protein amino acids have been identified as important determinants of aphid settling (Jördens-Röttger 1979) and as deterrents to the growth and development of Auchenorryncha (see Chapter 12). The amino acid analyses of the wheat leaves contained a group of seven non-protein amino acids which occurred regularly and in reasonable quantities and which reflected changes not linked to the protein amino acid metabolism. These were quantified and combined values for the levels which were found in the two crops are presented in Fig. 13.4. These amino acids are those which could be identified with reasonable certainty. Others occurred occasionally or were not possible to identify, with the exception of γ-amino butyric acid which was by far the most abundant non-protein amino acid present, but which followed the changes in

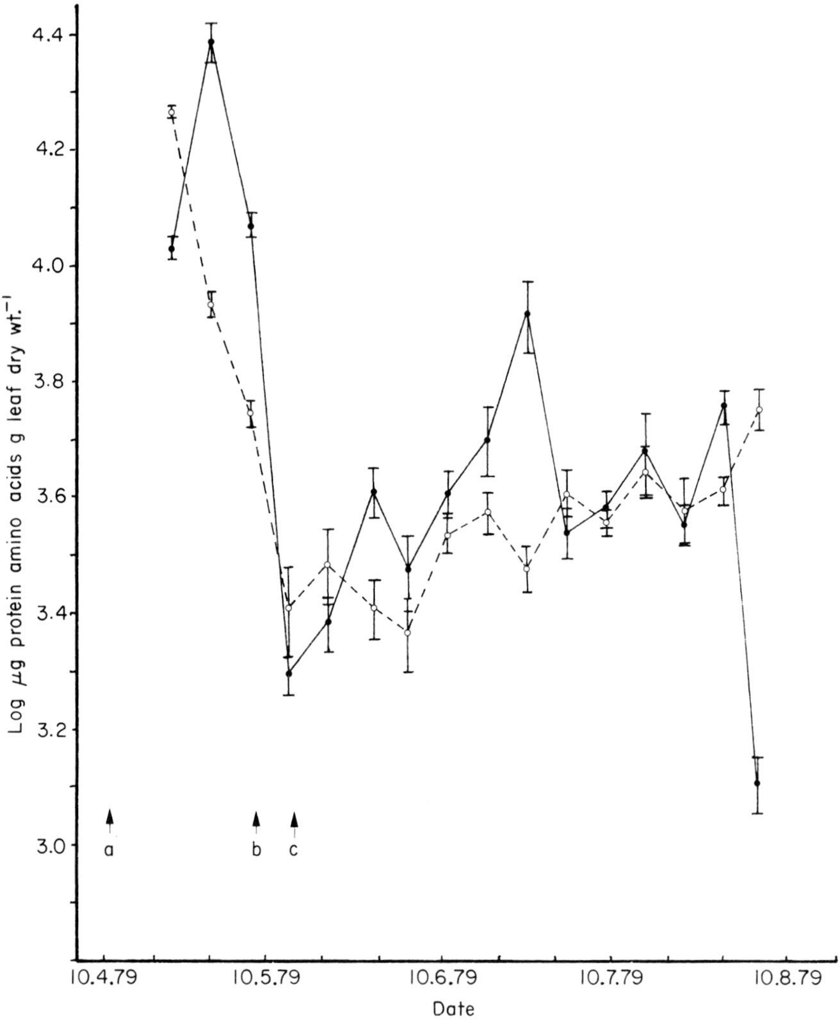

FIG. 13.3. Changes in the levels of free protein amino acids in the leaves of ●——● conventional and ○ - - - - ○ organic winter wheat. Bars represent ± one standard error. Chemical applications to the conventional crop were as follows: (a) 64.5 kg N ha^{-1}, (b) a mixture of Dicamba, Mecoprop and MCPA at 5 l ha^{-1}, (c) 1.59 kg Difenzoquat ha^{-1}.

free protein amino acid levels until mid-July. The seven non-protein amino acids were 2, 3 diaminopropionic acid, γ-amino and β-hydroxybutyric acid, anthranilic acid, δ-amino valeric acid, penicilamine dihydroxyphenylalanine and α-amino pimelic acid.

Once again, changes in these amino acid levels did not correlate with the

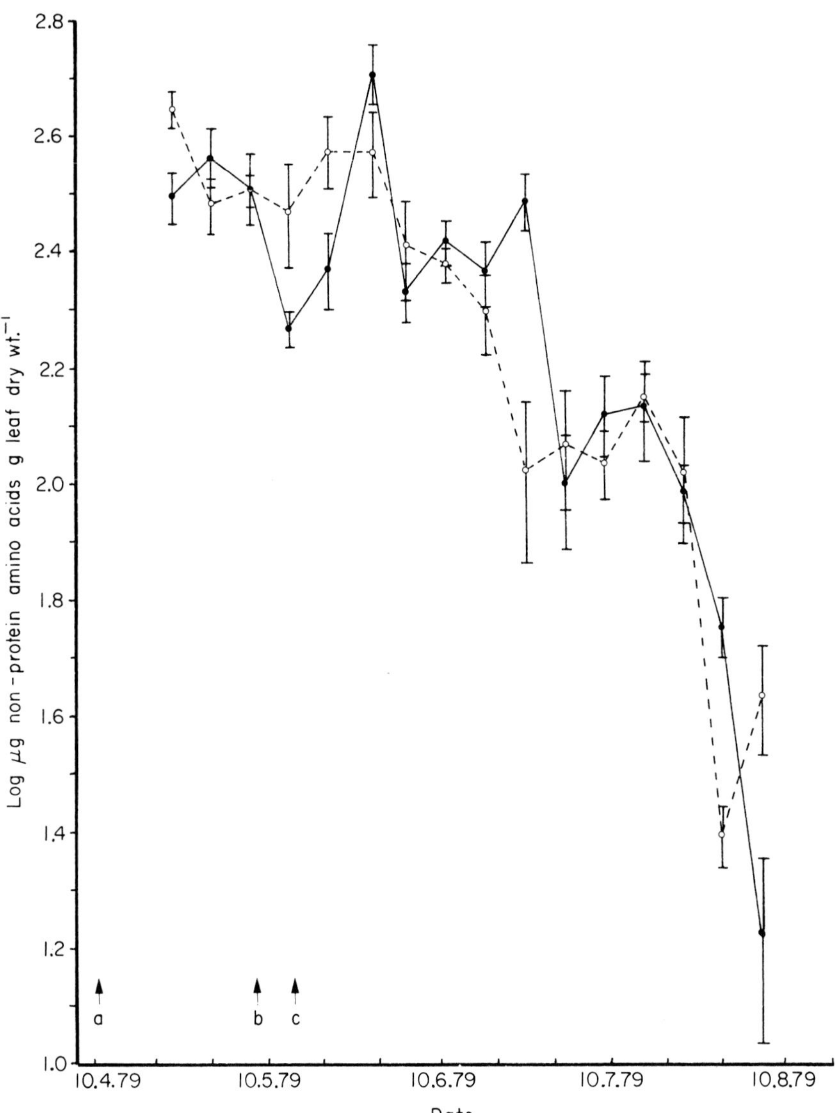

FIG. 13.4. Changes in the levels of non-protein amino acids in the leaves of ●——● conventional and ○ - - - - ○ organic winter wheat. Bars represent ± one standard error. Chemical applications to the conventional crop were as follows: (a) 64.5 kg N ha^{-1}, (b) a mixture of Dicamba, Mecoprop and MCPA at 5 l ha^{-1}, (c) 1.59 kg Difenzoquat ha^{-1}.

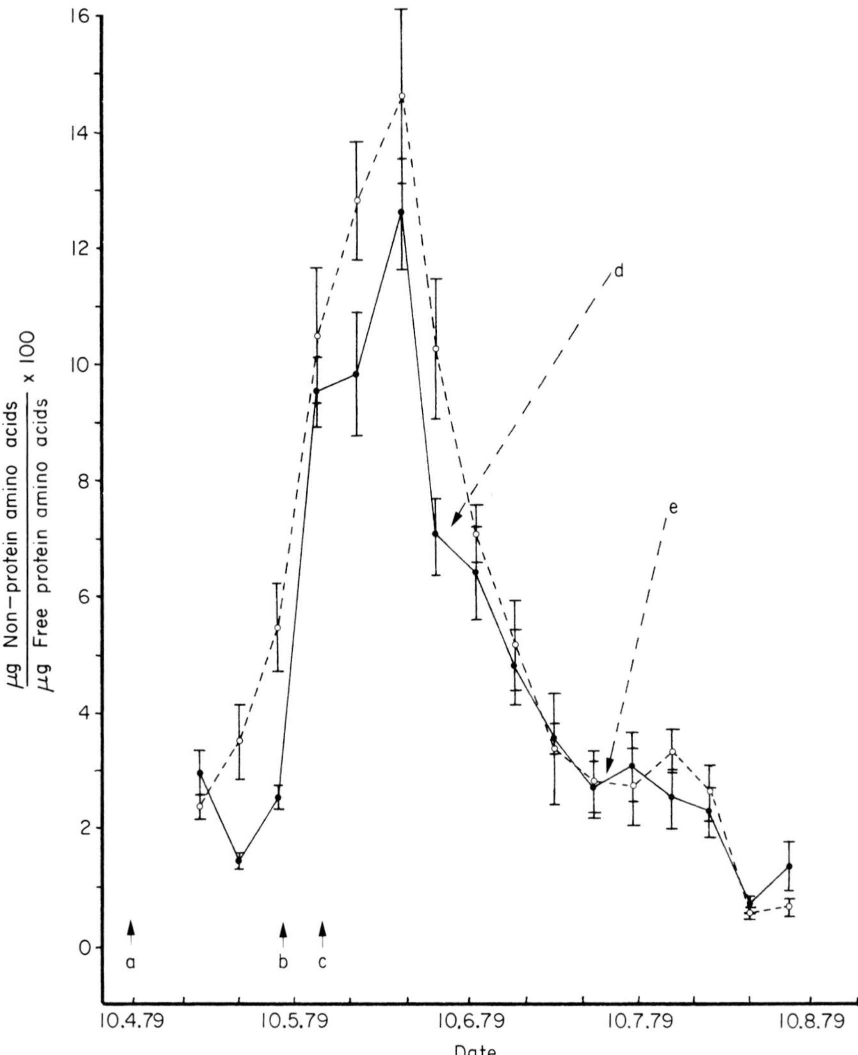

FIG. 13.5. Changes in the relative proportions of non-protein to protein amino acids in the leaves of ●——● conventional and ○- - - -○ organic winter wheat. Bars represent ± one standard error. Chemical applications to the conventional crop were as follows: (a) 64.5 kg N ha^{-1}, (b) a mixture of Dicamba, Mecoprop and MCPA at 5 l ha^{-1}, (c) 1.59 kg Difenzoquat ha^{-1}. Other points of interest: (d) The time of differentiation between the crops by *M. dirhodum*, (e) The time of differentiation between the crops by *S. avenae*.

respective aphid population development. Furthermore, there were less sustained differences between levels of these acids found in the respective crops than there had been for the protein amino acids. However, a comparison of the relative abundance of these non-protein to protein amino acids for the two crops was undertaken (see Chapter 12) and the results are presented in Fig. 13.5.

The first fact which emerges from such a comparison is that the week by week fluctuations so noticeable in Fig. 13.3 have virtually been eliminated. The graph for the organic crop is almost without perturbation, whereas the conventional, chemically treated crop does show some disjointed changes from week to week. Furthermore, the organic crop had a consistently higher proportion of non-protein to protein amino acids during the whole period when *M. dirhodum* was invading the crops. Indeed, during the period when the two respective populations were differentiated, and the final pattern of infestation was determined, the relative levels of non-protein to protein amino acids of the respective crops were also separated by a similar interval of time.

TABLE 13.1. Evaluation of parameters which contribute to overall yield for the two crops.

	Conventional crop	Organic crop	Difference	't' test
Tillers (m^2)	285.9 ± 7.5	270.3 ± 7.3	15.5	NS
Grains per ear	36.0 ± 1.1	44.3 ± 1.1	−8.3	$P < 0.001$
1000 grain wt. (g)	58.9 ± 1.0	60.7 ± 1.4	−1.8	NS
Calculated yield (tonne ha^{-1})	6.058 ± 0.457	7.263 ± 0.565	−1.205	NS

The growth stages of the two crops were never sufficiently different at any sampling event to warrant the allocation of a different decimal number (Tottman & Makepeace 1979). The performances of the crops as reflected in the three components which contribute to yield are summarized in Table 13.1 with the results of their respective 't' tests. The only significant difference is the number of grains per ear which was almost certainly reduced on the conventional crop by the Dicamba application (Tottman 1978). However, the reduction of grain number has a marked effect on grain weights (Bremner & Rawson 1978; Pinthus & Millet 1978). Thus the conventional crop probably compensated for losses in grain number by increasing grain weight, making any assessment of the impact of the different aphid populations impossible.

DISCUSSION

Studies of plant resistance to insect pests have shown that resistance varies with the age or growth stage of the plant (Dodd & van Emden 1979; Kishino & Ando 1979). This suggests that resistance is linked directly to the physiology of the plant (McMurtrey 1962) and thus any factor which affects the physiology of the plant may lead to changes in resistance to insect pests.

Cereal aphids of the species *Metopolophium dirhodum* and *Sitobion avenae* come into the cereal crops from the end of May until late June (Carter *et al.* 1980), although this immigration is subject to a wide variation (Taylor *et al.* 1980). Nevertheless, both species usually settle on the flag leaf and only *S. avenae* moves on to the ears after heading (Carter *et al.* 1980). If the proportion of non-protein to protein amino acids is a measure of antibiosis to these insects (see Chapter 12) then the decline of this proportion from the emergence of the flag leaf ligule (growth stage 29) onwards represents a falling antibiosis to which the aphids are likely to be responsive. Jördens-Röttger (1979) indicated that aphids were responsive to non-protein amino acids in leaf surfaces. Bond & Lowe (1975) showed that fewer adult *Aphis fabae* Scop. settled on bean plants (*Vicia faba* L.) of resistant varieties. The fact that aphid populations developed at the same rate on both treatments presented here, but that the date at which populations began to grow was later in the organic crop, is an indication that the response of the aphids was probably a behavioural one at the time of settling in the crop, in response to the proportion of non-protein to protein amino acids in the leaves.

The agency responsible for the different proportions of amino acids is most likely the nitrogen application of 11th April. Dewar (1980) also monitoring *M. dirhodum* in 1979 observed differences in populations on plots with different levels of applied nitrogen, whilst the rate of population development was the same. Similarly, Vereijken (1979) in laboratory trials observed similar rates of growth in *S. avenae* with different nitrogen treatments, although initial mortality/fecundity was related to the quantity of applied nitrogen. Hanisch (1980) confirmed higher reproduction of *S. avenae* in ear cages on wheat given greater nitrogen applications and also confirmed increased selection on the part of alate *S. avenae* of wheat plots given high rates of applied nitrogen.

The field observations of Henderson & Perry (1978), however, seem to contradict these findings as they suggested that higher aphid densities were associated with lower nitrate concentrations in plant sap. The observations at Haughley indicate that the semi-subjective assessment of the growth stage of the crop is a less reliable indication of the suitability for aphid colonization than physiological measurements. It is probable that the lower nitrate concentrations observed in the sap by Henderson & Perry were linked with a more advanced crop rather than reflecting the levels of nitrogen supplied to the crop.

The role of herbicides in determining aphid populations is more ambivalent. Vickerman (1974) concluded that differences caused by the herbicides Mecoprop and a mixture of Metoxuron and Simazine were purely secondary effects through their impact on natural enemies. Rautapää (1972) was unable to detect any effect of the herbicides MCPA, Mecoprop and a mixture of Mecoprop and Ioxynil on laboratory cultures of cereal aphids. He also tested the straw shortener chlormequat chloride with no observable interaction. However, Rautapää did observe a depressing effect of the herbicide Dinoseb on population development and Adams & Drew (1969) found that highest numbers of cereal aphids were always associated with a 2,4-D amine herbicide treatment. There is no indication that the application of dicamba, MCPA, Mecoprop and Difenzoquat to the conventional wheat at Haughley in 1979 had any influence on aphid population development, either directly or indirectly.

Despite the fact that *M. dirhodum* was over twice as abundant on the conventional crop compared with the organic crop, the real difference between the populations was a period of approximately seven days in the settling of immigrants. A crop of variety Hustler grown at Rothamsted Experimental Station and receiving two levels of nitrogen fertilizer (160 kg N ha^{-1} and 250 kg N ha^{-1}) in April 1979 also developed *M. dirhodum* populations which grew at the same rate but which were separated by approximately three days (Dewar 1980). *S. avenae* populations maintained in clip cages on wheat and treated with different nitrogen levels in the laboratory also reflected similar rates of growth but with a time lag of approximately one day (Vereijken 1979). This suggests that the effect of nitrogen declines as the quantity applied increases. Clearly these differences are extremely small in terms of the aphids' potential performance but nevertheless can have a profound effect upon final infestation levels and subsequent crop damage. Furthermore, the causal agency must also be relatively small, short lived and subtle. *S. avenae* was very late immigrating in 1979 (Taylor *et al.* 1980) and so came on to the ears rather than the leaves. Nevertheless the aphids showed a similar response to that of *M. dirhodum* on the two treatments despite the fact that amino acid pools in the leaves were virtually identical at that time. We must presume that pools of amino acids in the ears show a later but similar pattern of change to the pools in the leaves. This is true of perennial grasses (S. McNeill, pers. comm.).

Rabbinge & Vereyken (1980) pointed out that much of the damage caused by cereal aphids takes the form of early leaf senescence and secondary fungal infestations. Quite clearly the pattern of change of free protein amino acids at the end of July, as shown in Fig. 13.3, indicates the onset of flag leaf senescence earlier in the conventional crop. Furthermore, Table 13.1 shows that the conventional crop did not yield as well as its organic counterpart and that grain weights were lower, although they had probably compensated for loss in grain number which resulted from the herbicide application. An assessment of the full impact of the infestation on yield was not possible under these circumstances.

The observations made at Haughley enable us to speculate on the nature of resistance in cereals. Many authors have attempted to examine varietal and species resistance to cereal aphids (Markkula & Roukka 1972; Dewar 1977; Gill & Metcalfe 1977; Lowe 1978, 1980; Ba-Angood & Stewart 1980) and even to examine the interrelationships between feeding site, species of aphid and fecundity (Watt 1979). However, if resistance is manifested in the proportion of non-protein to protein amino acids and this changes continuously at all sites on the plant it becomes imperative to test the right site and growth stage with the appropriate species of aphid. If the interaction between variety or species, feeding site and growth stage can be examined from the point of view of the proportion of free amino acids then it may be possible to develop biochemical tests of the crop plant for resistance which remove the laborious live aphid trials to which recourse is now taken (Vickerman & Wratten 1979). Furthermore, it is likely that the prospect of resistance determined in this way would better represent the resistance likely to be achieved under field conditions.

With regard to population models and prognoses for the purpose of chemical applications to control infestations, the insect–plant interaction revealed in this study must pose some serious questions. Present chemical control recommendations for *S. avenae* are that an application is necessary when an average of 3–5 aphids per ear occur during the flowering of wheat (Wetzel, Freier & Heyer 1980) or five or more aphids can be found on average per tiller and the number rising at the beginning of flowering (George & Gair 1979). Quite clearly there is some scope for bioenvironmental control of aphid populations before the cereal reaches this stage and the data presented by George & Gair indicate that even in outbreak years there are crops which do not achieve levels of infestation which would necessitate chemical control. An investigation of such crops would perhaps be as effective as trying to work out the details of damage thresholds in order to prepare spraying recommendations. Moreover there is some evidence that the insect–plant interaction may also offset the effectiveness of insecticide treatments (Boness & El-Dessouki 1972) and this would also be worth studying in cereal aphids.

It would seem that a study of the interaction between the aphids and their primary plant hosts could be revealing with regard to the determination of outbreak years. Suter & Keller (1977) and Vickerman (1977) have drawn attention to the role of cold spring weather in delaying the aphid build-up on the primary host with the possible consequence of reduced natural enemy abundance which leads to higher numbers of immigrants and lower populations of natural enemies in the crop, although combinations of linear regression models of spring weather and summer aphid abundance have not proved very successful in predicting outbreak years (Carter *et al.* 1980). Clearly the role of weather in the interaction between soil nitrogen availability and the nitrogen status of the primary host would bear careful examination.

CONCLUSION

In conclusion it must be said that the study of cereal aphid populations in 1979 on winter wheat at Haughley did reveal a greater preponderance of the pest on chemically treated wheat, and that the organically grown cereal suffered less of an infestation because of increased resistance to aphid colonization as a result of receiving no soluble nitrogen top dressing in early April. The role of predators has not been fully established in this study, but it appeared to be proportional to the size of the infestation. This was probably a result of the small field sizes and the ley-arable system of farming, which has been shown elsewhere (Potts 1970) to retain natural enemy populations at higher levels than more intensive systems. This suggests that a comparative study of intensive systems with organic crops is likely to show an even greater divergence of pest incidence than the systems examined here. The study has served to emphasize the role of the insect–plant interaction as an important contributor to pest population dynamics and to further emphasize the role of free amino acids, in particular the relative proportions of non-protein to protein amino acids, in such insect–plant interactions. An understanding of the interaction between managerial practices and plant resistance to pest attack must contribute to the effective development of integrated control programmes.

ACKNOWLEDGEMENTS

The authors wish to thank Professor H. Vogtmann and Professor J. R. Thompson for their assistance in improving the methods of sample preparations for amino acid analysis by G.L.C. We are also grateful to Dr S. McNeill for his help, encouragement and criticisms of the study. The project was generously supported by Mr and Mrs J. A. Pye.

REFERENCES

Adams J.B. & Drew M.E. (1965) Grain aphids in New Brunswick. III. Aphid populations in herbicide-treated oat fields. *Canadian Journal of Zoology*, **43**, 789–94.

Adams J.B. & Drew M.E. (1969) Grain aphids in New Brunswick. IV. Effects of malathion and 2, 4-D amine on aphid populations and on yields of oats and barley. *Canadian Journal of Zoology*, **47**, 423–6.

Altieri M.A. & Whitcomb W.H. (1979) Manipulation of insect populations through seasonal disturbance of weed communities. *Protection Ecology*, **1**, 185–202.

Arnold A.J., Needham P.H. & Stevenson J.H. (1973) A self-powered portable insect suction sampler and its use to assess the effects of azinphos methyl and endosulfan on blossom beetle populations on oil seed rape. *Annals of Applied Biology*, **75**, 229–33.

Ba-Angood S.A. & Stewart R.K. (1980) Occurrence, development and distribution of cereal aphids on early and late cultivars of wheat, barley, and oats in South-western Quebec. *The Canadian Entomologist*, **112**, 615–20.

Balfour E.B. (1975) *The Living Soil and the Haughley Experiment*. Faber & Faber, London.

Baranyovits F. (1973) The increasing problem of aphids in agriculture and horticulture. *Outlook on Agriculture*, **7**, 102–8.

Boila R.J. (1977) *Amino acid metabolism by rumen papillae*. Unpublished Ph.D. thesis, University of Alberta, Canada.

Bond D.A. & Lowe H.J.B. (1975) Tests for resistance to *Aphis fabae* in field beans (*Vicia faba*). *Annals of Applied Biology*, **81**, 21–32.

Boness M. & El-Dessouki S.A. (1972) The influence of host plant species upon the effectiveness of insecticides especially in aphids. *Zeitschrift für Angewandte Entomologie*, **72**, 6–13.

Bremner P.M. & Rawson H.M. (1978) The weights of individual grains of the wheat ear in relation to their growth potential, the supply of assimilate and interaction between grains. *Australian Journal of Plant Physiology*, **5**, 61–72.

Byerly K.F., Gutierrez A.P., Jones R.E. & Luck R.F. (1978) A comparison of sampling methods for some arthropod populations in cotton. *Hilgardia*, **46**, 257–82.

Carter N., McLean I.F.G., Watt A.D. & Dixon A.F.G. (1980) Cereal aphids—a case study and review. In *Applied Biology* (ed. T.H. Coaker), Vol. 5, pp. 271–348. Academic Press, London & New York.

Daniels N.E. (1957) Greenbug populations and their damage to winter wheat as affected by fertilizer applications. *Journal of Economic Entomology*, **50**, 793–4.

Dempster J.P. & Pollard E. (1981) Fluctuations in resource availability and insect populations. *Oecologia* (Berlin) **50**, 412–16.

Dewar A.M. (1977) Assessment of methods for testing varietal resistance to aphids in cereals. *Annals of Applied Biology*, **87**, 183–90.

Dewar A.M. (1980) *Report of the Rothamsted Experimental Station 1979. Part 1*, pp. 17–21.

Dodd G.D. & van Emden H.F. (1979) Shifts in host plant resistance to the cabbage aphid (*Brevicoryne brassicae*) exhibited by Brussels sprout plants. *Annals of Applied Biology*, **91**, 251–62.

George K.S. & Gair R. (1979) Crop loss assessment on winter wheat attacked by the grain aphid *Sitobion avenae* (F.) 1974–77. *Plant Pathology*, **28**, 143–9.

Gill C.C. & Metcalfe D.R. (1977) Resistance in barley to the corn leaf aphid *Rhopalosiphum maidis*. *Canadian Journal of Plant Science*, **57**, 1063–70.

Hanisch H.C. (1980) The influence of increasing amounts of nitrogen applied to wheat on the reproduction of cereal aphids. *Zeitschrift für Pflanzenkrankheiten und Pflanzenschutz*, **87**, 546–56.

Henderson I.F. & Perry J.N. (1978) Some factors affecting the build-up of cereal aphid infestations in winter wheat. *Annals of Applied Biology*, **89**, 177–83.

Henderson I.F. & Whittaker T.M. (1977) The efficiency of an insect suction sampler in grassland. *Ecological Entomology*, **2**, 57–60.

Jördens-Röttger D. (1979) Behaviour of the black bean aphid *Aphis fabae* Scop. to chemical stimuli caused by leaf surfaces. *Zeitschrift für Angewandte Entomologie*, **88**, 158–66.

Kishino K. & Ando Y. (1979) Resistance of rice plant to the green rice leafhopper, *Nephotettix cincticeps* Uhler. II. Fluctuation of antibiosis with the growing stages of the resistant rice varieties. *Japanese Journal of Applied Entomology and Zoology*, **23**, 129–33.

Kolbe W. (1973) Studies on the occurrence of cereal aphids and the effect of feeding damage on yields in relation to infestation density levels and control. *Pflanschutz-Nachrichten Bayer*, **26**, 396–408.

Laresen I. (1977) *The influence of light and nitrogen supply on the amino acid metabolism and dry-matter production of Italian rye-grass* (Lolium multiflorum), *studied by means of pot experiments*. Stougaard Jensen, Copenhagen, Denmark.

Lawton J.H. & McNeill S. (1979) Between the devil and the deep blue sea: on the problem of being a herbivore. In *Population Dynamics* (eds R.M. Anderson, B.D. Turner & L.R. Taylor), Blackwell Scientific Publications, Oxford. pp. 223–44.

Lowe H.J.B. (1978) Detection of resistance to aphids in cereals. *Annals of Applied Biology*, **88**, 401–6.

Lowe H.J.B. (1980) Resistance to aphids in immature wheat and barley. *Annals of Applied Biology*, **95**, 129–35.

McMurtrey J.A. (1962) Resistance to alfalfa to spotted alfalfa aphid in relation to environmental factors. *Hilgardia*, **32**, 501–39.

McNeill S. & Southwood T.R.E. (1978) The role of nitrogen in the development of insect plant relationships. In *Biochemical Aspects of Plant and Animal Coevolution* (ed. J.B. Harborne), pp. 77–98. Academic Press, London & New York.

March J.F. (1975) A modified technique for the quantitative analysis of amino acids by gas chromatography using heptafluorobutyric n-propyl derivatives. *Analytical Biochemistry*, **69**, 420–42.

Markkula M. & Roukka K. (1972) Resistance of cereals to the aphids *Rhopalosiphum padi* (L.) and *Macrosiphum avenae* (F.) and fecundity of these aphids on Gramineae, Cyperaceae and Juncaceae. *Annales Agriculturae Fenniae*, **11**, 417–23.

Mayse M.A., Kogan M. & Price P.W. (1978) Sampling abundances of soybean arthropods; comparison of methods. *Journal of Economic Entomology*, **71**, 135–41.

Oelhaf R.C. (1978) *Organic Agriculture; Economic and Ecological Comparisons with Conventional Methods*. Allanheld, Osmun & Co., Montclair, New Jersey.

Perkins H.J. (1961) Note on chemical changes occurring in freeze-dried and fresh-frozen wheat leaves during storage. *Canadian Journal of Plant Science*, **41**, 689–91.

Pinthus M.J. & Millet E. (1978) Interactions among number of spikelets, number of grains and grain weight in the spikes of wheat (*Triticum aestivum* L.). *Annals of Botany*, **42**, 839–48.

Potts G.R. (1970) The effects of the use of herbicides in cereals on aphids. *Proceedings of the 10th British Weed Control Conference*, pp. 299–302.

Potts G.R. & Vickerman G.P. (1975) Arable ecosystems and the use of agrochemicals. In *Ecology of Resource Degradation and Renewal* (eds M.J. Chadwick & G.T. Goodman), pp. 17–29. Blackwell Scientific Publications, Oxford.

Prestidge R.A. (1980) *The influence of mineral fertilization on grassland leafhopper associations*. Unpublished Ph.D. thesis, University of London.

Rabbinge R. & Vereyken P.H. (1980) The effect of diseases or pests upon the host. *Zeitschrift für Pflanzenkrankheiten und Pflanzenschutz*, **87**, 409–22.

Rautapää J. (1972) Effect of herbicides and chlormequat chloride on host plant selection and population growth of *Macrosiphum avenae* (F.) (Hom., Aphididae). *Annales Agriculturae Fenniae*, **11**, 135–40.

Rautapää J. (1977) *Role of aphids in cereal production.* Academic dissertation of the Section of Mathematics and Natural Sciences, Faculty of Philosophy, University of Helsinki.

Rodriguez J.G. (1960) Nutrition of host and reaction to pests. In *Biological and Chemical Control of Plant and Pests* (ed. L.P. Reitz), pp. 149–67. American Association for the Advancement of Science, Washington, D.C.

Suter H. & Keller S. (1977) Okologische Untersuchungen an feldbaulich wichtigen Blattlausarten als Grundlage für eine Befallsprognose. *Zeitschrift für Angewandte Entomologie*, **83**, 371–93.

Taylor L.R., French R.A., Woiwod I.P., Tatchell G.M., Harrington R., Dupuch M.J. & Taylor M.S. (1980) *Report of the Rothamsted Experimental Station 1979, Part I*, pp. 81–2.

Tottman D.R. (1978) The effects of a dicamba herbicide mixture on the grain yield components of winter wheat. *Weed Research*, **18**, 335–9.

Tottman D.R. & Makepeace R.J. (1979) An explanation of the decimal code for the growth stages of cereals, with illustrations. *Annals of Applied Biology*, **93,** 221–34.

van Emden H.F. (1966) Studies on the relations of insects and host plant. III. A comparison of the reproduction of *Brevicoryne brassicae* and *Myzus persicae* (Hemiptera: Aphididae) on Brussel sprout plants supplied with different rates of nitrogen and potassium. *Entomologia Experimentalis et Applicata*, **9**, 444–60.

van Emden H.F. (1972) Aphids as phytochemists. In *Phytochemical Ecology* (ed. J.B. Harborne), pp. 25–43.

van Emden H.F. & Bashford M.A. (1971) The performance of *Brevicoryne brassicae* and *Myzus persicae* in relation to plant age and leaf amino acids. *Entomologia Experimentalis et Applicata*, **14,** 349–60.

Vereijken P.H. (1979) Feeding and multiplication of three cereal aphid species and their effect on yield of winter wheat. *Agricultural Research Report, Netherlands*, **888**, 1–58.

Vickerman G.P. (1974) Some effects of grass weed control on the arthropod fauna of cereals. *Proceedings of the 12th British Weed Control Conference*, pp. 929–39.

Vickerman G.P. (1977) Monitoring and forecasting insect pests of cereals. *Proceedings of 1977 British Crop Protection Conference—Pests and Diseases*, pp. 227–34.

Vickerman G.P. & Wratten S.D. (1979) The biology and pest status of cereal aphids (Hemiptera: Aphididae) in Europe: a review. *Bulletin of Entomological Research*, **69**, 1–32.

Watt A.D. (1979) The effect of cereal growth stages on the reproductive activity of *Sitobion avenae* and *Metopolophium dirhodum. Annals of Applied Biology*, **91**, 147–57.

Wetzel T., Freier B. & Heyer W. (1980) Model experiments on infestation-damage relations using insect pests of winter wheat. *Zeitschrift für Angewandte Entomologie*, **89**, 330–44.

Williams C.B. (1947) The logarithmic series and the comparison of island floras. *Proceedings of the Linnaean Society, London*, **158**, 104–8.

14. THE EFFECT OF FOLIAR NUTRIENTS UPON THE GROWTH AND FEEDING OF A LEPIDOPTERAN LARVA

G. R. W. WINT
Department of Zoology, South Parks Road, Oxford OX1 3PS

SUMMARY

It is suggested that levels of available leaf nitrogen, rather than total leaf nitrogen, are relevant to the growth and development of insect herbivores. A simple assay based on enzyme inhibition is described which estimates the levels of leaf protein complexing agents with host plant foliage. These estimates are then combined with total leaf protein values to give the levels of protein actually available to a feeding insect.

It is shown that available, rather than total, leaf protein levels are related to the performance of larval winter moths (*Operophtera brumata* L.), when they are fed on leaves from several host plant species.

Nitrogen availability is discussed with reference to the feeding biology of both sucking and chewing insect herbivores, and briefly with respect to the population levels of the winter moth.

INTRODUCTION

The relevance of the nutritional status of a food plant to the feeding and growth of phytophagous insects has been the subject of many studies during the past 30 years (Fraenkel 1953; House 1961; Feeny 1970; Southwood & McNeill 1978). Research has been concentrated largely on two categories of chemicals: nitrogen and its component amino acids (Slansky & Feeny 1977; Southwood & McNeill 1978), and those compounds, notably the tannins, which combine with plant proteins to reduce the availability of the latter to feeding herbivores (Feeny 1970; Swain 1976a; Bernays 1978).

The defensive function of plant phenolics, particularly against chewing insects, is well documented (Swain 1976a; Ibrahim, Axtell & Oswalt 1973; Reese & Beck 1976; Feeny 1970; Levin 1971; Oates, Swain & Zantovŝka 1977). However, there is evidence that the effects of phenolics upon herbivores in general, and insects in particular, are by no means simple. For example, tannins seem to have little effect upon the insects of *Eucalyptus* spp. (Fox & Macauley 1977), or upon the scale insect *Mycetapsis personatus* when feeding on varieties of *Mangifera indica* (Salama & Saleh 1972). Furthermore, the relative ineffectiveness of the hydrolysable tannins as a defence against the winter moth, reported by Feeny (1966), has been brought into question by evidence suggesting that they may be more effective than are the condensed tannins (Swain 1976a; Van Summere *et al.* 1975; Bernays 1978).

Nitrogen is well established in ecological lore as an important determinant of phytophagous insects' success (Southwood & McNeill 1978). Most of the clearest evidence comes from work on leaf-sucking insects. Thus high dietary nitrogen levels promote rapid development and increased growth in aphids (Dixon 1969) and can affect the distribution of several hemipteran species in the field (McNeill 1973; Gibson 1980). With chewing insects the evidence is less clear. Artificial diet experiments have shown that high nitrogen levels lead to larger lepidopteran larvae (House 1965), and some studies show that total plant nitrogen is positively correlated to an insect's feeding and growth (Fox & Macauley 1977). Much of the literature shows, however, that total nitrogen levels are unrelated to a chewing insect's growth (Slansky & Feeny 1977; Baker 1975), though they have distinct effects upon its digestion and feeding processes. Theoretical objections have also been raised to question the assumption that low nitrogen levels act as a defence against herbivores. Hamilton & Moran (1980) point out that low nutrient levels may well enhance insect consumption, so that, 'genes conferring lowered nutritive quality (in host plants) could even increase herbivore damage in some circumstances'.

Thus, total nitrogen as an indicator of diet quality or host plant defence strategy appears to be ambiguous: in some cases it is predictive while in others it is not. As a result, many studies have focussed on the components of dietary nitrogen—the amino acids (e.g. Prestidge & McNeill, p. 257 this volume). An alternative is to consider nitrogen availability. Many 'quantitative' plant defences (Feeny 1976) are assumed to act by complexing with leaf proteins to reduce their availability and yet most studies in the literature have looked at total nitrogen and the defence compounds as separate entities. As far as the author is aware, only Feeny (1970) makes any allowance for this interaction to provide estimates of the leaf proteins that are actually available to a feeding herbivore.

This paper reports on investigations of the associations between host plant quality and the performance of winter moth (*Operophtera brumata* L.) larvae when fed on the leaves of various ages from six common food plant species. Host plant quality was measured throughout the growing season by qualitative screening of foliar phenols, and by quantitative estimates of leaf toughness, leaf water content, total leaf nitrogen content and levels of leaf 'nutrient-reducing agents'. This last estimate was obtained from an enzyme assay and then combined with the foliar protein values to provide a measure of the amounts of protein available to the feeding larvae. Larval performance was assessed by monitoring 'overall' performance (pupal fresh weight, larval growth rate and weight increase) and mortality in 1976 and 1977, and feeding processes (consumption levels and feeding efficiencies) in 1977. Associations between the plant and animal parameters were then examined using standard correlation techniques.

METHODS

Analyses of leaf material

Six host plant species were selected: *Crataegus monogyna, Malus sylvestris* (cultivated apple), *Prunus spinosa, Corylus avellana, Fagus sylvatica* and *Quercus robur* (as a control). These species are all common winter moth host plants in the field which reduced the possibility of larval performance being affected by toxins and deterrents.

Leaves for chemical analyses and feeding trials were collected from the field into polythene bags which were pre-filled with nitrogen and generous quantities of ice to minimize chemical decay before analysis. All the material consisted of lower sun leaves and was collected in mid-morning.

Percentage leaf water was assessed gravimetrically using preweighed freeze-dried materal; leaf toughness was estimated using a modified penetrometer (Moreau 1965); percentage total nitrogen was measured from freeze-dried material using the micro-Kjeldahl method (Allen *et al.* 1974). Preliminary investigations of the leaf phenolics were performed by two-dimensional chromatography (Haslam 1966; Feeny 1966; Harborne 1973), using 2% acetic acid and butanol/acetic acid/water (12:3:5) as the first and second solvents respectively. These were developed with vanillin and $FeCl_3/K_3Fe(CN)_6$ sprays, and were examined under u.v. light with and without ammonia vapour. The analyses were carried out on leaves extracted in 70% aqueous acetone, stabilized by the addition of a little KCN, and stored in the dark at 4°C.

The available methods for quantitative estimates of nutrient-reducing agents like the phenolics (Brown, Love & Handley 1962; Feeny 1966; Haslam, pers. comm.) are all subject to several drawbacks. They often require extraction of the relevant compounds from the leaf tissues and there is no guarantee that these compounds remain unaltered by the extraction process (Brown, pers. comm.). Also, the identity of the substances extracted is frequently difficult to establish due, at least partly, to their heterogeneity.

Furthermore, the value of information concerning a single class of phenols such as the tannins, or even the phenols as a whole, is not only difficult to assess in terms of their effect upon the potential food plant quality, but may also represent one of a number of types of compound which can act as nutrient-reducing agents. Finally, the time needed to perform many of the standard methods prevents frequent screening of several plant species.

Professor G. C. Varley suggested that it might be possible to use the inhibition of bacterial amylase by tannins, observed by Feeny (1966), as the basis of a quantitative assay which avoids these problems. Such inhibition is well known and has been reported for a wide variety of enzymes including proteinases, invertases, lipases, phosphatases, decarboxylases, ureases and β-glucosidases, as well as amylases (Benoit & Starkey 1968).

There is also evidence that the level of inhibition of β-glucosidases and amylase is related to the amount of tannin present (Goldstein & Swain 1965; Boudet & Gadal 1965b). This suggests that any inhibition of amylase caused by leaf extract is paralleled by the effect of the leaf tissues upon the digestive enzymes of a feeding herbivore. Furthermore, the chemical similarity of nutrient proteins and enzymes suggests that an assay using enzyme inhibition mimics another major effect of nutrient-reducing agents—that is, the complexing between the nutrient proteins and compounds like tannins. Therefore, any inhibition detected by an assay of this type should reflect similar processes occurring in the gut of an herbivorous insect.

Examinations of the basic starch–amylase reaction revealed three relationships. First, when varying amounts of enzyme were added to constant amounts of starch, the end-point time (i.e. the isocolorimetric point as detected by an iodine indicator) was proportional to the inverse of the enzyme concentration. Second, aqueous extracts of the leaves of several host plant species (*Corylus, Crataegus* and a *Viburnum* sp.) were able to inhibit the starch digestion such that adding a fixed quantity of leaf extract to the enzyme solutions, prior to their admixture with starch, caused a delay in the end-point time which was linearly related to the inverse of the enzyme concentration. An example of this relationship using *Corylus* leaves is presented in Fig. 14.1a. The gradient of this line could provide a measure of the degree of inhibition of the starch digestion that was caused by the contents of the leaf extracts.

A third series of experiments showed that aqueous solutions of oak tannin extract (provided by B.R. Brown) inhibited the amylase in a manner identical to that shown by the leaf extracts. Also, the 'inhibition gradient' was linearly related to tannin concentration (Fig. 14.1b). Using this relationship, it was possible to describe the degree of enzyme inhibition caused by the leaf extracts in terms of the amount of oak tannin that produced an equivalent effect.

In addition, gallic acid, quercetin and (+)-catechin inhibited the starch–

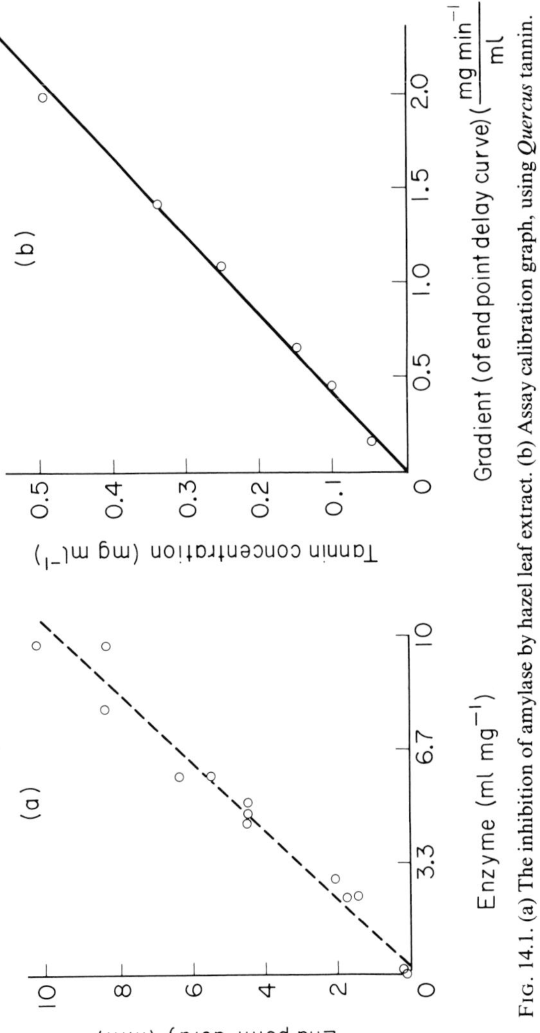

FIG. 14.1. (a) The inhibition of amylase by hazel leaf extract. (b) Assay calibration graph, using *Quercus* tannin.

amylase reaction, though the degree of inhibition was less than that produced by tannins. Thus gallic acid is about half (58%) as effective an inhibitor as the oak tannin extract, and quercetin and (+)-catechin are, respectively, 35% and 7% as effective. These inhibitory capacities parallel the reduction in pupal weight that is caused by the addition of these compounds to artificial diets of larval winter moth (Feeny 1966). This indicates that the effects of these compounds upon amylase activity closely mimics their effect upon larval growth.

As a result of this information, and after taking into consideration several additional factors—notably the complexing time between enzyme and extract, extract concentrations, the effect of leaf extracts on the starch iodine reaction and solution stabilities—a standard assay procedure was developed (see Appendix 1). This produced estimates of the levels of 'Inhibitors' within the host plant foliage and was calibrated in terms of 'Oak tannin equivalents'.

It has been shown that the degree of complexing between oak tannins and casein is related to the tannin/protein ratio (Feeny 1969). By calculating the ratio between the levels of leaf protein and the tannin equivalent (i.e. inhibitor) estimated by the enzyme assay, it is possible to obtain a measure which should reflect the amount leaf protein rendered unavailable by the inhibiting compounds. This allows calculation of the available protein levels.

The degree of tannin protein complexing also varies according to the type of tannin involved (Feeny 1969), and the calibration tannin extract contained both hydrolysable and condensed tannins. As the latter are much more effective at complexing with protein, it is likely that they produced most of enzyme inhibition caused by the calibration extract. On this assumption, the graph shown in Fig. 14.2, which represents the complexing characteristics of condensed tannins with casein, was used to calculate estimates of available protein. Only approximate levels of complexing were assessed due to the many uncertainties involved, so the reduction of protein availability was estimated to the nearest 5%. Total protein concentrations were obtained by multiplying the total nitrogen levels by 6.25 according to Allen *et al.* (1974).

Feeding trials

On eclosion, experimental animals were taken from laboratory stock in batches of at least 20, and fed on the experimental food plants. Feeding trials were conducted at 20°C, in 75% r.h. and 16:8 L/D, using leaves fresh from the field which were renewed every two days. Each larva was reared in numbered air-tight pots to reduce leaf water loss. Larvae were weighed daily from the fourth (penultimate) instar until they stopped feeding. Consumption and defaecation levels were estimated gravimetrically, and assimilation levels then

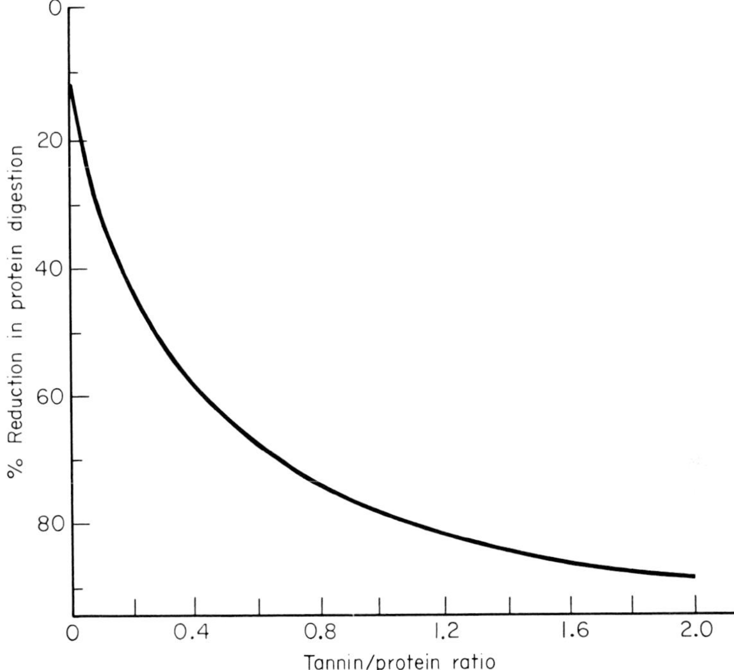

FIG. 14.2. Available protein estimation: the effect of the tannin/protein ratio upon the digestion of casein by trypsin (after Feeny 1966).

determined by difference (assimilation = consumption − faeces). Relative growth rates, consumption indices and feeding efficiencies (Assimilation Efficiency, AD; Growth Efficiency, ECD; Overall Conversion Efficiency, ECI) of the fifth instar larvae were defined and calculated according to Waldbauer (1964).

Data preparation and correlation

The mean values of each measure of larval performance were calculated for each feeding trial, using only those larvae which survived to pupation. The estimates of food plant quality relevant to each feeding trial were calculated from graphs of the type shown in Fig. 14.3, employing standard geometric methods. The values obtained were then corrected to allow for changes that were found to occur in the experimental food while it was being eaten. As no relationship was detected between leaf water content and any of the animal variables measured, both the larval performance data and the plant quality

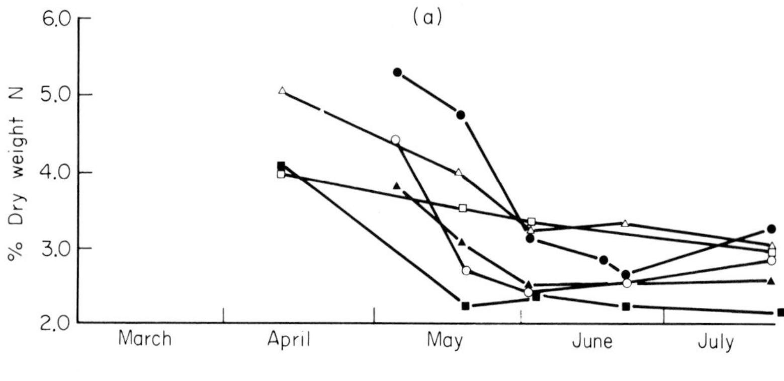

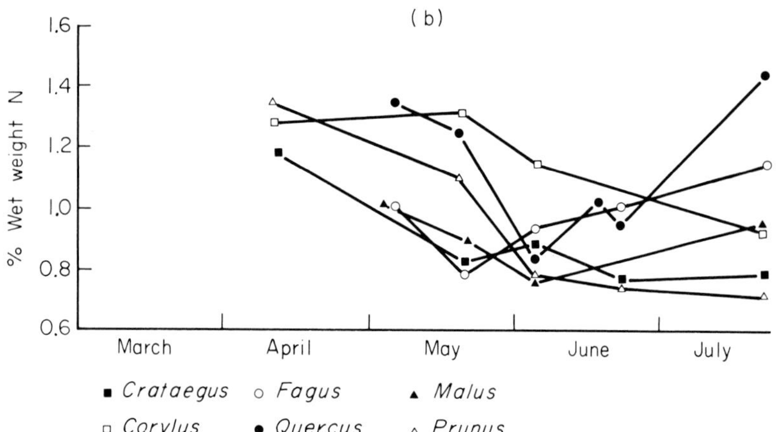

FIG. 14.3. Seasonal variations in % total leaf nitrogen for 1977. (a) % Dry weight. (b) % Wet weight.

values were calculated in terms of fresh weight throughout. Though less convenient, such values are likely to be of greater biological relevance than are the dry weight values. This is emphasized by a comparison of the dry and wet weight total nitrogen figures shown in Figs 14a & b. Though the range of values obtained for both variables shows a similar proportional variation, the pattern of general decline with time shown by the dry weight levels is not evident in the wet weight values. Indeed, in *Q. robur* leaves the July levels are higher than the April ones. Furthermore, the relative levels of wet and dry weight nitrogen levels in the different host plant species differ considerably.

Associations between the animal and plant variables were examined using both simple and partial correlation techniques (Snedecor & Cochran 1962).

The correlations were performed on the data from all the feeding trials together (transformed to give the best fit), on the assumption that a deductive approach is of more value than an inductive one, as well as to take advantage of the wider range of plant quality presented by the use of several plant species.

RESULTS

The enzyme bioassay

An example of the leaf inhibitor levels obtained from the enzyme assay is presented in Fig. 14.4a. It is conspicuous that the highest inhibitor levels were found in *Crataegus*, and the lowest in *Malus*. This agrees well with the number and intensity of the phenolic spots detected by the chromatographic analyses. In addition, there is good agreement between the maximum inhibitor level shown by the assay and the levels of tannins against which they were calibrated, as reported by Feeny (1970) using a specific chemical method. These points strongly suggest that the assay reacts quantitatively to the presence of phenols within leaf tissues. However, the data from the assays of *Corylus* foliage indicate that the assay detects compounds other than phenols. Thus the phenol content of 1977 *Corylus* leaves was considerably less than that of 1976, while the inhibitor content was higher. This implies that the assay

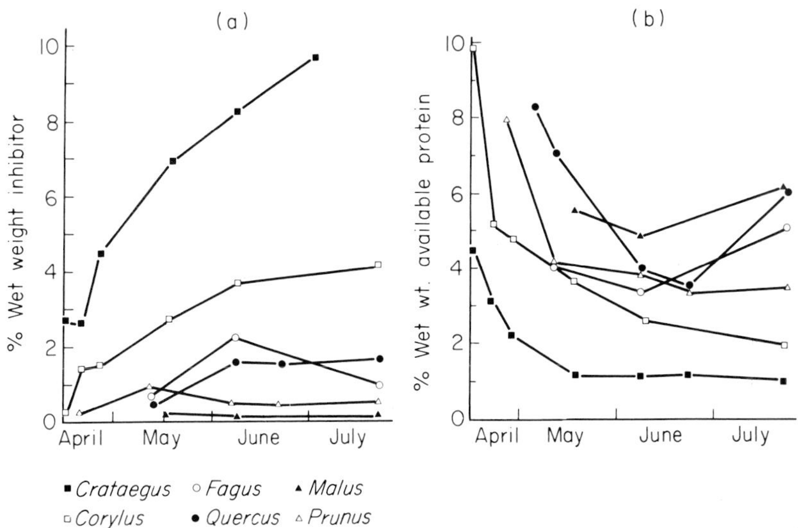

FIG. 14.4. Seasonal variations in leaf contents for 1977. (a) % Wet weight inhibitor. (b) % Wet weight available protein.

fulfills its desired role as an indicator of compounds, in addition to phenols, which may act as nutrient-reducing agents either by directly inhibiting herbivore enzymes or by complexing with foliar proteins.

This method has several advantages over the more chemical techniques. Firstly, as it uses similar principles to those involved in insect feeding (i.e. maceration, mixing then digestion), any inhibition detected is likely to be due to substances that also reduce the potential food quality of the leaves. Secondly, the assay can be performed on aqueous leaf extracts, so that the compounds detected are more likely to be identical to those present in the living leaf tissues than are those found in the acetone or methanol extracts which are necessary for the more chemical methods. Thirdly, this assay is sensitive not only to the presence of phenols, but is probably affected by other compounds with similar effects such as protein inhibitors (Buonocore, Petrucci & Silano 1977), and so it provides an estimate of all the nutrient-reducing agents within leaf tissues, regardless of chemical identity. Such a method is likely to have wider biological relevance than those which encompass a specific type of chemical. Lastly, this technique is easily and rapidly performed, which permits frequent analyses of several host plant species.

Available protein estimations

The value of these estimations can be seen by comparing the total leaf nitrogen and available protein levels for 1977 (Figs 14.3b & 4b). The available protein levels, in contrast to the total nitrogen values, fall below the 6.25% (equivalent to 1% nitrogen) levels suggested by some authors to be limiting to lepidopteran growth (Fox & Macauley 1977). This implies that larval performance may well be limited by nitrogen availability. In addition, the relative nutrient status of the six host plant species differs markedly according to the measure examined, particularly in the cases of *Crataegus* and *Corylus*. There is also a contrast in the phenologies of the two variables. The available protein levels tend to fall more rapidly in April and May, as well as rise less markedly in July, than do the corresponding total nitrogen contents.

The arguments used to develop these estimates are open to a number of criticisms. Firstly, there is no guarantee that the inhibitors detected act upon the feeding insect by reducing the available protein rather than directly inhibiting its digestive enzymes, and hence the assumption that all inhibitors detected act to reduce the available protein is speculative. Secondly, the use of a single class of phenolics and a simple protein to provide the conversion graph is open to censure—not only are many types of compound likely to be detected by the assay, but also there are many types of protein within the leaf tissues, most of which are more complex than casein. Thirdly, the conditions used by Feeny to produce the complexing curve bear little resemblance to conditions found either within a leaf, within an insect, or within the enzyme assay employed.

Associations between host plant quality and measure of larval performance

Overall larval performance

Of the independent plant variables, only the leaf inhibitor content is significantly correlated with pupal fresh weight, fifth instar weight increase and relative growth rate (Table 14.1a). The correlations are significant at the 1% level for the pupal fresh weights and weight increases, and at 2% for the relative growth rates: all are negative. These results suggest that while a rise in foliar inhibitor levels is related to a decrease in larval performance, changes in either leaf toughness or total leaf nitrogen contents are not reflected by changes in larval growth.

If, however, the interaction between leaf inhibitors and foliar nitrogen is considered, it becomes evident that nitrogen levels within the host plant leaves are of importance to the larval performance. The correlations between the animal variables and the available leaf protein levels are all significant at the 5% level or less, and are all positive, implying that an increase in larval growth accompanies a rise in available protein levels.

Using standard multiple regression (Steele & Torrie 1960), the proportion of the variation in pupal fresh weight explained by leaf toughness and available protein together (R^2) is 0.46 ($P<0.01$, $n=23$). Though significant, this suggests that many other factors are important in determining larval growth. While fifth instar weight increase and relative growth rate are measurements relevant to the fifth instar alone, the pupal fresh weight includes the effects of all five instars. 77% ($P<0.01$, $n=23$) of the variation in the last parameter is explained by the variation in the larval weight at the beginning of the fifth instar. Events leading to pupation may be triggered well in advance of the cessation of feeding (Varley, pers. comm.) and so may only be partially dependent upon variation in food quality after the 'trigger' has operated.

Larval consumption

As larval consumption rate (mg day^{-1}) is significantly related to the Averaged Mean Weight (Waldbauer 1964) of the larvae ($r=0.80$, $P<0.01$, $n=13$), correlations with the larval consumption index (that is the consumption rate

TABLE 14.1. Partial correlation coefficients between larval performance and host plant quality. (Significance levels in brackets. *⇒log transformation.)

Animal variables	Leaf inhibitor content	Total leaf nitrogen	Leaf toughness	Available leaf protein
	Plant variables			
(a) Overall measures of larval performance, 1976 & 1977 data, $n=23$.				
Pupal fresh weight	−0.6217*(0.01)	0.2744 (NS)	−0.3054 (NS)	0.5851*(0.01)
Fifth instar weight increase	−0.5587*(0.01)	0.1379 (NS)	−0.3775 (NS)	0.4905 (0.05)
Fifth instar growth rate	−0.5018 (0.02)	0.0423 (NS)	−0.2193*(NS)	0.4727*(0.05)
(b) Feeding processes, 1977 data, $n=13$.				
$\frac{C}{T.W.}$ (CI)	0.3867*(NS)	−0.2450*(NS)	−0.1766*(NS)	−0.3777*(NS)
G/A (ECD)	0.3167 (NS)	−0.5030 (NS)	−0.4349 (NS)	−0.2968*(NS)
A/C (AD)	−0.7397 (0.01)	0.7665 (0.01)	0.7087 (0.02)	0.6724*(0.05)
G/C (ECI)	−0.3966 (NS)	−0.0611 (NS)	−0.1273*(NS)	0.4563*(NS)

C = consumption (mg wet wt.) T = feeding times (days) W = averaged mean weight (mg wet wt.) G = weight increase (mg wet wt.) A = assimilation (mg wet wt.)

TABLE 14.2. Correlation coefficients between larval nutrient intake levels and measures of larval performance. (Significance levels in brackets; 1977 data, $n=13$, *⇒log transformation used.)

Animal variables	Total nitrogen (Partial correlation technique)	Inhibitor (Partial correlation technique)	Available protein (Simple correlation technique)
	Plant variables		
Fifth instar weight increase	0.4975* (NS)	−0.6261* (0.05)	0.6465* (0.05)
Pupal fresh weight	0.4805 (NS)	−0.6804* (0.02)	0.6933* (0.01)

corrected for larval size) were examined: none were significant (Table 14.1b). If the larval intake levels of nitrogen, inhibitor and available protein are calculated using the mean consumption levels for each feeding trial and the corresponding values of each plant parameter (Table 14.2), a different answer emerges. While the total nitrogen intake is not correlated to either fifth instar weight increase or to pupal fresh weight, the inhibitor and available protein intake levels are. This suggests that while diet quality does not influence larval consumption levels, consumption levels influence larval performance by virtue of their effect upon nutrient intake levels. Again the importance of leaf nitrogen only becomes apparent when its availability is considered.

This evidence is at variance with the published data (Tanton 1962; Iversen 1974; Reese & Beck 1976) which indicates that insect consumption levels respond to many plant variables. In the case of the winter moth, larval consumption appears to be primarily related to size which may mask any effects of plant quality.

Larval feeding efficiencies

The variation in leaf contents and in leaf toughness are not closely related to either larval ECI or ECD (see Table 14.1b). In contrast, the larval assimilation efficiency (AD) is significantly related to all four measures of host plant quality. This suggests that the major effect of host plant nutrient status upon larval digestion is the determination of the proportion of leaf tissues that can be digested. The highest correlation is with total leaf nitrogen level while the lowest is with available leaf protein content, though this may be due to the differences in correlation technique employed. The signs of the correlation coefficients show that larval AD increases in association with a rise in leaf nutrient availability, as represented by high nitrogen and available protein levels and a low inhibitor content. However, the larval assimilation efficiency also increases with rising leaf toughness. Tougher leaves are likely to contain more cellulose and lignin and therefore less nutrients. The non-significant relationship between larval consumption index and leaf toughness (see Table 14.1b) invalidates the possibility that rising toughness leads to reduced feeding rates and hence high assimilation efficiencies, as was reported by Soo Hoo & Fraenkel (1966) and Mehta & Saxena (1973). In the absence of further evidence, it seems possible that leaf toughness is related to a quantitative change within the leaves that itself affects the larval assimilation.

As the associations between the plant variables and the larval AD are not reflected by the Overall Conversion Efficiencies (ECI), it is probable that the larval ECD responds to entirely different parameters of food quality (e.g. the proportion of different amino acids), and that this contrast in response is evident as a lack of correlation between any of the measured plant variables and the larval ECI.

Larval mortality

The total larval mortality is the cumulative mortality of all five instars, and so examining it in relation to the diet quality during any one instar is meaningless. Such cumulative errors are reduced if the early instar mortalities are considered. First instar mortality may well be influenced by the egg quality and so the second instar cumulative mortality is the first which it is profitable to investigate. No chemical measure of food plant quality was correlated to this mortality (arcsine transformation), but leaf toughness was highly so ($r=0.74$, $P<0.001$, $n=23$). Hence, while leaf toughness exerts a direct effect upon early instar survival, the influence of the chemical aspects of diet quality is manifest by its effects on pupal fresh weight rather than larval mortality. Pupal fresh weight is closely correlated to subsequent adult survival and fecundity (Andrzejewska, in Gradwell 1973; Wint 1979), which suggests that any influence of nutrient availability upon the survival of the winter moth acts via reproductive success.

DISCUSSION

The evidence I have presented suggests that in terms of overall larval performance, it is the availability of nutrients in the host plant leaves, rather than other aspects of plant quality, that is most closely related to the growth of winter moth larvae. An increase in inhibitor levels, or a fall in available nitrogen content is associated with a decline in larval performance. Total nitrogen is, however, independent of larval growth. This last piece of evidence supports many of the reports in the literature concerning chewing insects, but it contrasts with the majority of evidence derived from leaf-sucking herbivores (Slansky & Feeny 1977; Southwood & McNeill 1978) as well as Fox & Macauley's (1977) assertion that the growth of *Eucalyptus* phytophages is closely allied to the foliar total nitrogen levels.

Such contradictions may be reconciled by considering the effects of nitrogen availability. In the case of Fox & Macauley's work, there is no correlation between the leaf tannin levels and the insects' growth. This suggests little or no interaction between the secondary compounds and either the insect digestion or the leaf proteins, and so nitrogen availability is unaffected by the defences. Insect growth is therefore closely related to total nitrogen concentrations. Many hemipterans extract their nitrogen directly from the host plant's vascular system and so avoid the majority of the quantitative defences which are located largely in the cells of the non-vascular tissues, and are present in the sap as inactive soluble glucosides (Siegler 1977). The plant proteins in the sap do not interact with the protein-complexing agents and so the insect's growth is closely related to the total (and available) nitrogen levels. The difference between sucking and chewing insects is further underlined if their digestion is considered. Proteolytic enzymes are used by chewing insects to hydrolyse the leaf-bound proteins to their constituent amino acids. These enzymes are likely to be affected by protein-complexing agents and are essential for the chewing insects' successful extraction of much of their dietary nitrogen. In the case of leaf-sucking insects, much of their nitrogen intake is already in small units and so their digestion is less vulnerable to inhibition.

If this is the case, both sucking and chewing insects are liable to be affected by the quantity (and by inference, the quality) of available nutrients, but only the latter are subject to many quantitative defences. The distribution of chemical defences is also likely to be an important factor, be they clumped, as in resin canals or essential oil glands, or evenly distributed throughout the non-vascular tissues. Such a dichotomy in the vulnerability of sucking and chewing insects to chemical plant defences may well be a contributing factor to the composition of phytophage communities on any specific host plant species.

The present evidence also suggests that plant nutrient levels affect the populations of chewing insects at the level of their reproductive capacity. In univoltine species such as the winter moth, this may mean that a host plant's nutrient status in any particular year may influence the insect population densities in the next. Conversely, the population levels in any particular year may be unrelated to the nutrient quality of the host plant tissues in that year. This concept is supported by preliminary evidence from a study now in progress (Southwood, Kennedy & Wint, unpublished data) in which the insect fauna of four oak species is being investigated alongside, among other factors, their nutrient status. Population levels of the major chewing herbivores (*Operophtera brumata* and *Tortrix viridana*) were estimated but were not related to the available protein content of the leaves on which they fed. Furthermore, no convincing evidence was found to relate the insect population densities to the initial nutrient content of the flushing leaves. Considerably more information is needed before any firm conclusions can be drawn.

ACKNOWLEDGEMENTS

I would like to thank the many people who contributed to this work in a multitude of ways. In particular, I am indebted to Professor G. C. Varley who was my supervisor; the many members of the Hope Department and the Zoology Department at Oxford who gave me invaluable help and advice; and

C. W. D. Gibson, C. E. J. Kennedy, V. C. Moran, N. J. Mills and J. Phillipson who criticized various drafts of the manuscript. The research was carried out during an SRC Studentship.

APPENDIX 1

THE QUANTITATIVE ESTIMATION OF THE NUTRITIVE INHIBITOR CONTENT OF EXPERIMENTAL HOST PLANT LEAVES: THE ENZYME ASSAY—REAGENTS AND PROCEDURES

THE BASIC STARCH–AMYLASE REACTION

The methodology of the basic starch–amylase reaction was modified from that used in undergraduate biochemistry practical experiments (P.C.J. Brunet, pers. comm.).

Reagents used

Stock starch reagent mixture

50 ml soluble starch solution (1 g in 500 ml water)
20 ml salt (NaCl) solution (1% w/v)
20 ml phosphate buffer, for pH 6.6
The buffer solution was made up from two solutions, in the ratio of 3:2, i.e. 150 ml solution 1 and 100 ml solution 2:

1. 2.269 g KH_2PO_4 in 250 ml water
2. 4.75 g Na_2HPO_4 $12H_20$ in 250 ml water

Enzyme solutions

An aqueous solution of α-amylase (Type III-A, from *Bacillus subtilis*, obtainable from Sigma London Chemical Co. Ltd) was prepared at 0.4 mg ml^{-1}. This was subsequently diluted to give four enzyme solutions—0.4 mg ml^{-1}, 0.3 mg ml^{-1}, 0.2 mg ml^{-1} and 0.1 mg ml^{-1}.

Iodine indicator solution

5 ml 0.2% I_2 in 5% KI
245 ml water

Standard reaction procedure

4.5 ml of the starch reagent mixture was added to 1 ml of each of the four enzyme solutions, held in a water bath at 20°C. Every 15 seconds, 0.25 ml of each reacting mixture was withdrawn and added to 4 ml of the iodine indicator solution until the end-point was reached, as indicated when no change in the colour of the iodine solution was produced by the addition of the enzyme–starch reactant mixture (i.e. the isocolorimetric point). This was originally determined by colorimetric comparison at 550 nm of the experimental solutions with 4 ml of indicator solution to which 0.25 ml of reagent blank solution had been added. It was subsequently discovered that reproducible end-point determinations could be obtained using simple visual comparisons, against a white background, in daylight and without fluorescent lighting. As the visual comparisons were considerably less cumbersome to perform, these were used in all the experimental enzyme assays.

THE ASSAY

Fresh leaves, collected as described under 'Methods', were macerated in distilled water, to give a standard extract concentration of 60 mg ml^{-1}. 0.25 ml of this was then added to 1 ml of each of the four enzyme solutions, which were left at 20°C to 'complex' for ten minutes precisely. 4.5 ml of the starch reagent mixture was then added to each extract enzyme mixture, and the resultant end-points determined as described above. However, as well as 0.25 ml of reagent blank, 0.25 ml of leaf extract was added to the control iodine indicator sample, to allow for any direct effects of the leaf extract upon the iodine–starch colour reaction.

A control assay was performed alongside the experimental assays in which water was substituted for the leaf extract. The end-points of this test were then subtracted from those of the leaf extract assays to give the delays in the end-point time of each enzyme–starch reaction that were caused by the addition of the leaf extracts. These figures were then plotted against the inverse of the relevant enzyme concentrations, and the slope of the resulting line calculated using linear regression. This gradient was used for subsequent calculations (see below) only if the regression was significant at $P<0.001$. If this condition was not fulfilled, the assay was repeated.

Leaf extracts were diluted when necessary to ensure that the levels of enzyme inhibition were equivalent to those of the calibration graph (Fig. 14.1b), and to ensure that the leaf extracts themselves did not affect the iodine–starch reaction.

The foliar levels of inhibitor were calculated from the gradients described above as follows:

$$\% \text{ wet weight inhibitor} = \frac{GC}{E} \times 100$$

G = gradient of the plot of end-point delay/enzyme.

C = gradient of the calibration curve. This was 0.25 in the 1976 and 1977 assays, but can vary according to the batch of enzyme used.

E = weight of leaf added to each enzyme solution (mg fresh wt.).

REFERENCES

Allen S.E., Grimshaw H.M., Parkinson J.A. & Quaimby C. (eds) (1974) *Chemical Analysis of Ecological Materials*. Blackwell Scientific Publications, Oxford.

Baker J.G. (1975) Protein utilization by larvae of the black carpet beetle, *Altagenus megatoma*. *Journal of Insect Physiology*, **21**, 613–21.

Benoit R.E. & Starkey R.L. (1968) Enzyme inactivation as a factor in the inhibition of decomposition of organic matter by tannins. *Soil Science*, **105**, 203–8.

Bernays E.A. (1978) Tannins: An alternative viewpoint. *Entomologia Experimentalis et Applicata*, **24**, 44–53.

Boudet A. & Gadal P. (1965b) Sur l'inhibition des enzymes par les tannins des feuilles de *Quercus sessilis*. Ehrh. Inhibition de la β-amylase. *Comptes Rendus de l'Académie des Sciences, Paris*, **260**, 4252–5.

Brown B.R., Love C.W. & Handley W.R.C. (1962) Protein fixing constituents of plants. III. *Report on Forest Research*, **62**, 90–3.

Buonocore V., Petrucci T. & Silano V. (1977) Wheat protein inhibitors of α-amylase. *Phytochemistry*, **16**, 811–20.

Dixon A.F.G. (1969) Quality and availability of food for a Sycamore aphid population. In *Animal Populations in Relation to the Food Resources* (ed. A. Watson). Blackwell Scientific Publications, Oxford.

Feeny P.P. (1966) Some effects on oak-feeding insects of seasonal changes in the nature of their food. *D. Phil. thesis*, University of Oxford.

Feeny P.P. (1969) Inhibitory effect of oak leaf tannins on the hydrolysis of proteins by trypsin. *Phytochemistry*, **8**, 2119–26.

Feeny P.P. (1970) Seasonal changes in oak leaf tannins and nutrients as a cause of spring feeding by winter moth caterpillars. *Ecology*, **51**, 565–81.

Feeny P.P. (1976) Plant apparency and chemical defence. *Recent Advances in Phytochemistry*, **10**, 1–40.

Fox L.R. & Macauley B.J. (1977) Insect grazing on *Eucalyptus* in response to variation in leaf tannins and nitrogen. *Oecologia* (Berlin), **29**, 145–62.

Fraenkel G. (1953) The nutritional value of green plants for insects. *Transactions of the 9th International Congress of Entomology, Amsterdam (1951)*. pp. 290–300.

Gibson C.W.D. (1980) Niche use patterns among some Stenodemini (Heteroptera: Miridae) of limestone grassland, and an investigation of the possibility of interspecific competition between *Notostira elongata* Geoffroy and *Megalocera recticornis* Geoffroy. *Oecologia* (Berlin), **47**, 352–64.

Goldstein J.L. & Swain T. (1965) The inhibition of enzymes by tannins. *Phytochemistry*, **4**, 185–92.

Gradwell G.R. (1973) The effect of defoliators on oak. In *The British Oak, Its History and Natural History* (eds M.G. Morris & F.H. Perring), pp. 182–93. Classey, Faringdon.

Hamilton W.D. & Moran N. (1980) Low nutritive quality as a defence against herbivores. *Journal of Theoretical Biology*, **86**, 247–54.

Harborne J.B. (1973) *Phytochemical Methods*. Chapman & Hall, London.

Haslam E. (1966) *The Chemistry of Vegetable Tannins*. Academic Press, London & New York.

House H.L. (1961) Insect nutrition. *Annual Review of Entomology*, **6**, 13–26.

House H.L. (1965) Effects of low levels of nutrient content of a food, and of nutrient imbalance on the feeding and nutrition of a phytophagous larva *Celerio euphorbiae* L. (Lep. Sphingidae). *Canadian Entomologist*, **97**, 62–8.

Ibrahim H.A., Axtell J.D., & Oswalt D.L. (1973) Effect of tannin in grain Sorghum on rat growth. *Journal of Animal Science*, **37**, 283–93.

Iversen T.M. (1974) Ingestion and growth in *Sericostoma personatum* (Trich) in relation to the nitrogen content of ingested leaves. *Oikos*, **25**, 278–82.

Levin D.A. (1971) Plant phenolics—an evolutionary perspective. *American Naturalist*, **105**, 157–81.

McNeill S. (1973) The dynamics of a population of *Leptoterna dolabrata* (Het. Miridae) in relation to its food resources. *Journal of Animal Ecology*, **40**, 495–507.

Mehta R.C. & Saxena K.N. (1973) Growth of the cotton spotted bollworm *Earias fabia* (lep. Noct) in relation to consumption, nutritive value, and utilisation of food from various plants. *Entomologia Experimentalis et Applicata*, **16**, 20–30.

Moreau J.P. (1965) A propos de la biologie D'*Arctia caja* L. (Lepidopteres, Arctiidae). *Proceedings of the XII International Congress of Entomology*, p. 539.

Oates J.F., Swain T. & Zantovska J. (1977) Secondary compounds and food selection by Colobus Monkey. *Biochemical Systematics and Ecology*, **5**, 317–21.

Reese J.C. & Beck S.D. (1976) Effects of allochemics on the black cutworm *Agrotis ipsilon* (Lep. Noct.): effects of catechol, L-Dopa, dopamine and chlorogenic acid on larval growth, development and utilization of food. *Annals of the Entomological Society of America*, **69**, 68–72.

Salama H.S. & Saleh M.R. (1972) Population of the scale insect *Mycetapsis Personatus* Constock. on different varieties of *Magnifera indica* L. *Zeitschrift für Angewandte Netomologie*, **70**, 328–32.

Siegler D.S. (1977) Primary roles for secondary compounds. *Biochemical Systematics and Ecology*, **47**, 209–28.

Slansky F. & Feeny P.P. (1977) Stabilisation of the rate of N_2 accumulation by larvae of the cabbage butterfly on wild and cultivated food plants. *Ecological Monographs*, **47**, 209–28.

Soo Hoo C.F. & Fraenkel G. (1966) The consumption, digestion and utilization of food plants by a polyphagous insect, *Prodenia eridania* (Cramer). *Journal of Insect Physiology*, **12**, 711–30.

Southwood T.R.E. & McNeill S. (1978) The role of nitrogen in the development of insect/plant relationships. In *Biochemical Aspects of Plant and Animal Coevolution* (ed. J.B. Harborne), *Phytochemical Society of Europe Symposium 15*, pp.77–98. Academic Press, London & New York.

Snedocor G.W. & Cochran W.G. (1962) *Statistical Methods*. Iowa State University Press, Ames, Iowa.

Steele R.G.D. & Torrie J.H. (1960) *Principles and Procedures of Statistics*. McGraw-Hill, London.

Swain T. (1976a) Angiosperm–reptile coevolution. In *Morphology and Biology of Reptiles* (eds A.D'A. Belliars & C.B. Cox), *Linnean Society Symposium Series*, pp. 107–44. Academic Press, London & New York.

Tanton M.T. (1962) Effect of leaf toughness on the feeding larvae of the Mustard Beetle *Phaedon cochlearia*. *Entomologia Experimentalis et Applicata*, **5**, 74–8.

Van Summere C.F., Albrecht J., Dedonder A., De Pooter H. & Pe I. (1975) Plant proteins and phenolics. In *The Chemistry and Biochemistry of Plant Proteins* (eds J.B. Harborne & C.F. Van Summere), pp. 211–64. Academic Press, London & New York.

Waldbauer G.P. (1964) The consumption, digestion and utilization of solenaceous and non-solenaceous plants by larvae of the tobacco hornworm *Protoparce sexta* (Johan) (Lepidoptera, Sphingidae). *Entomologia Experimentalis et Applicata*, **7**, 253–69.

Wint G.R.W. (1979) The effect of the seasonal accumulation of tannins upon the growth of lepidopteran larvae. D. Phil. thesis, University of Oxford.

15. NITROGEN IN DEFENCE AGAINST INSECTS

E. A. BERNAYS
Centre for Overseas Pest Research, College House, Wrights Lane, Kensington, London W8 5SJ

INTRODUCTION

The possible protective roles of secondary metabolites, including nitrogenous ones, is a topic which has recently attracted much attention, and been the subject of various generalizations and theories. Proof of the supposed protective roles is not always easy to obtain, and attempts to gather it have often failed. In this paper work is reviewed on the relationship between insect herbivores and the potentially deterrent or noxious compounds containing nitrogen which they may encounter in plants. What is the evidence for the protective role of such nitrogen-containing compounds in the biology and ecology of plants?

THE PLANTS

Several major groups of nitrogenous compounds from plants have toxic effects on man or domestic animals. Most important are the cyanogenic compounds, alkaloids, unusual amino acids and amines, and the glucosinolates, which are all loosely grouped as secondary metabolites. A number of unusual proteins and a few rarer chemicals which have been little studied to date may also affect animals. For the plant, there are two separate aspects of the occurrence of such chemicals. Firstly, they must be able to synthesize them and secondly they must have the capacity to store them. For example, most plants have the capacity to synthesize cyanide at least in minute amounts (Gewitz *et al.* 1974; Hegnauer 1976), but only certain species store cyanogenic compounds in measurable quantities. Storage of secondary metabolites generally involves spatial separation from active cytoplasm into vacuole or cell wall, or in cuticle, laticifers, glands and hairs (Schnepf 1976). They are also commonly retained in non-toxic chemical complexes (Müller 1976). Special biochemical tolerance mechanisms operate in some plants (Fowden & Lea 1979) and a further discussion of the relative importance of these mechanisms is given by McKey (1979).

Different types of compound tend to occur in different quantities—both in terms of absolute concentration, and as a percentage of the total leaf nitrogen. Alkaloids and glucosinolates are in low concentration, and invariably the nitrogen allocation as a proportion of plant nitrogen is very low. On the other hand, the non-protein amino acids commonly contain over 5% of the leaf nitrogen, while in seeds this may rise to 20% and occasionally very much more. Some selected examples are given in Table 15.1. In terms of cost to the plant, many alkaloids are large molecules with synthetic pathways rather distinct from primary metabolism, and the cost may be related to production *per se*, rather than the allocation of important resources.

Plants are conservative in their use of nitrogen generally and its supply

TABLE 15.1. Concentrations of nitrogenous secondary metabolites (% dry wt.) and nitrogen resource allocation for selected plants. L = leaves, S = seeds.

Plant	Typical high concentration	Estimated approximate % of N in compound	Reference
Cyanogenic glycosides			
Sorghum bicolor L	0.5	2.0	1
Trifolium repens L	0.2	1.5	2
Cynodon dactylon L	1.4	3.0	2
Manihot esculenta L	2.5	4.0	3, 4
Alkaloids			
Senecio vulgaris L	0.2	0.3	5
Lupinus luteus L	0.4	0.7	6
Lolium perenne L	0.05	0.08	7, 2
Lycopersicon esculenta L	2.0	0.6	8
Non-protein amino acids			
Leucaena leucocephala L	5.0	8.0	9
Indigofera spicata L	0.2	3.0	9
Canavalia ensiformis L	3.0	12.0	10
Vicia gigantea S	7.0	33.0	11
Wisteria floribunda S	12.0	56.0	11
Lathyrus cicera S	1.0	7.0	9
Glucosinolates			
Brassica nigra L	0.4	0.5	12
Trypsin inhibitor			
Lycopersicon esculenta L	0.6	3.0	13
Phytohaemagglutinin			
Phaseolus vulgaris S	2.0	5.0	14

1 Woodhead & Bernays 1977, 2 Bernays & Chamberlain, unpublished, 3 de Bruijn 1971, 4 Oomen 1964, 5 Rothschild *et al.* 1979, 6 Harley & Thorsteinson 1967, 7 Aasen *et al.* 1969, 8 Buhr, Toball & Schrieber 1958, 9 Rosenthal & Bell 1979, 10 Rosenthal 1972, 11 Rosenthal 1977b, 12 Blau *et al.* 1978, 13 Ryan 1979, 14 Janzen, Juster & Liener 1976.

may be a limiting factor in plant growth in many ecosystems (see Lee, Harmer & Ignaciuk, this volume). It is not surprising therefore that the accumulation of nitrogenous secondary compounds is generally greater in plant groups or species, or ecological situations where nitrogen is readily available. For example the Leguminosae, with associated nitrogen-fixing bacteria, is a family particularly rich in them. Nitrogen fertilization is known to increase the levels of a number of such compounds (de Bruijn, 1971; Dement & Mooney 1974). The common cyanogenic glycosides are synthesized from and readily converted back to four primary amino acids important in animal nutrition: valine, isoleucine, phenylalanine and tyrosine. There is certainly the possibility that apart from obvious defense, cyanogenic plants are making essential amino nitrogen as unavailable as possible by storing it as cyanide. Undoubtedly the quantity and quality of nutrient nitrogen is important for insect herbivores (McNeill & Southwood 1979; also Prestidge & McNeill, and van Emden, this volume). How important are the secondary metabolites in insect biology?

CYANOGENIC GLYCOSIDES

Although some cyanogenic compounds are lipoidal (Seigler 1977), most are stored in plants as vacuolar glycosides, over 20 of which have been identified to date, occurring in over 1000 plant species, and commonly present in concentrations of about 1% dry weight. Conn (1979, 1980) has recently reviewed the chemistry and occurrence of these compounds. Plants containing them also commonly contain a rather specific β-glycosidase which is spatially separated in the intact plant, but which is brought into contact with the glycoside when the tissue is damaged, with the result that HCN is released. Such release of cyanide will almost inevitably occur during chewing by a herbivore. Even in the absence of such an enzyme the glycosides may be hydrolysed after ingestion by gut enzymes.

Since cyanide is so universally toxic, its role in plants may seem obvious, but many insect species feed freely on cyanogenic plants (Jones 1972; Jones, Parsons & Rothschild 1962; Lane 1962; Scriber 1978), and suffer no ill effects. Some of these are probably specifically adapted to detoxify cyanide while others, such as the polyphagous caterpillar of *Spodoptera eridania*, are particularly effective at detoxification of a wide range of unusual compounds (Krieger, Feeny & Wilkinson 1971).

If cyanide is really protective one might hope to find evidence of selective grazing, and a potentially useful approach to the study of defence is the examination of cyanogenic and acyanogenic strains of species which are polymorphic and occur in the same habitat. Other biochemical differences are then to be assumed at a minimum. With *Lotus corniculatus* some of the mollusc species tested exhibited a preference for acyanogenic forms, and on Anglesey more slugs were found in places with high percentages of cyanogenic plants (Crawford Sidebothom 1971; Jones 1966; Jones, Keymer & Ellis 1978), but no relationship with insects was found. Also with *Trifolium repens*, Miller and his co-workers (1975) were unable to find any effects related to differential grazing on cyanogenic and acyanogenic forms. A complicating feature is the presence of interacting environmental parameters. Daday (1954), for example, showed that acyanogenic *T. repens* became dominant where winter temperatures dropped to low levels, although genes for both the glucoside, lotaustralin, and the enzyme, linamarase, are dominant. Later (1965) he showed that the cyanogenic form was more injured by frost and accounted for this in terms of linamarase activation and tissue death which measurably reduced fitness of cyanogenic plants at low temperature. More interesting in the present context, Daday found that, in a warm field site, growth of the cyanogenic form was superior, although there was apparently no selective grazing by a high grasshopper population. This indicates that there may be important advantages to the plants, related to cyanogenesis but unrelated to herbivore pressure, which are yet unknown, and is a most important observation in arguments concerning the presumed advantages of this characteristic. At present, we cannot assume that grazing provides the greater selection pressure.

To investigate the problem more fully, work at the Centre for Overseas Pest Research (COPR) has been aimed at separating the possible effects of the glycosides from the effects of the hydrolysis products. The pure glycosides have been tested with a number of insects and are generally not deterrent, at least at natural concentrations, while some are even phagostimulants to insects habitually feeding on plants containing them (Table 15.2). This is itself surprising since the acridids, at least, hydrolyse the glycosides after ingestion and cyanide is released in the gut (Bernays, unpublished data). Long-term experiments were therefore conducted to investigate the possibility of a harmful effect from the glycosides or their hydrolysis products as a result of their ingestion over a period. Pure amygdalin or linamarin was applied to wheat leaves and fed to *Locusta migratoria* over the length of the fifth nymphal instar. Growth was unaffected by linamarin and appeared to be improved by amygdalin (Table 15.3). Faecal analysis showed that approximately 70% of the glycoside had been hydrolysed in both cases. Thus the glycosides and the digestively released cyanide, acetone and benzaldehyde, from naturally occurring levels of glycoside were non-toxic, even in an insect which is not a specialist on cyanogenic plants. The amount of cyanide released in the higher concentrations amounted to a dose of about 500 mg kg body wt.$^{-1}$ day^{-1}:

10–100 times the dose which is lethal to livestock and other mammals (Jones 1973).

In conclusion, then, it may be assumed, at least for the present, that the glycosides themselves are generally neither deterrent nor deleterious to insects. The mechanism of this tolerance is unknown although rhodanese is

TABLE 15.2. The effect of cyanogenic glycosides at naturally occurring concentrations on insect feeding in the laboratory. 0 = no effect, + = phagostimulatory.

Insect	Glycoside	Effect	Reference
Malacosoma americana	amygdalin	+	1
Spodoptera eridiania	amygdalin	0	2
Pieris brassicae	prunasin	0	3
	amygdalin	0	4
Epilachna varivestis	linamarin	+	5
	lotaustralin	+	5
Locusta migratoria	dhurrin	0	6
	linamarin	0	7
	amygdalin	0	7
	prunasin	0	7
Schistocerca gregaria	linamarin	0	7
	amygdalin	0	7
	prunasin	0	7
Peregrinis maidis	dhurrin	+	8
Leptinotarsa decemlineata	amygdalin	0	4

1 Verschaeffelt 1910, 2 Rehr *et al.* 1973, 3 Jones & Firn 1979, 4 Schoonhoven & Jermy 1977, 5 Nayer & Fraenkel 1963, 6 Woodhead & Bernays 1977, 7 Bernays 1977, 8 Fisk 1980.

TABLE 15.3. Growth of *Locusta migratoria* over the fifth instar on wheat with and without added cyanogenic glycosides. Ten insects in each treatment.

Treatment	Concentration % dry wt.	% wt. increase day^{-1} mean ± s.e.	Length of instar, days
Amygdalin	2	11.4 ± 0.6	10
Control	0	10.0 ± 0.4	11
Amygdalin	3	13.8 ± 0.4	10
Control	0	10.4 ± 1.0	11
Linamarin	1	9.8 ± 0.4	13
Control	0	10.4 ± 0.3	13
Linamarin	4	8.0 ± 0.3	14
Control	0	8.3 ± 0.3	14

well known in insects adapted to feeding on cyanogenic plants (Jones, Parsons & Rothschild 1962; Parsons & Rothschild 1964), and this enzyme is responsible for the conversion of cyanide to the relatively non-toxic thiocyanate which may eventually be incorporated into intermediary metabolism. Various other possibilities are discussed by Brattsten (1979).

That cyanogenic glycosides can stimulate feeding certainly implies a lack of toxicity. In *Epilachna varivestis* those present in the food may be functioning as sign stimuli, indicating the presence of the host plant. Another possibility is that the cyanide is an additional source of readily available amino nitrogen.

CYANOGENESIS

Since cyanogenic glycosides are apparently not protective against insects, the next logical question was whether the hydrolysis products released during biting and chewing perhaps act peripherally and prevent feeding. The first evidence that this may be so was with bracken *Pteridium aquilinum*. Strains containing the glycoside only, the glycoside plus the enzyme, or neither, were tested for their palatability to the polyphagous locust *Schistocerca gregaria*. Only the truly cyanogenic form, containing the glycoside plus the enzyme, was totally unpalatable (Cooper-Driver *et al.* 1977). A similar result was obtained in laboratory tests with caterpillars of *Pieris brassicae* (Jones & Firn 1979). This indicates not only that cyanogenesis is effective but, again, that the glycoside itself is not.

Rehr, Feeny & Janzen (1973) examined herbivory on acacias and found that larvae of *Spodoptera eridania* performed poorly on the species normally without ants, compared with the species symbiotically housing ants. The latter had much lower levels of cyanogenic glycosides and of cyanogenesis. On the basis of experiments with artificial diet containing glycosides or extracted plant materials these authors assumed that there was no causal relationship, but it is likely that cyanide release was not adequately mimicked in the experiments.

In another approach to the role of cyanide Schroeder (1978) made an extensive survey of insect feeding on the highly cyanogenic black cherry, *Prunus serotina*, and found that insect feeding was only extensive when the glycoside levels and presumably cyanogenesis had dropped to well below the maximum.

These rather inconclusive experiments and observations suggested that a fuller study was needed. A very detailed study of *Sorghum bicolor* was undertaken at COPR. The unpalatability of seedling sorghum to the graminivorous *Locusta migratoria* was roughly correlated with levels of the cyanogenic glycoside in the plant, but when the glycoside dhurrin was extracted and purified, it was found not to be deterrent at natural concentrations, so the potential deterrent effects of hydrolysis products were investigated. A novel bioassay was developed for testing the volatile cyanide. Several days prior to testing, newly moulted fifth instar nymphs had a long cannula inserted through the labium, ending in the preoral cavity adjacent to the opening of the salivary duct. After 2–3 days allowed for recovery, a syringe was attached to the cannula. To test the insect, several microlitres of a solution of hydrocyanic acid were released into the mouth region soon after the beginning of feeding, and the reaction of the insect noted. The higher concentrations invariably prevented feeding (Table 15.4). To determine whether deterrent concentrations were released naturally from a leaf during chewing, estimates of cyanide release rates were obtained by measurement of the stoichometrically produced parahydroxybenzaldehyde, after crushing leaf fragments in buffer. The figures indicated that 0.5 mmol solutions of cyanide may be released during attempts at feeding on young sorghum, and that cyanide is therefore a very effective plant defence (Woodhead & Bernays 1977), and probably largely accounts for the unpalatability of field grown sorghum seedlings to acridids observed in West Africa (Launois-Luong 1976) and in India (Woodhead, Padgham & Bernays 1980).

The most complete study on a herbivore and a cyanogenic plant has been with an aposematic African grasshopper *Zonocerus variegatus* and the cassava plant *Manihot esculenta*. During the course of population studies of this insect in Nigeria, it was noted that only the later instars fed on cassava. Experiments with insects in net bags on the plant confirmed that only older, larger insects would feed on the growing plant. Moreover, even the larger insects would only feed freely and survive well if they were enclosed together in groups. Alternatively, individuals would feed on wilted but not on turgid leaves (Table 15.5), and indeed on the wilted leaves, growth and survival was better than on any other plants in the habitat (Bernays *et al.* 1977). The answer to this paradox lay in the release rates of cyanide. A method was developed for monitoring rapid release of cyanide in the field, and tests carried out on feeding followed by the immediate testing of the leaf for rapid cyanide release. It was found that, as the plant wilted, the release rates of cyanide dropped, and that the palatability simultaneously increased (Table 15.6). Total glycoside levels did not alter significantly so that the difference presumably lay in either the glycosidase activity or the greater mixing capacity of the turgid leaf tissue when damaged. Further evidence of the importance of cyanide was obtained by testing insects on different cultivars and different leaf ages having different release rates of cyanide. There was a significant negative correlation between feeding and cyanide release.

TABLE 15.4. The effect of different concentrations of hydrocyanic acid on feeding behaviour, after feeding on grass has started (partly from Woodhead & Bernays 1977).

		Behaviour		
Concentration	Number of tests	No effect	Feeding stops	Rapid withdrawal
0.1 M	5	0	0	5
0.01 M	20	0	9	11
1.0 mM	20	1	14	5
0.1 mM	20	10	10	0
Water	10	10	0	0
Air	20	20	0	0

TABLE 15.5. Consumption of cassava leaves by adult female *Zonocerus variegatus*. Five insects for each figure (after Bernays *et al.* 1977).

Insect age	Mean ingestion in mg $day^{-1} \pm$ s.d.	
(days)	Growing plants	Cut (wilted) leaves
1–2	25 ± 9	80 ± 3
3–5	49 ± 11	119 ± 21
6–10	46 ± 15	186 ± 20
11–13	38 ± 16	156 ± 18
14–18	35 ± 14	154 ± 15
19–20	28 ± 10	188 ± 16
21–23	26 ± 8	143 ± 26
24–26	25 ± 8	140 ± 20
27–30	20 ± 4	58 ± 9

TABLE 15.6. Percentage of insects feeding on leaves of cassava at intervals after picking the leaves from the growing plant, and the percentage of leaves producing an appreciable amount of cyanide within one second (from Bernays *et al.* 1977).

Time since picking (min)	% plants producing HCN in 1 s	% of insects feeding
0	100	0
20	88	20
30	55	48
50	45	65
70	25	79
80	15	95
90	10	98

There are two factors relating this study to the biology and pest status of *Z. variegatus* in southern Nigeria. Firstly, the insects are gregarious and tend to attack a plot over a very limited area. Although the gregariousness of the insects is probably related to the noxious nature of these aposematic insects, this behaviour pattern favours a situation where many insects attempt to feed on a single plant or leaf and cause wilting. This results in palatability. Secondly, during the dry season, a shortage of alternative host plants together with natural wilting of the cassava, will enhance the action of the group.

Finally, it is possible that the glycoside is favoured when cyanide release rates are not excessive. The bark, for example, which is extremely rich in linamarin is greatly favoured. In any case, cassava is classed as a very strongly cyanogenic plant, even when cyanide release rates are low (de Bruijn 1971). *Z. variegatus* may be ingesting as much as 1 g HCN kg body wt. day^{-1} as linamarin which is as much as 1000 times the mammalian toxicity. It is interesting, however, that cassava is a crop with relatively few pests, with *Z. variegatus* being exceptional, at least in West Africa.

The generalizations we can make at present are as follows:

1. Some insects are certainly deterred from feeding on cyanogenic plants and self-defence is likely to be one of the main values of cyanogenesis, although there are suggestions of other roles.
2. The mechanism of effectiveness is related to cyanide release rates during chewing and not to the levels of cyanogenic glycosides present. The biggest variable is probably enzyme activity levels.
3. Finally, insects are physiologically very tolerant of cyanide compared with mammals and the selection pressure provided by insects is likely to have been significant only where the plant is strongly cyanogenic.

ALKALOIDS

The alkaloids are perhaps the best known group of nitrogenous secondary metabolites. They are structurally diverse but generally contain nitrogen as part of a heterocyclic ring. There are many thousands of different structures and they probably occur in over 20% of vascular plant species. In a very extensive survey by Smolenski, Silinis & Farnsworth (1973, 1974a, 1974b, 1975), over 5000 species were examined, and approximately one third contained alkaloids. Numerous very novel chemicals appear to have arisen readily in some plant groups (McKey 1980). Biological activity is common, with many being extremely toxic at least to vertebrates. They act on diverse aspects of membrane transport, protein synthesis, DNA replication, enzyme activity and on receptor sites for endogenous chemicals. Very specific action on different parts of the nervous system is common, for example by blocking acetylcholine receptors or the activity of cholinesterase. Concentrations in plants are commonly between 0.1 and 1% dry weight, and an excellent recent review is given by Robinson (1979).

Circumstantial evidence of the defensive role of alkaloids is obtained by a study of their distribution within plants. This is reviewed by McKey (1974) but of particular significance are the facts that accumulation is usually in peripheral tissue, and that highest levels occur where herbivore attack would have greatest effect on plant fitness. Thus expanding leaves are richest in alkaloids during vegetative growth, while immature fruits tend to have highest concentrations overall. As far as it has been examined, a similar pattern exists for cyanogenic compounds and glucosinolates but not for non-protein amino acids.

The complexity of the field situation often precludes a clear demonstration of effectiveness of alkaloids in defence against insects, although there are a few studies of ecological correlates and cultivar differences in alkaloid concentrations. For example, the fruit fly *Oscinella frit* was shown to infest the grass *Phalaris arundinacea* more frequently when the alkaloid content was lower (Byers & Sherwood 1979). In *Lupinus angustifolius*, infestation by the aphid *Macrosiphum euphorbiae* was negatively correlated with alkaloid content (Brusse 1962). In the case of thrips on lupin populations there was not only a negative correlation: these insects were reported to be 100% accurate in infesting non-alkaloidal cultivars and were even suggested as a non-chemical means of selecting these cultivars in fodder breeding programmes (Forbes & Beck 1954; Gustafsson & Gadd 1965).

Most work has involved the family Solanaceae. High overall alkaloid levels in certain resistant crops are apparently effective against pests such as *Leptinotarsa decemlineata* (Schreiber 1958), *Manduca sexta* (Schoonhoven 1972) and *Empoasca fabae* (Dahlman & Hibbs 1967; Tingey, Mackenzie & Gregory 1978). The action of nicotine as an insecticide is well known, but the toxicity of the free alkaloid acting by contact is very much greater than the action of its salts ingested with the leaves of *Nicotiana* species. Nevertheless, the oral toxicity of tobacco alkaloids is well established and known to protect the plant from attack by most insects (Gordon 1961). Kuhn, Löw & Gauhe (1950) showed that the absence of certain alkaloids in species of Solanaceae determined the palatability of the host plant to *L. decemlineata*, although Hsiao & Fraenkel (1968) found that the egg-laying adult was much less particular than the young larvae. This resulted in the death of many larvae which hatched on the 'wrong' plants. Such a situation also occurs in other insects (Chew 1980). It is given a further ecological interpretation by Attsat & O'Dowd (1976), who suggest that some plants are acting as attractive decoys in a particular biome ensuring larval mortality and a generally reduced population of phytophagous insects.

In the laboratory, over 50 different plant alkaloids have been reported to be deterrent or toxic to a number of different kinds of insects. Very often, however, while the result is of physiological interest, its ecological relevance is obscure. For example, quinine has probably been most widely assayed (Bernays & Chapman 1977; Schoonhoven 1972), yet generally the insects involved are unlikely to encounter it in nature. The solanaceous alkaloids, have, however, been tested on a number of relevant insects, and the concentrations occurring naturally are such as to account for the unsuitability of Solanaceae for insects not normally feeding on this family. Some examples are given in Table 15.7 but more extensive lists are given by Hedin, Jenkins & Maxwell (1977) and Schoonhoven (1972).

By far the most extensive work has been with *L. decemlineata.* Buhr, Toball & Schreiber (1958) examined the effects of 33 solanaceous alkaloids and 30 other alkaloids from a variety of different plants. At natural concentrations many were deterrent. Many others were ingested but showed toxic effects. For most, these two effects were not separated but, as expected, the solanaceous alkaloids were less effective against this Solanaceae-feeding insect, than the non-solanaceous alkaloids (Table 15.8). In none of the non-solanaceous ones was there absolutely no effect. On the other hand, most of those naturally present in *Solanum tuberosum* had little or no effect. For *L.*

TABLE 15.7. Feeding deterrence (F) and/or toxicity (G) to various insects at naturally encountered concentrations of solanaceous alkaloids which could thus contribute to the unsuitability of the plant species as hosts. U = unfavourable, O = no measurable effect, NT = not tested.

Insect species	*Melanoplus bivittatus*[1]		*Locusta migratoria*[2]		*Pieris brassicae*[3]		*Dysdercus fulvoniger*[4]		*Myzus persicae*[5]	
Normal food	Mixed herbaceous		Gramineae		Cruciferae		Mixed, inc. Malvaceae		Mixed herbaceous	
Alkaloid	F	G	F	G	F	G	F	G	F	G
Atropine	NT	NT	U	NT	U	NT	U	NT	U	NT
Nicotine	U	O	U	U	U	NT	U	NT	NT	NT
Nornicotine	U	U	U	NT	NT	NT	NT	NT	NT	NT
Solanine	O	U	U	NT	U	NT	U	NT	U	U
Tomatine	O	U	U	U	U	NT	U	NT	O	O

[1] Harley & Thorsteinson 1967, [2] Bernays & Chapman 1977, Bernays, unpublished, [3] Ma 1972, Schoonhoven & Jermy, 1977, [4] Schoonhoven & Derksen-Koppers 1973, [5] Schoonhoven & Derksen-Koppers 1976.

TABLE 15.8. The percentage of alkaloids having a deleterious effect on *Leptinotarsa decemlineata* when added to a neutral diet at natural concentrations (from Buhr, Toball & Schreiber 1958).

Alkaloid	Little or no effect	Moderate effect or effect only at higher concentration	Very deterrent or toxic
Solanaceous alkaloids	35	35	30
non-solanaceous alkaloids	6	39	55

decemlineata, Hsaio & Fraenkel (1968) showed that toxicity is usually associated with deterrence but the relationship is by no means absolute, since low levels of nicotine may be ingested but rapidly cause mortality (Schreiber 1958).

It is difficult to generalize about chemicals of such a diverse group but concentrations in the diet which are deterrent, or deleterious if ingested, are lower for non-adapted insects than the concentrations which occur naturally. Janzen, Juster & Bell (1977) found that 0.1% by weight of most alkaloids tested in the diet of the bruchid beetle *Callosobruchus maculatus* were lethal. This insect normally feeds on a restricted range of legume seeds without alkaloids. Lower concentrations were not tested, but such chemicals are usually present at or above 0.1%. Much lower levels of most alkaloids were deterrent to the graminivorous *Locusta migratoria* (Bernays & Chapman 1977), and in a number of other isolated experiments with particular insects and alkaloids, it has been shown that, where they are effective against insects, the effective dose is relatively low compared with other secondary metabolites (Chapman 1974; Hedin, Jenkins & Maxwell 1977; Schoonhoven 1972).

There are as yet no surveys of general toxicity levels of alkaloids to insects, but calculations based on the work of Harley & Thorsteinson (1967) and unpublished work at COPR suggest that the LD_{50} for grasshoppers not specifically adapted for feeding on plants containing alkaloids, is extremely variable but generally higher than for mammals. Insects are very tolerant of changes in their internal milieu, including the presence of potentially noxious compounds and this is probably related to the particularly effective blood-brain barrier (Lane & Skaer 1980). Moreover, the excretory system takes up at least some toxic materials selectively (Maddrell 1981; Maddrell & Gardiner 1976).

That special detoxification methods occur in insects which readily feed on plants containing alkaloids must be assumed, and in some cases the detoxification mechanisms are known. Several excellent studies have elucidated the mechanisms involved in dealing with nicotine by tobacco-feeding insects (Self, Guthrie & Hodgson 1964; Yamamoto & Fraenkel 1960; Brattsten 1979). There is a combination of behavioural mechanisms, tolerance, excretion and biochemical detoxification processes. Tolerance is particularly well developed in some insect species which sequester plant alkaloids, which are then a means of protection from vertebrate predators. For example, all aposematic moths of the family Arctiidae which feed on plants containing pyrrolizidine alkaloids store these substances (Rothschild *et al.* 1979). There are now many other examples of this phenomenon (Duffey 1980), which may suggest that insects are better able to counter alkaloids than are vertebrates.

There can be little doubt of the protective role of alkaloids but it is very difficult to quantify in any way the impact in a given ecosystem. An interesting attempt has been made by Dolinger and his co-workers (1973), who made a study of a lycaenid caterpillar which feeds on the flowers of certain lupin species. They presented evidence which showed that plants produced fewer alkaloids and at low concentration where the environment was ecologically unfavourable for the insect. Where the situation was better for the butterfly the host plants produced not only more and greater variety of alkaloid types, but extreme variation occurred in the quantities produced by the individuals. This was interpreted as a means of reducing the effective pressure on the herbivore for counter-adaptations, in terms of physiological specialization to cope with the alkaloids.

A completely different approach has been taken by Levin (1976) and Levin & York (1978). In an attempt to look for geographical patterns they examined the amounts and relative toxicities of alkaloids from plants of different origin. They found that higher temperatures either in terms of latitude or altitude were correlated with higher alkaloid content (Table 15.9) and with increased toxicity as judged by known LD_{50} levels for different classes of alkaloids in mammals. This is assumed to have resulted from differences in the intensity of herbivory since a greater proportion of plant material is harvested by all classes of herbivore in tropical ecosystems than in temperate ones (Golley 1972). The argument is convincingly presented although there are inevitable exceptions and the overall problem is that at any one moment in time there will not be a complete evolutionary balance.

The following generalizations can be made at present:

1. Alkaloids are produced in quantity by plants primarily as a defence against other organisms including insects, with some insect species having adapted to individual alkaloid profiles.
2. Apart from the specifically adapted species, insects are physiologically sensitive to alkaloids, but generally less so than mammals.

TABLE 15.9. The mean percent dry weight of alkaloids in leaves followed by standard errors from plants in different regions (from Levin & York 1978).

Region	
Tropics	0.67 ± 0.69
Subtropics	0.60 ± 0.62
Temperate	0.48 ± 0.49
New Guinea	
Lowland rain forest	0.57 ± 0.18
Foothills rain forest	0.34 ± 0.11
Montane rain forest	0.22 ± 0.07

3. The heterogeneity in alkaloid structures in individual plant species, which is usual in alkaloid-bearing plants, may itself represent a particular defence strategy, or a state of transition necessitated by persistent adaptive measures taken by insect herbivores, with their selective advantage of a relatively short generation time.

NON-PROTEIN AMINO ACIDS

There are over 400 amino acids which are not used in protein formation, but which occur widely in plants in the free state, and they are particularly common in the Leguminosae (Bell 1972). Many have been shown to be directly neurotoxic to mammals, and in other cases the toxicity appears to be due to the fact that they are similar to common protein-forming amino acids and act as analogues in organisms not physiologically adapted for distinguishing them. Toxicity thus results from the formation of proteins not having appropriate properties. The details of the potential modes of action are discussed fully by Fowden, Lewis & Tristram (1967) and Fowden & Lea (1979). In some cases of mammalian toxicity such mechanisms have been demonstrated (Rosenthal & Bell 1979).

The case for the defensive role of at least some non-protein amino acids, particularly against vertebrates, is strong and has recently been argued by Rosenthal & Bell (1979). Many of these amino acids are general insect-feeding deterrents (Navon & Bernays 1978) when tested in the laboratory and many are toxic to particular insects (Janzen, Juster & Bell 1977; Rehr *et al.* 1973; Rosenthal 1977a; Rosenthal & Dahlman 1975). L-canavanine has been the most studied and its antimetabolic mode of action apparently depends on its substitution for arginine in protein synthesis, and it is potent against a number of insect species (Rosenthal & Bell 1979). On the other hand, the beetle

Caryodes brasiliensis is highly specific on the seeds of a tropical legume containing 13% dry weight of canavanine and possesses an arginyl-tRNA synthetase which discriminates against canavanine (Rosenthal, Janzen & Dahlman 1977).

Some subtle effects of β-cyanoalanine occurring, for example, in *Vicia* spp. have been found when it is placed in the diet of *Locusta migratoria*. Although it is deterrent at naturally occurring concentrations (Navon & Bernays 1978) insects will ingest artificial diet containing it if they are deprived of water or food. The overt physiological response is an inability to resorb water from the rectum which leads to death by dehydration (Schlesinger, Applebaum & Birk 1976). The exact mode of action is unknown but is assumed to be related in some way to the accumulation of β-cyanoalanine in the nervous system (Schlesinger, Applebaum & Birk 1977).

The non-protein amino acids occur in plants in higher concentration than alkaloids and their toxic or deterrent effects are generally seen at these higher levels (Bernays & Chapman 1977; Janzen, Juster & Bell 1977). It is interesting that, among leguminous plants, alkaloids are more commonly accumulated in large quantities in certain primitive tribes while non-protein amino acids are accumulated mostly in more advanced tribes. Bell (1980) reasons that this is evidence of the need for protective chemicals and that the non-protein amino acid accumulation is an evolutionary trend towards biochemical simplicity, in which a single amino acid solves both the problems of storage of nutrient and chemical protection, which would otherwise involve the synthesis of separate and more complex compounds in less advanced species.

The significance of non-protein amino acids as nitrogen stores cannot be disputed in respect of seeds where a high percentage of the nitrogen is often in such a compound and where it disappears during germination. High concentrations are not only found in storage organs, however, and it may be that, with these materials in the leaves, a further value accrues to the plant in terms of protection, other than by a direct deterrent or toxic action. The argument is as follows:

1. Nitrogen is usually at a premium for insect herbivores, and making it as unavailable as possible is an appropriate protective strategy for the plant (McNeill & Southwood 1978).
2. The nitrogen resource allocation to non-protein amino acids tends to be high (see Table 15.1), and it may provide the plant with its store of readily available amino nitrogen.
3. If this is so, and there is a concomitant lowering of amino acids normally considered essential nutrients, then the poverty of these nutrients is itself a defence against herbivores, which limits the rate of build-up of herbivore populations.

GLUCOSINOLATES

There are about 100 different glucosinolates which are characteristic of the family Cruciferae, although some also occur in other plant families. Both the glucosinolates and the products of hydrolysis (isothiocyanates) have known deleterious effects on mammals when they are ingested in large quantities. Amounts present in leaves vary from about 0.1 to 0.6% dry weight. Their chemistry and properties are reviewed by van Etten & Tookey (1979) and Tapper & Reay (1973).

This group of compounds is best known for the attractive properties of individual chemicals to particular insects which feed on plants containing them (Hedin, Jenkins & Maxwell 1977; Schoonhoven 1972). A large number of insects have adapted to the presence of glucosinolates, and they apparently then provide suitable sign stimuli indicating the presence of the host plant. Several studies have, however, shown that glucosinolates are deterrent or toxic to a number of non-adapted insects, indicating the protective role of such compounds (Table 15.10). How much of the effect is due to the glucoside and how much to the breakdown products is unknown since hydrolysis will occur during biting, by biting and chewing insects, and it certainly occurs after ingestion (Erickson & Feeny 1974).

The release of allylisothiocyanate during chewing on plants containing sinigrin occurs through cellular disruption, bringing myrosinase, which hydrolyses the glucoside, into contact with it (Tapper & Reay 1973). Thus the

TABLE 15.10. Examples of deterrence (F) or toxicity (G) of glucosinolates and isothiocyanates to insects, when presented at natural concentrations.

Species	Compound	Deterrence/ toxicity	Reference
Myzus persicae	sinigrin	F	1, 2
Plutella maculipennis	gluconasturtiin	G	3
	gluconappin	G	3
Manduca sexta	glucotropaeolin	F	4
Melanoplus sanguinipes	isothiocyanate	F	5
Locusta migratoria	sinigrin	F	6
	allylisothiocyanate	F	6
Papilio polyxenes	sinigrin	F,G	7
Leptinotarsa decemlineata	sinigrin	F	8

1 van Emden 1978, 2 Wearing 1968, 3 Nayer & Thorsteinson 1963, 4 Schoonhoven 1972, 5 Pawlowski Reigeita & Krymanski 1968, 6 Bernays & Chapman 1977, 7 Erickson & Feeny 1974, 8 Schoonhoven & Jermy 1977.

enzyme system is analagous to that of cyanogenesis. It is perhaps significant that the same degree of deterrence in *Locusta migratoria* is produced by allylisothiocyanate concentrations 50 times less than those of sinigrin (Bernays & Chapman 1977).

PROTEINS

Certain proteins are potentially important in plant defence. Best known are the glycoprotein phytohaemagglutinins, so called for their ability to agglutinate erythrocytes. They have, however, various other deleterious properties, at least partly associated with their affinity for carbohydrates. They are now generally known as lectins and occur very widely in the plant kingdom, especially in some seeds where they make up 2–10% of the protein. In a recent review, Liener (1979) lists the numerous proposed functions of these materials, few of which include defence. Little is known of insect responses to such materials, except that the lectin from *Phaseolus vulgaris* is toxic to the bruchid beetle *Callosobruchus maculatus* (Janzen, Juster & Liener 1976), and may be one reason why this insect is not found in seeds of this plant.

Secondly, there are proteinase inhibitors, which are themselves proteins or polypeptides. These are very widespread and have been proposed as protective agents for the plants, reducing the activity of insect digestive enzymes after feeding, and thus reducing growth (Applebaum 1964). Until recently, purification and identification of the material had not been achieved, so that critical experiments could not be carried out on their real effects *in vivo*. The situation is reviewed by Ryan (1979) whose work has given weight to the possibility that these materials are protective compounds, by showing that infestation and damage to solanaceous plants by *Leptinotarsa decemlineata* causes a rapid accumulation of proteinase inhibitor in the leaves (Green & Ryan 1972). It is an exciting possibility, but so far there is a shortage of experimental work with insects to show whether there is any deterioration of the digestive capacity as a result of the presence of such a trypsin inhibitor. The involvement of trypsin inhibitors in barley was suggested by Weiel & Hapner (1976). They demonstrated a negative correlation between cultivar levels of trypsin inhibitor and severity of grasshopper damage in North America. How such a short-term effect was mediated is not known since feeding experiments were inconclusive, but it could possibly be explained by an increased restlessness in insects which were nutritionally unsatisfied, leading to an accumulation of insects on more suitable food.

In a critical experiment with *Callosobruchus maculatus*, a relatively resistant cultivar of *Vigna unguiculata* was shown to have seeds containing a proteinase inhibitor which, when removed, allowed normal insect growth (Gatehouse *et al.* 1979).

CONCLUSIONS

No defence protects the plant from counter-adaptations by herbivores and the various theories of interactions and adaptations are most recently summarized by Fox (1981) and they do not conflict with the overall conclusion that secondary metabolites, including nitrogenous ones, are protective to the plants which possess them, particularly since natural insect herbivore pressure is now known to be much more effective than is apparent (Bentley, Whittaker & Malloch 1980; Lawton & McNeill 1979; Morrow & LaMarche 1978; Waloff & Richards 1977). Are there clear statements to be made on the overall impact of secondary metabolites?

For most insects examined with a wide range of plant extracts, deterrence by non-host plants is usual (Bernays & Chapman 1977; Jermy 1966; Schoonhoven & Jermy 1977), and the data presented in this paper summarizes the deterrent effects known concerning nitrogenous compounds. Species diversity on particular plants is not, however, influenced by plant chemistry (Lawton 1978; Strong 1979), which again demonstrates that the adaptive capabilities of the insects with respect to secondary metabolites are considerable.

Deterrent effects may signal toxicity. Additionally, or alternatively, they may indicate a plant to which the insect is less suited for other reasons. The majority of insect species are specialists—a strategy with certain theoretical advantages (van Emden 1978)—but it also means that the primary impact of feeding deterrents is to prevent most insects from feeding on the majority of plants and to minimize host plant switching.

Many nitrogenous compounds have been shown to be toxic to non-adapted insects, but even in adapted species, there is likely to be a cost involved in the process of detoxification. This is important since ideal food is commonly not readily available to insect herbivores (see Lawton & McNeill 1979). Ingestion of natural concentrations of alkaloids from the host plants of *Manduca sexta* has been shown to have a significant cost to the insect as measured by the efficiency of conversion of digested artificial diet (Schoonhoven & Meerman 1978). Thus even the 'adapted' oligophagous insect, deterred from feeding on the majority of plants, and able to cope with the secondary metabolites in its host plant, may nevertheless have its effective growth rate limited by these materials. Even small costs of this nature may have profound effects; phytophagous insect numbers are regulated largely by natural enemies but models predict that a small drop in population growth rate will result in a very large drop in the equilibrium population size (Lawton & McNeill 1979). Thus plant chemistry is likely to have a significant impact on the size of the population of individual species of herbivore.

It is likely that no further generalizations are possible at present. The direct effects act in concert with the more subtle and indirect effect of secondary metabolites in the biome. Plant-specific chemicals may attract parasitoids of plant-specific herbivores (Read, Feeny & Root 1970), and there are well-argued plant-positive aspects of herbivore grazing (Mattson & Addy 1975; Owen & Weigert 1976; Springett 1978) and cases of nitrogenous metabolites serving the herbivore in a more positive manner (van Emden 1978). What is now required are extensive surveys in both the experimental and ecological aspects of the problem.

REFERENCES

Aasen A.J., Culvenor C.C.J., Finnie E.P., Kellock A.W. & Smith L.W. (1969) Alkaloids as a possible cause of ryegrass staggers in grazing livestock. *Australian Journal of Agricultural Research*, **20**, 71–86.

Applebaum S.W. (1964) Physiological aspects of host specificity in the Bruchidae—1. General considerations of developmental compatibility. *Journal of Insect Physiology*, **10**, 783–8.

Attsatt P.R. & O'Dowd D.J. (1976) Plant defense guilds. *Science*, **193**, 24–9.

Bell E.A. (1972) Toxic amino acids in the Leguminosae. *Phytochemical Ecology* (ed. J.B. Harborne), pp. 165–77. Academic Press, London & New York.

Bell E.A. (1980) Non-protein amino acids in plants: their chemistry and possible biological significance. *Review Latinamerica Qulm*. **11**, 16–23.

Bentley S., Whittaker L.B. & Malloch A.J.C. (1980) Field experiments on the effects of grazing by a chrysomelid beetle (*Gastrophysa viridula*) on seed production and quality in *Rumex obtusifolius* and *Rumex crispus*. *Journal of Ecology*, **68**, 671–4.

Bernays E.A. (1977) Cyanogenic glycosides and their relevance in protection from insect attack. *International Organisation for Biological Control/Western Palearctic Regional Sector Bulletin*, **77 (3)**, 123–8.

Bernays E.A. & Chapman R.F. (1977) Deterrent chemicals as a basis of oligophagy in *Locusta migratoria*. *Ecological Entomology*, **2**, 1–18.

Bernays E.A., Chapman R.F., Leather E.M., McCaffery A.R. & Modder W.W.D. (1977) The relationship of *Zonocerus variegatus* (L.) (Acridoidea: Pyrogomorphidae) with cassava (*Manihot esculenta*). *Bulletin of Entomological Research*, **67**, 391–404.

Blau P.A., Feeny P., Contardo L. & Robson D.S. (1978) Allylglucosinolate and herbivorous caterpillars: a contrast in toxicity and tolerance. *Science*, **200**, 1296–8.

Brattsten L.B. (1979) Biochemical defence mechanisms in herbivores against plant allelochemicals. In *Herbivores: Their Interaction with Secondary Plant Metabolites* (eds G. A. Rosenthal & D. H. Janzen), pp. 200–70. Academic Press, London & New York.

Brusse M.J. (1963) Alkaloid content and aphid infestation in *Lupinus angustifolius* L. *New Zealand Journal of Agricultural Research*, **5**, 188–9.

Buhr H., Toball R. & Schreiber K. (1958) Die Wirkung von einigen pflanzenlichen Sonderstoffen, insbesondere von Alkaloiden, auf die Entwicklung der Larven des Kartoffelkafers (*Leptinotarsa decemlineata* Say). *Entomologie Experimentalis et Applicata*, **1**, 209–24.

Byers R.A. & Sherwood R.T. (1979) Differential reaction of *Phalaris arundinacea* to *Oscinella frit*. *Environmental Entomology*, **8**, 408–11.

Chapman R.F. (1974) The chemical inhibition of feeding by phytophagous insects: a review. *Bulletin of Entomological Research*, **64**, 339–63.

Chew F. (1980) Food plant preferences of *Pieris* caterpillars (Lepidoptera). *Oecologia*, **46**, 347–53.

Conn, E.E. (1979) Cyanide and cyanogenic glycosides. In *Herbivores: Their Interaction with Secondary Plant Metabolites* (eds G. A. Rosenthal & D. H. Janzen), pp. 387–412. Academic Press, London & New York.

Conn E.E. (1980) Cyanogenic glycosides. In *Secondary Plant Products* (eds E. A. Bell & B. V. Charlwood), pp. 461–92. Springer-Verlag, Berlin.

Cooper-Driver G., Finch S., Swain T. & Bernays E.A. (1977) Seasonal variation in secondary plant compounds in relation to the palatability of *Pteridium aquilinum*. *Biochemical Systematics and Ecology*, **5**, 177–83.

Crawford Sidebothom T.J. (1971) Studies of aspects of slug behaviour and the relation between molluscs and cyanogenic plants. Ph.D. thesis, University of Birmingham.

Daday H. (1954) Gene frequencies in wild populations of *Trifolium repens* L. I. Distribution by latitude. *Heredity*, **8**, 61–78.

Daday H. (1965) Gene frequencies in wild populations of *Trifolium repens* L. IV. Mechanism of natural selection. *Heredity*, **20**, 355–66.

Dahlman D.L. & Hibbs E.T. (1967) Responses of *Empoasca fabae* (Cicadellidae: Homoptera) to tomatine, solanine, leptine I, tomatidine, solanidine, and demissidine. *Annals of the Entomological Society of America*, **60**, 732–40.

de Bruijn G.H. (1971) A study of the cyanogenetic character of cassava. *Mededelingen Landbouwhogeschool Wageningen* 71–13, p. 140.

Dement W.A. & Mooney H.A. (1974) Seasonal variation in the production of tannins and cyanogenic glycosides in the chaparral shrub *Heteromeles arbutifolia*. *Oecologia*, **15**, 65–76.

Dolinger P.M., Ehrlich P.R., Fitch W.L. & Breedlove D.E. (1973) Alkaloid and predation patterns in Colorado lupine populations. *Oecologia*, **13**, 191–204.

Duffey S.S. (1980) Sequestration of plant natural products by insects. *Annual Review of Entomology*, **25**, 447–78.

Erickson J.M. & Feeny P. (1974) Sinigrin: a chemical barrier to the black swallowtail butterfly, *Papillo polyxenes*. *Ecology*, **55**, 103–11.

Fisk J. (1980) Effects of HCN, phenolic acids and related compounds in *Sorghum bicolor* on the feeding behaviour of the planthopper *Peregrinus maidis*. *Entomologia Experimentalis et Applicata*, **27**, 211–22.

Forbes L. & Beck E.W. (1954) A rapid biological technique for screening blue lupine populations for low alkaloidal plants. *Agronomy Journal*, **46**, 528–9.

Fowden L. & Lea P.J. (1979) Mechanism of plant avoidance of autotoxicity by secondary metabolites, especially by nonprotein amino acids. In *Herbivores: Their Interaction with Secondary Plant Metabolites* (eds G. A. Rosenthal & D. H. Janzen), pp. 135–60.

Fowden L., Lewis D. & Tristram H. (1967) Toxic amino acids: their action as antimetabolites. *Advances in Enzymology*, **29**, 89–163.

Fox L.R. (1981) Defence and dynamics in plant-herbivore ecosystems. *American Zoologist*, **21**, 853–64.

Gatehouse A.M.R., Gatehouse J.A., Dobie P., Kilminster A.M. & Boulter D. (1979) Biochemical basis of insect resistance in *Vigna unguiculata*. *Journal of the Science of Food and Agriculture*, **30**, 948–58.

Gewitz H.S., Lorimer G.H., Solomonson L.P. & Vennesland B. (1974) Presence of HCN in *Chorella vulgaris* and its possible role in controlling the reduction of nitrate. *Nature*, **249**, 79–81.

Golley F.B. (1972) Energy flux in ecosystems. In *Ecosystem Structure and Function* (ed. J. A. Wiens), pp. 69–90. Oregon State University Press, Corvallis.

Gordon H.T. (1961) Nutritional factors in insect resistance to chemicals. *Annual Review of Entomology*, **6**, 27–54.

Green T.R. & Ryan C.A. (1972) Wound-induced proteinase inhibitor in plant leaves: a possible defence mechanism against insects. *Science*, **175**, 776–7.

Gustafsson A. & Gadd I. (1965) Mutations and crop improvement II. The genus *Lupinus* (Leguminosae). *Hereditas*, **53**, 15–39.

Harley K.L.S. & Thorsteinson A.J. (1967) The influence of plant chemicals on the feeding behaviour, development and survival of the two striped grasshopper *Melanoplus bivattatus* (Say), Acrididae, Orthoptera. *Canadian Journal of Zoology*, **45**, 305–19.

Hedin P.A., Jenkins J.N. & Maxwell F.G. (1977) Behavioural and developmental factors affecting host plant resistance to insects. In *Host Plant Resistance to Insects* (ed. P. A. Hedin), pp. 231–75. American Chemical Society, Washington.

Hegnauer R. (1976) Accumulation of secondary products and its significance for biological systematics. *Nova Acta Leopoldina*, **7**, 45–76.

Hsiao T.H. & Fraenkel G. (1968) The role of secondary plant substances in food specificity of the Colorado potato beetle. *Annals of the Entomological Society of America*, **61**, 485–93.

Janzen D.H., Juster H.B. & Liener I.E. (1976) Insecticidal action of the phytohaemaglutinin in black beans of a bruchid beetle. *Science*, **192**, 795–6.

Janzen D.H., Juster H.B. & Bell E.A. (1977) Toxicity of secondary compounds to the seed-eating larvae of the bruchid beetle *Callosobruchus maculatus*. *Phytochemistry*, **16**, 223–7.

Jermy T. (1966) Feeding inhibitors and food preference in chewing phytophagous insects. *Entomologia Experimentalis et Applicata*, **9**, 1–12.

Jones C.G. & Firn R.D. (1979) Some allelochemicals of *Pteridium aquilinum* and their involvement in resistance to *Pieris brassicae*. *Biochemical Systematics and Ecology*, **7**, 187–92.

Jones D.A. (1966) On the polymorphism of cyanogenesis in *Lotus corniculatus*. I. Selection by animals. *Canadian Journal of Genetics and Cytology*, **8**, 556–67.

Jones D.A. (1972) Cyanogenic glycosides and their function. In *Phytochemical Ecology* (ed. J.B. Harborne), pp. 103–24. Academic Press, London & New York.

Jones D.A. (1973) Coevolution and cyanogenesis. In *Taxonomy and Ecology* (ed. V.H. Heywood), pp. 213–42. Academic Press, London & New York.

Jones D.A., Keymer R.J. & Ellis W.M. (1978) Cyanogenesis in plants and animal feeding. In *Biochemical Aspects of Plant and Animal Coevolution* (ed. J.B. Harborne), pp. 21–34. Academic Press, London & New York.

Jones D.A., Parsons J. & Rothschild M. (1962) Release of hydrocyanic acid from crushed tissues of all stages in the life cycle of species of Zygaeninae (Lepidoptera). *Nature*, **193**, 52–3.

Krieger R.J., Feeny P.P. & Wilkinson C.F. (1971) Detoxification enzymes in the guts of caterpillars: an evolutionary answer to plant defense? *Science*, **172**, 579–81.

Kuhn R., Löw I. & Gauhe A. (1950) The alkaloid glycoside of *Lycopersicon esculentum* var pruniforme and its action on the larvae of the potato beetle. *Chemische Berichte*, **83**, 448–52.

Lane C. (1962) Notes on the common blue, *Polyommatus icarus* (Rott.) egg-laying and feeding on the cyanogenic strains of bird's foot trefoil (*Lotus corinculatus* L.) *Entomologist's Gazette*, **13**, 112–16.

Lane N.J. & Skaer H.LeB., (1980) Intercellular junctions in insect tissues. *Advances in Insect Physiology*, **15**, 35–213.

Launois-Luong M-H. (1976) Étude comparée des préférences alimentaires de 26 acridiens du Sahel vis-à-vis des feuilles d'arachide, de haricôt, de mil et de sorgho. *Groupement d'Études et de Recherches pour le Développement de l'Agronomie Tropicale*. Ref. C/003/I.D.31.

Lawton J.H. (1978) Host-plant influences on insect diversity: the effects of space and time. In *Diversity of Insect Faunas* (eds L. A. Mound & N. Waloff), pp. 105–25. Blackwell Scientific Publications, Oxford.

Lawton J.H. & McNeill S. (1979) Between the devil and the deep blue sea; on the problem of being a herbivore. In *Population Dynamics* (eds R.M. Anderson, B.D. Turner & L.R. Taylor), pp. 223–44. Blackwell Scientific Publications, Oxford.

Levin D.A. (1976) The chemical defences of plants to pathogens and herbivores. *Annual Review of Ecology and Systematics*, **7**, 121–59.

Levin D.A. & York B.M. (1978) The toxicity of plant alkaloids: an ecogeographic perspective. *Biochemical Systematics and Ecology*, **6**, 61–76.

Liener I.E. (1979) Phytohemagglutinins. In *Herbivores: Their Interaction with Secondary Plant Metabolites* (eds G.A. Rosenthal & D.H. Janzen), pp. 567–98. Academic Press, London & New York.

Ma W.C. (1972) Dynamics of feeding responses in *Pieris brassicae* Linn. as a function of chemosensory input: a behavioural, ultrastructural and electrophysiological study. *Mededelingen Landbouwhogeschool Wageningen* 72–11.

McKey D. (1974) Adaptive patterns in alkaloid physiology. *American Naturalist*, **108**, 305–20.

McKey D. (1979) The distribution of secondary compounds within plants. In *Herbivores: Their Interaction with Secondary Plant Metabolites* (eds G.A. Rosenthal & D.H. Janzen), pp. 55–133. Academic Press, London & New York.

McKey D. (1980) Origins of novel alkaloid types: a mechanism for rapid phenotypic evolution of plant secondary compounds. *American Naturalist*, **115**, 745–9.

McNeill S. & Southwood T.R.E. (1978) The role of nitrogen in the development of insect/plant relationships. In *Biochemical Aspects of Plant and Animal Coevolution* (ed. by J.B. Harborne), pp. 77–98. Academic Press, London & New York.

Maddrell S.H.P. (1981) The functional design of the insect excretory system. *Journal of Experimental Biology*, **90**, 1–15.

Maddrell S.H.P. & Gardiner B.O.C. (1976) Excretion of alkaloids by Malpighian tubules of insects. *Journal of Experimental Biology*, **64**, 267–81.

Mattson M.J. & Addy N.D. (1975) Phytophagous insects as regulators of forest primary production. *Science*, **190**, 515–22.

Miller J.D., Gibson P.B., Cope W.A. & Knight W.E. (1975) Herbivore feeding on cyanogenic and acyanogenic white clover seedlings. *Crop Science*, **15**, 90–1.

Morrow P.A. & LaMarche V.C. (1978) Tree ring evidence for chronic suppression of productivity in subalpine *Eucalyptus*. *Science*, **201**, 1244–6.

Müller E. (1976) Principles in transport and accumulation of secondary products. *Nova Acta Leopoldina*, **7**, 123–8.

Navon A. & Bernays E.A. (1978) Inhibition of feeding in acridids by non-protein amino acids. *Comparative Biochemistry and Physiology*, **59A**, 161–4.

Nayer J.K. & Fraenkel G. (1963) The chemical basis of host selection in the Mexican bean beetle *Epilachna varivestis* (Coleoptera, Coccinellidae). *Annals of the American Entomological Society*, **56**, 174–8.

Nayer J.K. & Thorsteinson A.J. (1963) Further investigations into the chemical basis of insect–host relationships in an oligophagous insect, *Plutella maculipennis* (Curtis). *Canadian Journal of Zoology*, **41**, 923–9.

Oomen H.A.P.C. (1964) Vegetable greens, a tropical undevelopment. *Chronica Horticulturae*, **4**, 3–5.

Owen D.F. & Weigert R.G. (1976) Do consumers maximise plant fitness? *Oikos*, **27**, 488–92.

Parsons J. & Rothschild M. (1964) Rhodanese in the larvae and pupae of the common blue butterfly *Polyommatus icaris* (Rott.) (Lepidoptera). *Entomologist's Gazette*, **15**, 58–9.

Pawlowski S.H., Reigert P.W. & Krymanski J. (1968) Use of grasshoppers in bioassay of thioglycosides in rape seed (*Brassica napus*). *Nature*, **220**, 174–5.

Read D.P., Feeny P. & Root R.B. (1980) Habitat selection by the aphid parasite *Diaeretiella rapae* (Hymenoptera: Braconidae) and hyperparasite *Charips brassicae* (Hymenoptera: Cynipidae). *Canadian Entomologist*, **102**, 1567–78.

Rehr S.S., Feeny P.P. & Janzen D.H. (1973) Chemical defense in Central American non-ant-acacias. *Journal of Animal Ecology*, **42**, 405–16.

Rehr S.S., Bell E.A., Janzen D.H. & Feeny P.P. (1973) Insecticidal amino acids in legume seeds. *Biochemical Systematics*, **1**, 63–7.

Robinson T. (1979) The evolutionary ecology of alkaloids. In *Herbivores: Their Interaction with Secondary Plant Metabolites* (eds G.A. Rosenthal & D.H. Janzen), pp. 413–48. Academic Press, London & New York.

Rosenthal G.A. (1972) Investigations of canavinine biochemistry in the jack bean plant *Canavalia ensiformis* (L.)D.C. II. Canavanine biosynthesis in the developing plant. *Plant Physiology*, **50**, 328–31.

Rosenthal G.A. (1977a) The biological effects and mode of action of L-canavanine, a structural analogue of L-arginine. *Quarterly Review of Biology*, **52**, 115–78.

Rosenthal G.A. (1977b) Nitrogen allocation for L-canavanine synthesis and its relationship to chemical defense of the seed. *Biochemical Systematics and Ecology*, **5**, 219–20.

Rosenthal G.A. & Bell E.A. (1979) Naturally occurring, toxic nonprotein amino acids. In *Herbivores: Their Interactions with Secondary Plant Metabolites* (eds G.A. Rosenthal & D.H. Janzen), pp. 353–85. Academic Press, London & New York.

Rosenthal G.A. & Dahlman D.L. (1975) Non-protein amino acid–insect interactions II. Effects of canaline urea cycle amino acids on growth and development of the tobacco horn worm, *Manduca sexta* L (Sphingidae). *Comparative Biochemistry and Physiology*, **52**, 105–8.

Rosenthal G.A., Janzen D.H. & Dahlman D.L. (1977) Degradation and detoxification of canavanine by a specialised seed predator. *Science*, **196**, 658–60.

Rothschild M., Aplin R.T., Cockrum P.A., Edgar J.A., Fairweather P. & Lees R. (1979) Pyrrolizidine alkaloids in arctiid moths (Lep.) with a discussion on host plant relationships and the role of these secondary plant substances in the Arctiidae. *Biological Journal of the Linnean Society*, **12**, 305–26.

Ryan C.A. (1979) Proteinase inhibitors. In *Herbivores: Their Interaction with Secondary Plant Metabolites* (eds G.A. Rosenthal & D.H. Janzen), pp. 559–656. Academic Press, London & New York.

Schlesinger H.M., Applebaum S.W. & Birk Y. (1976) Effect of β-cyanoalanine on the water balance of *Locusta migratoria*. *Journal of Insect Physiology*, **22**, 1421–5.

Schlesinger H.M. Applebaum S.M. & Birk Y. (1977) Comparative uptake, tissue permeability and turnover of β,γ-diaminobutyric acid and β-cyanoalanine in locusts. *Journal of Insect Physiology*, **23**, 1311–14.

Schnepf E. (1976) Morphology and cytology of storage spaces. *Nova Acta Leopoldina*, **7**, 23–44.

Schoonhoven L.M. (1972) Secondary plant substance and insects. In *Structural and Functional Aspects of Phytochemistry* (eds C.V. Runekless & T.C. Tso), pp. 197–224.

Schoonhoven L.M. & Derkson-Koppers I. (1973) Effects of secondary plant substances on drinking behaviour in some Heteroptera. *Entomologia Experimentalis et Applicata*, **16**, 141–5.

Schoonhoven L.M. & Derksen-Koppers I. (1976) Effects of some allelochemicals on food uptake and survival of a polyphagous aphid, *Myzus persicae*. *Entomologia Experimentalis et Applicata*, **19**, 52–6.

Schoonhoven L.M. & Jermy T. (1977) A behavioural and electrophysiological analysis of insect feeding deterrents. In *Crop Protection Agents—Their Biological Evaluation* (ed. N.R. McFarlane), pp. 133–45. Academic Press, London & New York.

Schoonhoven L.M. & Meerman J. (1978) Metabolic cost of neutralising allelochemics. *Entomologia Experimentalis et Applicata*, **24**, 689–93.

Schreiber K. (1958) Über einige Inhaltsstoffe der Solanacean und ihre bedeutung für die Kartoffelkäferresistenz. *Entomologia Experimentalis et Applicata*, **1**, 28–37.

Schroeder L.A. (1978) Consumption of black cherry leaves by phytophagous insects. *American Midland Naturalist*, **100**, 294–306.

Scriber J.M. (1978) Cyanogenic glycosides in *Lotus corniculatus*: their effect upon growth, energy budget, and nitrogen utilisation of the southern army worm *Spodoptera eridania*. *Oecologia*, **34**, 143–55.

Seigler D.S. (1977) The naturally occurring cyanogenic glycosides. In *Progress in Phytochemistry*, Vol. 4 (eds L. Reinholt, J.B. Harborne & T. Swain), pp. 83–120. Pergamon Press, Oxford.

Self L., Guthrie F. & Hodgson E. (1964) Metabolism of nicotine in tobacco-feeding insects. *Nature*, **204**, 300–1.

Smolenski S.J., Silinis H. & Farnsworth N.R. (1973) Alkaloid screening III. *Lloydia*, **36**, 359–89.

Smolenski S.J., Silinis H. & Farnsworth N.R. (1974a) Alkaloid screening. IV. *Lloydia*, **37**, 30–61.

Smolenski S.J., Silinis H. & Farnsworth N.R. (1974b) Alkaloid screening V. *Lloydia*, **37**, 506–36.

Smolenski S.J., Silinis H. & Farnsworth N.R. (1975) Alkaloid screening VI. *Lloydia*, **38**, 411–41.

Springett B.P. (1978) On the ecological role of insects in Australian *Eucalyptus* forests. *Australian Journal of Ecology*, **3**, 129–39.

Strong D.R. (1979) Biogeographic dynamics of insect-host plant communities. *Annual Review of Entomology*, **24**, 89–119.

Tapper B.A. & Reay P.F. (1973) Cyanogenic glycosides and glucosinolates. In *Chemistry and Biochemistry of Herbage*, Vol. I (eds T.W. Butler & R.W. Bailey), pp. 447–77. Academic Press, London & New York.

Tingey W.M., Mackenzie J.D. & Gregory P. (1978) Total foliar glycoalkaloids and resistance of wild potato species to *Empoasca fabae* (Harris). *American Potato Journal*, **85**, 577–85.

van Emden H.F. (1978) Insects and secondary plant substances—an alternative viewpoint with special reference to aphids. In *Biochemical Aspects of Plant and Animal Coevolution* (ed. J.B. Harborne), pp. 309–26. Academic Press, London & New York.

van Etten C.N. & Tookey H.L. (1979) Chemistry and biological effects of glucosinolates. In *Herbivores: Their Interaction with Secondary Plant Metabolites* (eds G. A. Rosenthal & D. H. Janzen), pp. 471–500. Academic Press, London & New York.

Verschaeffelt E. (1910) Cause determining the selection of food in some herbivorous insects. *Proceedings of the Academy of Science, Amsterdam*, **13**, 536–42.

Waloff N. & Richards O.W. (1977) The effect of insect fauna on growth, mortality, and natality of broom, *Sarothamnus scoparius*. *Journal of Applied Ecology*, **14**, 787–98.

Wearing C.H. (1968) Responses of aphids to pressure applied to liquid diet behind parafilm membrane. *New Zealand Journal of Science*, **11**, 105–21.

Weiel J. & Hapner K.D. (1976) Barley proteinase inhibitors: a possible role in grasshopper control? *Phytochemistry*, **15**, 1885–7.

Woodhead S. & Bernays E.A. (1977) Changes in the release rates of cyanide in relation to palatability of *Sorghum* to insects. *Nature*, **270**, 235–6.

Woodhead S., Padgham D.E. & Bernays E.A. (1980) Insect feeding on different sorghum cultivars in relation to cyanide and phenolic acid content. *Annals of Applied Biology*, **95**, 151–7.

Yamamoto R.T. & Fraenkel G. (1960) The suitability of tobacco for the growth of the cigarette beetle, *Lasioderma serricorne*. *Journal of Economic Entomology*, **53**, 318–84.

16. LARGE HERBIVORES AND FOOD QUALITY

MALCOLM COE
Animal Ecology Research Group, Department of Zoology, South Parks Road, Oxford OX1 3PS

SUMMARY

1. The mean body size of ungulates changes along a rainfall gradient from semi-arid to humid climates.
2. Rumen size is also related to body size, but food intake increases as a percentage of body mass with increased body size.
3. Large ruminants are classified into bulk roughage feeders, concentrate selectors and intermediate feeders. This classification is based upon their stomach structure which influences intake and throughput.
4. Grazing ungulates coexist in a habitat by a variety of means, a paramount feature of which is the selection of plant parts rather than just species. This mode of selection leads to degrees of feeding facilitation or grazing succession. Browsing mammals or concentrate selectors show little overlap in their preferred species between seasons.
5. Cokes hartebeest and female buffalo show high energy deficits in the dry season.
6. Ungulates show a high degree of intake compensation with varying levels of crude protein in the diet down to a level of 5% crude protein.
7. Faecal protein appears to be a good predictor of dietary crude protein above 5%.
8. Morphological indices of condition show significantly lower body condition in territorial compared with non-territorial Cokes hartebeest. Bone marrow fat, kidney fat indices, and physiological measurements of condition are described.
9. In addition to selection for energy and protein, large ungulates also appear to be able to select for minerals such as sodium and calcium.
10. In terms of the time budgets of foraging ungulates the greater kudu increases its foraging time in dry season. With increased foraging time there is a corresponding decrease in the time devoted to other miscellaneous motor activities.
11. Models for buffalo and Cokes hartebeest are discussed in relation to selection for nutrients and the control of food intake.

INTRODUCTION

The modern ungulates comprise 94 genera (Walker 1964) which, although first emerging as the Condylarths in the Palaeocene period, did not show really rapid radiation until the Miocene (Young 1950). This rapid evolution may be related to the occurrence of a somewhat drier climate during this period. It seems likely that the expansion of the plains habitats represented a strong selection pressure in favour of the evolution of grazing mammals which developed deep hypsodont molars which could withstand the abrasive action of grasses with high silica levels. Indeed, when we observe that at the present time high herbivore densities lead to the selection of procumbent morphs of many grass species on the Serengeti Plains (McNaughton 1979), it becomes apparent that the evolution of the savanna grasses must have coevolved with the rapidly radiating large herbivores both during and since the Miocene.

Vrba (1980) has described the rapid radiation of the *Alcelaphini* (blesbok, hartebeest and wildebeest) and the *Aepycerotini* (impala) during the last six million years. During this period the alcelaphines, which are specialist bulk and roughage feeders (grazers), had some of the most advanced stomach forms of all the large herbivores (Hoffman 1973), and evolved up to 27 species, of which seven forms still exist. These creatures were widespread on the African continent in the drier and more open parts of the habitat spectrum. By contrast, the *Aepycerotini* only appear to have evolved two species during the same period, of which only the impala (*Aepyceros melampus* Lichtenstein) is still extant. This herbivore has the longest virtually unchanged history of all the groups mentioned above, a fact which is probably related to its occupation of an ecotone boundary, resulting in its present almost continuous distribution as a single species from the Equator to South Africa (Coe 1980a). This distribution pattern may in turn be related to the fact that it is capable of utilizing a wide range of different food plants ranging from almost entirely grass to predominantly browse in different parts of its range (Hoffman & Stewart 1972).

Today the African continent still has the most diverse ungulate fauna in the world, and it is the study of these mammals that has provided us with the most detailed information on feeding and the importance of food quality to wild populations.

BODY SIZE, HABITAT AND FOOD INTAKE

When we examine the whole spectrum of African ungulates, we are dealing with creatures that range in size from the diminutive Royal antelope

(*Neotragus pygmaeus* L.) which weighs up to 2.5 kg, to the African elephant (*Loxodonta africana* Blumenbach) with a unit weight of 1725 kg, though large bulls may exceed 6000 kg. Since the energy demands of mammals may be related to their body mass (basal metabolic rate BMR = 70 kg$W^{0.75}$Kcal day^{-1}) the smaller species have a much higher BMR in relation to their body size due to their large surface area/volume ratio (Kleiber 1961). Total energy expenditure by large herbivores has been calculated as BMR × 3 (Lamprey 1964; Eltringham 1974), but a more likely estimate for free-living ungulates is BMR × 1.5 (Moen 1973).

Along a gradient from semi-arid savannas, with their high temperatures and low rainfall, to areas of high rainfall we pass through habitats dominated by low scrub, open grassland, and woodland to various types of forest. In such a transition we note that the annual primary production/biomass ratio changes from 0.30 (30%) in grassland to 0.06 (6%) in forest habitats (Coe 1980a). This change in turnover time may be related to the high standing crop biomass of wood in forested areas, compared with the low levels achieved in grassland. The effect of these differences is that whereas a high percentage of the standing crop is available as food for large herbivores in grassland, in woodland and forest a large part of the annual production is potentially unavailable as food (in mixed woodland and grassland savanna intermediate turnover rates are to be expected). Thus on the open long and short grassland of the Serengeti Plains, the white bearded wildebeest (*Connochaetes taurinus* Burchell) may remove up to 85% of the grass production on the western edge of the plains (McNaughton 1976). They will at the same time leave up to 140 g m^{-2} behind as dung, which will be rapidly recycled and the nutrients released for further plant growth by dung beetles in the wet season and by termites in the dry season (Coe 1980a).

Coe (1980b) in a study of African large herbivore communities has demonstrated that mean body mass increases along a rainfall gradient from arid habitats with little rainfall to those receiving up to 1000 mm. Above this level of precipitation we encounter forest environments where the mean herbivore body mass falls rapidly, but there is little reliable community data available on herbivore densities/biomass for really humid environments. The observed increase in mean body mass is due to an increasing number of larger species, although the actual species composition changes very little over the lower part of the range.

Thackeray (1980) has demonstrated for the ungulates of South West Africa that mean ungulate body mass is linearly related to primary production, which in turn may be related to net above-ground primary production for 16 datum points whose rainfall varied from 50 mm year^{-1} to just over 500 mm year $^{-1}$. Thus mean ungulate body mass may be calculated from the expression:

$$MUB = (0.26 \times MAR) + (1.05 \times MAT) + 18.6 \text{ where}$$

MUB = mean ungulate body mass
MAR = mean annual rainfall
MAT = mean annual temperature.

The significance of increased body size along a presumed stability gradient from arid to humid environments must be sought in the ecology of the habitats located along its length. In the more arid zones large herbivores have evolved complex means of conserving water and withstanding heat stress, but although larger body size would confer advantages in being able to range over larger areas in search of food (Pennycuick 1979), the generally low and erratic availability of food makes it impossible to support a really large body size.

If we ask ourselves why body size should increase to a peak in moderate rainfall that will support a wooded savanna, and then fall rapidly in forest environments, we must seek the probable answer in the life history strategies of both the animals and the plants. We have seen that the environments dominated by large herbivores occur at the lower end of the rainfall spectrum. Above these levels an increasing quantity of nutrients is tied up in woody material, which must be an important contributing factor in explaining why the more humid environments support low standing crop biomasses of large mammals and which also attain a lower mean unit body mass.

In the same way that forest vegetation has a slower turnover time than grassland, increasing body size in large herbivores will also be accompanied by longer generation times (Western 1979). In consequence their production to biomass ratio will also fall, so that while a small gazelle population may be expected to turnover 78% of their biomass each year, the eland (*Taurotragus oryx* Pallas) will turnover 16%, while the elephant will turnover 7% (Coe 1980b & c). Thus in low rainfall areas whose vegetation has a rapid turnover time a significant proportion of the available nutrients is immobilized in large herbivores with low turnover rates while in high rainfall areas the opposite appears to be the rule.

Since BMR is a function of body size, we might expect to find that the amount of food consumed by a large herbivore is also related to body size. Hoffman (1973) in a detailed study of the ruminant stomach gives details of the average body mass and the stomach volume of 27 species of East African ruminant (Table 16.1). If we regress body mass on stomach volume we obtain a significant linear relationship ($n = 27$; $r = 0.96$; $P = 0.001$). Stomach volume may be calculated from the expression:

$$Y \text{ (stomach volume)} = 0.1616 \times \text{(body mass)} + 4.0514.$$

If we are to be able to determine the effect that large herbivores have on available primary production it is important to estimate the amount of food they consume. Sinclair (1974, 1975) in his studies of the African buffalo (*Syncerus caffer* Sparrman) has calculated that these mammals need the equivalent of 2% of their body mass per day. This figure of 2% is equivalent to 9% of their metabolic weight ($W^{0.75}$), so he used this figure to calculate the daily food intake of the Serengeti ungulate species. This calculation assumes that the ratio of food consumed to metabolic weight remains constant over the whole size range.

In studies of defaecation in large ungulates the author has obtained measurements of daily faecal output (dry wt.) for seven species of domestic

TABLE 16.1. Body mass and stomach volume of 27 species of East African ruminant. The species groups are divided on the left into their major feeding categories (after Hoffman 1973).

	Ungulate species	Average body size (kg)	Total stomach value (l)
Concentrate selectors, fruit and dicotyledon foliage selectors	Guenthers dikdik	4	0.85
	Kirks dikdik	5	1.05
	Suni	6	1.45
	Klipspringer	11	2.81
	Grey duiker	14	3.41
	Harveys duiker	16	5.49
	Bushbuck	49	7.75
Concentrate selectors, tree and shrub foliage eaters	Giraffe	750	110.00
	Lesser kudu	90	14.10
	Greater kudu	213	50.15
	Gerenuk	43	6.86
Bulk and roughage eaters (grass eaters)	Buffalo	599	119.00
	Oribi	16	4.39
	Bohor reedbuck	45	10.56
	Uganda kob	79	10.81
	Waterbuck	200	44.70
	Wildebeest	182	44.30
	Kongoni (Cokes hartebeest)	138	25.60
	Jacksons hartebeest	174	44.60
	Topi	119	34.00
	Mountain reedbuck	24	7.97
	Oryx	181	39.10
Intermediate feeders	Impala	55	13.10
	Thomsons gazelle	21	6.17
	Eland	465	59.50
	Grants gazelle	55	14.10
	Steenbok	10	2.68

and wild ungulate. If we assume that ruminants have a gross assimilation efficiency of 75% (five species), the zebra 50% (Louwe, pers. comm.) and the elephant 22% (Rees 1977), we may calculate the dry weight of food consumed per day for these species (Table 16.2). These data bear a strong positive log linear correlation ($r=0.977$; $P=0.001$) so that the amount of food consumed may be calculated for a whole range of large ungulates by the expression:

$$Y \text{ (food consumed kg dry wt. day}^{-1}) = 1.33 \log_e W - 5.75.$$

The estimates obtained from this expression fall close to those given in the middle weight range of Sinclair's (1975) Serengeti ungulates. At the lower end

TABLE 16.2. Body size and daily faecal output (dry wt.) for seven species of wild and domestic ungulate.

Species	Body mass (kg)	Dung production (kg dry wt. day^{-1})	Food intake (kg dry wt. day^{-1})
White tailed deer	26	0.13	0.30
Sheep	50	0.48	0.64
Blesbok	50	0.43	0.61
Oryx	119	1.38	1.56
Zebra	200	4.12	4.71
Cow	376	4.15	5.42
Elephant	2482	20.32	142.41

of the weight scale, however, they are only one third of those calculated as 9% $W^{0.75}$, while at the upper end of the scale (elephant) they are three times larger. It is interesting to note that there is a strong indication that the food consumed as a percentage of W increases with body size, being 1.15% for the white tailed deer (*Odocoileus virginianus* Boddaert) (26 kg) and 5.73% for the African elephant (2482 kg). This latter figure is close to that calculated by Laws, Parker & Johnstone (1975) which fell between 4.2 and 5.6%.

FOOD SELECTION BY LARGE HERBIVORES

Until Hoffman (1973) studied the stomach of East African ruminants the feeding habits of large herbivores had been considered to be broadly divisible into groups of browsers, grazers and intermediate feeders (Hoffmann & Stewart 1972). These authors concluded that the East African ruminants could more accurately be divided into the following categories:

1. Bulk and roughage feeders (i.e. grazers)

(a) roughage grazers
(b) fresh grass grazers dependent upon water
(c) dry region grazers
2. Selectors of juicy, concentrated herbage
(a) tree and foliage feeders
(b) fruit and dicotyledon (tree, shrub and forb) foliage selectors
3. Intermediate feeders
(a) preferring grasses
(b) preferring forbs, shrubs and tree foliage

Hoffman (1973) demonstrated that the structure of the rumen closely reflected an animal's feeding habits, so that bulk and roughage feeders had a capacious rumen with uneven papillation and an omasum whose surface mucosa was greatly enlarged. These features result in a maximum delay of their coarse food. By contrast, the selectors of juicy concentrated herbage possess a simple rumen and a small but muscular omasum which allows rapid food passage. Although intermediate feeders show features of both the other categories in their stomach structure, they show variation in this structure both with age and season. Thus the structural components of the stomach determine the physical limitation of food intake. Maloiy, Kay & Goodall (1968) showed that red deer (*Cervus elephas* L.) need to compensate for greater roughage levels in their diet by passing more food through the gut. This need was reflected in the fact that although the red deer was only 10–25% heavier than a domestic sheep, the poorer quality of its forage required that its intake be up to 50% greater.

Lamprey (1963) in an early study of 14 ungulate species in the Tarangire Game Reserve in Tanganyika counted these animals on transects that passed through a wide variety of habitats. As a result he was able to derive a pattern whereby these species could coexist. He concluded that separation was achieved by means of their preferred habitats, selection of food types, adaptation to different feeding levels (related to body size), seasonal variation in habitat occupation, and the occupation of feeding refuges in the dry season which may be assumed to be the major period of stress for all savanna herbivores.

The selection of different plant species, however, still does not answer the major problem of coexistence of several ungulate species of similar size in the same habitat. Since it seems likely that there must be at least a small degree of overlap between their preferred food plants it was necessary to examine whether several species could coexist by using the same plant species in different ways through the selection of different plant parts. Vesey-Fitzgerald (1960) was the first to suggest that the feeding habits of one species could modify the vegetation in such a way that it became suitable for other species to feed on it. He found that where elephants invaded and fed on 2 m tall swamp grasses, the resulting regeneration was utilized by buffalo, while their effects through both grazing and trampling modified the habitat such that topi (*Damaliscus lunatus* Burchell) could then graze on patches where short grasses were available.

This pattern of feeding facilitation has been termed a grazing succession by Gwynne & Bell (1968), and Bell (1970) who have demonstrated that while there is considerable overlap between the species of plant taken by zebra (*Equus burchelli* Gray), wildebeest and Thomsons gazelle (*Gazella thomsoni* Gunther), they show a considerable degree of separation on the plant parts taken. Bell (1970) suggests that the non-ruminant zebra is the first species to move into long grass habitats because they can cope with the highly lignified vegetation better than the wildebeest and the Thomsons gazelle. The zebra therefore removed the coarse top stems which made the lower leaves available to wildebeest, while after this species had lowered the grass height still further the Thomsons gazelle could feed on the short nutritious grass regrowth.

Since this time McNaughton (1976) has shown that Thomsons gazelle feed on the dense short grass whose growth has been stimulated by grazing wildebeest in long grass areas. McNaughton (1979) has also proposed a grazing model which incorporates factors such as patterns of seasonal rainfall, species composition and structure, and the genetic properties of the plants. This model takes into account the increased production due to the stimulation of growth through grazing. Maddock (1979) has suggested, however, that this grazing succession is not applicable all over the plains since the seasonal distribution of migratory species does not show complete overlap, but she does confirm that ecological separation can on the whole be explained in terms of differences in feeding.

The selection of different species, plant parts or habitats may explain the coexistence of a wide variety of herbivore species, but Sinclair (1974) has shown that it is not necessary for more than a small percentage of one species to overlap with another to cause a severe shortage of food. In this study of buffalo he observed that even though only 7.1% of the wildebeest population entered the riverine grasslands, their pressure on the available grass was responsible for a marked reduction in the food available and consequently in the condition of buffalo. Eltringham (1974) has shown a similar effect following the removal of hippo from grasslands of the Queen Elizabeth National Park (now the Rwenzori National Park) in Uganda in 1957. As an apparent consequence of the removal of hippo, the additional food made available to buffalo led to a six-fold increase in their numbers by 1968.

The problems faced by the selectors of concentrated foliage are very different from those of roughage feeders (grazers). The availability of leafy shoots differs between species, and green material is likely to be available on herbaceous and woody plants for a variable period after the quality of grass has been reduced, so that we might expect these herbivores to show considerable seasonal changes in their pattern of species selection. Hall-Martin & Basson (1975) in a study of the giraffe (*Giraffa camelopardalis* L.) in the low veldt of the Transvaal, South Africa, have shown the manner in which this species preference changes with season. During the hot wet season from November to March they concentrate on new leaf growth of *Acacia nigrescens* Oliv., *A. exuvialis* Verdoorn, *A. nilotica* (L) Del., *Combretum apiculatum* Sond., *C. zeyheri* Sond., *Terminalia prunoides* Laws., and *Ziziphus mucronata* Willd. In the cool dry season from April to July they feed on *Colophospermum mopane* (Kirk ex Benth) Kirk ex J. Leon, *Bolusanthus speciosus* (H. Bol.) Harms, *Combretum hereoense* Schinz, *C. imberbe* Wawra, *Acacia senegal* (L) Willd. and *A. hereroensis* Engl., while in the hot dry season from August to October at a time of drastic leaf fall they switch to a wide variety of evergreen leaves including *Euclea undulata* Thumb., *Colophospermum mopane*, *Maytenus senegalensis* (Lam.) Exell., *Schotia brachypetala* Sond., *Diospyros mespiliformis* Hochst ex A. DC., and *Albizia harveyi* Fourn. It will be noted that throughout the period of species selection described above only one species (*Colophospermum mopane*) shows any overlap from one seasonal period to another.

By contrast the small body size of the dikdik (*Modoqua* spp.), suni (*Nesotragus moschatus* Van Duben), the klipspringer (*Oreotragus oreotragus* Zimmerman), the duikers (*Cephalophus* spp.) and the bushbuck (*Tragelaphus scriptus* Pallas) which range in size from 4–49 kg, fall into Hoffman's (1973) classification of concentrate feeders (2b) which live largely in thick bush or forest. As we might predict from their smaller body size and consequently high metabolic rate, they are very selective in their diet, taking energy and nutrient-rich seeds, pods and fruits as well as leaves. This pattern of selection is reflected in their simple 'primitive' rumen, reflecting the more easily fermented food and its relatively fast passage time. When we noted earlier that body size decreased with increasing rainfall, it was indeed this group of herbivores that became the predominant herbivores in forest environments.

FOOD SELECTION FOR QUALITY

One of the profound differences between herbivores feeding on grass and those on browse lies in the basic difference in the anatomy of monocotyledons and dicotyledons. The primary meristem of grasses is characteristically intercalary, while that of browse plants (dicotyledons) is apical. Thus during the growing season a relatively small amount of a browse plant is likely to be growing compared with a grass, but in consequence the grass is likely to produce a somewhat more homogeneous food source than that of browse. The consequence of these differences in growth pattern is that the apices of a browse plant do not regrow, while the meristem of a grass will continue to grow when the plant is grazed (Jarman 1974). It is, however, true that plants which are favoured elements of a browse diet are likely to produce adventitious shoots more rapidly than those which are not, so that favoured browse plants are also 'stimulated' by being eaten by herbivores.

Herbivores feeding on fruits, pods or leafy shoots are likely to be more selective in their diet than a grazer since the items that they favour are more widely spaced between relatively inedible woody material, while most grass species tend to grow close together and the level of searching for the next food item is likely to be less. In this sense we might predict that time-partitioning may be much more of a critical problem for a browser than a grazer.

Jarman (1974) in discussing food selection amongst antelopes has described the potential differences in the food quality that will be available to a grazer feeding on a typical grass plant. Those feeding on terminal leaves may only obtain up to 15% crude protein (dry wt.) while those taking terminal leaves and the sheath below will only obtain 7% crude protein, and those that remove the whole plant including the stem would only obtain an average of 4%. This phenomenon explains graphically the nature of the facilitation described by Gwynne & Bell (1968), when they showed the importance of selection for plant parts by the zebra, wildebeest and Thomsons gazelle.

Although most research on optimal foraging has been directed at predators and their prey, which comprise discrete nutrient packages of generally high digestibility, natural selection must also have selected modes of food harvesting by hervibores which maximize their food intake efficiency. Belovsky (1978) in a recent study of the moose (*Alces alces* Gray) has analysed this herbivore's feeding strategy in terms of the contrast between the maximization of energy intake and the minimization of feeding time. Thus while a herbivore might be expected to use a feeding strategy that provides it with its energy needs, it will also be constrained to reduce its feeding time (by maximizing its food selection) in order to reduce its exposure to predators while feeding, and will also allow more time for behavioural activities like mate selection and care of offspring, etc. In terms of their energy requirements, Belovsky has also demonstrated that the age at weaning and maximum body size are determined by their ability to obtain sufficient energy in terms of their rumen size and the speed with which they can pass food through the gut.

Stanley-Price (1974, 1978) examined the effect of energy limitation on the Cokes hartebeest or kongoni (*Alcelaphus buselaphus cokei* Gunther.) by comparing the food intake of shot samples with those of captive hartebeest and sheep. He estimated the metabolizable energy (ME) available for production by comparing the ME intake above maintenance cost minus the estimated ME cost of activity (kcal kg $W^{0.73}$ day^{-1}). He concluded that the hartebeest had a deficit of 5.9 kcal kg $W^{0.73}$ day^{-1} in April (the end of the short dry season and the beginning of the long rains) and a much larger deficit of 62.3 kcal kg $W^{0.73}$ day^{-1} in September, at the height of the long dry season (May–November) (Table 16.3). Metabolic weight in this study was expressed as $W^{0.73}$ since this function is more commonly used in studies of cattle than the more usual $W^{0.75}$. When his calculations for ME intake were compared with data obtained from captive hartebeest and sheep (70.1 kcal kg$^{0.73}$ day^{-1}) he found that in wild hartebeest in September it fell 36.5% below this level, while in January, April and July it exceeded the level of captive animals by 92%, 44% and 118% respectively. These results clearly indicate that while there is little seasonal variation in the gross energy levels of the diet (3955–4039 cal g^{-1}), factors such as the influence of grass moisture content on the bulk of dry matter intake and varying crude fibre levels have a profound effect on an animal's gross energy budget.

TABLE 16.3. Estimated metabolizable energy (ME) intake for shot Cokes hartebeest in relation to the calculated energy available for production after deduction of the cost of maintenance and activity (after Stanley-Price 1974).

Month	ME intake kcal kg $W^{0.73}$ day^{-1}	ME intake above maintenance cost of kcal kg $W^{0.73}$ day^{-1}	ME cost of activity kg $W^{0.73}$ day^{-1}	ME available for production, kcal
January	134.4	64.3	36.7	27.6
April	100.9	30.8	36.7	−5.9
July	153.2	83.1	36.7	46.4
September	44.5	−25.6	36.7	−62.3

Sinclair (1977) has also shown that a female buffalo is likely to experience an energy deficit throughout the dry season, based upon his estimates of their energy requirement and food availability on the Serengeti Plains. Even at high levels of energy availability a female will experience a deficit in her breeding cycle between March and May, while at intermediate food levels of available energy this deficit will last from March to September. Since the largest energy demand is likely to occur during the period of lactation, we might expect this to coincide with the availability of younger green grass, which would be a period of maximum energy intake. Sinclair (op cit.) suggests, however, that lactation continues throughout the dry season when the female experiences a large energy deficit. It therefore seems likely that the female is building up her fat reserves during the period from December to February as a store to be drawn on during the dry season lactation period. A similar picture emerges for reproducing wildebeest which experience an energy deficit at and following the birth peak (February–April) and during the dry season, following conception, from June to October.

In terms of a large herbivore's need to grow and reproduce it is clearly also important that it should be able to maximize its protein intake as well as energy and minerals, etc. Since in old senescent vegetable matter the level of crude fibre increases, its digestibility is inversely related to crude fibre content (Blaxter, Wainman & Wilson 1961). In terms of both their energy and protein needs it is thus perhaps not surprising that small ungulates, which feed selectively on nutrient-rich food units such as shoots and fruits, are taking materials that are low in crude fibre. In order, however, to obtain the necessary protein levels ungulates are capable of increasing their dry matter intake as the protein level in its diet falls. Putman (1981) has shown that captive fallow deer (*Dama dama* L.) fawns increased their dry matter intake linearly with percentage protein levels in their diet between 7.33 and 5.41%. When, however, this level fell to 5%, although the dry matter intake still rose, it did so proportionately less, indicating that the animal's ability to maintain its protein intake at a nearly constant level had reached its limit. The activity of the rumen microflora is inhibited at crude protein levels below 5% (Chalmers 1961). Bredon, Harker & Marshall (1963) have also shown that the digestibility coefficient (DC) is related to the crude protein percentage (CP) in the diet of cattle ($DC = 100.89 \log CP - 44.45$), which indicates a zero digestibility at a crude protein level of 2.76% (Sinclair 1977). This phenomenon of increasing intake compensation in response to variation in forage quality is well documented for domestic ruminants (Blaxter, Wainman & Wilson 1961; Conrad, Pratt & Hibbs 1964).

One of the major problems facing the field ecologist studying large mammals is monitoring exactly which plants or their parts are being eaten. These difficulties can be reduced by using tame animals which can be followed and their feeding closely observed (Field 1970; Napier-Bax & Sheldrick 1963), although the problem of assessing the quality of food intake for a population at large remains. Sinclair (1974) tackled this problem by calculating the percentage of leaf, sheath and stem in the diet of buffalo from the analysis of rumen contents in different seasons. Crude protein levels were also analysed for the same grass parts in grass species that were commonly eaten by buffalo. Thus by multiplying the proportions of plant parts in the diet by their crude protein levels it was possible to estimate the animal's crude protein intake. These results indicated a range of crude protein intake of between 5.06 and 9.87% between November and June, with a steady decline from 5.22 to 2.18% in the dry season between July and October. This decline was closely followed by a decreasing percentage of grass leaf and an increasing percentage of stem in the diet, together with an overall decline in the crude protein levels of all three grass components. During the period (March–October) the percentage of crude protein in grass leaf decreased from 12.7 to 4.0%, while that of the stem component fell from 5.2 to 1.6%. The fact that the stem component with its low crude protein levels increased proportionately in the diet from 11% in March to 42% in October is a clear indication of not simply reduced crude protein levels in the plant components, but also the availability of the preferred leaf items on the sward. During such periods of nutrient stress the time a large herbivore would have to spend searching for scarce items of high food quality would rapidly become inefficient and result in an even greater energy and protein deficit.

Unless the field biologist is able to monitor both the nutrient availability in both the food plants and the rumen it is impossible to monitor changes in the quality of food intake. A number of authors have, however, investigated the relationship between faecal crude protein (or nitrogen) and dietary crude protein (Bredon, Harker & Marshall 1963; Erasmus, Penzhorn & Fairall 1978; Grobler 1978; Sinclair 1977; Stanley-Price 1974). These authors have shown that providing the level of dietary crude protein (CP) is above 5%, measurements of faecal crude protein (FP) bear a positive linear correlation with dietary crude protein. Sinclair (1977) using the regression $CP = 1.677\ FP - 6.93$ established by Bredon, Harker & Marshall for cattle, showed that these data for buffalo (another bovid) closely followed this regression. The regression for wildebeest (Sinclair 1977) is $CP = 2.61\ FP - 11.90$, demonstrating that these animals are achieving somewhat higher nitrogen levels than cattle or buffalo.

Grobler (1978) has established a similar regression for metabolic faecal nitrogen (MFN) and metabolic crude protein (MCP) for the sable antelope (*Hippotragus niger* Harris) with the expression $MCP = 3.179\ MFN - 2.83$ ($r = 0.69$). While Stanley-Price (1974) has demonstrated that rumen nitrogen and faecal nitrogen in Cokes hartebeest are highly correlated ($r = 0.771$; $P = 0.001$, d.f. 28) and has in addition shown that dry matter intake (DMI) may be estimated from faecal nitrogen (FN) measurements by the expression $DMI = 109.46\ FN - 61.89$ ($r = 0.681$, $P = 0.001$, d.f. 28). Hence although it is necessary to establish a regression for each species or group of species, faecal crude protein or nitrogen may be used as a good predictor of the quality of dietary intake.

Once a herbivore has satisfied its basic maintenance requirements, any surplus may be used for growth, the deposition of energy reserves and reproduction. Whether an animal is succeeding in satisfying these basic needs will be reflected in its body condition, though clearly a deterioration in its condition will not appear immediately a deficit is experienced, since it will be able to call on accumulated reserves before a reduction in body condition can be expected to manifest itself. In semi-arid regions with large variations in annual rainfall, during periods of drought when one or several rainy seasons may fail, the slow reduction in body condition followed by mass mortality is not an uncommon feature. Phillipson (1975) showed that the distribution of elephant mortality in the Tsavo National Park (East) during the drought of 1971 could be directly related to the effect of rainfall on available primary production.

The concept of body condition may be assessed as a series of categories reflecting the animal's external appearance as good, poor or bad (Albl 1971; Riney 1955, 1960; Stanley-Price 1974). A more widely employed method of assessing condition is to measure the animal's fat reserves in either the bone marrow or fat deposits associated with the kidney (Brooks, Hanks & Ludbrook 1977; Grobler 1978; Melton 1978; Ransom 1965; Riney 1955; Sinclair 1977; Sinclair & Duncan 1972). A further technique employs the measurement of physiological parameters, notably those associated with the blood (Bandy *et al.* 1957; Franzmann 1972; Melton 1975). Since, however, most of these measures vary with sex and status as well as with season they have failed to provide a really reliable basis for estimating nutrient limitation on a whole population.

The method of visually assessing body condition of known individuals (territorial males) was employed by Stanley-Price (1974) and was based upon periodic photographs taken from the side. His classification of condition was based on the appearance of the body outline in relation to anatomical features such as the sacral vertebrae, the iliac process, the scapula and the ribs. All these four major features become more visible as the animal's condition fell through four grades from excellent to poor condition. When the percentage of territorial males in poor condition (Category 4) and the ratio of precipitation to monthly evapotranspiration (P/Et) were analysed by the Spearmann rank correlation, he showed that a significant relationship existed between them and that the highest correlation occurs with the cumulative rainfall of the previous four months. Additionally he showed that the proportion of non-territorial males in excellent condition was always significantly greater than that of territorial males. This may be accounted for by assuming either that territorial males occupy poorer areas, that territorial males' food intake is reduced by the amount of time available for feeding, or that if food intake of territorial males is the same as that of non-territorial animals their increased activity accounts for their loss of condition.

The use of bone marrow fat as an index of condition has been reviewed by Brooks, Hanks & Ludbrook (1977). This index is based upon a sample of marrow removed from the centre of one of the long bones which is dried to constant weight at 100°C. Fat content is then estimated by Soxhlet extraction. Sinclair & Duncan (1972) showed that the percentage of dry mass was a good estimator of the percentage of fat present, and Brooks, Hanks & Ludbrook (1977) have shown for eight species of African ruminant that % marrow fat = % dry mass − 7. Over a large series of analyses these authors have demonstrated that the fat of the bone marrow is mobilized from the humerus and femur before the other long bones, after which the suggested order of mobilization is the metacarpus and metatarsus before the radius and ulna.

Sinclair (1977) has shown that the bone marrow fat in wildebeest lactating females fell dramatically in April when the newborn young were lactating. This month should have been the highest rainfall month which would guarantee a good supply for lactation, but in 1971 the rains were late and consequently the marrow fat reserves were mobilized, although when the rain did eventually fall in May the animals' condition rapidly improved and they regained their marrow fat. The commonest alternative measure of fat reserves in the ungulate body is that of the kidney fat index (KFI) which is measured as the ratio of perinephric fat/kidney fat × 100 (Riney 1955; Ransom 1965). Brooks, Hanks & Ludbrook (1977) discussed the use of this index as a measure of condition, and have shown that in buffalo, eland and impala the perinephric fat is mobilized before that of the bone marrow, indicating that the KFI index is likely to be a good early indicator of nutrient stress but, when this index falls below 40, bone marrow fat is a more valuable continuing estimator of falling condition. This conclusion is similar to that of Sinclair & Duncan (1972) who showed that bone marrow fat was the last fat deposit to be mobilized during periods of shortage and the first to be replaced.

These indices are therefore clearly of considerable value in studying condition, but during periods of nutritional stress they fall from high to low values very quickly. Clearly, the storage of fat must be considered to be a strategy used by large herbivores as a preparation for periods of excessive energy demands such as lactation in females or the rut in males. Thus although the animals cannot avoid every period of nutritional stress in a climatically variable environment with great seasonal extremes, this strategy must be considered one of minimizing rather than eliminating stress.

In relation to the importance of nitrogen or crude protein Melton (1978), in a study of waterbuck (*Kobus ellipsiprymnus* Ogilby), favoured the analysis of blood urea nitrogen (BUN) and serum total protein (TP) as these both reflect the level of nitrogen in the diet (Harper 1975). Melton showed that the BUN (mg. dl^{-1}) showed a rapid decline as the winter progressed in Natal, South Africa. BUN was shown to bear a significant linear relationship to total protein (g. dl^{-1}), $BUN = 0.73 + 2.19 \log_e TP$ ($r = 0.64$; $P = 0.01$).

DISCUSSION

We have seen that the body size of large herbivores has a profound influence on both their mode of feeding and food intake, so that an African elephant needs to spend up to 15 hours a day feeding (Wing & Buss 1970), while a 3 kg hyrax (*Procavia* spp; *Heterohyrax* spp.) will obtain all its food requirements in little more than one hour a day (Sale 1965). The time spent feeding each day in ruminants also varies seasonally in relation to food quality (Owen-Smith 1979). Sinclair (1977) showed that poor quality food in the dry season and a consequent decrease in food intake, resulted in buffalo spending a longer time grazing and ruminating. This need for an extra grazing period in the dry season necessitated the animals feeding in the middle of the day when they are faced with problems of heat stress as well as nutrient stress.

Although an ungulate can adjust the amount of food that passes through its stomach in relation to the energy and protein content of its forage, Belovsky (1978) has shown that the size at weaning (67–76 kg) of the moose is determined by stomach size (and presumably buccal anatomy) and its consequent inability to take in enough food independently below this limit, while the upper size limit of a bull (645 kg) is also limited in the same manner. These minimum and maximum body sizes are determined by the points at which the curve of net energy (NE) intersects that of energy metabolism (MR). The points at which the NE curve exceeds that of the MR curve by the greatest distance defines the optimum body size which for a bull moose is 250 kg and a barren cow 307 kg.

In discussing the importance of food quality it is important to remember that although it is essential that the level of digestible energy and nitrogen intake should be at such a level that the animal should maintain itself and be able to grow and reproduce, there are many other essential nutrients that all terrestrial animals must obtain through their food. It is therefore very difficult to consider a single element of food quality that the animal must be able to monitor. Westoby (1974) has constructed a model of diet selection by large generalist herbivores. This author points out that it is necessary for these animals to choose foods which give them a balanced nutrient intake across the whole spectrum of essential materials. The manner in which a large mammal may select those materials may either be through examination and rejection before ingestion or after ingestion where the nutritional properties and their effect on the animal's internal physiological state monitor the rate and level of intake. These two mechanisms may be termed, respectively, sensory and nutritional properties.

Belovsky (1978) in his study of the optimization of energy intake in the moose has also examined the manner in which this herbivore obtains its sodium needs, and has shown that the food plants that are best suited to supply their energy demands are not the same as those that satisfy their demand for sodium. Thus time expanded in satisfying energy (and protein) must be balanced with time spent ingesting materials that provide a satisfactory level of sodium.

The manner in which a large mammal satisfies its demand for a nutrient like sodium is well illustrated by a study of elephants (Weir 1971, 1972) in the Wankie National Park, Zimbabwe. This study showed that a number of elephants visiting water-holes bears a strong positive linear correlation to sodium levels in the water. If the sodium balance of an elephant is calculated, it can be shown that an elephant weighing 2041 kg (Dougall & Sheldrick 1964) will lose up to 30 g day^{-1} through its faeces, and up to 120 g day^{-1} through its urine, a total daily loss of 150 g day^{-1} which needs to be replaced if not exceeded. Weir estimates that an elephant feeding randomly on food plants containing an average sodium level would obtain no more than 14–20 g day^{-1}, while by selecting food items with high sodium levels it could increase its intake to 3.4 g h^{-1}. Since, however, the favoured water-holes in Wankie have high levels of sodium, animals drinking at these would obtain all their sodium requirements with ease, while at water-holes with low levels it would be necessary for them to supplement their sodium intake from vegetable food or mineral licks. This postulate is confirmed by the observation that while elephant numbers are positively correlated with sodium levels in the water, the number of salt licks around water-holes is strongly negatively correlated with elephant numbers (Weir 1971, 1972). We must therefore assume that the elephant is internally monitoring its sodium intake after the manner proposed by Westoby (1974).

In a similar manner McCullagh (1973) has studied the potential importance of essential fatty acids (linoleic and linolenic acid) in the diet of elephants and has suggested that elephant damage to trees in the Murchison Falls (now Kabalega Falls) National Park, Uganda and the Tsavo National Park, Kenya may be related to high levels of these nutrients in *Terminalia* species in the former area and baobab (*Adansonia digitata* L.) in the latter. Napier-Bax & Sheldrick (1963) have also pointed out that high calcium levels in trees commonly damaged by elephants in Tsavo may account for these animals' preference for these species. Coe (1980b) has also observed that the large amounts of calcium immobilized in the bones of large mammals in Tsavo and low levels of soil calcium may be a factor responsible for the apparent selection of calcium by elephants in their diet.

In a recent study of the greater kudu (*Tragelaphus strepsiceros* Pallas) in South Africa, Owen-Smith (1979) has assessed the feeding efficiency of large herbivores in terms of accepted food abundance, expressed as feeding time achieved per unit distance covered, and food ingestion rate expressed as the proportion of foraging time devoted to feeding. He showed that in a favoured hill slope base ecotone habitat the daylight time devoted to foraging increased from 54.1% in the late wet season (January–March) to 63.5% in the late dry season, while in *Acacia nigrescens* savanna, a less favourable habitat, the daylight foraging time in the late dry season was significantly greater (77.1%). There was, as might be expected, no difference in the foraging time in the late wet season between the two habitats. If we presume that the time available for activities other than feeding may be counter-productive in a period of potential nutrient stress in the late dry season we would expect to find that the time devoted to activities other than feeding should be significantly lower at this time. Owen-Smith (1979) showed that in the transition from the late wet season (January–March) to the height of the dry season (September) the amount of time spent moving (searching time) increased from 15.3 to 22.1% with a proportional decrease in the time devoted to miscellaneous activities, such as standing alert and grooming, from 15 to 8.9% in the hill-slope base ecotone. In the *A. nigrescens* savanna habitat the time devoted to miscellaneous activities declined from 15.1 to 7.1% over the same period. Hence during the dry season there was apparently a considerable increase in the energetic costs of foraging. The difference between the preferred habitat and the less favourable habitat is well illustrated by the difference in the movement rate while foraging, which between the wet and dry season was 8.8–13.8 steps per minute in the hill-slope base ecotone and averaged 19.4 steps per minute in the *A. nigrescens* savanna habitat.

Freeland & Janzen (1974) have drawn attention to the problems herbivorous mammals face in dealing with secondary plant compounds, and suggest that the unique nature of these materials must place further constraints on their patterns of food selection. While it is probable that the smaller species of herbivorous mammal with high metabolic rates and specialized feeding selection are required to solve these problems, it seems likely from the studies of Belovsky (1978), Owen-Smith (1979) and Sinclair (1974) that the protein and minerals are of much greater importance to large herbivorous mammals than that of secondary plant compounds in the evolution of their feeding strategies.

If we accept that the feeding strategies of a large herbivore are directed towards maximizing food intake and minimizing the time spent feeding, then it should be possible to observe the manner in which nutrients may operate as a regulating factor on such populations. Sinclair (1974, 1975, 1977) has modelled the many factors responsible for the natural regulation of a buffalo population. This model shows the complicated web of cause and effect which operates in limiting a population's potential, but paramount of all these factors is its food supply which is determined either directly through the effect of climate (mainly rainfall or, more correctly, evapotranspiration), or by seasonal competition with other species such as the wildebeest in riverine woodland. Sinclair's study of the relationship between the buffalo of the Serengeti Plains and Mount Meru and their food supply graphically illustrates the danger of assuming that studies of a species in one habitat can be extrapolated for that species throughout its range. The Mount Meru habitat was located in montane pasture at 1800 m altitude and the quantity of available food was measured directly. Estimates of the minimum food requirement of the buffalo population indicated that available grass growth was less than the populations requirement in the dry season of 1968 and 1969, although at this altitude grass quality remained high and did not therefore show the marked seasonal variation in quality recorded on the Serengeti. Thus although the pattern of nutrient availability was very different in the two areas, the ultimate effect of intake limitation was the same, suggesting that resource limitation may therefore be a common feature for buffalo in East Africa, where its range extends from semi-arid savanna with barely 500 mm rainfall per year to montane forest where rainfall may exceed 2500 mm per year.

We have seen that selection for crude protein is closely related to the main growth season of the herbivore's food plants. These seasonal changes in protein level are virtually the same whether we are considering temperate zone large mammals or tropical species. Rushworth (1975) has noted that the peaking of crude protein levels in browse plants in the Wankie Game Reserve, Zimbabwe is later than that of grasses but persists longer, a fact that may have a profound influence on the timing of breeding in grazing and browsing species. The problems faced by the large herbivore in obtaining a balance between its energy, protein and other nutrient requirements, requires a flexible strategy which will allow it to detect, select, and monitor all these requirements, although, clearly, unless it can satisfy its energy and protein requirements as a primary aim of its feeding strategy, it will be unable to achieve a suitable maintenance level and in consequence be unable to grow or reproduce.

The model proposed by Stanley-Price (1974, 1977, 1978) for the Cokes hartebeest attempts to indicate the relationship of the environment with the herbivore's dry matter intake and the digestive tract (Fig. 16.1). The water content and the particle specific gravity of savanna grasses declines with the progression of the dry season, and the food retention time increases. Total dry matter in the rumen, however, increases as the nitrogen level decreases, a fact which is of great importance to the herbivore in maintaining a fairly constant level of nitrogen in the rumen (Putman 1981) since rumen microorganism

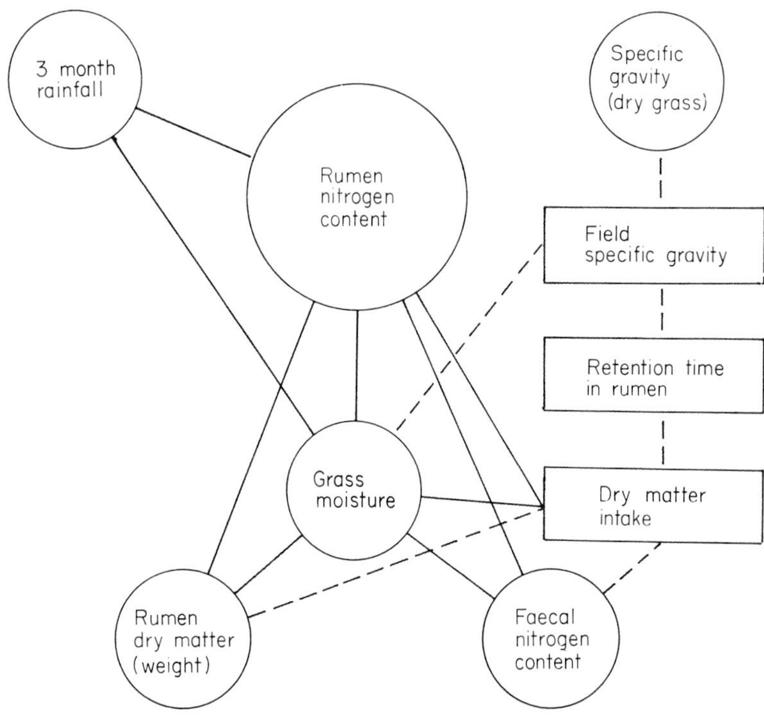

FIG. 16.1. Model of some of the environmental influences on the Cokes hartebeest nutrition and associated parameters (after Stanley-Price 1974).

density is correlated with the nitrogen or protein level. Hence we can see that environmental factors such as rainfall will have a profound effect on the animal's dietary intake, since precipitation leads to new grass growth, and new vegetation has both a high moisture content and a high protein level. Since dry matter intake varies with the moisture content of the sward, through the effect of a lowered retention time, the animal will actually eat more when the grass is nutritious. Thus Stanley-Price concludes that since dry matter intake is inversely related to retention time, intake control may be considered to be an attribute of the food rather than the requirements of the animal. The effect of varying levels of grass moisture in dry matter intake might be expected to be a feature of semi-arid and tropical rather than temperate environments since grass moisture in the former may vary seasonally from 5 to 70%. In relation to the animal's intake of protein, the concentration of nitrogen in the rumen is correlated with dry matter intake and grass moisture, so an animal feeding on herbage with short retention time should have a high rumen nitrogen. Thus in spite of the fact that herbivores are able to show intake compensation, this model for the hartebeest indicates that the nitrogen content of the rumen is more closely related to levels of dry matter intake than to varying levels of protein in the food.

Although there is an extensive literature on domestic ruminants we are only just beginning to understand the complex factors which enable wild herbivores to maximize their nutritional efficiency both in terms of seasonal variation in quality, selection and the regulation of nutrient intake.

ACKNOWLEDGEMENTS

I should like to express my sincere thanks to my many colleagues in Kenya, South Africa, Tanzania, Uganda, the UK and Zimbabwe who have over 25 years shared with me their knowledge and enthusiasm for the large mammals of the African continent. Without the work of Drs Richard Bell, Patrick Duncan, John Hanks, Tony Sinclair, Mark Stanley-Price and others too numerous to mention, our knowledge of wild ungulates would be so much the poorer.

REFERENCES

Albl P. (1971) Studies on assessment of physical condition in African elephants. *Biological Conservation*, **3 (2)**, 134–40.

Bandy P.J., Kitts W.D., Wood A.J. & Cowan I. McT. (1957) The effect of age and the plane of nutrition on the blood chemistry of the Columbian black-tailed deer (*Odocoileus hermionus columbianus*). *Canadian Journal of Zoology*, **35**, 283–9.

Bell R.H.V. (1970) The use of the herb layer by grazing ungulates in the Serengeti. In *Animal Populations in Relation to Their Food Resources*, (ed. A. Watson), pp. 111–23. Blackwell Scientific Publications, Oxford.

Belovsky G.E. (1978) Diet optimisation in a generalist herbivore: The Moose. *Theoretical Population Biology*, **14**, 105–34.

Blaxter K.L., Wainman F.W. & Wilson R.S. (1961) The regulation of food intake by sheep. *Animal Production*, **3**, 51–61.

Bredon R.M., Harker K.W. & Marshall B. (1963) The nutritive value of grasses grown in Uganda when fed to Zebra cattle. *Journal of Agricultural Science* (Cambridge) **61**, 101–4.

Brooks P.M., Hanks J. & Ludbrook (1977) Bone marrow as an index of condition in African ungulates. *South African Journal of Wildlife Research*, **7**, 61–6.

Chalmers M.I. (1961) Protein synthesis in the rumen. In *Digestive Physiology and Nutrition in the Ruminant* (ed. D. Lewis), pp. 205–22. Butterworth, London.

Coe M. (1980a) African mammals and savanna habitats. In *Habitats and their Influences on Wildlife*. pp. 83–109. Endangered Wildlife Trust, Pretoria. (Mimeographed.)

Coe M. (1980b) The role of modern ecological studies in the reconstruction of palaeoenvironments in sub-Saharan Africa. In *Fossils in the Making* (eds A.K. Behrensmeyer & A.P. Hill), pp. 55–71. University of Chicago Press, Chicago.

Coe M. (1980c) African wildlife resources. In *Conservation Biology* (eds M. E. Soule & B. A. Wilcox), pp. 273–302. Sinauer Associates, Sunderland, Massachusetts.

Conrad H.R., Pratt A.D. & Hibbs J.W. (1964) Regulation of feed intake in dairy cows 1. Changes in importance of physical and physiological factors with increasing digestibility. *Journal of Dairy Science*, **47**, 54–62.

Dougall H.W. & Sheldrick D.L.W. (1964) On the chemical composition of a day's diet of an elephant. *East African Wildlife Journal*, **2**, 51–9.

Eltringham S.K. (1974) Changes in the large mammal community of Mweya Peninsula, Rwenzori National Park, Uganda, following the removal of hippopotamus. *Journal of Applied Ecology*, **11**, 855–66.

Erasmus T., Penzhorn B.L. & Fairall N. (1978) Chemical composition of faeces as an index of veld quality. *South African Journal of Wildlife Research*, **8**, 19–24.

Field C.R. (1970) Observations on the food habits of tame wart hog and antelope in Uganda. *Journal of Applied Ecology*, **7**, 273–94.

Franzmann A.W. (1972) Environmental sources of variation of big horn sheep physiologic values. *Journal of Wildlife Management*, **36**, 924–32.

Freeland W.J. & Janzen D.H. (1974) Strategies in herbivory by mammals: the role of secondary plant compounds. *American Naturalist*, **108**, 269–89.

Grobler J.H. (1978) Population dynamics of Sable in Rhodesia. D. Phil. thesis, University of Rhodesia.

Gwynne M.D. & Bell R.H.V. (1968) Selection of grazing components by grazing ungulates in Serengeti National Park. *Nature*, **220**, 390–3.

Hall-Martin A.J. & Basson W.D. (1975) Seasonal chemical composition of the diet of Transvaal lowveld giraffe. *Journal of the South African Wildlife Management Association*, **5**, 19–22.

Harper H.A. (1975) *Review of Physiological Chemistry*. Lange Medical Publications, Los Altos, California.

Hoffman R.R. (1973) *The Ruminant Stomach*. East African Monographs in Biology, 2. East African Literature Bureau, Nairobi.

Hoffman R.R. & Stewart D.R.M. (1972) Grazer or browser: a classification on the stomach structure and feeding habits of East African ruminants. *Mammalia*, **36**, 226–40.

Jarman P.J. (1974) The social organisation of antelope in relation to their ecology. *Behaviour*, **48**, 215–66.

Kleiber M. (1961) The fire of life. In *An Introduction to Animal Energetics*. John Wiley & Sons, New York.

Lamprey H.F. (1963) Ecological separation of large mammal species in the Tarangire Game Reserve, Tanganyika. *East African Wildlife Journal*, **1**, 63–92.

Lamprey H.F. (1964) Estimation of the large mammal densities, biomass and energy exchange in the Tarangire Game Reserve and the Masai Steppe in Tanganyika. *East African Wildlife Journal*, **2**, 1–46.

Laws R.M., Parker I.S.C. & Johnstone P. (1975) *Elephants and Their Habitats. The Ecology of Elephants in North Bunyoro, Uganda*. Oxford University Press, Oxford.

McCullagh K.G. (1973) Are African elephants deficient in essential fatty acids? *Nature*, **242**, 267–8.

McNaughton S.J. (1976) Serengeti migratory wildebeest: Facilitation of energy flow by grazing. *Science*, **191**, 92–4.

McNaughton S.J. (1979) Grassland–herbivore dynamics. In *Serengeti: Dynamics of an Ecosystem* (eds A.R.E. Sinclair & M. Norton-Griffiths), pp. 46–81. University of Chicago Press, Chicago.

Maddock L. (1979) The 'migration' and grazing succession. In *Serengeti: Dynamics of an Ecosystem* (eds A.R.E. Sinclair & M. Norton-Griffiths), pp. 104–29. University of Chicago Press, Chicago.

Maloiy G.M.O., Kay R.N.B. & Goodall E.D. (1968) Studies on the physiology of digestion and metabolism of the red deer (*Cervus elephus*). In *Comparative Nutrition of Wild Animals* (ed. M.A. Crawford). 21st Symposium of the Zoological Society of London, pp. 101–8.

Melton D.A. (1978) Ecology of Waterbuck (*Kobus ellipsiprymnus*) (Ogilby 1833) in Umfolozi Game Reserve. D.Sc. thesis, University of Pretoria, South Africa.

Moen A.N. (1973) *Wildlife Ecology: An Analytical Approach.* Freeman, San Francisco.

Napier-Bax P. & Sheldrick D.L.W. (1963) Some preliminary observations on the food of elephant in the Tsavo Royal National Park (East) of Kenya. *East African Wildlife Journal*, **1,** 40–53.

Owen-Smith N. (1979) Assessing foraging efficiency of a large herbivore, the Kudu. *South African Journal of Wildlife Research*, **9**, 102–10.

Pennycuick C.J. (1979) Energy cost of locomotion and the concept of 'foraging radius'. In *Serengeti: Dynamics of an Ecosystem* (eds A.R.E. Sinclair & M. Norton-Griffiths), pp. 164–84. University of Chicago Press, Chicago.

Phillipson J. (1975) Rainfall, primary production and "carrying capacity" of Tsavo National Park (East), Kenya. *East African Wildlife Journal*, **13**, 171–201.

Putman R.J. (1981) Consumption, protein and energy intake in Fallow deer fawns on diets of differing nutritional quality. *Acta Theri Logica*, **25**, 403–13.

Ransom A.B. (1965) Kidney and marrow fat as indicators of white tailed deer condition. *Journal of Wildlife Management*, **29**, 397–8.

Rees P. (1977) Defaecation in captive African elephants. Honours project, University of Manchester. (Mimeographed.)

Riney T. (1955) Evaluating condition of free-ranging Red deer (*Cervus eltephas*) with special reference to New Zealand. *New Zealand Journal of Science and Technology*, **26 (5)**, 429–63.

Riney T. (1960) A field technique for assessing field condition of some ungulates. *Journal of Wildlife Management*, **24**, 92–3.

Rushworth J.E. (1975) The floristic, physiognomic and biomass structure of Kalahari sand scrub vegetation in relation to fire and frost in Wankie National Park, Rhodesia. M.Sc. thesis, University of Rhodesia.

Sale J.B. (1965) Some aspects of the behaviour and ecology of the rock hyraces (Genera *Procavia* and *Heterohyrax*) in Kenya. Ph.D. thesis, University of London.

Sinclair A.R.E. (1974) The natural regulation of buffalo populations in East Africa. IV. The food supply as a regulating factor, and competition. *East African Wildlife Journal*, **12**, 291–311.

Sinclair A.R.E. (1975) The resource limitation of trophic levels in tropical grassland ecosystems. *Journal of Animal Ecology*, **44**, 497–520.

Sinclair A.R.E. (1977) *The African Buffalo: A Study of Resource Limitation of Populations.* University of Chicago Press, Chicago.

Sinclair A.R.E. & Duncan P. (1972) Indices of condition in tropical ruminants. *East African Wildlife Journal*, **10**, 143–50.

Stanley-Price M.R. (1974) The feeding ecology of Coke's Hartebeest, *Alcelaphus buselaphus cokei* Gunther in Kenya. D.Phil. thesis, University of Oxford.

Stanley-Price M.R. (1977) The estimation of food intake, and its seasonal variation in the Hartebeest. *East African Journal*, **15**, 107–24.

Stanley-Price M.R. (1978) The nutritional ecology of Cokes Hartebeest *Alcelaphus buselaphus cokei* in Kenya. *Journal of Applied Ecology*, **15**, 35–49.

Thackeray J.F. (1980) New approaches in interpreting archaeological faunal assemblages with examples from Southern Africa. *South African Journal of Science*, **76**, 216–24.

Vesey-Fitzgerald D.F. (1960) Grazing succession amongst East African game animals. *Journal of Mammology*, **41**, 160–70.

Vrba E.S. (1980) Evolution, species and fossils: How does life evolve? *South African Journal of Science*, **76**, 61–84.

Walker E.P. (1964) *Mammals of the World.* Vol. 2, p. 1500. The Johns Hopkins University Press, Baltimore.

Weir J.S. (1971) The effect of creating additional water supplies in a central African National Park. In *The Scientific Management of Animal and Plant Communities for Conservation* (eds E. Duffey & A.S. Watt), pp. 367–85. Blackwell Scientific Publications, Oxford.

Weir J.S. (1972) Spatial distribution of elephants in an African National Park in relation to environmental sodium. *Oikos*, **23**, 1–13.

Western D. (1979) Size, life history and ecology in mammals. *African Journal of Ecology*, **17**, 185–204.

Westoby M. (1974) An analysis of diet selection by large generalist herbivores. *American Naturalist*, **108**, 290–304.

Wing L.D. & Buss I.O. (1970) Elephants and forests. *Wildlife Monographs*, **19**, 1–92.

Young J.Z. (1950) *The Life of Vertebrates*. Clarendon Press, Oxford.

17. CHANGES IN NITROGEN COMPOUNDS IN FISH AND THEIR ECOLOGICAL CONSEQUENCES

ZOFIA FISCHER
Institute of Ecology, Polish Academy of Sciences, Dziekanow Lesny

SUMMARY

The role of fish in the nitrogen budgets of lakes is considered in relation to current work in Poland on the nitrogen economy of intensive fish-rearing techniques.

INTRODUCTION

In order to understand fully the multiple interrelationships between fish and their environment in terms of nitrogen economy, it is useful to ask how the efficiency of energy utilization by fish varies. Nitrogen compounds are both builders of matter and carriers of energy, and a study of the loss of matter by respiration will enable us to gain some knowledge of nitrogen transformations.

Intensive fish culture is becoming more and more common throughout the world and it is frequently carried out in natural bodies of water. This type of intensive rearing raises the question, 'How does the aquatic ecosystem cope with the large amounts of extra matter accumulating at each trophic level?' In addition to considering such cycles of matter we can go even further and look at what happens to a single biogenic element such as nitrogen.

There is a great number of papers in the literature where the economy of biogenic elements in various kinds of fish is discussed, but it is only recently that such studies have considered the circulation of such substances in whole aquatic ecosystems. Many such papers are now being published in *Polish Archives of Hydrobiology*, and represent a new approach in showing effects of intensive fish culture on natural waters, for example Korzeniewski *et al.* (1982). Penczak *et al.* (1982) investigated the intake by fish of biogenic elements and their release into the lake while the fish were kept in floating cages and fed intensively. Cumulative budget methods were used to quantify these effects.

EFFECT ON THE NITROGEN BALANCE OF THE LAKE

It is of interest to know how much of the biogenic elements (particularly nitrogen) are released by one generation of fish, e.g. rainbow trout (Fig. 17.1). During 16 months of culturing, 21.5 tons of food were supplied, representing 2.5 tons of nitrogen. This figure is the total amount of organic matter introduced into the lake, some part of which is never found by the experimental fish. The non-consumed food stayed in the water and served as food for wild fish or invertebrates, or else it decomposed.

The major part of the food is consumed and finally excreted as liquid or solid faeces, or alternatively burned in the metabolic process. In this way 1114 kg nitrogen are released by the organisms into the lake. Each type of foodstuff is, of course different, but there is, in any case, a heavy inflow into the lake of biogenic elements which ultimately cause eutrophication. Some of the fish are removed for sale and, in this way, part of the biogenic elements found in fish biomass are removed from the system.

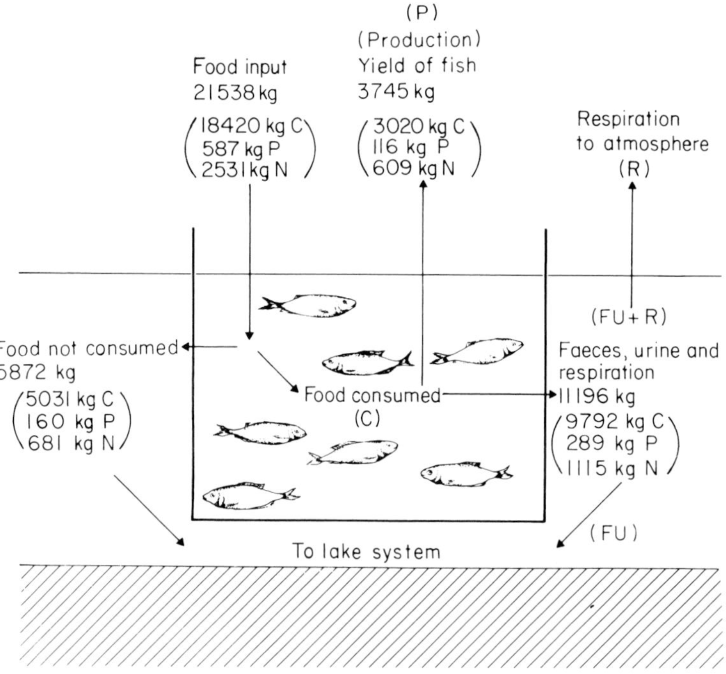

FIG. 17.1. Nutrient budget for cage-bred rainbow trout. This represents the addition of 0.67 kg C, 0.02 kg P and 0.083 kg N to the lake system per kg live wt. of fish cropped from the cage. All figures are in kg dry wt. (Compiled from Penczak *et al.* (1982) and Penczak, unpublished data.)

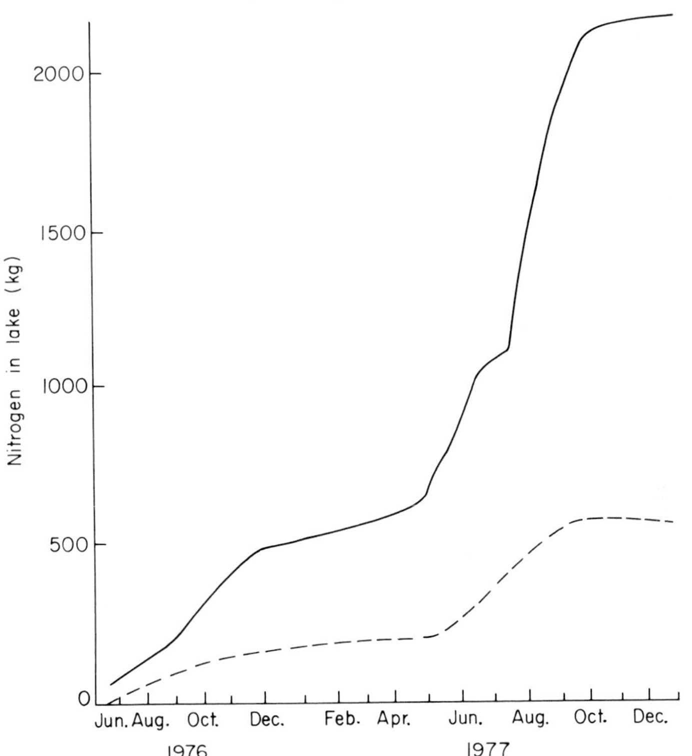

FIG. 17.2. Nitrogen load in lake.
—— Urine input
- - - - - Faeces input

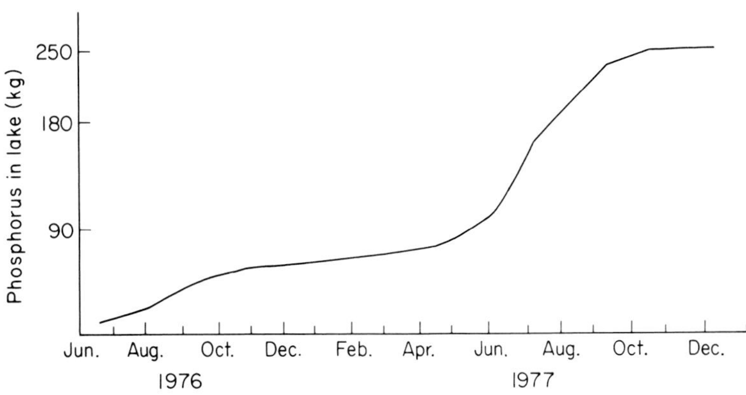

FIG. 17.3. Phosphorus load in lake.

In order to analyse the input of biogenic elements by fish in cages in a given lake, it is necessary to take a closer look at how they excrete nitrogen, carbon and phosphorus. Fig. 17.2 shows the cumulative nitrogen input from urine and faeces, and if this is compared to Fig. 17.3 (cumulated phosphorus) it can be seen that the ratio of nitrogen to phosphorus in the lake is now changed.

This would suggest that if a lake receives its main input of nutrient elements from fish culture there may be a change in the major limiting factor to productivity in that lake.

NITROGEN ECONOMY OF THE FISH

Fish are capable of playing a very important role in a body of water as a nitrogen converter and certainly do often play this role. Hence a question arises as to the internal nitrogen economy of fish, namely, 'What is the strategy of fish in dealing with this element?'

The crucial point in the nitrogen economy of fish is the waste material, comprising undigested protein waste and the soluble nitrogenous end products of protein catabolism. We know that proteins are usually assimilated by fish to a high degree and that they usually occur in considerable quantities in faeces. However, protein is not fully utilized by all fish species from all kinds of food; for instance, the grass carp that feeds with equal readiness on both plant and animal food excretes high levels of protein when the food is of plant origin. The amount of protein excreted depends to a high degree on the food quality; in unfavourable feeding conditions the grass carp fry may lose as much as 30% of the total excreted nitrogen in the form of protein.

The end products of nitrogen metabolism in fish are ammonia, urea, traces of purines, amino acids, creatine, trimethylamine, and others, with ammonia and urea being the main ones (Smith 1929). According to this classic work, urea accounts for about 15% of ammonia nitrogen in cyprinids. This value is not, however, constant.

Chalupova & Blažka (1960) proved that the level of urea excretion depends on the animal's activity. In conditions of enforced activity the fish excretes twice as much ammonia, and it has to be remembered here that enforced activity involves an increased oxygen consumption. This is a point which will be returned to later.

The question then arises as to what determines nitrogen excretion patterns, its total level and the proportion of particular compounds. Fischer (1970, 1977a, 1979) demonstrated a tendency in cyprinids to excrete more nitrogen when there was an increased supply in the food, but this ratio was not

1:1; with two kinds of food pellets differing by a factor of 1.8 in nitrogen content, excretion differed by only a factor of 1.5.

The amount of excreted nitrogen also depends on the quality of protein taken in by the fish. Urban (in press) emphasizes, however, the dependency on the weight of the fish as the essential factor. Urban experimented with grass carp fry weighing between 2 mg and 25 mg and fed them with natural plant, animal, and mixed animal and plant foods. The growth of the fry was good with the mixed food, fair with an animal diet, and very poor, with occasional deaths, when only plant food was given. It was found, however, that irrespective of food quality, the total excreted nitrogen depended solely on the weight of fish. This relationship could be expressed as $Q=6W^{0.8}$, where Q is micrograms of nitrogen excreted by a single fish in 24 hours and W is the weight of the fish in milligrams (Fig. 17.4).

This result suggests that the relationship between the excreted total

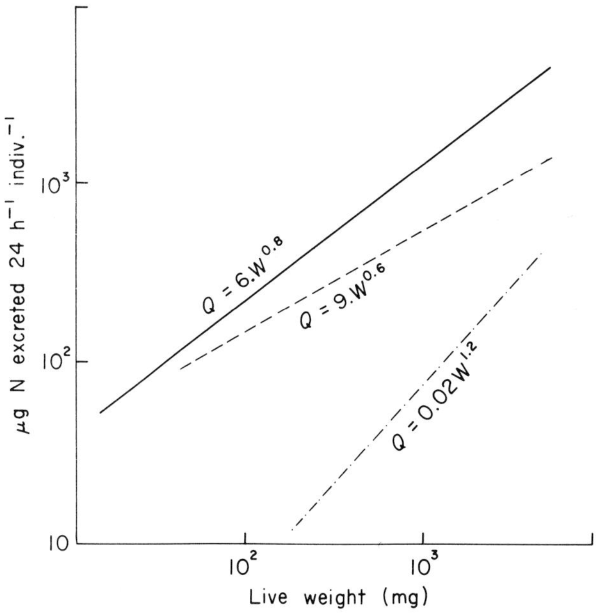

FIG. 17.4. Excretion of nitrogen as dependent on body weight in grass carp fry (after Urban, in press).
—— Total nitrogen.
– – – – – Ammonium-N excreted by fish fed with mixed plant-animal fodder (good growth rates).
–·–·– Ammonium-N excreted by fish fed with plant fodder (poor growth rates).

nitrogen and fish weight is similar to that between respiration and weight, which also has an exponent of 0.8 (Winberg 1961, 1965). This indicates that the essential role in nitrogen excretion is played by the end products of the respiratory protein combustion rather than by undigested or non-assimilated proteins, amino acids or others compounds. A similar relationship between the total excreted nitrogen and fish weight can be found by analysing the data of Wood (1958).

Urban (in press) in her experiments on grass carp found a similar relationship between fish body weight and the excretion rate of urea with a regression coefficient of 0.778. Urban suggested that urea excretion rate should be further investigated in other species and different-sized fish. Urea is synthesized in the liver, a process therefore requiring large amounts of energy. Although it is an ideal waste product, readily soluble in water and non-toxic, excretion of urea is probably very much dependent on the type of food consumed and the condition of the animal (Baranowski 1959; Vellas & Serfaty 1967). This type of relationship, however, applies only to normal conditions with proper food levels, normal temperatures, etc. Experiments on eels starved for a prolonged time, i.e. 330 days (Fischer 1977) revealed a reverse relationship: eels excreted more nitrogen as they got lighter in weight. Obviously the origin of the phenomenon is completely different to normal metabolism described above. A starving eel is metabolizing its fat reserves, and after a long deprivation of food it starts to use its own body protein.

The excretion of ammonia nitrogen seems to be strongly dependent both on food and on the environment of the fish. Pequin & Serfaty (1963) report extensive changes in ammonia excretion by carp at various seasons of the year and at a range of temperatures. Accoding to them, when the termperature falls from 20°C to 7°C the amount of excreted ammonia is more than halved.

Urban (in press) reports that in grass carp fry the relationship between ammonium nitrogen excretion and body weight differs from that for total nitrogen excretion or oxygen consumption. The regression coefficent between excreted ammonia and fish weight is high for plant food (1.2), while for mixed food the coefficient is much lower, i.e. 0.6 (Fig. 17.4).

Let us consider further the problem of the proportion of ammonia nitrogen in the total excreted nitrogen. Data from Fischer & Lipka (in press) and from Urban (in press) are collected in Table 17.1. It can be seen that for fast-growing fish in good condition the share of excreted ammonia reaches 40% of the total nitrogenous wastes; at lower growth rates it amounts to 30%, while in poor feeding conditions the proportion of excreted ammonia nitrogen drops steeply to less than 10%. A similar tendency can be seen in the data of Smith (1929).

Fig. 17.5 shows how important the role of waste matter (faeces) is and how

TABLE 17.1. A comparison of the percentage of ammonia in total excreted nitrogen for selected fish species.

Species	Weight (g)	Growth rates	Food	%NH_3 in total N	Reference
Carp	70	good	animal	40	Fischer 1977a
Grass carp	0.6	good	animal	35	Urban (in press)
Carp	100	fair	Super	32	Fischer & Lipka (in press)
Grass carp	0.5	fair	animal	31	Urban (in press)
Grass carp	30	poor	mixed	26	Fischer 1977a
Carp	50	poor	animal	23	Fischer 1977a
Grass carp	0.1	poor	animal	23	Urban (in press)
Grass carp	0.06	poor	animal	23	Urban (in press)
Carp	100	poor	Standart*	22	Fischer & Lipka (in press)
Grass carp	0.2	very poor	plant	7	Urban (in press)
Grass carp	0.02	negative	plant	5	Urban (in press)
Eel	160	negative	starved for 30 days	20	Fischer 1977b
Eel	130		starved for 90 days	10	Fischer 1977b

* Artificial carp pellets

it varies from one fish species to another. This figure presents the energy budgets of a number of fish species and it is easy to see how large the differences are in excretion between species.

The conversion of nitrogen in animals is one of the key issues—it is the only possible way of understanding the partitioning of food protein intake into that proportion built into the body tissues or used metabolically and that excreted as waste. Fig 17.6 shows examples of a protein budget for five carp specimens. These budgets are not complete since not all the end products of metabolic protein combustion were able to be measured by the available analytical methods. Incomplete as they are, they serve to give a picture of the order of magnitude in the nitrogen economy of a common fish species, the common carp.

CONCLUSION

In conclusion it is worth noting how great the flexibility of the nitrogen economy of fish is in relation to variable feeding conditions. A good example would be the budget of grass carp fed plant or animal food and the results of

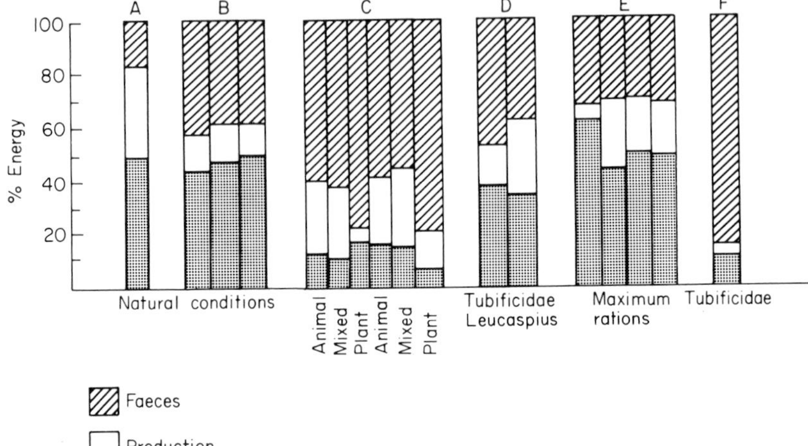

FIG. 17.5. Example of energy budgets in fish derived from a number of authors.
A. *Cyprinus carpio*
B. *Carassius auratus*
C. *Ctenopharyngodon idella*
D. *Perca fluviatilis*
E. *Salmo trutta*
F. *Anguilla anguilla*

such a study are shown in Table 17.2. The coefficient K_1 for protein measures the utilization of consumed protein for body growth; the higher K_1, the greater is the fraction of food protein incorporated into the body, and the fish is thus a more efficient converter of food protein into fish protein. The coefficient K_2 for protein is the fraction of the digested assimilated protein that is utilized for body growth, the remainder being excreted as respiratory nitrogen end products. A low value of K_2 means that the organism is not utilizing the majority of its protein intake for growth but is utilizing it as an energy source, especially for movement. The coefficient U^{-1} for protein measures the assimilation efficiency, high values indicating better quality food protein which is readily digested and assimilated. Some important observations resulted from these experiments. When the grass carp obtains pure animal food (*Tubifex tubifex*) it wastes much of the nitrogen, assimilating only 26% of it from the food while, on the other hand, when it is fed plant material (lettuce) it assimilates a greater proportion of the nitrogen, about 40%, and when mixed animal and plant food is supplied, the grass carp derives as much as 80% of its protein from the animal material.

The few examples discussed here throw some light on the nitrogen flow

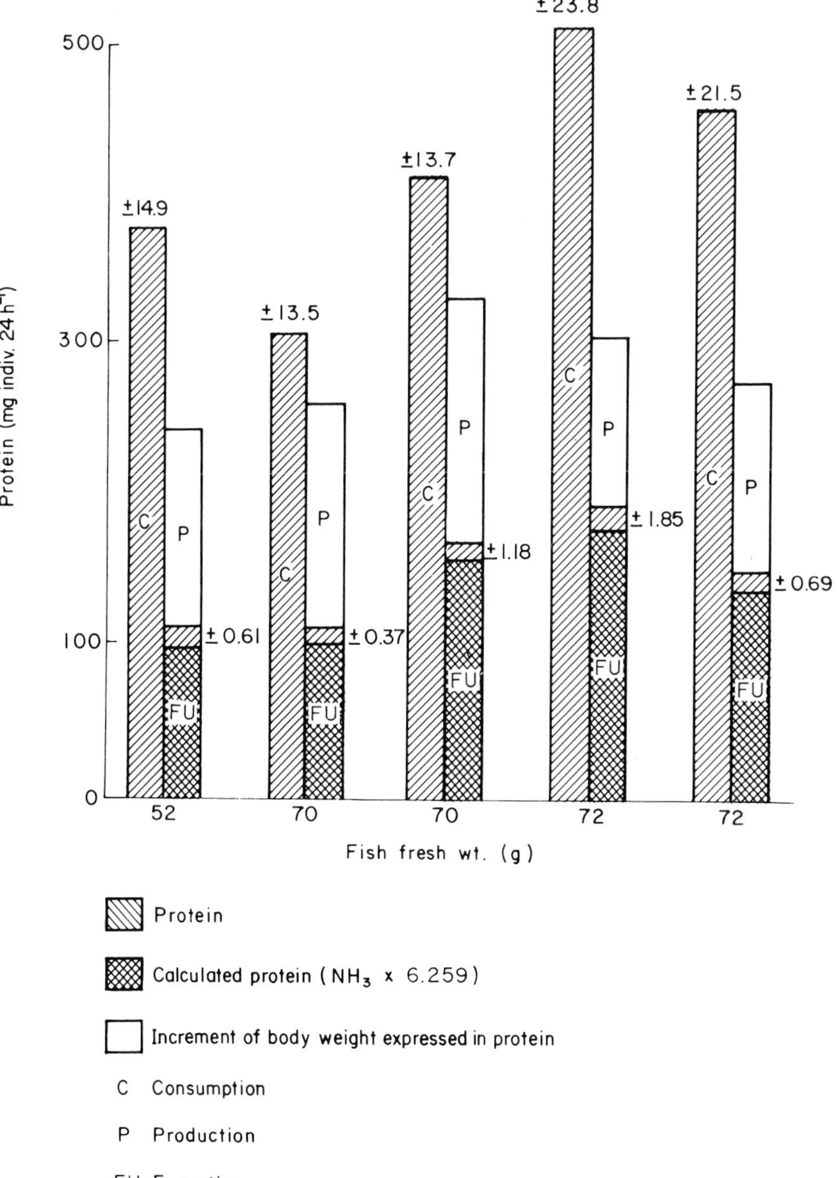

FIG. 17.6. Protein budget in carp. (Standard error of consumption (C) estimate in mg; standard error in P & FU column is amount (mg) by which budget does not balance.)

TABLE 17.2. The share of proteins, carbohydrates and lipids in diet conversion of grass carp. (For details, see text.)

Coefficient	Proteins	Lipids	Carbohydrates
		Animal food	
U^{-1}	26.1	44.3	92.3
K_1	15.2	13.9	30.1
K_2	60.1	30.1	0.32
		Plant food (cellulose-free)	
U^{-1}	41.0	36.6	75.4
K_1	6.5	23.8	0.003
K_2	15.9	39.2	0.003

through a fish population in a body of water; they suggest that the role of fish in the nitrogen economy of a freshwater habitat depends strongly on the ecological efficiencies of the population with regard to food protein. Wild herbivorous fish populations may intensify the nitrogen turnover in aquatic ecosystems in a similar way to the high density fish cultures discussed here.

REFERENCES

Baranowski T. (1959) *Concise Handbook of Physiological Chemistry*. PZWL, Warsaw.

Chalupova Z. & Blažka P. (1960) Protein metabolism in fish. *Babak's Collection*, **16**, 139–42.

Fischer Z. (1970) Some remarks about the food ration. *Polish Archives of Hydrobiology*, **17**, 177–82.

Fischer Z. (1977a) Nitrogen conversion in grass carp *Ctenopharyngodon idella* Val. *Polish Archives of Hydrobiology*, **24**, 203–14.

Fischer Z. (1977b) Nitrogen changes in starved eel *Anguilla anguilla* L. *Polish Archives of Hydrobiology*, **24**, 289–98.

Fischer Z. (1979) Selected problems of fish bioenergetics. From *Proceedings of World Symposium on Finfish Nutrition and Fishfeed Technology*, Hamburg, 20–23 June, 1978. Vol. I, pp. 17–55.

Fischer Z. & Lipka J. (in press) The role of amino acids in nitrogen conversion in carp fed with artificial food. *Polish Archives of Hydrobiology*, **30.**

Korzeniewski K., Trojanowski J. & Mrozek B. (1982) Effect of intensive trout culture on contents of nutrients in water. *Polish Archives of Hydrobiology*, **29**, 625–32.

Penczak T., Galicka W., Molinski M., Kusto E. & Zalewski M. (1982) The enrichment of a mesotrophic lake by carbon, phosphorus and nitrogen from the cage aquaculture of rainbow trout *Salmo gairdneri*. *Journal of Applied Ecology*, **19**, 371–93.

Pequin L. & Serfaty A. (1963) L'excretion of ammonia cale chez un Teleostee dulcicole: *Cyprinus carpio* L. *Comparative Biochemistry and Physiology*, **10**, 315–24.

Smith H. (1929) The excretion of ammonia and urea by the gills fish. *Journal of Biological Chemistry*, **81**, 727–42.

Urban E. (in press) The elements of energy balance in fry of grass carp. *Polish Journal of Ecology*, **32**.

Vellas F. & Serfaty A. (1967) Sur l'excretion ureique de la carpe *Cyprinus carpio*. *Archives des Sciences Physiologiques*, **21**, 185–92.

Winberg G.G. (1961) New data on intensity of metabolism in fish. *Voprosy Ikhtiologii*, **1 (18)**, 157–66.

Winberg G.G. (1965) Biotic balance of matter and energy and the biological productivity of water basins. *Gidrobiologicheskii Zhurnal*, **1**, 25–32.

Wood J.D. (1958) Nitrogen excretion in some marine teleosts. *Canadian Journal of Biochemistry and Physiology*, **36**, 1237–42.

18. THE UTILIZATION OF NITROGEN RESOURCES BY TERMITES (ISOPTERA)

N. M. COLLINS*

International Centre of Insect Physiology and Ecology, P.O. Box 30772, Nairobi, Kenya

SUMMARY

Termites accumulate nitrogen either through dietary input or by fixation of atmospheric nitrogen. There is no evidence for selection of food-plant taxa for high nitrogen content, except perhaps by *Hospitalitermes*, which forages over long distances for lichens. State of decay is an important basis for food selection, but there is little evidence that nitrogenous compounds are attractive. Live plants are rarely consumed even though their nitrogen content is high. Fresh dead litter is consumed by certain termites, some of which have alternative nitrogen strategies. Rotten litter is preferred by many termites. There is evidence that fungi improve the nitrogenous quality of litter.

A wide variety of termites contain nitrogen-fixing bacteria in their guts, but levels of fixation are unpredictable.

Termites conserve waste nitrogen by cannibalism and necrophagy, by bacterial metabolism of uric acid and its derivatives, or by recycling faeces. Utilization of the waste products of gut symbionts has not been thoroughly investigated.

Nitrogen budgets for two species of the fungus-growing genus *Macrotermes* show that dietary nitrogen input is efficiently used in termite production. Fungal digestion and translocation of nitrogenous material in the faecal fungus combs increases the available nitrogen. Since the system is in effect an external digestive system, large amounts of material can be processed, and production/biomass ratios are consequently high.

INTRODUCTION

Termites feed on plant material, sometimes living, but more usually dead and in various stages of decay. Such substrates are generally poor in nitrogen content and it has long been debated how termites subsist, and indeed thrive,

* Present address: IUCN Conservation Monitoring Centre, 219c Huntingdon Road, Cambridge CB3 0DL.

on them. Quantitatively, this question remains largely unanswered, but it is apparent that termites use free-living and symbiotic fungi, gut bacteria and protozoa to assist in metabolizing and conserving nitrogenous compounds. Termites obtain nitrogen either through their diet, or by fixing atmospheric nitrogen. Conservation of nitrogen is by internal recycling of nitrogenous waste from the termites or their gut symbionts, by recycling of faeces, or by cannibalism and necrophagy.

UTILIZATION OF DIETARY NITROGEN

The lower termites are predominantly xylophagous (Cleveland 1926), a habit considered primitive by Grassé & Noirot (1959). The higher termites (Termitidae) embrace a more catholic range of usually more decomposed foodstuffs, which may contain substantial amounts of fungal or other extraneous material. Under normal circumstances each species adheres to a fairly well-defined diet (Wood 1978), although adaptations may occur under stress. For example, *Trinervitermes trinervius* (Rambur) normally feeds on grass, but has been known to switch to dead wood during a drought (Lepage 1974). Similarly, during a drought in Kenya the litter-feeding *Macrotermes michaelseni** (Sjöstedt) began to attack succulents (J.P.E.C. Darlington, pers. comm.).

Within one habitat different species of termites have a choice of plant species or groups of species, a choice of different organs or tissues, a choice of dead, moribund or live material, and a choice of material in various stages of decay (Bouillon 1970). There is evidence that some of these categories may be advantageous in terms of nitrogen availability.

Selection for food condition

Live plants

With a few notable exceptions (e.g. *Mastotermes, Cryptotermes, Coptotermes* and *Hospitalitermes*), termites do not generally include living vegetation in their diet, even though live material would often have a higher nitrogen content than freshly dead material. Trees have efficient recycling mechanisms and dead wood has lost most of its nitrogen prior to abscision (Cosling & Merrill 1966). The physiological reasons why termites do not usually eat live wood have not been clarified. Of the relatively few live trees attacked by termites, many are exotic species (Dhanarajan 1969; Tho 1974; Harris 1969).

* Classification of *Macrotermes* spp. follows Ruelle (1970, 1977).

Sound plant litter

The majority of termites feed on dead plant material in some form, ranging from fresh herbaceous or woody litter to the finest remnants of soil humus (Wood 1978). Such materials are typically low in nitrogen content. Herbaceous tissues range from 0.5–5% nitrogen by dry weight, while woody tissues usually contain only 0.03–0.10% (Cowling & Merrill 1966).

Some of the lower termites, particularly the Kalotermitidae, are able to live on apparently sound and fungus-free dead wood (Sands 1969; Pence 1956). However, even the dry-wood termite *Incisitermes minor* (Hagen) has been found to be associated with 17 genera of free-living fungi, which may chemically improve its diet (Hendee 1933). As La Fage & Nutting (1978) have noted, the role of symbiotic gut microorganisms in metabolism of dietary nitrogen is not at all clear. Bacteria-free *Reticulitermes santonensis* has a reduced capacity to utilize ^{14}C-labelled glucose in amino acid synthesis, but no important changes have been noted in *Nasutitermes nigriceps* (Haldemann) (Speck, Becker & Lenz 1971). Similarly, *Coptotermes formosanus* Shiraki maintained normal amino acid levels in its tissues three weeks after all protozoans and most bacteria were removed (Mauldin & Smythe 1973). The importance of gut symbionts in recycling waste termite nitrogen is discussed below.

Relatively fresh litter is widely utilized by termites in the dry grasslands, savannas and deserts of the world, but this cannot be regarded as selective, since rotting litter is a rarity in these environments. Fresh litter is widely consumed by the Macrotermitinae in West and East Africa (Wood 1976; Collins 1981a, 1982), and by *Coptotermes, Schedorhinotermes, Amitermes* and *Nasutitermes* in Australia (Lee & Wood 1971a). Some of these taxa are known to utilize alternative nitrogen strategies. The Macrotermitinae maintain a symbiotic fungus in their nests and thus bypass the need for free-living fungi, while *Nasutitermes* and *Coptotermes* have nitrogen-fixing gut bacteria (see below). Such fauna may be contrasted with those of rain forests, where the majority of termites take rotting litter in some form. In rain forest in Sarawak, 62–68% of termite species feed on rotten wood and only 12–26% on dry wood (Collins, in press). Once again, this may be more a question of the state of available food, rather than selection. However, there is no doubt that in the Isoptera as a whole, most species require some degree of pre-rotting to bring litter, particularly wood, to a palatable state (Lee & Wood 1971a).

Rotten plant litter

Fungi affect termite nutrition in two ways—firstly by adding or removing compounds which attract, repel or poison termites, and secondly, by consuming and digesting wood (La Fage & Nutting 1978). Only this second category is of importance in terms of nitrogen utilization.

The advantages conferred on soil fauna by the pre-rotting of litter have been considered at the ecosystem level by Fittkau & Klinge (1973). They consider that fungi play a decisive role in concentrating otherwise limited nutrient resources in the Amazonian rain forest. The consumer food chains are based largely on a soil fauna (in which termites are a major component), which itself depends on fungi to break down plant litter and immobilize nutrients. The immobilization of nitrogen by fungi was first demonstrated by Hungate (1940), who found that despite a weight loss of over 30% in rotten wood, the amount of nitrogen per unit volume remained constant. In some cases it has been demonstrated that rotting wood accumulates nitrogen (Hungate 1941; Swift 1977), either by transfer from adjacent litter along fungal hyphae (Bocock 1963; Gosz, Likens & Bormann 1973), or by nitrogen fixation (Sharp & Millbank 1973).

Despite a wealth of observational data on the preferences of termites for rotten wood, experimental data on the advantages conferred in terms of nutrient quality are few. Becker (1971) found that termites cultured in decayed wood showed significant increases in survival and biomass over those in sound wood. The termites perferred wood which had lost 5–15% of its weight due to attack by a brown-rot fungus. Both *Nasutitermes corniger* and *Rhynchotermes perarmatus* Snyder have been found to feed on litter significantly higher in nitrogen content than similar but unutilized litter in Costa Rica (Prestwich, Bentley & Carpenter 1980). Some of the nitrogen in the litter was the result of nitrogen fixation at rates varying from 0.04–8.98 μg N g^{-1} (fresh wt.) day^{-1}, although the termites did not select food on the basis of its nitrogenase activity. The possibility that faeces or imported microorganisms had caused nitrogen increases in the food was not eliminated. It is known that termites in Malaysian rain forest import decomposer microorganisms into wood litter (Abe 1980).

In some cases the biochemical impact of fungi on wood may be correlated with the digestive capabilities of termites. There is evidence that fungi mobilize the nitrogenous compounds in plant material and translocate them to fungal material in forms more readily digestible by termites (Martin 1979; Levi, Merrill & Cowling 1968). Analysis of various woods has shown that although 50% of total nitrogen is recovered in the amino or ammonia form, 15% remains bound in cellulose or lignin hydrolysis residues (Scurfield & Nicholls 1970; Lincoln & Mulay 1929). This is generally attributed to lignin-bound protein, but Whitehead & Quicke (1960) found that lignin from some grasses contains nitrogen in the methylated form $-NCH_3$. The evidence for lignin digestion by termites is contradictory (La Fage & Nutting 1978). Several authors have found that lignin passes through termites with little or no change (Moore 1969; Hungate 1938; Leopold 1952; Nel, Hewitt & Joubert 1970). Conversely, Seifert & Becker (1965) found 2–26% of lignin to be digested by *Kalotermes flavicollis* (F.) and 14–40% by *Heterotermes indicola* (Wasmann). High results for *Reticulitermes* and *Nasutitermes* were equivocal due to observed coprophagy, but still indicated lignin digestion. Digestion of ^{14}C lignin has been demonstrated in *Nasutitermes exitiosus* (Butler & Buckerfield 1979), and some lignin is degraded by *Microcerotermes edentatus* Wasmann, although this is more complete if wood is previously subjected to fungal attack (Kovoor 1964a & b, 1966).

White-rot fungi attack lignin with extracellular oxidases requiring aerobic conditions (Schubert 1965), while cellulose breakdown in both lower and higher termites is an anaerobic fermentation (Lee & Wood 1971a). Hence, unless the degradation of lignin in termites proceeds by a pathway different from that in fungi, it must be possible for anaerobic and aerobic decomposition to occur simultaneously (Lee & Wood 1971a). This important matter requires clarification since lignin digestion may be important in releasing nitrogenous compounds. The fungus-growing Macrotermitinae appear to be dependent upon their symbiotic white-rot fungus to break down dietary lignin, a subject further discussed below. Since white-rot fungi are capable of utilizing a wide range of nitrogenous compounds for protein synthesis (Levi, Merrill & Cowling 1968), it might be expected that termites would be attracted to white-rotted wood. Once again the data are contradictory. *Nasutitermes exitiosus* (Hill) prefers *Eucalyptus* wood attacked by the white-rot *Fomes lividus* (Kalch.) Sacc. (Ruyooka 1979), despite the termite's apparent ability to digest lignin (Butler & Buckerfield 1979), whereas white rot inhibits the attack of timbers by *Reticulitermes flavipes* (Amburgey & Beal 1977).

As a component of fungal tissue, protein is second only to carbohydrate, and is usually present as 20–40% of dry weight, although 60% has been recorded (Martin 1979). These proteins consist of the normal amino acids and, along with free amino acids, are presumably freely absorbed in the midgut of all termites (Noirot & Noirot-Timothée 1969). Proteolytic enzymes have been demonstrated in the guts of *Heterotermes indicola* (Krishnamoorthy 1960; Rao 1962) and are probably a usual component of termite digestive systems. The concentration of partially bound amino acids (PBAAs) is somewhat higher in rotten wood (31.7 μmol g^{-1}) than in sound wood (19.3 μmol g^{-1}) (Carter & Smythe 1973), as might be expected from observations on the conservation of nitrogen by wood-rotting fungi (Hungate 1941; Swift 1977). The molar ratios of the various PBAAs in the woods also differ, but this has no effect on the PBAAs of *Reticulitermes flavipes* (Kollar) fed on them (Carter & Smythe 1973). These authors found that some variation in the free amino acids of the termites was detectable, but noted that FAA composition varies with metabolic changes as well as diet.

The polyphenolic compounds (lignin, tannins) which complex with proteins in the cell walls of higher plants and may reduce their digestibility to insects (Feeny 1976; Bernays 1978) are absent from fungi (Aronson 1965; Bartnícki-Garcia 1970). The cell walls of fungi are composed of chitin, a β (1,4) polysaccharide derived from acetyl-glucosamine, and mixture of β-D-(1,3)- and β-D-(1,6)-glucans (Martin 1979). In addition to containing its own molecular nitrogen, chitin in nature is always chemically combined with proteins (Waterhouse, Hackman & McKellar 1961). There is some evidence that chitinase activity may be widespread in termites (Table 18.1). Indeed, since the moulting process involves enzymatic degradation of chitin, the capacity of the epidermal cells of the integument to produce chitinase may be universal in insects (Martin 1979). Chitinase has been detected in the wood-feeding termites *Coptotermes lacteus* (Froggatt), *Nasutitermes exitiosus* (Tracey & Youatt 1958; Waterhouse, Hackman & McKeller 1961) and *Zootermopsis angusticollis* (Hagen); in the grass-feeding *Trinervitermes trinervoides* (Sjöstedt) and in the fungus-growing *Macrotermes ukuzii* Fuller (Rohrmann & Rossman 1980). In this last species, fungal chitin content has been estimated at 0.8% and 1.0% in old and new comb respectively, and 2.7% in the fungal conidiophores. Chitinase has also been found in the soil-feeding genus *Cubitermes* (Rohrmann & Rossman 1980). Since hexosamines from chitinous remains of arthropods and fungi make up 5–13% of total soil nitrogen (Bremner & Shaw 1954), an ability to digest this material could be an important nutrient strategy (Rohrmann & Rossman 1980). Martin (1979) considers that chitinase enzymes are crucial to any mycophage and has

Table 18.1. Records of termites with a chitinase capacity. *Hodotermes mossambicus* has been found not to have chitinase (Retief & Hewitt 1973).

Species	Reference
Hodotermitidae	
Zootermopsis angusticollis	Rohrmann & Rossman 1980
Rhinotermitidae	
Coptotermes lacteus	Waterhouse, Hackman & McKellar 1961
Termitidae	
Cubitermes sp.	Rohrmann & Rossman 1980
Macrotermes ukuzii	Rohrmann & Rossman 1980
Trinervitermes trinervoides	Rohrmann & Rossman 1980
Nasutitermes exitiosus	Tracey & Youatt 1958

reviewed data demonstrating chitinase in roaches, ants, beetles and other invertebrates.

Urea is commonly accumulated in the higher fungi, but it is not clear whether insects can utilize it for protein synthesis (Martin 1979). The symbiotic gut bacteria in *Reticulitermes flavipes* have a facultative uricase capacity, but the uric acid substrate is excretory waste held in the fat tissue (Potrikus & Breznak 1980a & b). It is not clear whether dietary fungal urea may be similarly utilized, nor how widespread such bacteria are in other groups of termites.

To summarize, it is apparent that pre-rotting of food by fungi may be important in increasing the amount of nitrogenous nutrients available for metabolism by termites. The influence of white-rot fungi in decomposing lignin appears to release proteinaceous and other nitrogenous substances which might otherwise be unavailable to termites. Immobilized nitrogen in fungal proteins, chitin and urea are available at least to those termites known to have the necessary enzymes.

Mammalian dung is an important food source for many termites inhabiting pastures, and may be regarded as a special form of rotting litter (Ferrar & Watson 1970; Wood 1978). Dung may be a rich source of utilizable nitrogen, but there is no information on this subject relating to termites. Moisture may be an equally important acquisition.

Selection for food-plant taxa

There is a wide literature on preferred species of food, particularly with regard to building timbers. Such studies are beyond the scope of this paper since they are usually associated with hardness or the presence of noxious substances, and elemental analyses are rarely given.

Several studies on grass-feeding termites have demonstrated distinct preferences, often for the less common species. *Hodotermes mossambicus* Hagen prefers grasses to shrubs, and 16 species of grass have been arranged in order of preference (Hewitt & Nel 1969; Nel & Hewitt 1969; Nel, Hewitt & Joubert 1970). *Themeda triandra* Forsk. is the favourite, but green grass, whether fresh or dried, is toxic to laboratory colonies (Botha & Hewitt 1978). Dougall, Drysdale & Glover (1964) estimated the crude protein content of 41 species of grass and obtained a mean of 11.51% (dry wt.). By dividing by a factor of 6.25 they obtained a mean proteinaceous nitrogen content of 1.84%. By this calculation, *T. triandra* whole flowering plant was only 0·7% nitrogen, while even fresh growth after burning was below average at 1.74%. Grass litter consisting mostly of *T. triandra* is only 0.52% nitrogen (M.A. Arshad, pers.

comm.) and it therefore seems unlikely that nitrogen content is the trigger for *Hodotermes*' preference for *Themeda.*

Trinervitermes geminatus (Wasmann) was shown to prefer *Andropogon gayanus* Kunth. to five other grasses tested, even though its density in the field was less than 0.5% of that of *Hyparrhenia dissoluta* (Hochst.) Chiov. (Ohiagu 1978). Laboratory work with the same termite revealed preferences for the smaller, fine-leaved species of grass (Sands 1961). Unfortunately, no elemental analyses are available, but food size and coarseness seem to be the most important factors for *Trinervitermes geminatus.*

Conversely, preliminary results for the grass and litter feeder *Macrotermes subhyalinus* Rambur in Kenya suggest that the woody perennial grass *Pennisetum* is preferred to other softer and finer genera (Collins 1981a, 1982). According to Dougall, Drysdale & Glover (1964) the crude protein levels of *Pennisetum* are well below average, and the woody dead parts preferred by *Macrotermes* would probably be of particularly poor quality. *Macrotermes* is a fungus-growing genus which may not be concerned with initial food quality, depending on its fungus combs to translocate and concentrate nitrogen into fungal tissue.

In some cases a rich nitrogenous diet appears to be disadvantageous or repellent to termites. It has been demonstrated that the desert termite *Gnathamitermes tubiformans* (Buckley) achieves maximal growth and survival on low amino acid diets (Spears & Ueckert 1976). High dietary nitrogen causes abnormal gut faunation and inhibition of nitrogenase activity in *Coptotermes* (Breznak 1975). Similarly, *Amitermes hastatus* (Haviland) refused a series of 16 amino acids when added to the normal diet in small quantities (Skaife 1955). This suggests that the nitrogenous requirements were more suitably fulfilled in the normal diet (Bouillon 1970). In short, there is little evidence that termite preferences for plant species are the result of nitrogen content.

One possible exception is the oriental nasute genus *Hospitalitermes*, which habitually feeds on living mosses and lichens, with a minor input of bark, wood, algae and fungi (Kalshoven 1958; Jander & Daumer 1974; Collins 1979a). Lengthy surface forays of up to 300 m are necessary to gather the lichens from the rain forest canopy. Since an abundance of mosses and other acceptable epiphytic materials are easily available near the nest, it seems that the lichens may be of special dietary significance. The genera of lichens collected by *Hospitalitermes umbrinus* (Haviland) are *Melaspilea, Phaeographis* and *Phaeotrema*, all of which contain blue-green algal phycobionts (Collins 1979a; N. Sammy, pers. comm.). Lichens with a blue-green phycobiont are the only ones known to fix atmospheric nitrogen and have higher nitrogen contents than lichens with green phycobionts (Millbank &

Kershaw 1974; Forman 1975). Hence, the lichens may be an important nitrogen source for these termites, and worth travelling long distances to collect. *Odontotermes* has been observed feeding, perhaps incidentally, on the epiphytic lichen *Parmelia andina* Mull. Arg. in Kenya (Collins, unpublished data; B. Coppins & T.D.V. Swinscow, pers. comm.), but this lichen contains a green phycobiont, and other species in the genus are not particularly high in nitrogen content, e.g. *P. physodes* 0.49%, *P. sulcata* 0.96% (Hitch 1971).

FIXATION OF ATMOSPHERIC NITROGEN

The earliest suggestion that termites might be able to fix nitrogen was made by Cleveland (1925), whose colonies of *Zootermopsis* showed a 40-fold weight increase after 18 months on a diet of apparently pure cellulose. However, Roessler (1932) found that *Zootermopsis* only just survived on such a diet, while Hendee's (1935) colonies lost weight or died. Hungate (1941) then observed that the filter papers used by Cleveland were 0.03% nitrogen and the idea of nitrogen fixation was discredited.

The development of the acetylene reduction method for assay of nitrogen fixation (Hardy *et al.* 1968) has led to renewed assessments. In 1973 Breznak *et al.* and Benemann independently demonstrated that *Zootermopsis* nymphs and workers can indeed fix nitrogen. Levels were low and, by inference from earlier work, would be insufficient to support the colonies. Nitrogen fixation has since been demonstrated in species of the lower termite families Mastotermitidae, Kalotermitidae, Hodotermitidae and Rhinotermitidae, and in the higher termite subfamilies Termitinae and Nasutitermitinae (Table 18.2). The cockroach *Cryptocercus punctulatus*, believed to be closely related to termites, has also been found to fix nitrogen (Breznak 1975). No nitrogenase activity has been found in three species of Macrotermitinae (Rohrmann & Rossman 1980; N.M. Collins & P.G. McDowell, unpublished data; G.D. Prestwich & B.L. Bentley, pers. comm.), nor in representatives of the Coleoptera, Diptera or Hymenoptera (Benemann 1973; Breznak *et al.* 1973).

The degree of nitrogenase activity by groups of termites varies within the very wide limits of a mere trace (<0.01 μg N g^{-1} (dry wt.) h^{-1}) to 5.71 μg g^{-1} h^{-1}. A recent study using whole colonies of *Nasutitermes corniger* has produced an even higher figure of 12.92 μg g^{-1} h^{-1} (range 7.61–22.69, Prestwich & Bentley 1981), although the possibility of fixation in nest material was not ruled out. This high variability of results may be caused by two factors. Firstly, it seems that the acetylene assay method, although simple to operate, poses some technical problems regarding termite and bacterial behaviour. It has been found that rates of nitrogen fixation by groups of

TABLE 18.2. Estimated levels of nitrogen fixation by termites, converted where necessary to a common base of μg N fixed g^{-1} (dry wt.) h^{-1}. Figures in μmol C_2H_4 have been converted by multiplying by 9.3 (Prestwich, Bentley & Carpenter 1980); assumed fresh/dry weight ratios of 5 (workers) or 4 (soldiers) have been used where dry weights were not given. Nymphs from six species tested by Benemann (1973), Breznak *et al.* (1973) and Rohrmann & Rossman (1980) were all zero or below 0.01 μg^{-1} h^{-1} except *Cryptotermes brevis* 0.08 μg g^{-1} h^{-1}.

Species	Nitrogen fixation rates (μg N g^{-1} (dry wt.) h^{-1}) Worker	Soldier	Workers + soldiers	Reference
Mastotermitidae				
Mastotermes darwiniensis	0–0.89			French, Turner & Bradbury 1976
Kalotermitidae				
Incisitermes minor	1.60	Trace		Benemann 1973
Cryptotermes brevis			Trace	Benemann 1973
Hodotermitidae				
Zootermopsis angusticollis			Trace	Benemann 1973
Zootermopsis sp.	0.01			Breznak *et al.* 1973
Rhinotermitidae				
Heterotermes sp.	0.26	0		Sylvester-Bradley, Bandeira & de Oliveira 1978
Reticulitermes flavipes	0.01	Trace		Breznak *et al.* 1973
Coptotermes formosanus	0.03	0.01		Breznak *et al.* 1973
Coptotermes lacteus	0.08–0.39			French, Turner & Bradbury 1976
Termitidae: Termitinae				
Cubitermes sp.		Trace		Rohrmann & Rossman 1980
Gnathamitermes tubiformans			1.83–5.71	Schaefer & Whitford 1981
Amitermes sp. A	0.10–0.23	0.13–0.37		Sylvester-Bradley, Bandeira & de Oliveira 1978
Amitermes sp. B	0	0		Sylvester-Bradley, Bandeira & de Oliveira 1978
Termes sp.	0.19	0		Sylvester-Bradley, Bandeira & de Oliveira 1978
Neocapritermes sp. A			0	Sylvester-Bradley, Bandeira & de Oliveira 1978
Neocapritermes sp. B	0.17	0		Sylvester-Bradley, Bandeira & de Oliveira 1978
Termitidae: Nasutitermitinae				
Trinervitermes trinervoides	0.29	0.19		Rohrmann & Rossman 1980
Nasutitermes corniger	4.93[1]	1.4–1.8[1]	12.92[2]	[1]Prestwich, Bentley & Carpenter 1980; [2]Prestwich & Bentley 1981
Nasutitermes exitiosus	0–1.17			French, Turner & Bradbury 1976
Nasutitermes sp. A			0.31–5.09	Sylvester-Bradley, Bandeira &
Nasutitermes sp. C	0.15–2.99	0.11–2.07		Sylvester-Bradley, Bandeira &
Nasutitermes sp. D			2.36	Sylvester-Bradley, Bandeira &
Nasutitermes sp. F	0.43–0.83	0.68–1.94		Sylvester-Bradley, Bandeira &
Nasutitermes sp. G	0.06			Sylvester-Bradley, Bandeira &

TABLE 18.2. (*cont.*)

Nasutitermes sp. H	3.69			de Oliveira 1978
Cornitermes sp. A			0.29	
Cornitermes sp. B	0.07	0.05		
Constrictotermes sp.	0.04			
Rhynchotermes perarmatus	0.04	0.96		Prestwich, Bentley & Carpenter 1980
Syntermes sp.	0.17	0.01		
Grigiotermes sp.	0–0.04	0	0	
Spinitermes sp.	0	0		Sylvester-Bradley, Bandeira & de Oliveira 1978
Armitermes sp.	0	0		
Armitermes sp.			0.40	
Labiotermes sp.	0.04			
Termitidae:				
Macrotermitinae				
Macrotermes ukuzii	0	0		Rohrmann & Rossman 1980
Macrotermes michaelseni	0	0		N.M. Collins & P.G. McDowell, unpublished data
Macrotermes natalensis	0	0		G.D. Prestwich & B.L. Bentley, pers. comm.

Nasutitermes corniger are ten times lower than weight-specific rates for whole colonies, suggesting that physical disturbance and manipulation reduce nitrogenase activity (Prestwich & Bentley 1981). Nitrogen fixation also drops in specimens held for 2–24 hours between collection and assay, and declines after more than 2–3 hours in the assay chamber (Prestwich, Bentley & Carpenter 1980; Prestwich & Bentley 1981; Sylvester-Bradley, Bandeira & de Oliveira 1978). It appears that 20% acetylene in the assay chamber is eventually sufficient to poison the nitrogenase pathways. Experiments with radioactive isotopes of nitrogen offer an alternative method. Nitrogenase-positive bacterial isolates from termites grown on nitrogen-free agar slants in an atmosphere of $^{15}N_2$ incorporated radioactive nitrogen within one week (French, Turner & Bradbury 1976). Experiments using whole nests in an atmosphere with radioactive nitrogen are currently under way (Prestwich & Bentley 1981).

The second reason for variability in fixation rates is that diet is apparently important in regulating the rate of nitrogen fixation. *Coptotermes formosanus* Shiraki fed on filter paper containing no added nitrogen source has over ten times greater nitrogenase activity than termites fed on wood (Breznak *et al.* 1973). However, addition of nitrogenous compounds to the filter papers reduces nitrogen fixation to below the levels found in wood-fed termites.

Similarly, *Rhynchotermes perarmatus* Snyder, which feeds on leaf litter of relatively high nitrogen content, has a lower fixation rate than the wood-feeding *Nasutitermes corniger* (as *ephratae*) (Prestwich *et al.* 1980). Rates of fixation by termites in Amazonian primary and secondary forest are lower than in pasture (Sylvester-Bradley, Bandeira & de Oliveira 1978). Although elemental analyses of foodstuffs are lacking, it seems likely that a higher nitrogen content diet would be available in the forest, where decomposer microorganisms may be more active.

The fixation of nitrogen by termites has been attributed to several species of facultative anaerobic bacteria (Enterobacteriaceae) in the termite hindgut. De-gutted termite bodies do not fix nitrogen, whereas excised guts fix nitrogen at a reduced level (Breznak *et al.* 1973). Termites fed on filter papers containing antibacterial pharmaceuticals lose the ability to fix nitrogen. *Enterobacter agglomerans* has been identified as the nitrogen-fixing agent in *Coptotermes formosanus* (Potrikus & Breznak 1977), and *Citrobacter freundii* (Braak) Werkman & Gillen is the active species in *Mastotermes darwiniensis* Froggatt, *Coptotermes lacteus* (Froggatt) and *Nasutitermes exitiosus* (French *et al.* 1976). In addition to the records in Table 18.2, *Enterobacter* species capable of growth on a nitrogen-free medium in the presence of gaseous nitrogen have been isolated from *Mastotermes darwiniensis, Cryptotermes primus* Hill, *Heterotermes ferox* (Froggatt), *Coptotermes acinaciformis* (Froggatt), *C. lacteus, Schedorhinotermes intermedius* (Brauer) (all lower termites) and *Nasutitermes graveolus* (Hill) (a higher termite) (Eutick, O'Brien & Slayton 1978). No activity was found in *Nasutitermes walkeri* (Hill) or in *N. exitiosus* (French, Turner & Bradbury 1976).

In general, nitrogen fixation estimates are too variable for extrapolation to a field population or community of termites. However, Schaefer & Whitford (1981) have guardedly calculated that a population of 80 colonies ha^{-1} of *Gnathamitermes tubiformans* (Buckley) (10^4 termites per colony, biomass 25 g dry wt.) fixed 32–100 g N ha^{-1} $year^{-1}$ in a Chihuahuan desert. The mean value of 66 g fixed ha^{-1} $year^{-1}$ represents 13% of total nitrogen output in the form of faecal carton galleries (100 g N ha^{-1} $year^{-1}$) and losses to predation (410 g N ha^{-1} $year^{-1}$).

Various authors have calculated approximate times for doubling of termite nitrogen content by assuming constant fixation rates. *Incisitermes minor* would take 7–8 years at a fixation rate of 139–212 μg g^{-1} (fresh wt.) $month^{-1}$, or 2.5 years at the highest observed rate of 566 μg g^{-1} $month^{-1}$ (Benemann 1973). At a fixation rate of 0.28–5.56 μg N g^{-1} (dry wt.) h^{-1}, *Amitermes wheeleri* (Desneux) and *Reticulitermes tibialis* Banks have doubling times of 0.25–6 years (Schaefer & Whitford 1979). At the very high mean rate of 12.92 μg g^{-1} (dry wt.) h^{-1}, colonies of *Nasutitermes corniger* could double their nitrogen content in one year (Prestwich & Bentley 1981). Given an approximate production/biomass ratio of 3 : 1 for nasutes (see Table 18.8), this fixation rate would account for 33% of the annual nitrogen requirement. It has been estimated that nitrogen fixation by another nasute, *Trinervitermes trinervoides*, would permit a doubling time of 8–12.5 years (Rohrmann & Rossman 1980).

RECYCLING OF WASTE NITROGENOUS MATERIAL

There are three basic ways in which waste nitrogenous material may be conserved by recycling:

1. Internal recycling of nitrogenous waste from termites or their gut symbionts.
2. Recycling of voided waste material (faeces).
3. Recycling of wounded, superfluous or dead nest-mates.

Internal recycling of waste from termites or their gut symbionts

The digestive physiology of termites is dominated by trophallaxis (nutrient exchange), and by the digestive collaboration of symbiotic microorganisms in the proctodeum (Noirot & Noirot-Timothée 1969). The lower families of termites contain flagellate protozoans which are spread by proctodeal feeding. Higher termites have no flagellates and do not usually exhibit proctodeal feeding, but the gut contains a complex and variable association of bacteria (Grassé & Noirot 1959). In most, if not all, cases, these gut symbionts are essential for efficient digestion. Termites themselves can degrade simple compounds like soluble sugars and protein, but the symbionts are generally required for cellulose digestion (see reviews by Lee & Wood 1971a; Honigberg 1970; Noirot & Noirot-Timothée 1969). There is, however, recent evidence of cellulase production by the lower termite *Coptotermes lacteus* (O'Brien *et al.* 1979) and by the higher termites *Trinervitermes trinervoides* (Potts & Hewitt 1973), *Microcerotermes edentatus* (Kovoor 1970) and *Nasutitermes exitiosus* (O'Brien *et al.* 1979).

Nitrogenous waste from the gut symbionts may be of two types—dead symbionts or excretory products. It has been claimed that the protozoa in the lower termites may, upon death, supply their hosts with significant amounts of assimilable nitrogen, but this has not been verified. Mitotic flares of the symbionts appear to be limited to the post-moult period and the individual flagellates are said to be very long-living. They would not therefore die in sufficient quantities to represent a significant nitrogen flux (Honigberg 1970). Mauldin & Smythe (1973) found that defaunation of *Coptotermes formosanus* resulted in few differences in amino acid content of the termites. They concluded that the termites are able to maintain their protein levels without protozoa, that dead protozoa probably do not furnish needed nitrogenous substances, and that symbiotic protozoa do not fix nitrogen.

The most abundant waste product of protozoa is ammonia. There are no data on whether it may be used as a nitrogen source by termites, but Roessler (1932) observed some apparent utilization of ammonium compounds by *Zootermopsis*. La Fage & Nutting (1978) have drawn attention to the fact that ruminants depend on bacteria for biosynthesis of protein from non-protein sources such as ammonia and urea. It is now known that some termites contain uricolytic bacteria (see below) but it is not known whether protozoan waste products may also be utilized by these bacteria.

The normal end products of nitrogen metabolism by all terrestrial insects are uric acid and its derivatives. Leach & Granovsky (1938) have suggested that termites might be able to reutilize these compounds, either directly, or with the help of uricolytic gut symbionts. Quantities of urates have been found in the fat-bodies of many termites (Jucci 1921) and uric acid has been detected in the faeces of *Zootermopsis* (Hungate 1941) and *Nasutitermes exitiosus* (Moore 1969). An apparent absence of uric acid in the faecal fungus combs of *Macrotermes falciger* (Gerstäcker) and *Odontotermes badius* (Haviland) has been reported (Cmelik & Douglas 1970), but uric acid was found in the comb of *Macrotermes michaelseni* by using a more sensitive technique (Abo-Khatwa 1977). Uric acid is present as 1–45% (dry wt.) of the lower termites *Reticulitermes flavipes, R. virginicus* (Banks), *Coptotermes formosanus, Marginitermes hubbardi, Paraneotermes simplicicornis* (Banks) and *Cryptotermes cavifrons* (Banks) (Potrikus & Breznak 1980a). This amounts to 4–69% of the termites' total nitrogen. In *Reticulitermes flavipes* almost all the uric acid is associated with fat-body tissue, while faeces contain less than 0.2%. This species contains dense populations of *Streptococcus* sp., *Bacteroides termitidis* and *Citrobacter* sp. in its hindgut (Potrikus & Breznak 1980b). All are able to use uric acid anaerobically as an energy source, but none have an absolute requirement for it. It has been proposed that the uric acid produced by the termites as an excretory product is broken down by the bacteria to products utilizable by termites as carbon, nitrogen and/or energy sources, very little uric acid being voided in the faeces (Potrikus & Breznak 1980b). Similarly, the cockroach *Periplaneta americana* (L.) stores uric acid internally and uses it as a metabolic reserve when placed on nitrogen-deficient diets (Mullins & Cochran 1975a & b).

It has yet to be demonstrated whether or not uricolytic bacteria are present in the guts of higher termites. However, large populations of an apparently pure culture of bacteria have been found in the mixed segment between the midgut and first proctodeal segment of the Termitinae (Noirot & Noirot-Timothée 1969). These bacteria are well situated to make use of the wastes of nitrogen metabolism, since the Malpighian tubules open into the gut at the anterior end of the mixed segment. If uricolytic activity is a common characteristic of the symbiotic gut microbiota of termites, then the urea in fungi may be another important source of nitrogen for termites feeding on rotten wood.

Although the metabolism of uric acid and its derivatives may play a highly significant role in termite productivity, it is not known whether the uricolytic bacteria are widespread in termites, nor how much nitrogen may be conserved via this pathway.

Recycling of faeces

Interspecific coprophagy

The recycling of termite faeces occurs both within and between species. Interspecific coprophagy is best exemplified by two Australian genera, *Ahamitermes* and *Incolitermes*, which subsist entirely on the faecal carton in the nurseries of *Coptotermes* nests (Calaby 1956; Gay & Calaby 1970). The inquiline colonies have no access to the outside, and alates mix with the host alates at swarming time. Clearly the carton must be sufficiently nitrogenous to allow growth of the inquilines, and it has been found that the faeces of *Coptotermes lacteus* contain minute fragments of undigested plant tissue (Gay *et al.* 1955). It is likely that interspecific consumption of faeces is common since facultative nest-sharing is frequently encountered, particularly in the Termitinae, all of which build their nests from faeces. In an area of rain forest in Cameroun, 31% of nests were occupied by species other than the original builders (Collins 1977a, 1980). Well-established nest-sharing, such as the use of *Hospitalitermes* nests by *Termes*, may be based on recycling of host faeces by the guest, which offers extra defence in return.

The brittle carton nests built by many wood-feeding Nasutitermitinae do not appear to be suitable for recycling. Few of the nasute nests found in rain forest in Cameroun, Ecuador and Sarawak contain guest species (Collins, unpublished data). It has been found that whereas *Nasutitermes exitiosus* can survive on the faeces of *Coptotermes lacteus*, the reverse is not so (Gay *et al.* 1955). Although nests of *Nasutitermes exitiosus* contain 0.095–0.14% nitrogen, microbial breakdown of organic matter in abandoned mounds is extremely slow due to lack of carbohydrate substrate and the presence of toxic humic acids (Lee & Wood 1971a & b). Similarly, the assimilation efficiency of *Nasutitermes ephratae* is so high (75–85%), that few extractable nutrients remain in the faeces (Seifert & Becker 1965).

Intraspecific coprophagy

Intraspecific recycling of faeces through proctodeal feeding is a characteristic of the lower termites, and is essential for the exchange of symbiotic flagellates (Noirot & Noirot-Timothée 1969). Undoubtedly, proctodeal feeding is also a nutrient conservation strategy, but there are no data on this matter. The higher termites do not generally practise proctodeal feeding, but it seems that many nasutes, including *Nasutitermes exitiosus*, reingest carton during nest construction and expansion, thus digesting some of the nutrients not decomposed previously (Lee & Wood 1971a).

The best-known examples of intraspecific coprophagy are in the higher termite subfamily Macrotermitinae. These termites build their faeces into comb-like structures upon which fungi of the basidiomycete genus *Termitomyces* are cultured. Some authors considered these fungus combs to be made of masticated foodstuff, but the main advocate, P.P. Grassé, has recently agreed to at least a partially proctodeal origin (Grassé 1978). Rohrmann (1978) also concluded, mainly from chemical analyses, that combs were constructed from food material with some faeces added. However, Sands' (1960) observation of a solely faecal origin for combs has been confirmed by many field workers in West and East Africa (Wood 1978; Josens 1971; Ruyooka 1980). The fungal symbiosis appears to be essential to the long-term survival of colonies (Sands 1956; Ausat *et al.* 1962), although a *Macrotermes* colony has been kept alive for 18 months without combs (Grassé 1959).

The fungus combs are dynamic structures, constantly being eaten away from the underside while new material is added above. The termites are stimulated to feed on the comb when it reaches a certain state (Alibert 1964), which may take 5–8 weeks (Collins 1977b; Josens 1971). There is evidence that the combs help to regulate nest temperature (Lüscher 1951) and humidity (Rohrmann 1977). In addition, the fungus produces a cellulolytic enzyme which is acquired by the termites as they feed on the comb (Martin & Martin 1978, 1979; Abo-Khatwa 1978). Although some cellulose digestion occurs in the first passage of food through the gut (Sands 1969; Seifert & Becker 1965), the lignin/cellulose ratios in the faeces are similar to plant tissue and quite different from those reported for the faecal carton of other termites (Lee & Wood 1971a). Grassé & Noirot (1957, 1958) believed that lignin breakdown was the most important function of the combs. Since *Termitomyces* is a white-rot basidiomycete, it does indeed have this capability (Cowling 1961; Wilcox 1973; Schubert 1965). New comb and faeces stain strongly for lignin (Grassé 1959), which is degraded by the fungus, thus exposing cellulose which can be digested by hindgut bacteria (Grassé & Noirot 1958). It seems likely that in this process of lignin degradation, some bound proteins are released, but this has not been demonstrated.

The importance of the fungus combs as a strategy for nitrogen metabolism and conservation has been independently suggested by several workers (Collins 1977b; Matsumoto 1976; Rohrmann 1977). The influence of free-living fungi in translocating and immobilizing nitrogen from dead wood applies equally to *Termitomyces*. Estimation of respiration rates of a population of *Macrotermes bellicosus* has shown that the annual respiratory budget of the fungus combs is over 5.5 times that of the termites (Collins 1977b). This clearly indicates that the major part of the metabolism of food brought in by the termites is carried out by the combs. A similar conclusion was drawn for a population of *Macrotermes ukuzii* (Rohrmann 1977). The high losses of respiratory CO_2 from the combs causes a decrease in the carbon/nitrogen ratios, which is reflected in the comb nitrogen content (Table 18.3). The temperature of *Macrotermes bellicosus* nests is maintained at 30–31°C, which is optimal for fungal metabolism (Collins 1977b).

Estimates of the fungal tissue content of combs range from 27–33% in *Macrotermes ukuzii* (Rohrmann & Rossman 1980) to 15% in *Odontotermes badius* (Haviland) (Cmelik & Douglas 1970). The nitrogen content of the combs (average 1.4%) and particularly of the fungal conidiophores (7.3%) is high compared to the plant litter that the Macrotermitinae consume (see Table 18.3). Much of this is proteinaceous nitrogen, some of which may be bound to polyphenols. The conidiophores on the combs have been estimated as 38% protein and the fungus is able to supply all the amino acids determined from proteins of *Macrotermes falciger* alates (Rohrmann & Rossman 1980). The conidiophores are an important source of nitrogenous material for growing larvae and nymphs, which may feed on them either directly or

TABLE 18.3. Nitrogen content of Macrotermitinae fungus combs.

Macrotermes species	Whole comb	New comb	Middle comb	Old comb	Conidio-phores	Reference
M. bellicosus	0.82 } 0.96	0.85	0.80	0.82	6.68	Thomas 1981
M. bellicosus	1.10	—	—	—	—	Hesse 1957
M. michaelseni	1.20	1.22	1.11	1.18	7.90	Abo-Khatwa 1977
M. ukuzii	1.43	1.52	—	1.34	—	Rohrmann & Rossman 1980
M. natalensis	1.18	1.24	—	1.11	7.13	
M. carbonarius	1.70	—	—	—	7.30	Matsumoto 1976
M. malaccensis	1.50	—	—	—	7.40	
M. falciger	1.50 } 1.55	—	—	—	—	Hesse 1957
M. falciger	1.60	—	—	—	—	Cmelik & Douglas 1970
M. subhyalinus	2.00	—	—	—	—	Hesse 1957
Odontotermes badius	1.29	—	—	—	—	Cmelik & Douglas 1970
O. redemanni	1.33	—	—	—	—	Joachim & Kandiah 1940
Mean	1.41	1.21	0.96	1.11	7.28	

trophallactically. The conidiophores are even attractive to the doryline ant *Aenictus* sp., which has been seen to strip fungus combs during a raid on a colony of *Macrotermes michaelseni* (J.P.E.C. Darlington, pers. comm.).

Chitin comprises 2.7% of fungal conidiophores, 0.8% of old comb and 1.0% of new comb in *Macrotermes ukuzii* (Rohrmann & Rossman 1980). Chitinase activity has been detected in the termite guts, but it is not known whether this is a capacity of the termites themselves or of their gut microbiota (Rohrmann & Rossman 1980).

The concentration of uric acid in combs of *Macrotermes michaelseni* drops from 0.66 mg g^{-1} (dry wt.) in new comb to 0.33 mg g^{-1} in old comb, and this reduction has been assumed to be due to uricase activity by the fungus (Abo-Khatwa 1977). No reference has been found to this ability in fungi, which are known to store urea (Martin 1979). In addition, it is not known whether the Macrotermitinae themselves have gut bacteria capable of uricase activity.

To conclude this section, it appears that the main function of the symbiotic fungus *Termitomyces* in the nests of the Macrotermitinae is as an external digestive system capable of providing both the carbohydrate and amino acid requirements of the termites. In some species, the building of complex nest structures facilitates heating and humidification of the nest atmosphere to levels suitable for both the termites and the fungus (Collins 1979b).

Cannibalism and necrophagy

Examples of cannibalism and necrophagy are given in Table 18.4. Three main functions of these behaviour patterns have been suggested:

1. Nest sanitation.
2. Caste proportion control.
3. Conservation of nitrogenous compounds.

Although Moore (1969) considered controlled cannibalism to be an important factor in the nitrogen economy of termites, there is evidence of cannibalism when nutrition is apparently adequate (Hendee 1935; Grassé 1949; Nutting 1969). *Zootermopsis angusticollis* is regularly cannibalistic, even on a diet of rotting wood similar to that devoured in nature (Hendee 1935). Moreover, attacked individuals appear quite healthy, unlike the wounded individuals attacked by *Reticulitermes* and *Cubitermes* (Buchli 1950; Williams 1959). Nevertheless, increased cannibalism occurs in *Zootermopsis* colonies fed on sound, fungus-free wood, or on diets lacking nitrogen (Hendee 1935; Andrew 1930; Cook & Scott 1933). Similarly, protein starvation and severe injury lead to increased cannibalism in *Reticulitermes lucifugus* var. *santonensis* (Feytaud) (Dhanarajan 1978). The stimulus to attack has been found to be haemolymph, and in colonies under feeding stress a higher proportion of individuals are damaged through soliciting behaviour (Dhanarajan 1978). There have been no reports of species in which cannibalism occurs *only* when under feeding stress, but it is apparent that this behaviour would serve to conserve nutrients to some extent.

It appears to be general practice for termites to kill and eat individuals of a caste which are superfluous to the needs of the colony (La Fage & Nutting 1978). Supernumerary reproductives are eliminated by *Kalotermes flavicollis* (F.) (Grassé & Noirot 1960; Lüscher 1952) and by *Neotermes jouteli* (Banks) (Nagin 1972). In *Kalotermes*, fighting between reproductives leads to injuries, and injured parties are eaten by larvae and nymphs (Ruppli & Lüscher 1964). This agrees with observations that haemolymph stimulates cannibalism (Dhanarajan 1978).

The stimulus for necrophagy in *Reticulitermes*, and in certain ants, is fatty acids (Dhanarajan 1978). Badly decomposed cadavers are normally walled up in blind chambers. *Amitermes laurensis* Mjoberg stores cadavers in peripheral nest cavities and Gay & Calaby (1970) consider these to be food stores. In

TABLE 18.4. Cannibalism and necrophagy in termites.

Species	Victims	Reference
Kalotermitidae		
Kalotermes flavicollis	excess reproductives	Lüscher 1952 Ruppli 1969
Neotermes jouteli	excess reproductives and soldiers	Grassé & Noirot 1946, 1960
Pterotermes occidentis	at least soldiers	W.L. Nutting, pers. comm.
Incisitermes minor	at least soldiers	W.L. Nutting, pers. comm.
Paraneotermes simplicicornis	at least soldiers	W.L. Nutting, pers. comm.
Hodotermitidae		
Zootermopsis angusticollis	wounded	Castle 1934
Rhinotermitidae		
Reticulitermes lucifugus	brood & wounded	Dhanarajan 1978 Buchli 1950
Reticulitermes flavipes	brood	Snyder & Popenoe 1932
Coptotermes lacteus	alates	Ratcliffe, Gay & Greaves 1952
Termitidae		
Amitermes laurensis	cadavers	Gay & Calaby 1970
Trinervitermes bettonianus	cadavers	Bouillon 1970
Cubitermes ugandensis	brood & wounded	Williams 1959
Macrotermes michaelseni	cadavers	J.P.E.C. Darlington, pers. comm.

nature, *Macrotermes bellicosus* (Smeathman) and *M. michaelseni* are only known to wall up cadavers following extensive mortalities (Collins, unpublished data; J.P.E.C. Darlington, pers. comm.). Soldier head capsules are commonly found in faecal dumps and abandoned galleries of the Arizona dry-wood termites *Pterotermes occidentis* (Walker), *Incisitermes minor* and *Paraneotermes simplicicornis*, but it is not known whether these are cannibalized or eaten as cadavers (W.L. Nutting, pers. comm.). Although cadavers of foreign species are normally walled up (Dhanarajan 1978), gnawed head capsules of *Hodotermes* soldiers have been found in the bottom of a nest of *Macrotermes michaelseni*, presumably having been eaten after a battle with the residents (J.P.E.C. Darlington, pers. comm.).

Proteinaceous nitrogen from the bodies of nest-mates would presumably be digested by the termites, and there is evidence that chitin and uric acid might also be recycled. Chitin is 7% nitrogen and chitinase activity has been demonstrated in species from most of the important families (see Table 18.1). It has been estimated that chitin comprises 2% (dry wt.) of *Macrotermes falciger* alates (Rohrmann & Rossman 1980).

THE IMPACT OF NITROGEN UTILIZATION STRATEGIES ON TERMITE POPULATIONS AND COMMUNITIES

The only nitrogen strategy for which there are enough data for an impact analysis on the population level is the recycling of faeces by the fungus-growing Macrotermitinae. These termites do not use their faeces for building nests, nor for lining galleries, as is so commonly found in other groups. Some fungus combs in the nests of *Macrotermes bellicosus* may consist of old faecal dumps (Collins 1977b), and small amounts of amorphous waste material have been found in *M. michaelseni* nests (J.P.E.C. Darlington, pers. comm.). However, waste material only leaves *Macrotermes* nests by death of the whole colony, or by predation on foragers. It might therefore be supposed that utilization of foodstuffs is very efficient. Unless sufficient excretory uric acid is lost from the colony via predatory pathways, it must be postulated that either the termite gut microbiota or the fungus combs have a uricolytic capacity. This hypothesis has yet to be tested.

Macrotermes bellicosus feeds on wood and leaf litter in the West African savannas, while *Macrotermes michaelseni* consumes grass litter and dung in similar environments in East Africa. In Table 18.5, data on caste proportions, individual biomass and nitrogen content have been used to calculate the average nitrogen content of 100 adult neuters of each species. In Table 18.6, the annual productivity and the proportion of dietary nitrogen accountable in

TABLE 18.5. Calculation of the nitrogen content of 100 individuals of *Macrotermes bellicosus* and *M. michaelseni*, taking caste proportions and caste nitrogen content into account.

	M. bellicosus[1]				M. michaelseni[3]			
Caste	Major soldier	Minor soldier	Major worker	Minor worker	Major soldier	Minor soldier	Major worker	Minor worker
Caste proportions of adult neuters (%)	1.58	2.33	24.45	71.64	1.37	2.55	34.25	61.83
Dry weight per individual (mg)	8.767	1.498	2.609	1.008	23.77	3.60	4.26	2.36
Proportional weight in 100 individuals (mg)	13.85	3.49	63.79	72.21	32.56	9.18	145.91	145.92
Nitrogen content per caste (%)[2]	7.9	7.7	6.1	4.5	7.9	7.7	6.1	4.5
Proportional weight of nitrogen in 100 individuals (mg)	1.09	0.27	3.89	3.25	2.57	0.71	8.90	6.57

[1] From Collins 1977b.
[2] Buhlmann's (1977) data for *M. michaelseni* have been used for both species.
[3] Caste proportions and biomass of *M. michaelseni* from J.P.E.C. Darlington (pers. comm.).

that productivity are calculated. Production of neuters is calculated from the populations of larvae and their development time from hatching, as estimated from laboratory incipient colonies (*M. bellicosus*), or by removal of queens from nests and resampling at a later date (*M. michaelseni*, J.P.E.C. Darlington, pers. comm.). It is assumed that the entire larval population reaches maturity and is turned over once during a developmental period. The adult neuters of *M. bellicosus* are turned over 15.2 times per year, giving an average lifespan of only 3–4 weeks. In the larger species, *M. michaelseni*, a longer developmental time for larvae and a lower larval proportion (45% as opposed to 58.77%) give an adult neuter turnover of 6.6 times per year and a lifespan of 7–8 weeks. The nitrogen content of the sterile castes recruited annually is estimated as 686 g ha^{-1} for *M. bellicosus*, and 6376 g ha^{-1} for *M. michaelseni*, which has far larger populations. Alate production accounts for a a further 73 g N ha^{-1} $year^{-1}$ in *M. bellicosus* (Collins 1981b) and 1085 g ha^{-1} $year^{-1}$ in *M. michaelseni*, giving total values for nitrogen required as 759 g ha^{-1} $year^{-1}$ and 7461 g ha^{-1} $year^{-1}$ respectively.

The field estimate of consumption by *M. bellicosus* is 241 kg ha^{-1} $year^{-1}$ of wood and leaf litter, and is complete except for a small amount of tree bark (Collins 1981d). The nitrogen content of wood litter at the site has been

TABLE 18.6. Calculations of the proportion of dietary nitrogen input used in production by *Macrotermes bellicosus* and *M. michaelseni*.

	M. bellicosus	*M. michaelseni*
Larval development time	34.25 days	45 days[1]
Turnover of larvae year^{-1}	10.66	8.11
Population of larvae ha^{-1}	757×10^3[5]	4193×10^3[1]
Number of adults developing year^{-1}	8070×10^3	$34\,005 \times 10^3$
Biomass per adult sterile	1.533 mg	3.336 mg
N content per adult sterile	0.085 mg (5.54%)	0.1875 mg (5.62%)
N content of annual adult sterile recruits	685 g	6376 g
Number of alates produced ha^{-1} year^{-1}	31 570	151 000[1]
Biomass of alates produced ha^{-1} year^{-1}	1140 g	16 960 g
N content of annual alate recruits	73 g (6.4%)[6]	1085 g (6.4%)[6]
N flux in production ha^{-1} year^{-1}	758 g	7461 g
Estimated consumption ha^{-1} year^{-1}	241 kg[5]	1250–1600 kg[2]
N content of food	0.28%[4]	0.52%[3]
N content of annual food consumed	603–747 g	6.5–8.32 kg
% of dietary N used in production	101–126%	90–115%

[1] J.P.E.C. Darlington, pers. comm.
[2] Lepage's (1979) data for the non-drought year 1977 do not include consumption of dung.
[3] M.A. Arshad, pers. comm.
[4] This value is for wood litter and makes no allowance for leaf litter consumed, which would have a higher N content (Thomas 1981).
[5] Collins 1981b, 1981d.
[6] Buhlmann 1977.

estimated at 0.28%±0.03 (Thomas 1981). The nitrogen content of leaf litter, which constituted 26% of the diet of *M. bellicosus*, is unfortunately not known, but would probably be slightly higher than for wood litter. Therefore in Table 18.6, where 0.25–0.31% has been used as an overall range for food nitrogen content, the value of 101–126% utilization of nitrogen is a slight overestimate. A nitrogen content of 0.315% for food consumed would suggest 100% utilization.

Field consumption of grass by *M. michaelseni* has been estimated at 1250–1600 kg ha^{-1} year^{-1} in a non-drought year (Lepage 1979). The grass litter consumed is estimated to be 0.52% nitrogen (M.A. Arshad, pers. comm.), and the annual dietary nitrogen input from grass is therefore 6.5–8.3 kg ha^{-1} year^{-1}. Unfortunately, the annual consumption of dung, a minor food source for this species, has not been estimated. The calculation in Table 18.6 of 90–115% utilization of dietary nitrogen is therefore an overestimate. Although it is not possible to refine these figures further, the data clearly support the conclusion that 90–100% of dietary nitrogen input is utilized in production by these species.

There is only one other comparable study of nitrogen assimilation by termites, in which the lower termite *Zootermopsis nevadensis* (Hagen) assimilated about 50% of its dietary nitrogen (Hungate 1944). This study was on individuals in the laboratory rather than populations in the field, and although it was considered to be a very efficient level of utilization, it is clearly low compared with these remarkable estimates for *Macrotermes*.

The impact of such efficiency in *Macrotermes* is far-reaching, since very high production/biomass ratios result. Calculations in Table 18.7 show nitrogen-based production/biomass ratios of 13.3 for *M. bellicosus* and 7.2 for *M. michaelseni*. These results are slightly different to the more usual energetic-based results of 13.2 and 7.4, respectively, given in Table 18.8. The figure for *M. bellicosus* appears to be rather high compared to other species, possibly due to underestimated adult neuter populations. This would not affect the calculations of nitrogen utilization.

It is apparent from Table 18.8 that the annual production/biomass ratios for fungus growers is 2–4 times higher than for the rhinotermitid and nasute species, clearly emphasizing the value of the fungus combs. Production by groups other than the Macrotermitinae is typical of other invertebrates, at

TABLE 18.7. Calculation of nitrogenous production/biomass ratios for *Macrotermes bellicosus* and *M. michaelseni*.

	M. bellicosus	*M. michaelseni*[2]
Population of all steriles ha^{-1}	1288×10^3	9317×10^3
Population of adult neuters ha^{-1}	531×10^3	5124×10^3
Biomass of all steriles ha^{-1}	1.062 kg	18.490 kg
Biomass of larvae ha^{-1}	0.248 kg	1.396 kg
Biomass of adult neuters ha^{-1}	0.814 kg	17.094 kg
Total N content of population ha^{-1}	57 g	1029 g
N content of neuters ha^{-1}[1]	45 g	961 g
N content of larvae ha^{-1}	12 g	68 g
Biomass of sterile recruits ha^{-1} $year^{-1}$	12.371 kg	113.441 kg
Biomass of alate recruits ha^{-1} $year^{-1}$	1.14 kg	16.960 kg
Total biomass recruits ha^{-1} $year^{-1}$	13.511 kg	130.401 kg
N content of sterile recruits ha^{-1} $year^{-1}$	685 g	6375 g
N content of alate recruits ha^{-1} $year^{-1}$	73 g	1085 g
Total N used in production ha^{-1} $year^{-1}$	758 g	7460 g
Nitrogenous production/biomass ratio	13.3	7.2
Alate production/sterile biomass ratio	1.3	1.0

[1] N content of larvae = 4.9% (Buhlmann 1977), mean N content of adult neuters = 5.54% (*M. bellicosus*) or 5.62% (*M. michaelseni*) (see Table 18.5).
[2] Population and biomass data for *M. michaelseni* from J.P.E.C. Darlington (pers. comm.).

TABLE 18.8. Energetic production/biomass ratios for termites. Values for species other than *Macrotermes bellicosus* and *M. michaelseni* calculated by Wood & Sands (1978). Energetic values calculated from average figures for neuters 21.8 kJ g^{-1}, and alates 31.0 kJ g^{-1} (Wood & Sands 1978) , except *M. bellicosus* measured directly (Collins 1981b).

Species	Population kJ m^{-2}	Production kJ m^{-2} $year^{-1}$ Neuters	Alates	Production/ biomass ratio
Rhinotermitidae				
Psammotermes hybostoma	0.720	2.394	0.029	3.4
Macrotermitinae				
Ancistrotermes cavithorax	0.975	9.207	0.745	10.2
Odontotermes smeathmani	0.720	2.415	1.456	5.4
M. subhyalinus	2.088	9.684	3.842	6.5
M. bellicosus	2.137	24.605	3.670	13.2
M. michaelseni	40.31	247.3	52.58	7.4
Nasutitermitinae				
**Trinervitermes geminatus*	0.067	0.130	0.088	3.3
**Trinervitermes geminatus*	3.022	3.139	1.318	1.5
Trinervitermes trinervius	0.239	0.393	0.557	3.9

* Josens (1982) has recently estimated a P/B ratio of 2.62 for laboratory colonies of *T. geminatus*, supporting these earlier calculated values.

about 3:1 (Phillipson 1973). The Macrotermitinae are not only efficient in reducing nitrogenous plant material to a form available to the termites, but since the combs are in effect an external digestive system, the amount of material processed is greatly increased. This is reflected in very high consumption rates by Macrotermitinae (Collins 1981c). For their part, the termites have developed nests suitable for accommodation, protection and tending of the fungus combs. The highly complex nests built by *Macrotermes*, arguably the most advanced built by any invertebrate, are witness to this coevolution (Collins 1979b).

In the relatively dry savannas of Africa, the activity of most decomposers is limited by the slow breakdown of litter by free-living microbes and fungi (Collins 1981d; Wood 1976). The Macrotermitinae, however, with their symbiotic fungus maintained in the humid atmosphere of the nest, are able to process large amounts of litter and, in effect, turn it into termites. *Macrotermes bellicosus* is only one of about ten species of Macrotermitinae in the Southern Guinea savanna of Nigeria. Between them they take 60% of annual wood litter production, 60% of grass litter production and 3% of leaf litter production (49% of leaves are burnt in fires) (Collins 1981d; Ohiagu & Wood 1979). This amounts to 35% of total annual litter produced (Collins 1981d). As a result of their high productivity, fungus-growing termites are central to many predatory food webs in African savannas. Ants are particularly important predators (Longhurst 1978; Longhurst *et al.* 1978), but many species of opportunistic birds and mammals, including man, also take termites (Wood & Sands 1978). In Africa, termites are the only insects which entirely support one large mammal, the aardvark (*Orycteropus afer* (Pallus)), and are the main foodstuff of at least five others, the aardwolf (*Proteles cristatus* Sparrman) and four species of pangolins (*Manis* spp.) (Dorst & Dandelot 1972; Kruuk & Sands 1972).

ACKNOWLEDGEMENTS

I am grateful to M. A. Arshad, B. L. Bentley, B. Coppins, P. G. McDowell, W. L. Nutting, G. D. Prestwich, N. Sammy, D. A. Schaefer, T. D. V. Swinscow and W. G. Whitford for personal communications and permission to use unpublished material. I particularly thank J. P. E. C. Darlington who has so generously allowed me to use her unparalleled data on *Macrotermes michaelseni*. The field work on *M. bellicosus* was done while I was with the Centre for Overseas Pest Research, London.

REFERENCES

Abe T. (1980) Studies on the distribution and ecological role of termites in a lowland rain forest of West Malaysia (4) The role of termites in the process of wood decomposition in Pasoh Forest Reserve. *Revue d'Écologie et de Biologie du Sol*, **17**, 23–40.

Abo-Khatwa N. (1977) Natural products from the tropical termite *Macrotermes subhyalinus*: chemical composition and function of 'fungus gardens'. In *Natural Products and the Protection of Plants: Pontificiae Academiae Scientiarum Scripta Varia*, **41**, 447–67, Elsevier, Amsterdam.

Abo-Khatwa N. (1978) Cellulase of fungus-growing termites: A new hypothesis on its origin. *Experientia*, **34**, 559–60.

Alibert J. (1964) L'évolution dans le temps des meules à champignons construites par les termites. *Comptes Rendus Hebdomadaires des Séances de l'Académie des Sciences, Paris*, **258**, 5260–3.

Amburgey T.L. & Beal R.H. (1977) White rot inhibits termite attack. *Sociobiology*, **3**, 35–8.

Andrew B.J. (1930) Method and rate of protozoan refaunation in the termite *Termopsis angusticollis* Hagen. *University of California Publications in Zoology*, **33**, 449–70.

Aronson J.M. (1965) The cell wall. In *The Fungi* (eds G.C. Ainsworth & A.S. Sussman), pp. 49–76. Academic Press, London & New York.

Ausat A., Cheema P.S., Koskhi T., Petri S.L. & Ranganathan S.K. (1962) Laboratory culturing of termites. In *Proceedings of the New Delhi Symposium 1960, Termites in the Humid Tropics*, pp. 121–5. UNESCO, Paris.

Bartnicki-Garcia S. (1970) Cell wall composition and other biochemical markers in fungal phylogeny. In *Phytochemical Phylogeny* (ed. J.B. Harborne), pp. 81–103. Academic Press, London & New York.

Becker G. (1971) Physiological influences on wood-destroying insects of wood compounds and substances produced by micro-organisms. *Wood Science and Technology*, **5**, 236–46.

Benemann J.R. (1973) Nitrogen fixation in termites. *Science*, **181**, 164–5.

Bernays E.A. (1978) Tannins: an alternative viewpoint. *Entomologia Experimentalis et Applicata*, **24**, 244–53.

Bocock K.L. (1963) Changes in the amount of nitrogen in decomposing leaf litter of sessile oak (*Quercus petraea*). *Journal of Ecology*, **51**, 555–66.

Botha T.C. & Hewitt P.H. (1978) Influence of diets containing green material on laboratory colonies of *Hodotermes mossambicus*. *Phytophylactica*, **10**, 93–8.

Bouillon A. (1970) Termites of the Ethiopian region. In *Biology of Termites* (eds K. Krishna & F.M. Weesner), Vol. 2, pp. 153–280. Academic Press, London & New York.

Bremner J.M. & Shaw K. (1954) Studies on the estimation and decomposition of amino-sugars in soil. *Journal of Agricultural Science*, **44**, 152–9.

Breznak J.A. (1975) Symbiotic relationships between termites and their intestinal microbiota. In *Symbiosis* (eds D.H. Jennings & D.L. Lee), pp. 559–80. Cambridge University Press, Cambridge.

Breznak J.A., Brill W.J., Mertins J.W. & Coppell H.C. (1973) Nitrogen fixation in termites. *Nature*, **244**, 577–80.

Buchli H. (1950) Recherche sur la fondation et le développement des nouvelles colonies chez le termite lucifuge (*Reticulitermes lucifugus* (Rossi)). *Physiologia Comparata et Oecologia*, **2**, 145–60.

Buhlmann G. (1977) Termite physiology and caste differentiation. *1976 Annual Report of the International Centre of Insect Physiology & Ecology,* Nairobi, Kenya, pp. 55–6.

Butler J.H.A. & Buckerfield J.C. (1979) Digestion of lignin by termites (*Nasutitermes exitiosus*). *Soil Biology and Biochemistry*, **11**, 507–14.

Calaby J.H. (1956) The distribution and biology of *Ahamitermes* (Isoptera). *Australian Journal of Zoology*, **4**, 111–24.

Carter F.L. & Smythe R.V. (1973) Effect of sound and *Lenzites*-decayed wood on the amino-acid composition of *Reticulitermes flavipes*. *Journal of Insect Physiology*, **19**, 1623–9.

Castle G.B. (1934) The damp-wood termites of western United States, genus *Zootermopsis* (formerly *Termopsis*). In *Termites & Termite Control* (ed. C.A. Kofoid), 2nd edn, pp. 273–310. University of California Press, Berkeley.

Cleveland L.R. (1925) The ability of termites to live perhaps indefinitely on a diet of pure cellulose. *Biological Bulletin,* Marine Biological Laboratory, Woods Hole, **48**, 289–93.

Cleveland L.R. (1926) Symbiosis among animals with special reference to termites and their intestinal flagellates. *Quarterly Review of Biology*, **1**, 51–60.

Cmelik S.H.W. & Douglas C.C. (1970) Chemical composition of 'fungus gardens' from two species of termites. *Comparative Biochemistry and Physiology*, **36**, 493–502.

Collins N.M. (1977a) Oxford Expedition to the Edea-Marienberg Forest Reserve, United Republic of Cameroon 1973. *Bulletin, Oxford University Exploration Club*, New Series **3**, 5–15.

Collins N.M. (1977b) The population ecology and energetics of *Macrotermes bellicosus* (Smeathman), Isoptera. Unpublished Ph.D. thesis, University of London.

Collins N.M. (1979a) Observations on the foraging activity of *Hospitalitermes umbrinus* (Haviland) in the Gunong Mulu National Park, Sarawak. *Ecological Entomology*, **4**, 231–8.

Collins N.M. (1979b) The nests of *Macrotermes bellicosus* (Smeathman) from Mokwa, Nigeria. *Insectes Sociaux*, **26**, 240–6.

Collins N.M. (1980) Inhabitation of epigeal termite (Isoptera) nests by secondary termites in Cameroun rain forest. *Sociobiology*, **5**, 47–54.

Collins N.M. (1981a) The impact of feeding by *Macrotermes* on semi-arid pastures. *1980 Annual Report of the International Centre of Insect Physiology & Ecology,* Nairobi, Kenya.

Collins N.M. (1981b) Populations, age structure and survivorship of colonies of *Macrotermes bellicosus* (Smeathman) (Isoptera: Macrotermitinae). *Journal of Animal Ecology*, **50**, 293–311.

Collins N.M. (1981c) Consumption of wood by artificially isolated colonies of the fungus-growing termite *Macrotermes bellicosus*. *Entomologia Experimentalis et Applicata*, **29**, 313–20.

Collins N.M. (1981d) The role of termites in the decomposition of wood and leaf litter in the Southern Guinea savanna of Nigeria. *Oecologia (Berlin)*, **51**, 389–99.

Collins N.M. (1982) The interaction and impact of domestic stock and termites in a Kenyan rangeland. In *The Biology of Social Insects* (eds M.D. Breed, C.D. Michener & H.E. Evans), pp. 80–4. Westview Press, Boulder.

Collins N.M. (in press) The termites (Isoptera) of the Gunung Mulu National Park, with a key to the genera known from Sarawak. *Sarawak Museum Journal Supplement.*

Cook S.F. & Scott K.G. (1933) The nutritional requirements of *Zootermopsis angusticollis*. *Journal of Cellular and Comparative Physiology*, **4**, 95–110.

Cowling E.B. (1961) Comparative biochemistry of the decay of sweetgum sapwood by white-rot and brown-rot fungi. *United States Department of Agriculture Technical Bulletin, 1258.*

Cowling E.B. & Merrill W. (1966) Nitrogen in wood and its role in wood deterioration. *Canadian Journal of Botany*, **44**, 1539–54.

Dhanarajan G. (1969) The termite fauna of Malaya and its economic importance. *Malaysian Forester*, **22**, 274–8.

Dhanarajan G. (1978) Cannibalism and necrophagy in a subterranean termite (*Reticulitermes lucifugus* var. *santonensis*). *Malayan Nature Journal*, **31**, 237–51.

Dorst J. & Dandelot P. (1972) *A Field Guide to the Larger Mammals of Africa*, 2nd edn, Collins, London.

Dougall H.W., Drysdale V.M. & Glover P.E. (1964) The chemical composition of Kenya browse and pasture herbage. *East African Wildlife Journal*, **2**, 86–121.

Eutick M.L., O'Brien R.W. & Slaytor M. (1978) Bacteria from the gut of Australian termites. *Applied and Environmental Microbiology*, **35**, 823–8.

Feeny P.P. (1976) Plant apparency and chemical defence. In *Biochemical Interaction Between Plants and Insects* (eds J.W. Wallace & R.L. Mansell). pp. 1–40. Plenum Press, New York.

Ferrar P. & Watson J.A.L. (1970) Termites (Isoptera) associated with dung in Australia. *Journal of the Australian Entomological Society*, **9**, 100–2.

Fittkau E.J. & Klinge H. (1973) On biomass and trophic structure of the central Amazonian rain forest ecosystem. *Biotropica*, **5**, 2–14.

Forman R.T.T. (1975) Canopy lichens with blue-green algae: a nitrogen source in a Colombian rain forest. *Ecology*, **56**, 1176–84.

French J.R.J., Turner G.L. & Bradbury J.F. (1976) Nitrogen fixation by bacteria from the hind gut of termites. *Journal of General Microbiology*, **95**, 202–6.

Gay F.J. & Calaby J.H. (1970) Termites from the Australian region. In *Biology of Termites* (eds K. Krishna & F.M. Weesner), pp. 393–448. Academic Press, London & New York.

Gay F.J., Greaves T., Holdaway F.G. & Wetherby A.H. (1955) Standard laboratory cultures of termites for evaluating the resistance of timber, timber preservatives and other materials to termite attack. *Bulletin of the Commonwealth Scientific and Industrial Research Organisation, Melbourne, 277.*

Gosz J.R., Likens G.E. & Bormann F.H. (1973) Nutrient release from decomposing leaf and branch litter in the Hubbard Brook Forest, New Hampshire. *Ecological Monographs*, **47**, 173–91.

Grassé P.P. (1949) Ordre des Isoptères ou Termites. In *Traité de Zoologie*, (ed. P.P. Grassé), Vol. **9**, pp. 408–544. Masson, Paris.

Grassé P.P. (1959) Un nouveau type de symbiose: la meule alimentaire des termites champignonnistes. *Nature (London)*, **3293**, 385–9.

Grassé P.P. (1978) Sur la véritable nature et le rôle des meules à champignons construites par les termites Macrotermitinae (Isoptera Termitidae). *Comptes Rendus Hebdomadaires des Séances de l'Académie des Sciences, Paris* (Série D) **287**, 1223–6.

Grassé P.P. & Noirot C. (1946) La production des sexués néoténiques chez le termite à cou jaune

(*Calotermes flavicollis* (F.)): inhibition germinale et inhibition somatique. *Comptes Rendus Hebdomadaires des Séances de l'Académie des Sciences*, **223**, 869–71.

Grassé P.P. & Noirot C. (1957) La signification des meules à champignons des Macrotermitinae (Ins., Isoptères). *Comptes Rendus Hebdomadaires des Séances de l'Académie des Sciences, Paris*, **244**, 1845–50.

Grassé P.P. & Noirot C. (1958) La meule des termites champignonnistes et sa signification symbiotique. *Annales des Sciences Naturelles (Zoologie)*, Série 11, **20**, 113–28.

Grassé P.P. & Noirot C. (1959) L'évolution de la symbiose chez les Isoptères. *Experientia*, **15**, 365–72.

Grassé P.P. & Noirot C. (1960) Rôle respectif des mâles et des femelles dans la formation des sexués néoténiques chez *Calotermes flavicollis*. *Insectes Sociaux*, **7**, 109–23.

Hardy R.W.F., Holsten R.D., Jackson E.K. & Burns R.C. (1968) The acetylene-ethylene assay for N_2 fixation: laboratory and field evaluation. *Plant Physiology*, **43**, 1185–207.

Harris W.V. (1969) *Termites as Pests of Crops and Trees*, Commonwealth Institute of Entomology, London.

Hendee E.C. (1933) The association of the termites, *Kalotermes minor, Reticulitermes hesperus* and *Zootermopsis angusticollis* with fungi. *University of California Publications in Zoology*, **39**, 111–34.

Hendee E.C. (1935) The role of fungi in the diet of the common damp-wood termite *Zootermopsis angusticollis*. *Hilgardia*, **9**, 499–525.

Hesse P.R. (1957) Fungus combs in termite mounds. *East African Agricultural Journal*, **23**, 104–5.

Hewitt P.H. & Nel J.J.C. (1969) Toxicity and repellancy of *Chrysocoma tenuifolia* (Berg.) (Compositae) to the harvester termite *Hodotermes mossambicus* (Hagen) (Hodotermitidae). *Journal of the Entomological Society of Southern Africa*, **32**, 133–6.

Hitch C.J.B. (1971) A study of some environmental factors affecting nitrogenase activity in lichens. Unpublished M.Sc. thesis, University of Dundee.

Honigberg B.M. (1970) Protozoa associated with termites and their role in digestion. In *Biology of Termites* (eds K. Krishna & F.M. Weesner), Vol. 2, pp. 1–36. Academic Press, London & New York.

Hungate R.E. (1938) Studies on the nutrition of *Zootermopsis* II. The relative importance of the termite and the protozoa in wood digestion. *Ecology*, **19**, 1–25.

Hungate R.E. (1940) Nitrogen content of sound and decayed coniferous woods and its relation to loss in weight during decay. *Botanical Gazette*, **102**, 389–92.

Hungate R.E. (1941) Experiments on nitrogen economy of termites. *Annals of the Entomological Society of America*, **34**, 467–89.

Hungate R.E. (1944) Termite growth and nitrogen utilisation in laboratory cultures (*Zootermopsis*). *Proceedings and Transactions of the Texas Academy of Science*, **27**, 91–8.

Jander R. & Daumer K. (1974) Guide-line and gravity orientation of blind termites foraging in the open (Termitidae: *Macrotermes, Hospitalitermes*). *Insectes Sociaux*, **21**, 45–69.

Joachim A.W.R. & Kandiah S. (1940) Studies on Ceylon soils XIV. A comparison of soils from termite mounds and adjacent land. *The Tropical Agriculturist and Magazine of the Ceylon Agricultural Society*, **95**, 333–8.

Josens G. (1971) Le renouvellement des meules à champignons construites par quatre Macrotermitinae (Isoptères) des savanes de Lamto-Pakobo (Côte d'Ivoire). *Comptes Rendus Hebdomadaires des Séances de l'Académie des Sciences, Paris*, Série D, **272**, 3329–32.

Josens G. (1982) Le bilan énergétique de *Trinervitermes geminatus* Wasmann (Termitidae, Nasutitermitinae) 1. Mesure de biomasses, d'équivalents énergétiques, de longévité et de production en laboratoire. *Insectes Sociaux*, **29**, 297–307.

Jucci C. (1921) Sulla presenza di depositi uratici nei tessuto adipose dei termitidi. *Atti dell'Accademia Nazionale dei Lincei. Memorie, Classe di Scienze Fisiche, Matematiche e Naturali*, **30**, 213–15.

Kalshoven L.G.E. (1958) Observations on the black termites, *Hospitalitermes* spp. of Java and Sumatra. *Insectes Sociaux*, **5**, 9–30.

Kovoor J. (1964a) Modifications chimiques d'une sciure de bois de peuplier sous l'action d'un termitidé: *Microcerotermes edentatus* Wasmann. *Comptes Rendus Hebdomadaires des Séances del'Académie des Sciences, Paris*, **258**, 2887–9.

Kovoor J. (1964b) Modifications chimiques provoquées par un termitidé (*Microcerotermes edentatus* Was.) dans du bois de peuplier sain ou partiellement dégradé par les champignons. *Bulletin Biologique de la France et de la Belgique*, **98**, 491–510.

Kovoor J. (1966) Contribution à l'étude de la digestion chez un termite supérieur (*Microcerotermes edentatus* Was., Isoptera, Termitidae). Unpublished doctoral thesis, University of Paris.

Kovoor J. (1970) Présence d'enzymes cellulolytiques dans l'intestin d'un termite supérieur, *Microcerotermes edentatus* Was. *Annales des Sciences Naturelles (Zoologie)*, **12**, 65–71.

Krishnamoorthy R.V. (1960) The digestive enzymes of the termite *Heterotermes indicola. Journal of Animal Morphology & Physiology*, **7**, 156–61.

Kruuk H. & Sands W.A. (1972) The aardwolf (*Proteles cristatus* Sparrman, 1783) as predator of termites. *East African Wildlife Journal*, **10**, 211–27.

La Fage J.P. & Nutting W.L. (1978) Nutrient dynamics of termites. In *Production Ecology of Ants and Termites* (ed. M.V. Brian), pp. 165–232. Cambridge University Press, Cambridge.

Leach J.G. & Granovsky A.A. (1938) Nitrogen in the nutrition of termites. *Science*, **87**, 66–7.

Lee K.E. & Wood T.G. (1971a) *Termites and Soils*. Academic Press, London & New York.

Lee K.E. & Wood T.G. (1971b) Physical and chemical effects on soils of some Australian termites and their pedological significance. *Pedobiologia*, **11**, 376–409.

Leopold B. (1952) Studies on lignin XIV. The composition of Douglas fir wood digested by the West Indian dry-wood termite (*Cryptotermes brevis* Walker). *Svensk Papperstidning*, **5**, 784–6.

Lepage M.G. (1974) Recherches écologiques sur une savane sahelienne du Ferlo septentrional, Sénégal: influence de la sécheresse sur le peuplement en termites. *Terre et la Vie*, **28**, 76–94.

Lepage M.G. (1979) La récolte en strate herbacée de *Macrotermes* aff. *subhyalinus* dans un ecosystème semi-aride (Kajiado-Kenya). *Comptes Rendus, Union Internationale pour l'Etude des Insectes Sociaux, Section Française-Lausanne 1979*, pp. 145–51.

Levi M.P., Merrill W. & Cowling E.B. (1968) Role of nitrogen in wood deterioration VI. Mycelial fractions and model nitrogen compounds as substrates for growth of *Polyporus versicolor* and other wood-destroying and wood-inhabiting fungi. *Phytopathology*, **58**, 626–34.

Lincoln F.B. & Mulay A.S. (1929) The extraction of nitrogenous material from pear tissues. *Plant Physiology*, **4**, 233–50.

Longhurst C. (1978) Behavioural, chemical and ecological interactions between West African ants and termites. Unpublished Ph.D. thesis, University of Southampton.

Longhurst C., Johnson R.A. & Wood T.G. (1978) Predation by *Megaponera foetens* (Fabr.) (Hymenoptera:Formicidae) on termites in the Nigerian southern Guinea savanna. *Oecologia*, **32**, 101–7.

Lüscher M. (1951) Significance of fungus gardens in termite nests. *Nature (London)*, **167**, 34–5.

Lüscher M. (1952) Die Produktion und Elimination von Ersatzgeschlechstieren bei der Termite Kalotermes flavicollis (Fabr.). *Zeitschrift für Vergleichende Physiologie*, **34**, 123–41.

Martin M.M. (1979) Biochemical implications of insect mycophagy. *Biological Reviews*, **54**, 1–21.

Martin M.M. & Martin J.S. (1978) Cellulose digestion in the mid-gut of the fungus-growing termite *Macrotermes natalensis*. The role of acquired digestive enzymes. *Science*, **199**, 1453–5.

Martin M.M. & Martin J.S. (1979) The distribution and origins of the cellulolytic enzymes of the higher termite *Macrotermes natalensis. Physiological Zoology*, **52**, 11–21.

Matsumoto T. (1976) The role of termites in an equatorial rain forest ecosystem of West Malaysia.

I. Population density, biomass, carbon, nitrogen and calorific content and respiration rate. *Oecologia (Berlin)*, **22**, 153–78.

Mauldin J.K. & Smythe R.G. (1973) Protein-bound amino-acid content of normally and abnormally faunated Formosa termites. *Journal of Insect Physiology*, **19**, 1955–60.

Millbank J.W. & Kershaw K.A. (1974) Nitrogen metabolism. In *The Lichens* (eds V. Ahmadjian & M.E. Hale), pp. 289–307. Academic Press, London & New York.

Moore B.P. (1969) Biochemical studies in termites. In *Biology of Termites* (eds K. Krishna & F.M. Weesner), pp. 407–32. Academic Press, London & New York.

Mullins D.E. & Cochran D.G. (1975a) Nitrogen metabolism in the American cockroach. I. An examination of positive nitrogen balance with respect to uric acid stores. *Comparative Biochemistry and Physiology*, **50**, 489–500.

Mullins D.E. & Cochran D.G. (1975b) Nitrogen metabolism in the American cockroach. II. An examination of negative nitrogen balance with respect to mobilisation of uric acid stores. *Comparative Biochemistry and Physiology*, **50**, 501–10.

Nagin R. (1972) Caste determination in *Neotermes jouteli* (Banks). *Insectes Sociaux*, **19**, 39–61.

Nel J.J.C. & Hewitt P.H. (1969) A study of the food eaten by a field population of the harvester termite *Hodotermes mossambicus* (Hagen) and its relation to population density. *Journal of the Entomological Society of Southern Africa*, **32**, 123–31.

Nel J.J.C., Hewitt P.H. & Joubert L. (1970) The collection and utilisation of redgrass *Themeda triandra* (Forsk.) by laboratory colonies of the harvester termite *Hodotermes mossambicus* (Hagen) and its relation to population density. *Journal of the Entomological Society of Southern Africa*, **33**, 331–40.

Noirot C. & Noirot-Timothée C. (1969) The digestive system. In *Biology of Termites* (eds K. Krishna & F.M. Weesner), Vol. 1, pp. 49–88. Academic Press, London & New York.

Nutting W.L. (1969) Flight and colony foundation. In *Biology of Termites* (eds K. Krishna & F.M. Weesner), Vol. 1. pp. 233–82. Academic Press, London & New York.

O'Brien G.W., Veivers P.C., McEwen S.E., Slaytor M. & O'Brien R.W. (1979) The origin and distribution of cellulase in the termites *Nasutitermes exitiosus* and *Coptotermes lacteus*. *Insect Biochemistry*, **9**, 619–26.

Ohiagu C.E. (1978) Laboratory tests on food preferences of *Trinervitermes geminatus* (Isoptera, Nasutitermitinae). *Entomologia Experimentalis et Applicata*, **23**, 110–14.

Ohiagu C.E. & Wood T.G. (1979) Grass production and decomposition in Southern Guinea savanna, Nigeria. *Oecologia (Berlin)*, **40**, 155–65.

Pence R.J. (1956) The tolerance of the drywood termite *Kalotermes minor* (Hagen) to desiccation. *Journal of Economic Entomology*, **49**, 553–4.

Phillipson J. (1973) The biological efficiency of protein production by grazing and other land-based systems. In *The Biological Efficiency of Protein Production* (ed. J.G.W. Jones), pp. 217–35. Cambridge University Press, London.

Potrikus C.J. & Breznak J.A. (1977) Nitrogen-fixing *Enterobacter agglomerans* isolated from guts of wood-eating termites. *Applied and Environmental Microbiology*, **33**, 392–9.

Potrikus C.J. & Breznak J.A. (1980a) Uric acid in wood-eating termites. *Insect Biochemistry*, **10**, 19–28.

Potrikus C.J. & Breznak J.A. (1980b) Uric acid degrading bacteria in guts of termites (*Reticulitermes flavipes* (Kollar)). *Applied and Environmental Microbiology*, **40**, 117–24.

Potts R.C. & Hewitt P.H. (1973) The distribution of intestinal bacteria and cellulase activity in the harvester termite *Trinervitermes trinervoides* (Nasutitermitinae). *Insectes Sociaux*, **20**, 215–20.

Prestwich G.D. & Bentley B.L. (1981) Nitrogen fixation by intact colonies of the termite *Nasutitermes corniger*. *Oecologia (Berlin)*, **49**, 249–51.

Prestwich G.D., Bentley B.L. & Carpenter E.J. (1980) Nitrogen sources for neotropical nasute termites, fixation and selective foraging. *Oecologia*, **46**, 379–401.

Rao K.P. (1962) Occurrence of enzymes for protein digestion in the termite *Heterotermes indicola*. In *Termites in the Humid Tropics, Proceedings of the New Delhi Symposium 1960*, pp. 71–2. UNESCO, Paris.

Ratcliffe F.N., Gay F.J. & Greaves T. (1952) *Australian Termites. The Biology, Recognition and Economic Importance of the Common Species*. Commonwealth Scientific and Industrial Research Organisation, Melbourne.

Retief L.W. & Hewitt P.H. (1973) Digestive B-glycosidases of the harvester termite, *Hodotermes mossambicus*: properties and distribution. *Journal of Insect Physiology*, **19**, 1837–47.

Roessler E.S. (1932) A preliminary study of the nitrogen needs of growing *Termopsis*. *University of California Publications in Zoology*, **36**, 357–68.

Rohrmann G.F. (1977) Biomass, distribution and respiration of colony components of *Macrotermes ukuzii* Fuller. *Sociobiology*, **2**, 283–95.

Rohrmann G.F. (1978) The origin, structure and nutritional importance of the comb in two species of Macrotermitinae (Insecta, Isoptera). *Pedobiologia*, **18**, 89–98.

Rohrmann G.F. & Rossman A.Y. (1980) Nutrient strategies of *Macrotermes ukuzii* (Isoptera: Termitidae). *Pedobiologia*, **20**, 61–73.

Ruelle J.E. (1970) A revision of the termites of the genus *Macrotermes* from the Ethiopian region (Isoptera: Termitidae). *Bulletin of the British Museum (Natural History), Entomology*, **24**, 365–444.

Ruelle J.E. (1977) *Macrotermes michaelseni*: new names for *Macrotermes mossambicus* (Isoptera: Termitidae). *Journal of the Entomological Society of Southern Africa*, **40**, 1077.

Ruppli E. (1969) Die Elimination überzähliger Ersatzgeschlechtstiere bei der Termite *Kalotermes flavicollis* (Fabr.). *Insectes Sociaux*, **16**, 235–48.

Ruppli E. & Lüscher M. (1964) Die Elimination überzähliger Ersatzgeschlechtstiere bei der Termite *Kalotermes flavicollis* (Fabr.) (Vorläufige Mitteilunge). *Revue Suisse de Zoologie*, **71**, 626–32.

Ruyooka D.B.A. (1979) Associations of *Nasutitermes exitiosus* (Hill) (Termitidae) and woodrotting fungi in *Eucalyptus regnans* F. Muell. And *Eucalyptus grandis* W. Hill ex Maiden: choice-feeding, laboratory study. *Zeitschrift für Argewandte Entomologie*, **87**, 377–88.

Ruyooka D.B.A. (1980) Food handling and fungus comb turnover in laboratory and field *Macrotermes michaelseni* colonies. *1979 Annual Report of the International Centre of Insect Physiology & Ecology*, Nairobi, Kenya, pp. 79–80.

Sands W.A. (1956) Some factors affecting the survival of *Odontotermes badius*. *Insectes Sociaux*, **3**, 531–6.

Sands W.A. (1960) The initiation of fungus comb construction in laboratory colonies of *Ancistrotermes guineensis* (Silvestri). *Insectes Sociaux*, **7**, 251–9.

Sands W.A. (1961) Foraging behaviour and feeding habits in five species of *Trinervitermes* in West Africa. *Entomologia Experimentalis et Applicata*, **4**, 277–88.

Sands W.A. (1969) The association of termites and fungi. In *Biology of Termites* (eds K. Krishna & F.M. Weesner), Vol. 1, pp. 495–524. Academic Press, London & New York.

Schaefer D.A. & Whitford W.G. (1979) Nitrogen fixation in desert termites. *Bulletin of the Ecological Society of America*, **60**, 128.

Schaefer D.A. & Whiford W.G. (1981) Nutrient cycling by the subterranean termite *Gnathamitermes tubiformans* in a Chihuahuan desert ecosystem. *Oecologia (Berlin)*, **48**, 277–83.

Schubert W.J. (1965) *Lignin Biochemistry*. Academic Press, London & New York.

Scurfield G. & Nicholls P.W. (1970) Amino-acid composition of wood proteins. *Journal of Experimental Botany*, **21**, 857–68.

Seifert K. & Becker G. (1965) Der chemische Abbau von Laub- und Nadelholzarten durch verschiedene Termiten. *Holzforschung*, **19**, 105–11.

Sharp R.F. & Millbank J.W. (1973) Nitrogen fixation in deteriorating wood. *Experientia*, **29**, 895–6.

Skaife S.H. (1955) *Dwellers in Darkness*, p. 74. Longmans Green, London.

Snyder T.E. & Popenoe E.P. (1932) The founding of new colonies by *Reticulitermes flavipes* (Kollar). *Proceedings of the Biological Society of Washington*, **45**, 153–8.

Spears B.M. & Ueckert D.N. (1976) Survival and food consumption by the desert termite *Gnathamitermes tubiformans* in relation to dietary nitrogen source and levels. *Environmental Entomology*, **5**, 1022–5.

Speck U., Becker G. & Lenz M. (1971) Ernährungsphysiologische Untersuchungen an Termiten nach selecktiver medikamentöser Ausschaltung der Darmsymbionten. *Zeitschrift für angewandte Zoologie*, **58**, 475–91.

Sylvester-Bradley R., Bandeira A.G. & de Oliveira L.A. (1978) Fixação de nitrogênio (redução de acetileno) em cupins (Insecta: Isoptera) da Amazônia Central. *Acta Amazonica*, **8**, 621–7.

Swift M.J. (1977) The roles of fungi and animals in the immobilisation and release of nutrient elements from decomposing branch-wood. *Ecological Bulletin (Stockholm)*, **25**, 180–92.

Tho Y.P. (1974) The termite problem in plantation forestry in Peninsular Malaysia. *Malaysian Forester*, **37**, 278–83.

Thomas R.J. (1981) Ecological studies on the symbiosis of *Termitomyces* Heim with Nigerian Macrotermitinae. Unpublished Ph.D. thesis, University of London.

Tracey M.V. & Youatt G. (1958) Cellulase and chitinase in two species of Australian termites. *Enzymologia*, **19**, 70–2.

Waterhouse D.F., Hackman R.H. & McKellar J.W. (1961) An investigation of chitinase activity in cockroach and termite extracts. *Journal of Insect Physiology*, **6**, 96–112.

Whitehead D.L. & Quicke G.V. (1960) The nitrogen content of grass lignin. *Journal of the Science of Food & Agriculture*, **3**, 151–2.

Wilcox W.W. (1973) Degradation in relation to wood structure. In *Wood Deterioration and its Prevention by Preservative Treatments*, Vol 1, (ed. D.D. Nicholas), pp. 107–48. Syracuse University Press, Syracuse.

Williams R.M.C. (1959) Colony development in *Cubitermes ugandensis* Fuller (Isoptera: Termitidae). *Insectes Sociaux*, **6**, 291–304.

Wood T.G. (1976) The role of termites (Isoptera) in decomposition processes. In *The Role of Terrestrial and Aquatic Organisms in Decomposition Processes* (eds J.M. Anderson & A. Macfadyen), pp. 145–68. Blackwell Scientific Publications, Oxford.

Wood T.G. (1978) The food and feeding habits of termites. In *Production Ecology of Ants and Termites*, (ed. M.V. Brian), pp. 55–80. Cambridge University Press, Cambridge.

Wood T.G. & Sands W.A. (1978) The role of termites in ecosystems. In *Production Ecology of Ants and Termites* (ed. M.V. Brian) pp. 245–92. Cambridge University Press, Cambridge.

19. INTERACTIONS BETWEEN SOIL ARTHROPODS AND MICROORGANISMS IN CARBON, NITROGEN AND MINERAL ELEMENT FLUXES FROM DECOMPOSING LEAF LITTER

J. M. ANDERSON AND P. INESON

Wolfson Ecology Laboratory, Department of Biological Sciences, University of Exeter

SUMMARY

The release of nitrogen (and other limiting nutrients) from litter for root uptake primarily depends upon the balance between mineralization and immobilization processes. Soil animals do not appear to contribute to these processes directly in temperate forest soils but have important local effects on net mineralization rates through interactions with fungal and bacterial populations. These effects vary qualitatively and quantitatively between major groups of soil animals.

Macroarthropods (e.g. woodlice and millipedes) can enhance microbial respiration but feeding intensities above optimal values decrease microbial activity and result in gross shifts in the balance of fungal and bacterial populations. This effect appears to be related to the sensitivity of the fungal thallus to disruption and the favourable environment of saprotroph guts for bacterial growth.

Microarthropods such as collembola feed selectively on fungi and the inhibitory effects of grazing may be more marked than those of litter-feeding animals. However, the effects of fungal food quality on collembola populations, and the relationships between litter microhabitat complexity and the availability of hyphae, have important modifying influences on grazing effects.

Leaching rates of cations (calcium, potassium and sodium) from leaf litter are enhanced by soil animal feeding activities but the effects are proportionally much smaller than the impact of animals on nitrogen mineralization as ammonium. Woodlice (*Oniscus*) and millipedes (*Glomeris*) enhanced ammonium leaching rates up to six times control values without animals. The response was proportional to the number of *Oniscus* but increasing numbers of *Glomeris* above optimum values reduced ammonium leaching rates. These effects are not directly related to nitrogen excretion by the animals and qualitative influences of the animals on the litter microflora are implicated.

It is concluded that soil animal feeding activities can significantly affect nitrogen mineralization rates through locally reducing fungal immobilization. Some nitrogen from food materials may be transferred to bacterial biomass during gut passage but continued growth in faecal material is limited by available carbon and energy resources so that excess nitrogen can be taken up by roots and mycorrhizas as ammonium. Protozoa and nematodes may be involved in nitrogen release from the residual bacterial biomass.

INTRODUCTION

The importance of animal microbial interactions in decomposition and nutrient cycling processes has been demonstrated for aquatic systems both in the field (Cooper 1973, Flint & Goldman 1975; Nixon, Oviatt & Hale 1976) and laboratory (Barsdate, Fenchel & Prentki 1974; Fenchel & Harrison 1976; McDiffett & Jordan 1978) but little is known about equivalent processes in soils.

Microbial biomass and respiration exceed that of the soil fauna by one or two orders of magnitude and it is frequently assumed, with the exception of termites (Wood & Sands 1977) and burrowing earthworms, that the fauna has little significance in soil processes. Over recent years, however, there has been increasing appreciation that the observed levels of microbial activity are the sum effects of phenomena operating in a mosaic of microsites in litter, soil and the rhizosphere which are highly variable in time and space. At this scale of organization the fauna can be seen to affect soil processes indirectly through the formation of microsites (litter comminution, burrowing activities and faecal aggregates) as well as through direct effects of feeding on fungal and bacterial populations.

Earthworms in temperate, agricultural soils have been shown to qualitatively and quantitatively modify the microflora (Parle 1963), affect patterns of root growth (Aldag & Graff 1974; Edwards & Lofty 1978), increase the availability of nitrogen, phosphorus and mineral nutrients (Graff 1970; Syers, Sharpley & Keeney 1979) and influence transport pathways of dissolved nutrients (Sharpley, Syers & Springett 1979). A large earthworm biomass is not, however, characteristic of most soil types and the effects of other temperate invertebrate groups on soil processes are difficult to quantify under field conditions.

Patten & Witkamp (1967) used laboratory microcosms containing tree seedlings, litter, soil, millipedes and microbial components to demonstrate that the animals were integral to the transfers of ^{137}Cesium from litter back to the plants. Similar results were obtained by Witkamp & Frank (1969, 1970) for transfers of ^{137}Cesium as well as potassium and magnesium. The addition of millipedes or snails to decomposing leaf litter reduced microbial ^{137}Cesium immobilization from 36.0 to 10.8% of the initial litter content through the consumption of the litter microflora (Witkamp & Frank 1969). This process was considered critical by Ausmus, Edwards & Witkamp (1976) for the turnover of nutrients in a warm temperate, deciduous forest. The immobilization of nitrogen, phosphorus and potassium in microbial biomass was highest in summer and autumn (fall) and lowest in spring during the period of maximum root growth. The nutrient transfers to the smaller faunal biomass were highest during this period but excess nutrients to saprotroph demands were not lost to ground water because of active root uptake. The soil fauna was predominantly mycophagous and was calculated to consume 86% of fungal production in this site (McBrayer 1974). During periods of high rainfall and low root growth the immobilization of limiting nutrients in the soil biota (including mycorrhizas) was considered to be a major process preventing leaching from the rooting zone of the soil profile. Similar processes have been quantified for simulated rhizosphere systems in which bacterially immobilized phosphorus was mobilized through the feeding activities of amoebae (Cole *et al.* 1978).

Thus there is evidence that soil faunas affect nutrient fluxes in soils through their influence on microbial mineralization and immobilization. We therefore briefly review these basic processes before considering the mechanisms involved in the direct and indirect effects of soil animals on nutrient flux rates and pathways.

MINERALIZATION AND IMMOBILIZATION

The twin processes of mineralization and immobilization are invariably involved in decomposition since catabolism provides the energy and nutrients required for the maintenance, growth and reproduction of saprotrophic organisms (Swift, Heal & Anderson 1979). The balance of these processes (net mineralization) determines the nutrient supply to higher plants and relates directly to the availability of that element to the organisms (fungi, bacteria and animals) involved in organic matter decomposition.

Sodium and potassium are usually available in excess of saprotroph requirements. These elements are predominantly present as inorganic ions and are leached rapidly from decaying litter. The loss rates generally decrease with time as concentrations in the litter approach those for microbial tissues (Gosz, Likens & Bormann 1973).

Under laboratory conditions microbial immobilization of potassium may be up to 80% higher than the retention in leached, sterile-litter controls (Witkamp & Barzansky 1968). This suggests that the ion exchange complex of litter (as opposed to humus) is less efficient than microorganisms for the retention of these mobile cations.

All nitrogen and sulphur, some calcium and most phosphorus, are present in organic compounds in plant and animal remains and are mobilized through enzyme activity. The availability of nitrogen and phosphorus limits secondary production in the soil, just as they limit primary production in most natural terrestrial ecosystems, and these elements are therefore efficiently annexed and conserved by microorganisms.

During the course of decomposition the carbon/nitrogen ratio decreases (as well as ratios for other limiting nutrients), and microbial tissues represent an increasing proportion of the resource mass. Swift (1973) found that sawdust lost 39% of the initial weight after 15 weeks' incubation in the laboratory but 39% of the remaining material was fungal mycelium. Nitrogen is not theoretically mobilized until the carbon/nitrogen ratio of the resource complex approaches that of microbial tissues (about 10/20:1, depending upon species and growth conditions); similar principles underlie the release of other microbially bound nutrients. In acid, organic soils, such as moorlands, tundra, coniferous forests and mor-humus forms in deciduous forests, the carbon/nitrogen ratio of soil organic matter may be considerably above the threshold value where nitrogen mineralization theoretically occurs.

A critical step in nutrient cycling processes is therefore the release of nutrients from microbial tissues. This involves abiotic processes such as freezing/thawing and wetting/drying cycles which occur in soils subject to variable extremes of weather and climate (Witkamp 1969; Witkamp & Frank 1970) but biotic processes are probably more important under temperate conditions. Microbial lysis and autolysis undoubtedly contribute to the turnover of bacterial and fungal tissues (Mitchell & Alexander 1963; Shields *et al.* 1973) but the result of microcosm experiments suggests that soil fauna feeding activities are quantitatively more important. The feeding activities of soil fauna are also widely stated to stimulate microbial growth and activity through the grazing of senescent tissues. The balance of bacterial and/or fungal growth and fauna consumption is therefore an important variable in determining net mineralization rates.

The soil fauna may therefore affect the availability of inorganic nitrogen in three ways: by the enhancement of carbon mineralization and hence the reduction of litter carbon/nitrogen ratios, by the stimulation or reduction of microbial biomass and the excretion of nitrogen compounds. The evidence for these phenomena is reviewed before considering their possible contribution to nitrogen fluxes in the soil.

Effects of soil animals on microbial respiration

Van der Drift & Witkamp (1960) and Nicholson, Bocock & Heal (1966) suggested from a comparison of microbial respiration rates on arthropod faeces, mechanically ground litter and intact leaves that the main contribution of soil animals to decomposition processes was the short-term enhancement of microbial activity through litter comminution. A criticism of these experiments is that they involved the 'static' effects of a single gut passage rather than the continuous 'dynamic' effects of coprophagy and direct grazing on microbial populations. Thus Nicholson, Bocock & Heal (1966) found no significant differences in weight losses after one year in the field between millipede faeces and intact leaves enclosed in fine-mesh litter bags. Standen (1978), however, demonstrated that enchytraeid worms or Diptera larvae enclosed in fine-mesh litter bags significantly increased carbon mineralization through microbial respiration. In addition, Addison & Parkinson (1978) have shown that microbial respiration was enhanced in cores from tundra soils containing low numbers of mycophagous collembola compared with defaunated samples.

The dynamic effects of macroarthropod feeding activities on microbial respiration were investigated by Hanlon & Anderson (1980). Fragmented (2–4 mm) and mechanically ground (0.1–0.2 mm) oak leaf litter was inoculated with a mixed culture of fungi and bacteria and incubated in the laboratory. After 40 days both treatments showed similar levels of microbial respiration. Various numbers of woodlice (*Oniscus asellus* L.) or millipedes (*Glomeris marginata* Villers) were then added and respiratory rates were measured for a further 40 days.

Microbial respiration in fragmented litter was initially increased to twice the control rates by four *Oniscus* and to 1.6 times control rates by six *Glomeris* but subsequently declined to rates slightly above controls. However, cultures containing ten *Oniscus* showed respiratory rates below those of controls after 20 days. Cumulative carbon dioxide measurements, corrected for animal respiration, showed that microbial activity was decreased by grazing pressures higher than optimum values (Fig. 19.1). Animals feeding on previously ground litter produced a similar, reduced response suggesting that litter comminution was the main factor contributing to the enhancement of carbon mineralization but confirming that secondary effects were involved in the effects of the animals on microbial populations.

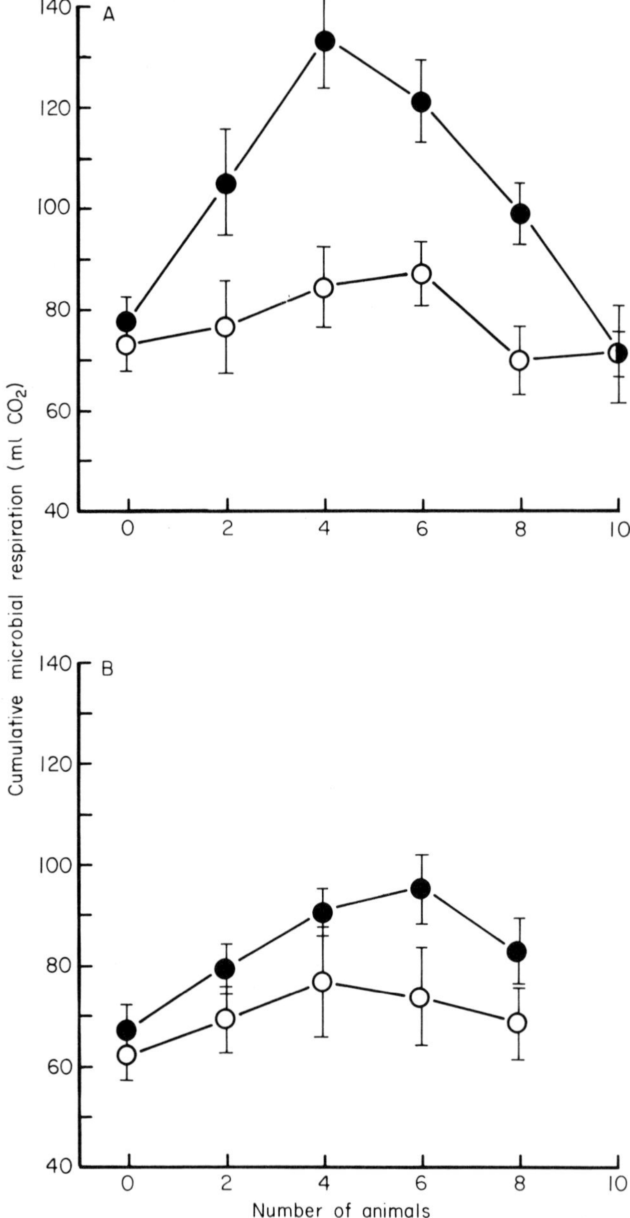

FIG. 19.1. Cumulative microbial respiration over 40 days (mean values ± 95% confidence limits) for different numbers of *Oniscus asellus* (A) and *Glomeris marginata* (B) feeding on 0.5 g oak leaf fragments (●) or ground oak leaves (○). (After Hanlon & Anderson 1980.)

The effects of fungal grazing in the absence of litter comminution by the animals were investigated in experiments using mycophagous collembola (*Folsomia candida* Willem) to graze *Coriolus versicolor* (L.ex Fc) growing on ground leaf litter (Hanlon & Anderson 1979). Five collembola per 0.1 g leaf litter stimulated fungal respiration for approximately 14 days while higher numbers of animals, up to 20 individuals, produced no stimulatory effect or inhibited fungal respiration. All treatments were significantly below controls three weeks after the introduction of the animals. Total cumulative respiration showed a non-linear response to animal numbers similar to that shown in Fig. 19.1.

Hargrave (1970) has shown similar bell-shaped curves for the effects of amphipods on bacterial activity in lake sediments. In these experiments microbial activity was not inhibited until amphipod respiration (30 animals) accounted for 65% of community respiration. In the experiments reported here *Oniscus* produced an inhibitory response at only 13% of total respiration while the more selectively mycophagous collembola inhibited fungal activity at less than 5% of total respiration. These effects therefore fall within the animal densities and activity ranges found in natural soils.

While it can be concluded that optimum levels of soil animal feeding activities do enhance carbon mineralization rates by microorganisms, it is also evident that high feeding intensities can inhibit microbial activity. Investigation of this phenomenon has revealed gross changes in the balance of bacterial and fungal populations affected by animal feeding activities.

INTERACTIONS BETWEEN ANIMAL AND MICROBIAL POPULATIONS

Analysis of changes in bacterial and fungal populations in the above experiments (using direct count methods) revealed that reduced microbial respiration resulting from superoptimal levels of grazing by macroarthropods and collembola was associated with a reduction of fungal standing crop and an increase in bacteria (Hanlon & Anderson 1979, 1980). Results for *Oniscus* grazing fragmented litter are shown in Fig. 19.2. Fungal standing crop was reduced by all levels of animal feeding, the effects being particularly pronounced over the first three days of the experiment. After 35 days fungal standing crop was reduced by the animals to approximately one third of control levels. Bacterial standing crop, however, increased proportional to feeding intensity and after 35 days litter with six *Oniscus* contained ten times more bacteria than controls. Similar effects were detected in ground litter and in litter grazed by collembola and *Glomeris* (Table 19.1.)

Thus it is evident that not only slow-growing basidiomycetes such as

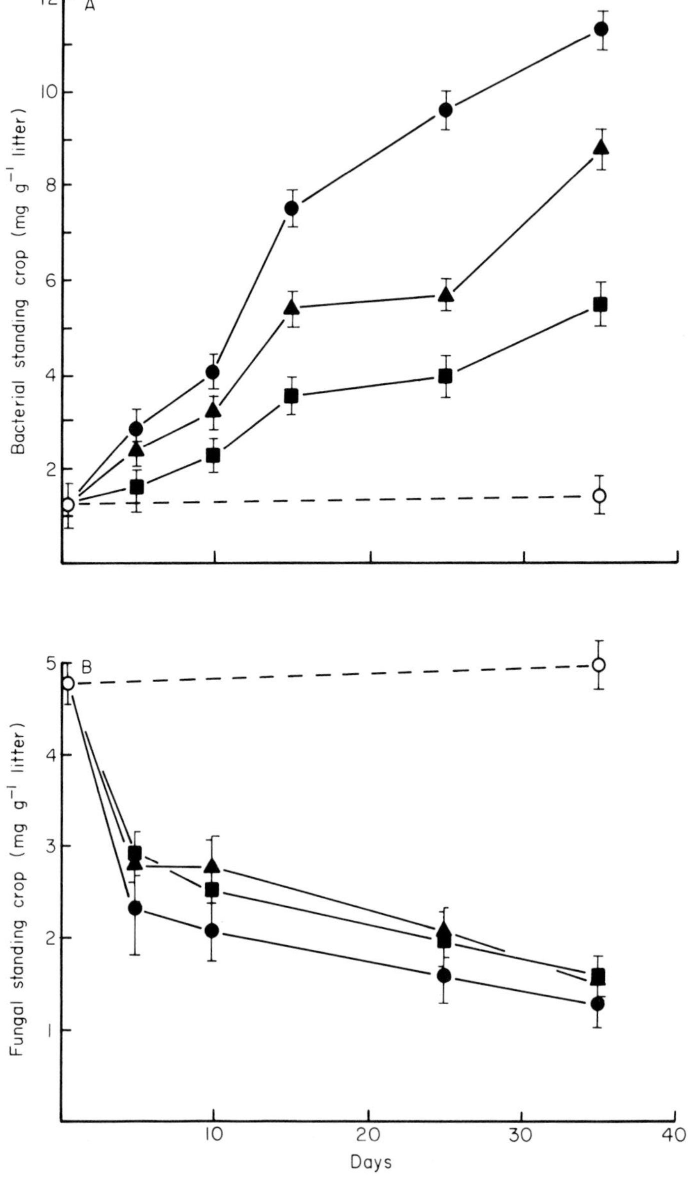

FIG. 19.2. Bacterial (A) and fungal (B) standing crops in 0.5 g fragmented oak leaf litter grazed by different numbers of *Oniscus asellus*. Mean values ($\pm 95\%$ limits) are shown for control chambers without animals (○) and for chambers, containing 2 (■), 4 (▲) or 6 (●) animals. Microbial standing crops were determined by direct counts using membrane filters: 3 replicate chambers, 4 filters per replicate and 10 fields per filter ($n = 120$) for each estimate. Results are expressed as milligram of fungi or bacteria per gram dry weight litter. The small fiducial limits are typical of results using the filter technique. (After Hanlon & Anderson 1980.)

TABLE 19.1. Bacteria in the food, gut contents and faeces of *Glomeris marginata* and *Oniscus asellus*. Dilution series of litter, gut contents and faecal homogenates were prepared in 0.1% peptone water. Counts of viable bacteria were made using pour plates (Oxoid Nutrient Broth powder 13 g l^{-1}; agar 15 g l^{-1}). Nystatin (50 μg cm^{-3}) was included to suppress fungal growth. Homogenates of gut contents were filtered through weighed Millipore filters (0.45 μm pore size), dried in a vacuum oven for 48 hours and then reweighed to determine the dry weight of gut contents. Results are expressed as counts per gram dry weight material.

Sample	Replicate no.	Viable bacteria (10^8 g^{-1})	Mean (10^8 g^{-1})
Litter	1	3.4	
	2	6.7	4.3
	3	2.8	
Glomeris	1	13.0	
gut	2	11.9	22.8
	3	43.5	
Glomeris	1	695.5	
faeces	2	306.9	404.1
	3	209.9	
Oniscus	1	8.9	
gut	2	65.3	28.2
	3	10.3	
Oniscus	1	437.5	
faeces	2	386.0	349.6
	3	225.0	

C. versicolor but also the faster-growing microfungi, such as *Penicillium*, *Mucor* and *Trichoderma* which were dominant in the macroarthropod grazed litter, are extremely sensitive to animal feeding activities. The guts of macroarthropods are a favourable environment for bacterial growth (Reyes & Tiedje 1976; Anderson & Bignell 1980) and increased counts in gut contents and faeces of *Oniscus* and *Glomeris* are an evident consequence of gut passage (Table 19.2). These data simply refute the statement by Boyle & Michell (1979) that the guts of various crustacea, including *Oniscus asellus*, are sterile. Reyes & Tiedje (1976) observed that although the net effect of gut passage in *Tracheoniscus rathkei* Brandt (Isopoda) was an increase in bacteria over the food litter, lysis and digestion of bacterial cells also occurred. The same phenomenon has been established for *Glomeris* and 53.2%, 53.4%, and 28.9% of tritiated *Pseudomonas syringae*, *Erwinia herbicola* and *Escherichia coli* were assimilated (Anderson & Bignell, unpublished data).

The effects of animal grazing on microbial populations are further

TABLE 19.2. Effect of *Glomeris* numbers on bacteria populations in oak leaf litter from the experimental microcosms.
(a) Total viable counts using pour plates (see Table 19.1).
(b) Ammonifying bacteria estimated by the most probable number technique (MPN) using 0.2% tryptone medium (three series of five replicates). Results are expressed as counts per gram dry weight of litter.

Treatment (no. animals)	(a) Total viable (10^9 g^{-1}) Replicate	Count	Mean	(b) Ammonifiers (10^5 g^{-1}) Series	Mean MPN	Overall mean MPN
0	1	2.6		1	0.04	
	2	1.9	1.87	2	0.01	0.04
	3	1.1		3	0.08	
4	1	2.4		1	0.7	
	2	0.9	2.17	2	0.3	2.47
	3	3.2		3	6.4	
8	1	12.0		1	4.0	
	2	2.0	6.00	2	4.0	4.00
	3	4.0		3	4.0	

complicated by the influence of resource quality on microbial populations and the quality of microbial tissues as a food resource for the animals. These interactions are further modified by the availability of microorganisms to the animals in soil and litter microhabitats.

Booth & Anderson (1979) demonstrated that the growth rates and reproduction of collembola (*Folsomia candida*) feeding on *Coriolus versicolor* were affected by the nitrogen content (as asparagine) of the fungal culture medium and hence the protein and amino acid content of the fungus. Egg laying rates at 200 μg l^{-1} N in the culture medium were over three times higher than at 2 μg l^{-1} N. A less marked reproductive response was found for another litter basidiomycete *Hypholoma fasiculare* Huds grown under the same conditions. These experiments have been repeated by Leonard (see p. 441, this volume) using *Mucor plumbeus* Bon. grown in two-dimensional and three-dimensional bead matrices at different nitrogen concentrations. The two-dimensional surface of fine glass beads exposes the fungus to attack by the collembola, and at 20 or 200 μg l^{-1}N the growth response of the animal population resulted in the elimination of the fungus. At 2 μg l^{-1} N an equilibrium was established because both the fungus and the collembola were nitrogen-limited but the fungus was apparently better adapted to these low nutrient conditions. In the three-dimensional matrix the effect of the higher nitrogen levels was reversed. The beads formed a microhabitat refuge for the fungal hyphae, where they were protected against grazing, and the fungus proliferated under the same nutrient conditions where it had previously been reduced (Fig. 19.3). Experiments using more natural soil and litter systems are in progress but these results serve to illustrate that the feeding activities and

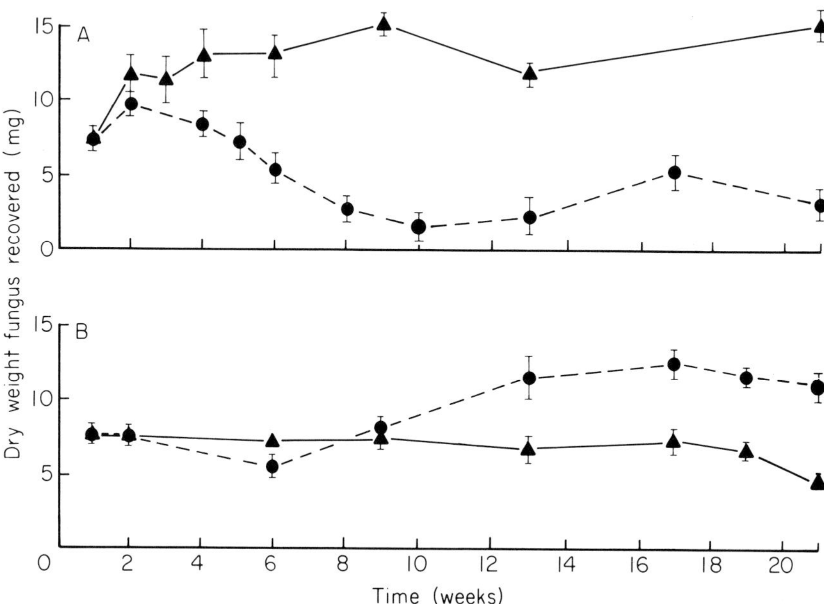

FIG. 19.3. Effect of spatial heterogeneity on the growth of *Mucor plumbeus* in controls (▲) and in cultures grazed by *Folsomia candida* (●). The fungus was grown in liquid medium containing 20 mg l^{-1} N (as asparagine) on a two-dimensional surface formed from fine glass chromatography beads (A) or in a three-dimensional matrix of ceramic column packing (B). The collembola were added seven days after the fungal inoculum. Fungal mycelium was recovered by washing and sonic separation from the matrix. Results are expressed as mean weight per culture $\pm$ 1 standard error ($n=5$). (See also Leonard p. 441, this volume.)

effects of microarthropods on microbial populations are influenced by the structural complexity of soil and litter microhabitats as well as by the nutritional status of the litter for fungal growth.

Effects of animals on nutrient fluxes from decomposing leaf litter

The majority of studies emphasize the role of soil animals in carbon mineralization processes and flux rates of non-limiting, mineral elements including radio isotopes. There have been no detailed studies of animal–

microbial interactions in nitrogen flux pathways despite the recognized importance of nitrogen in limiting ecosystem processes and the large immobilized pools of organic nitrogen in most natural soils. Ausmus, Edwards & Witkamp (1976) have suggested that the soil fauna is critical for the mobilization of nitrogen in microbial tissues but there is no direct evidence for this phenomenon.

Current research in this laboratory is directed towards an understanding of the role of arthropods in nitrogen flux pathways in forest soils, particularly bacterially mediated steps in nitrogen transformations.

Investigations of these processes in the field are hindered by the biological and physico-chemical complexity of soils and their variability in time and space. At the other extreme, laboratory experiments may be difficult to extrapolate to field conditions because important synergistic effects may be excluded by an oversimplification of the system. Microcosms offer a means of bridging these extremes if a holistic approach is adapted to their design and analysis, while retaining the analytical precision of laboratory experiments and the ability to replicate them. The field system can be approached by either using microcosms of increasing complexity and evaluating the kinetics of compartments as they are added to the model, an approach developed by Patten & Witkamp (1967), or by using intact soil cores which are manipulated in the laboratory and their dynamics compared with field samples of lysimeters. The first generation of experiments reported here were not intended as analogues of natural systems but to form a link between the various phenomena described above and more complex systems.

Experiments were carried out to investigate the effects of macroarthropods on nutrient losses from decomposing oak (*Quercus robur* L.) leaf litter. Approximately 3 g of leaf litter fragments (2–4 mm) were placed in containers whose bases were covered with fine stainless steel mesh. The litter was inoculated with a homogenate of decomposing leaf litter and incubated for three weeks at 15°C before 0, 2, 4, 6 or 8 *Glomeris* or *Oniscus* were added. The containers were leached at intervals of seven days using distilled water and the leachates were analysed for calcium, potassium, sodium, ammonium, and nitrate.

Cation losses from leaf litter showed similar trends for treatments with *Glomeris* and *Oniscus* and are illustrated by the data for potassium shown in Fig. 19.4. Losses of potassium from controls followed an approximately negative exponential pattern. Treatments with *Glomeris* showed similar trends to the controls but cumulative release rates, although proportional to grazing intensity and generally significantly higher than controls ($P<0.01$), showed only small increases in mineral nutrient mobilization in the presence of animals (see Fig. 19.5).

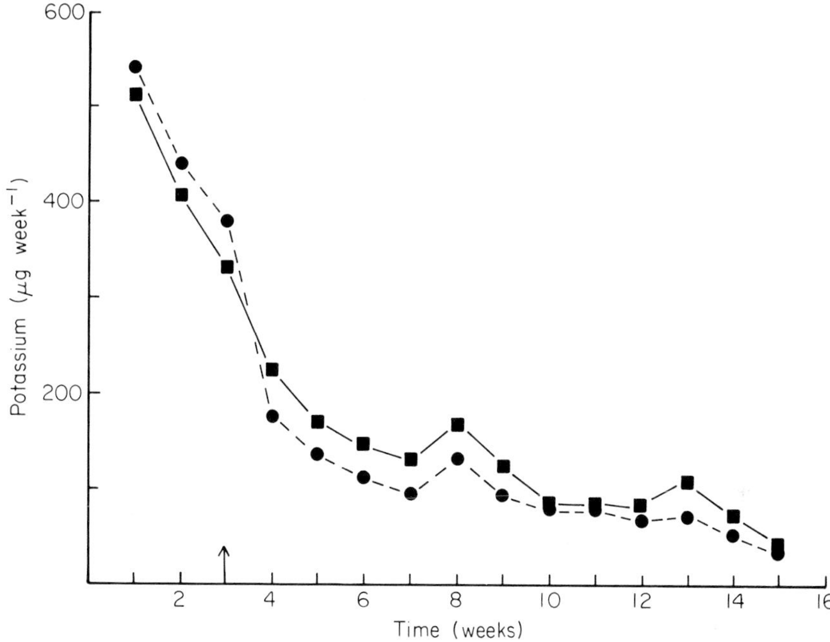

FIG. 19.4. The effect of *Glomeris* feeding activities on potassium fluxes from decomposing oak leaf litter. Aliquots of 3 g (dry wt.) fragmented leaf litter (3–5 mm) were added to microcosm chambers and leached in distilled water for 24 hours. The drained samples were inoculated with a litter suspension and incubated for three weeks at 15°C before 0, 2, 4, 6 or 8 animals were added (↑). Nine replicates of each treatment were prepared. The samples were leached with 60 ml distilled water every seven days and the mineral nutrient content of the leachates determined using an atomic absorption spectrophotometer (Pye Unicam SP9-800). Mean potassium concentrations in leachates are shown for chambers without animals (●) and for chambers containing 8 *Glomeris* (■). Results for treatments with 2, 4 or 6 animals were intermediate between these values and are not shown for clarity of presentation. Standard errors are obscured by the symbols in all cases.

Small losses of nitrate were detected in leachates from all treatments. This is in accordance with expectations that mineral nitrogen fluxes in acid litter would be predominantly in the form of ammonium.

The pattern of ammonium losses from controls showed a similar trend to cation leaching though initial loss rates were lower (Fig. 19.5). The effect of adding animals, however, was dramatically different to the pattern shown to the other nutrients. Ammonium losses from the litter were considerably enhanced by the presence of animals and after 16 weeks cumulative losses of ammonium were over three times higher than controls. Both instantaneous release rates and cumulative losses showed a non-linear response to the numbers of *Glomeris* reflecting the pattern shown for microbial respiration (see Fig. 19.1). Ammonium concentrations in leachates were increased by the

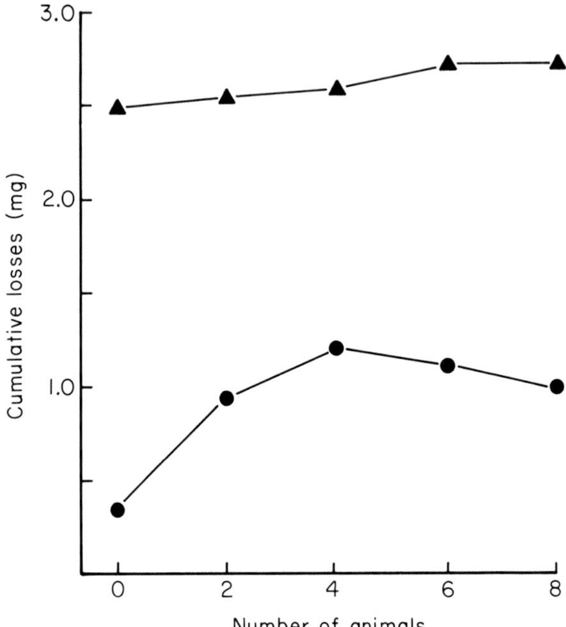

FIG. 19.5. Cumulative losses of potassium (▲) and ammonium (●) ions in leachates from chambers containing different numbers of *Glomeris* over 15 weeks.

addition of up to four animals but higher numbers reduced nitrogen losses as ammonium; ammonium concentrations in leachates from all treatments with animals were, however, considerably above those of the controls (Fig. 19.5). The removal of animals resulted in ammonium loss rates falling to approximately control levels over a period of 2–3 weeks (Fig. 19.6).

The addition of *Oniscus* to microcosms produced similar results to *Glomeris* in terms of cation and ammonium mobilization rates but, unlike the experiments with *Glomeris*, ammonium fluxes were proportional to the number of animals in the microcosms.

The mechanisms responsible for the different patterns of ammonium release by *Glomeris* and *Oniscus* have not yet been determined. They are not, however, simply related to the different nitrogenous excretory products of the two groups of animals. Most terrestrial arthropods, including millipedes, are generally assumed to excrete uric acid as an adaptation to water conservation (Chapman 1975). The principal nitrogenous excretory product of terrestrial isopods, however, is ammonia (Wieser & Schweizer 1970) reflecting their close physiological affinities to aquatic crustacea. It was initially hypothesized that the breakdown of uric acid in millipede faecal material might account for the continued release of ammonium from the litter for three weeks after the

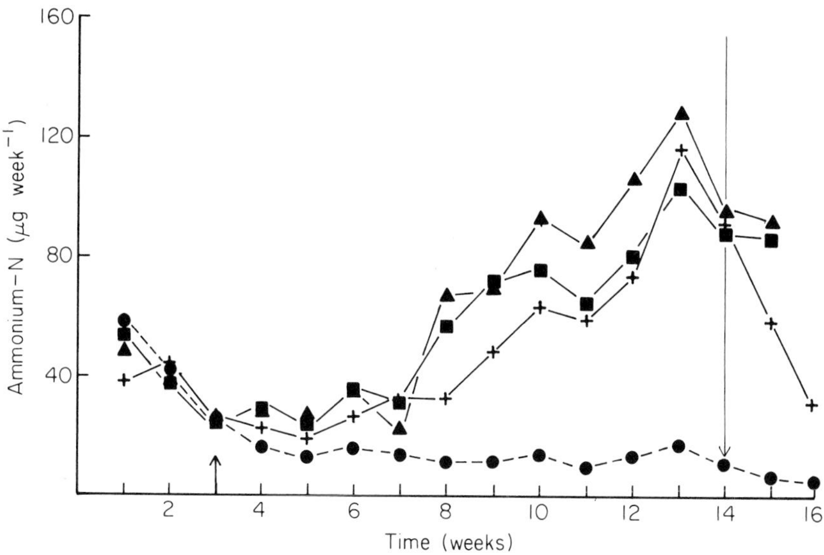

FIG. 19.6. Effect of *Glomeris* feeding activities on ammonium fluxes from decomposing oak leaves. Ammonium content of leachates was determined by Technicon Autoanalyser using the phenol method. Results are presented for mean ammonium losses from chambers containing 2 (+), 4 (▲) or 8 (■) *Glomeris* and for controls (●) without animals. Results for six animals were intermediate between those for four and eight animals and are not presented. Animals were removed from chambers containing two *Glomeris* at 14 weeks and ammonium concentrations in leachates, in the absence of feeding activities, were measured for a further two weeks. (See Fig. 19.4. for further details of methods.)

animals were removed. Faeces, gut contents, gut tissues and eviscerated bodies of *Glomeris* and *Oniscus* were assayed for uric acid. Homogenates were treated with pig liver uricase and measured spectrophotometrically at 292 nm (Uric Acid Diagnostic Kit No. 292-uv; Sigma Chemical Co., St Louis). No uric acid was detected in any of the preparations except for the body tissues of *Glomeris* which were found to contain up to 2.5% uric acid by weight. A similar phenomenon has been described by Mullins & Cochran (1976) who found that uric acid was absent from cockroach faeces but was sequestered in the fat bodies. These nitrogen reserves could be utilized by the animal as a response to low protein diets or oviposition. Potrikus & Breznak (1980a,b) have shown that uric acid can form up to 45% of the dry body weight of a wood-feeding termite and that the mobilization and re-utilization of the nitrogen is mediated by gut bacteria. Paradoxically, we have found uricolytic bacteria present at similar densities in the guts of *Glomeris* and *Oniscus*. Inorganic nitrogen available in the gut is in excess of that required by the proliferating bacterial flora and fresh faeces contain high concentrations of ammonium nitrogen (Table 19.3) as well as an increase in bacterial biomass through the gut passage (see Table 19.2).

TABLE 19.3. Changes in the nitrogen content of hazel litter eaten by *Glomeris* (from Bocock 1963).

Material	Amino-N ($\mu g\ g^{-1}$)	NH_4^+ ($\mu g\ g^{-1}$)	As % of total N	Total N ($\mu g\ g^{-1}$)
Food	193	175	2.7	13 600
Uneaten food	86	335	3.2	13 100
Faecal pellets	103	1385	9.5	15 700

Experiments with these two macroarthropods reveal that cation leaching rates are not substantially increased through litter comminution but the animals appear to have a significant role in nitrogen mobilization and mineralization. This effect is not simply a function of the nitrogen excretion by the animals, since ammonium losses from the microcosms did not increase step-wise with the addition of animals, but involves qualitative and quantitative interactions with the litter microflora (see Table 19.2).

CONCLUSIONS

Soil animals may enhance or inhibit microbial activity, act differentially on bacterial and fungal populations, and directly or indirectly affect nutrient fluxes. These effects have primarily been demonstrated in litter systems but are not unrelated to field conditions since soil biological processes, including mycorrhizal activity, are mainly associated with the top 5 cm of organic soils. The influences of soil animals on microbial populations and nutrient fluxes may be related to major functional groups of the fauna and the structural characteristics of soil and litter habitats.

The microfauna (protozoa and nematodes) have small body sizes and short generation times which promote the regulation of bacterial populations through specific feeding activities and rapid population responses to changes in bacterial biomass. Bacterial populations in mineral soils are primarily energy-limited (Hisset & Gray 1976) and have mean generation times in the order of days rather than hours. In the rhizosphere, particularly in grasslands, microbial activity is promoted by root exudates, sloughed cells and the rapid turnover of root hairs. Competition for nutrients between the roots and rhizosphere microflora can occur under these conditions but Cole *et al.* (1978) have shown that the bacteriophagous microfauna can increase phosphorus turnover to rates comparable with aquatic systems.

The role of microfauna in acid, forest soils, where fungal biomass predominates, has not been quantified.

The mesofauna do not contribute significantly to the primary processes of litter comminution in forest soils and most collembola and mites are more cryptic than fossorial. The structure of soil and litter horizons therefore influences the availability of fungal hyphae as food resources (Anderson 1975). Fungal biomass increases during the early phases of litter decay if the leaves remain uncomminuted, but microarthropod gut contents generally contain decreasing amounts of hyphae after the winter months. This appears to be the result of fungi penetrating and ramifying the leaf tissues and thus reducing the amount of the mycelium exposed to attack on leaf surfaces. In addition, limiting nutrients are translocated to hyphal tips during growth and many of the hyphae bridging soil cavities are highly vacuolated and a poor quality food resource for microarthropods. The high structural complexity of the surface layers of organic soils (Anderson 1978) modifies microarthropod population growth rates and their influences on nutrient mobilization through grazing effects.

The feeding activities of macrofauna modify these interactions through litter comminution, changing both the physical environment for fungal growth and the litter structural complexity, as well as altering the balance of fungal and bacterial populations. The selective destruction of the fungal component of the soil flora initially results in reduced soil microbial activity (Hanlon & Anderson 1980), followed by a 'flush of decomposition' as dead organisms are utilized by the proliferating bacteria. This flush can be measured in terms of carbon and nitrogen mineralization, and likened to the effects of fumigation (Jenkinson 1976). Macroarthropods significantly affect nitrogen mineralization rates through the combined effect of feeding, gut passage and excretion, and in the experiments reported here ammonium leaching was enhanced. Under natural conditions root and/or mycorrhizal uptake would form an ammonium sink and it was expected that in these experiments the high bacterial biomass in macroarthropod faeces would immobilize a major proportion of the available nitrogen. The observed ammonium concentrations in leachates implies, however, that bacterial growth is limited by resources other than nitrogen. This is in accordance with the view that the carbon and energy sources of structural polysaccharides in litter are unavailable to most bacteria but are attacked by many fungi (Eklund & Gyllenberg 1974). The feeding activities of the fauna may therefore temporarily shift the balance between mineralization and fungal immobilization so that mineral nitrogen is available to roots. Baath *et al.* (1978) have shown, using pot experiments, that the growth rate of pine seedlings was inhibited by treatments with glucose and nitrogen because nitrogen deficiency was induced by microbial growth and immobilization.

The importance of these nitrogen mineralization processes under field conditions is unknown but it only requires the mobilization of a small proportion of the nitrogen pool in organic soils, which can be as high as 100 t ha^{-1} under mature conifers (Weetman & Webber 1972), to maintain the production of a climax woodland.

The role of grazing herbivores for nutrient cycles in grasslands is generally accepted by agriculturalists (Whitehead 1970; Floate 1970; Swift *et al.* 1979). It is reasonable to expect soil animals to have similar roles in decomposer systems where invertebrate biomass is usually one or two orders of magnitude higher per square metre than that of herbivores in the same ecosystem.

ACKNOWLEDGEMENTS

Financial support for these studies by the Natural Environment Research Council is gratefully acknowledged.

REFERENCES

Addison J.A. & Parkinson D. (1978) Influence of collembolan feeding activities on soil metabolism at a high arctic site. *Oikos*, **30**, 529–38.

Aldag R. & Graff O. (1974) Einfluss der Regenwurm-tätigkeit auf Proteingehalt und Proteinqualität junger Haferpflanzen. *Zietschrift für Landwirt-schaftliche Forschung*, **31**, 277–84.

Anderson J.M. (1975) Succession, diversity and trophic relationships of some soil animals in decomposing leaf litter. *Journal of Animal Ecology*, **44**, 475–96.

Anderson J.M. (1978) Inter- and intra-habitat relationships between woodland Cryptostigmata species diversity and the diversity of soil and litter microhabitats. *Oecologia*, **32**, 341–8.

Anderson J.M. & Bignell D.E. (1980) Bacteria in the food, gut contents and faeces of the litter feeding millipede *Glomeris marginata* (Villers). *Soil Biology and Biochemistry*, **12**, 251–4.

Ausmus B.S., Edwards N.T. & Witkamp M. (1976) Microbial immobilisation of carbon, nitrogen, phosphorus and potassium: implications for forest processes. In *The Role of Terrestrial and Aquatic Organisms in Decomposition Processes* (eds J. M. Anderson & A. Macfadyen), pp. 397–416. Blackwell Scientific Publications, Oxford.

Bååth E., Lohm U., Lundgren B., Rosswall T., Söderström B., Sohlenius B. & Wirén A. (1978) The effect of nitrogen and carbon supply on the development of soil organism populations and pine seedlings: a microcosm experiment. *Oikos*, **31**, 153–63.

Barsdate R.J., Fenchel T. & Prentki R.T. (1974) Phosphorus cycle of model ecosystems: significance for decomposer food chains and effects of bacterial grazers. *Oikos*, **25**, 239–51.

Bocock K.L. (1963) The digestion of food by *Glomeris*. In *Soil Organisms* (eds J. Doeksen & J. van der Drift), pp. 85–91. Elsevier-North Holland, Amsterdam.

Booth R.G. & Anderson J.M. (1979) The influence of fungal food quality on the growth and fecundity of *Folsomia candida* (Collembola: Isotomidae). *Oecologia*, **38**, 317–23.

Boyle P.J. & Mitchell R. (1979) Absence of micro-organisms in Crustacean digestive tracts. *Science*, **200**, 1157–9.

Chapman R.F. (1975) *The Insects: Structure and Function*. English Universities Press, London.

Cole C.V., Elliot E.T., Hunt H.W. & Coleman D.C. (1978) Trophic interactions in soils as they affect energy and nutrient dynamics. V. Phosphorus transformations. *Microbial Ecology*, **4**, 381–7.

Cooper D.C. (1973) Enhancement of net primary productivity of herbivore grazing in aquatic laboratory microcosms. *Limnology and Oceanography*, **18,** 31–7.

Edwards C.A. & Lofty J.R. (1978) The influence of arthropods and earthworms upon root growth of direct drilled cereals. *Journal of Applied Ecology*, **15,** 789–95.

Eklund E. & Gyllenberg H.G. (1974) Bacteria. In *The Biology of Plant Litter Decomposition* (eds C. H. Dickinson & G. J. F. Pugh), pp. 245–68. Academic Press, London & New York.

Fenchel T. & Harrison P.G. (1976) The significance of bacterial grazing and mineral cycling for the decomposition of particulate detritus. In *The Role of Terrestrial and Aquatic Organisms in Decomposition Processes* (eds J. M. Anderson & A. Macfadyen), pp. 285–99. Blackwell Scientific Publications, Oxford.

Flint R.W. & Goldman C.R. (1975) The effects of a benthic grazer on the primary productivity of the littoral zone of Lake Tahoe. *Limnology and Oceanography*, **20,** 935–44.

Floate M.J.S. (1970) Mineralisation of nitrogen and phosphorus from organic materials of plant and animal origin and its significance in the nutrient cycle of grazed upland and hill soils. *Journal of the British Grassland Society*, **25,** 295–302.

Gosz J.R., Likens G.E. & Bormann F.H. (1973) Nutrient release from decomposing leaf and branch litter in Hubbard Brook Forest, New Hampshire. *Ecological Monographs*, **47**, 173–91.

Graff O. (1970) Phosphorus content of earthworm casts. *Landbau Forschung Völkenrode*, **20**, 33–6.

Hanlon R.D.G. & Anderson J.M. (1979) The effects of collembola grazing on microbial activity in decomposing leaf litter. *Oecologia*, **38**, 93–9.

Hanlon R.D.G. & Anderson J.M. (1980) The influence of macroarthropod feeding activities on microflora in decomposing oak leaves. *Soil Biology and Biochemistry*, **12**, 255–61.

Hargrave B.T. (1970) The effect of a deposit-feeding amphipod on the metabolism of benthic microflora. *Limnology and Oceanography*, **15**, 21–30.

Hissett R. & Gray T.R.G. (1976) Microsites and time changes in soil microbe ecology. In *The Role of Terrestrial and Aquatic Organisms in Decomposition Processes* (eds J. M. Anderson & A. Macfadyen), pp. 23–39. Blackwell Scientific Publications, Oxford.

Jenkinson D.S. (1976) The effects of biocidal treatments on metabolism in soil. IV. The decomposition of fumigated organisms in soil. *Soil Biology and Biochemistry*, **8**, 203–8.

McBrayer J.F., Reichle D.E. & Witkamp M. (1973) *Energy flow and nutrient cycling in a cryptozoan food web*. Oak Ridge National Laboratory. EDFB-IBP-73-8.

McDiffett W.F. & Jordan T.E. (1978) The effects of an aquatic detritivore on the release of inorganic N and P from decomposing leaf litter. *American Midland Naturalist*, **99**, 36–44.

Mitchell R. & Alexander M. (1963) Lysis of soil fungi. *Canadian Journal of Microbiology*, **9**, 169–77.

Mullins D.E. & Cochran D.G. (1976) A comparative study of nitrogen excretion in twenty-three cockroach species. *Comparative Biochemistry and Physiology*, **53A**, 393–9.

Nicholson P.B., Bocock K.L. & Heal O.W.H. (1966) Studies on the decomposition of the faecal pellets of a millipede (*Glomeris marginata* Villers). *Journal of Ecology*, **54**, 755–66.

Nixon S.W., Oviatt C.A. & Hale S.S. (1976) Nitrogen regeneration and the metabolism of coastal marine bottom communities. In *The Role of Terrestrial and Aquatic Organisms in Decomposition Processes* (eds J.M. Anderson & A. Macfadyen), pp. 269–83. Blackwell Scientific Publications, Oxford.

Parle J.N. (1963) A microbiological study of earthworm casts. *Journal of General Microbiology*, **31**, 13–22.

Patten B.C. & Witkamp M. (1967) System analysis of ^{134}Cesium kinetics in terrestrial microcosms. *Ecology*, **48**, 814–24.

Potrikus C.J. & Breznak J.A. (1980a) Uric acid in wood eating termites. *Insect Biochemistry*, **10**, 19–27.

Potrikus C.J. & Breznak J.A. (1980b) Uric acid degrading bacteria in guts of termites. *Applied and Environmental Microbiology*, **40**, 117–24.

Reyes V.G. & Tiedje J.M. (1976) Ecology of the gut microbiota of *Tracheoniscus rathkei* (Crustacea, Isopoda). *Pedobiologia*, **16**, 67–74.

Sharpley A.N., Syers J.K. & Springett J.A. (1979) Effect of surface-casting earthworms on the transport of phosphorus and nitrogen in surface run-off from pasture. *Soil Biology and Biochemistry*, **11**, 459–62.

Sheilds J.A., Paul E.A., Lowe W.E. & Parkinson D. (1973) Turnover of microbial tissue in soil under field conditions. *Soil Biology and Biochemistry*, **5**, 753–64.

Standen V. (1978) The influence of soil fauna on decomposition by microorganisms in blanket bog litter. *Journal of Animal Ecology*, **47**, 25–38.

Swift M.J. (1973) The estimation of mycelial biomass by determination of the hexosamine content of wood tissue decayed by fungi. *Soil Biology and Biochemistry*, **5**, 321–32.

Swift M.J., Heal O.W. & Anderson J.M. (1979) *Decomposition in Terrestrial Ecosystems*. Blackwell Scientific Publications, Oxford.

Syers J.K., Sharpley A.N. & Keeney D.R. (1979) Cycling of nitrogen by surface casting earthworms in a pasture ecosystem. *Soil Biology and Biochemistry*, **11**, 181–5.

Van der Drift J. & Witkamp M. (1960) The significance of the breakdown of oak litter by *Eniocyla pusilla* Burm. *Archives Néerlandaises de Zoologie*, **13**, 486–92.

Weetman G.F. & Webber B. (1972) The influence of wood harvesting on the nutrient status of two spruce stands. *Canadian Journal of Forest Research*, **2**, 351–69.

Whitehead D.C. (1970) *The role of nitrogen in grassland productivity*. Bulletin 48, Commonwealth Agricultural Bureau, Farnham Royal, Bucks.

Wieser W. & Schweizer, G. (1970) A re-examination of the excretion of nitrogen by terrestrial isopods. *Journal of Experimental Biology*, **52**, 267–74.

Witkamp M. (1969) Environmental effects on microbial turnover of some mineral elements. Part I – Abiotic factors. *Soil Biology and Biochemistry*, **1**, 167–76.

Witkamp M. & Barzansky B. (1968) Microbial immobilisation of ^{137}Cs in forest litter. *Oikos*, **19**, 392–5.

Witkamp M. & Frank M.L. (1969) Cesium-137 kinetics in terrestrial microcosms. In *Proceedings of the Second National Symposium on Radioecology* (eds D. J. Nelson & F. C. Evans), pp. 635–43. US AEC Conference–670503, US Department of Commerce, Springfield, Virginia 22151.

Witkamp M. & Frank M.L. (1970) Effects of temperature, rainfall and fauna on transfer of ^{137}Cs, K, Mg and mass in consumer-decomposer microcosms. *Ecology*, **51**, 465–74.

Wood T.G. & Sands W.A. (1977) The role of termites in ecosystems. In *Production Ecology of Ants and Termites* (ed. M. V. Brian), pp. 245–92. Cambridge University Press, Cambridge.

20. ABSTRACTS OF POSTER PAPERS

Seasonal patterns in the availability of mineral nitrogen and comparisons with above-ground nitrogen uptake

S. R. TROELSTRA AND R. WAGENAAR *Institute for Ecological Research, Oostvoorne, The Netherlands*

Mineral-nitrogen determinations and mineralization experiments were carried out monthly at six rather dissimilar sites in an old inner-dune area with a rolling relief and a natural grassland vegetation on the island of Goeree (The Netherlands). The primary aims were as follows:
1. To obtain an impression of the relative importance of ammonium nitrogen and nitrate nitrogen.
2. To assess nitrate concentrations in the soil solution.
3. To assess 'plant-available' mineral nitrogen flushes during the season.
4. To compare these flushes with nitrogen uptake in the above-ground vegetation.

Since the parameters in question are strongly dependent on environmental factors and microbial activity, and are unquestionably influenced by the experimental conditions, the results must be seen as semi-quantitative estimates and expected trends or orders of magnitude.

Measurements were confined to the 0–10 cm layer (pH-H_2O: 4–5) at all sites and the following experimental conditions required:
1. Monthly sampling within a grid (60 × 60 cm).
2. Adjacent displacement of the grid during the season.
3. Sampling, field incubation, and analysis of undisturbed 0–10 cm soil cores after removal of the above-ground vegetation.
4. Clipping at ground level of the total vegetation at regular intervals in five adjacent 0.18 m^2 plots.

During the season, initial ammonium levels were almost always higher than those of nitrate at all sites and on the whole never higher than 4 μg N g^{-1} and 2 μg N g^{-1} (oven-dry soil), respectively. At very wet (lower bulk density) and relatively high sites, higher values were recorded, particularly at the latter for ammonium-N (up to *c.* 20 μg g^{-1}).

In general, estimated nitrate concentrations in the soil solution were not higher than 0.30 mmol and 3.0 mmol for the wet and intermediate/dry sites, respectively. However, these concentrations varied widely during the season and some extreme values (*c.* 20 mmol) were found for very high sites. These high nitrate concentrations coincided with unfavourable nitrate transport conditions. Mean seasonal values were 0.15 mmol, 0.95 mmol, and 3.0 mmol for wet, intermediate, and dry sites, respectively.

At the six locations, net production of mineral nitrogen in the 0–10 cm soil layer (field-incubation technique) was predominantly in the form of ammonium, particularly at the wet sites, and showed a distinct seasonal pattern. Total net production amounted to 5–11 g N m^{-2} year $^{-1}$.

The amount of nitrogen taken up by the above-ground vegetation was compared with the net production found in the field-incubation experiments at four sites. The degree of agreement was very satisfactory (whether total amounts or seasonal patterns were compared): 10/8.4/4.8/9.4 and 8.8/6.1/8.4/3.4 g N m^{-2} for total net production and total uptake, respectively. Differences between these values can be partially explained by other soil factors, e.g. soil moisture (nutrient transport) and total nitrogen distribution in the soil profile (uptake of nitrogen from the subsoil).

The field-incubation techniques used in the same place during two or three successive years yielded reproducible yearly values for total net production of mineral nitrogen and for the ammonium/nitrate ratio of this net production. This procedure can therefore provide reliable estimates of 'plant-available' mineral nitrogen flushes during the season. However, the technique may be less suitable for wet locations, especially with respect to the measurement of nitrate production. Such sites may be flooded during part of the year, which might give rise to denitrification processes.

Growth and nitrogen utilization of two *Plantago* species of different environments

I. STULEN, H. LAMBERS, L. LANTING, F. POSTHUMUS, S. J. VAN DE DIJK AND R. HOFSTRA *University of Gröningen, Biological Centre, Department of Plant Physiology, Kerklaan 30, P.O. Box 14, Haren (GN), The Netherlands*

In a comparative study we investigated energy and nitrogen metabolism of *Plantago lanceolata* and *P. major* L. ssp. *major* as dependent on the supply of mineral nutrients and a sudden change therein. *P. lanceolata* is a grassland species found in environments with a relatively low supply of mineral nutrients whereas *P. major* ssp. *major* requires an adequate supply of water and mineral nutrients.

Both species were grown in a concentrated ($\frac{1}{4}$ strength Hoagland; 3.75 mM NO_3^-) and in a diluted nutrient solution (2% of the concentrated solution; 0.075 mM NO_3^-). Part of the plants was transferred from a concentrated into a diluted, or from a diluted into a concentrated solution.

Transfer of *P. lanceolata* plants immediately affected the shoot to root ratio while the ratio was rather constant in *P. major* ssp. *major*. Growth data of both species revealed that growth after a switch from a concentrated to a diluted nutrient solution probably occurred at the expense of stored ions. Since nitrate content declined rapidly while soluble sugar content increased, and since at the same time respiration data suggested a decreased availability of respiratory substrate, it is very likely that sugars are pumped into the vacuole to replace the osmotic function of nitrate. Support for this hypothesis was obtained from leakage experiments which showed that vacuoles of low-nutrient plants contained less nitrate and more sugars than vacuoles of high nutrient plants.

In *P. lanceolata* both the activities of nitrate reductase, glutamate dehydrogenase and glutamine synthetase and the reduced nitrogen content were correlated with the level of the nutrient supply. The data on *P. major* ssp. *major* showed that both the activities of the enzymes and the reduced nitrogen content were rather independent of the nutrient supply. The results obtained with the nitrate reductase assay reflected the actual nitrate reduction in both species. On a whole plant basis both species reduced more nitrate in the root than in the shoot. In *P. lanceolata* the V_{max} of uptake system I for nitrate was sufficient to account for the amount of nitrate actually taken up, but in *P. major* ssp. *major* system II may be operative as well.

A comparison of growth, energy and nitrogen metabolism of *P. lanceolata* and *P. major* ssp. *major* gives the impression that *P. lanceolata* reacts more flexibly to a change in the nutrient supply than *P. major* ssp. *major*. These physiological characteristics may be related to the differences in nutrient supply in the natural environments of the two species. The flexibility of *P. lanceolata* may be of advantage when the nutrient supply becomes severely limiting for a relatively long period whilst *P. major* ssp. *major* may have some advantage if the nutrient supply is limited for a relatively short period only. Since nitrogen metabolism is rigid *P. major* ssp. *major* would not have to readjust enzyme levels to a renewed supply of nutrients.

Nitrate uptake kinetics and nitrogen uptake capacity of species from habitats differing in nitrogen supply

S. J. VAN DE DIJK, L. LANTING, H. LAMBERS, F. POSTHUMS, I. STULEN AND R. HOFSTRA *University of Gröningen, Biological Centre, Department of Plant Physiology, Kerklaan 30, P.O. Box 14, Haren (GN), The Netherlands*

The species *Urtica dioica*, *Plantago major* ssp. *major*, *Plantago lanceolata*, *Hypochaeris radicata* ssp. *radicata* and *Hypochaeris radicata* ssp. *ericetorum* (mentioned in the order from high to low nitrogen supply in their habitats) were grown under high and low nutrient conditions ($\frac{1}{4}$ Hoagland and 2% of $\frac{1}{4}$ Hoagland, further called the 100% and 2% treatment, containing 3.75 mM NO_3^- and 0.075 mM NO_3^-, respectively). After a a certain period half of the plants were switched from low to high and from high to low nutrients.

The nitrate uptake kinetics in the range of system I of the five species grown under the different nutrient conditions were measured during a three-week experimental period. The nitrate uptake of the species showed Michaelis–Menten kinetics. Under low nutrient conditions the V_{max} of *U. dioica* per gram dry root was lower than under high nutrient conditions. For *H. radicata* ssp. *radicata* and for *H. radicata* ssp. *ericetorum* the reverse was found. The values of V_{max} in the two treatments of *P. major* ssp. *major* were almost equal. In young plants of *P. lanceolata*, V_{max} was higher in the 100% treatment; in old plants it was the reverse. The results are explained in relation to the relative growth rate, the shoot to root ratio and the natural environment of the species. The K_m values were not influenced by the different treatments. Differences in K_m between the species, if any, are very small.

It is suggested that the V_{max} is more important for the distribution of plant species in the field than is the K_m. However, the V_{max} of *U. dioica* measured at 2% nutrients was still higher than that of both subspecies of *H. radicata* at 2% nutrients. The V_{max} of both subspecies of *H. radicata* was essentially the same under the applied nutrient conditions.

It seemed relevant to compare the nitrate uptake capacities of these (sub)species at extremely low nitrate concentrations since the nutrient conditions had been shown to influence the kinetic parameters of nitrate uptake. A new growth technique was applied for this comparison. Plants were grown in a situation of severe competition for nitrate (or ammonium). With an infusion pump a fixed amount of nitrate or ammonium was supplied continuously to a culture vessel supporting a fixed number of seedlings, at a rate limiting the uptake rate of the plants. At a concentration of about 1 μM the uptake rate of the plants equalled the rate of supply.

Plants were grown in monocultures (as a control) or in mixed cultures (two species per culture vessel) in the following combination:
1. *H. radicata* ssp. *radicata* and *U. dioica* supplied with nitrate.
2. *H. radicata* ssp. *radicata* and *H. radicata* ssp. *ericetorum* supplied with nitrate.
3. *H. radicata* ssp. *radicata* and *H. radicata* ssp. *ericetorum* supplied with ammonium.
In the mixed culture *H. radicata* ssp. *radicata* accumulated significantly more nitrogen at the expense of nitrogen-accumulation of *U. dioica*. Under high nutrient conditions, on the other hand, *U. dioica* accumulated much more nitrogen than *H. radicata* ssp. *radicata*. Uptake of both species seems related to their natural environment.

The two subspecies of *H. radicata* did not differ in nitrogen accumulation with either nitrate or ammonium as a nitrogen source. Apparently, other factors than nitrogen uptake capacity play a part in the distribution of the two subspecies in the field.

Growth of *Plantago* under various well-defined regimes of suboptimal nitrate supply

A. H. J. FREIJSEN AND H. OTTEN *Institute for Ecological Research, Oostvoorne, The Netherlands*

INTRODUCTION

In Europe, *Plantago lanceolata* L. occurs in various natural and man-made grasslands. This species can grow on poor soils, e.g. on coastal dunes. *Plantago major* L. ssp. *major* is a plant of trampled places such as roadsides and intensively used meadows. Habitats of *P. major* do not seem to be poor in nutrients; competition is on a low level. We hypothesized that *P. lanceolata* has a lower demand for nutrients, particularly nitrate, than does *P. major*. To test this hypothesis, both species were grown in culture experiments in the laboratory under various well-defined regimes of suboptimal nitrate supply. Most of these experiments are described in detail in Freijsen & Otten (1981).

CULTURE TECHNIQUES

The following three types of suboptimal nitrate supply were applied:
1. A 25 μmol concentration of nitrate, which is of the same order of magnitude as K_m values found for a number of species.
2. Intermittent supply of 750 and 25 μM: alternately one day with, and one or two days without, nitrate.
3. Exponentially increasing supply of nitrate: the daily addition equals a constant percentage (e.g. 5%) of the amount of nitrogen already taken up by the plants.
The concentration experiments (1 & 2) were carried out with a simple flowing-culture system. Depletion around the roots was avoided by circulation of the nutrient solution through the system and stirring of the solution in the culture vessels. A rapid fall of the bulk concentration of the solution was avoided by the use of relatively large amounts

of nutrient solution. In the experiments with exponentially increased addition of nitrate (3), this nutrient was supplied to the solution by an infusion pump working at a constant rate, and the concentration of the stock solution was increased daily. This procedure was developed by Ingestad & Lund (1979).

RESULTS AND CONCLUSIONS

The lower uptake rate at the external concentration of 25 μM NO_3^-, the reduction being due to an unsaturated uptake mechanism, was easily compensated for by enlargement of the root system and a more economical use of nitrogen by the plant. No significant reduction of the growth rate was found. This is in agreement with findings made by other authors (Clement, Hopper & Jones 1978). No differences between the *Plantago* species were detectable under these conditions.

Intermittent supply led to marked effects. The shoot/root ratio and the nitrogen content showed a greater decrease. After two days without nitrate supply, the accumulated nitrate was almost completely depleted. The relative growth rate was reduced. Because *P. major* was affected more strongly than *P. lanceolata*, we concluded that the latter species is indeed better adapted to suboptimal nitrate availability, particularly when the supply is intermittent.

When the exponential nitrate addition amounted to 5%, the nitrate concentration in the culture vessel was *c.* 1 μM. Below this level, no nitrate was absorbed. It seems likely that greater specific differences will be found with the Ingestad technique than in concentration experiments. Furthermore, the former seems to provide a good simulation of natural soil conditions.

REFERENCES

Clement C.R., Hopper M.J. & Jones L.H.P. (1978) The uptake of nitrate by *Lolium perenne* from flowing nutrient solution. I. Effect of NO_3^- concentration. *Journal of Experimental Botany*, **29**, 453–64.

Freijsen A.H.J. & Otten H. (1981) Nitrate uptake and growth of *Plantago lanceolata* L. and *Plantago major* L. ssp. *major* at continuous and intermittent supply of low and high concentrations of nitrate. *Physiologia Plantarum* (submitted).

Ingestad T. & Lund A.B. (1979) Nitrogen stress in Birch seedlings. I. Growth technique and growth. *Physiologia Plantarum*, **45**, 137–48.

The ionic balance of *Plantago* species and its relation to nitrogen nutrition

S.R. TROELSTRA, A. H. J. FREIJSEN AND W. SMANT *Institute for Ecological Research, Oostvoorne, The Netherlands*

Plants of the genus *Plantago*, collected in greenhouse experiments, experimental plots, and their natural habitats, were analysed for total nitrogen and the quantitatively important inorganic constituents of the ionic balance (NH_4^+, Na^+, K^+, Ca^{++}, Mg^{++}, NO_3^-, $H_2PO_4^-$, SO_4^{--}, Cl^-). The difference between accumulated amounts of cations and inorganic anions, designated as C minus A, is a measure of the organic-anion content of the plant tissue. This C – A level is partially determined by the form in which nitrogen is taken up, and is lower for ammonium than for nitrate supply.

Analysis of many plants grown in greenhouse and field experiments has already made it possible to establish a rough sequence of organic-anion contents (m-equiv. kg shoot dry wt.$^{-1}$) within the group of *Plantago* species: > 1000 (*P. media, P. major*), *c.* 1000 (*P. lanceolata, P. coronopus*), and < 1000 (*P. maritima*). This sequence of C − A levels was confirmed by the analysis of plants from many natural habitats (m-equiv. kg shoot dry wt.$^{-1}$): 2100 (*P. media*), 1500 *P. major*), 1150 (*P. lanceolata*), 1050 (*P. coronopus*), and 700 (*P. maritima*).

Plantago lanceolata occurs in the field at a wide range of pH vaues but more often at relatively low than at high pH values. Habitats of *P. media* and *P. major* generally have higher pH values, which might be related to the higher organic-anion content of *P. media* and *P. major* (i.e. nitrogen more available in the nitrate form). *P. coronopus* and *P. maritima* also occur frequently at relatively high pH values, but here salinity leads to relatively low C − A levels.

These findings led to the hypothesis that C − A levels in *Plantago* are specific and reflect adaptations to the natural environment (pH, nitrogen source, chloride availability). For *P. media* and *P. major* this suggests a more or less obligatory preference for nitrate-nitrogen. Water culture experiments with five *Plantago* species and five nitrate/ammonium ratios are in progress to test this hypothesis.

Results of preliminary experiments with *P. media*, *P. major*, and *P. lanceolata* and three nitrate/ammonium ratios (100 : 0, 50 : 50, 0 : 100) at nitrogen levels of 1–4 mM and an average pH of 6.0–6.5, suggest that these three species do not require a specific C − A level for normal growth. The C − A levels in the shoots of ammonium-grown plants were 60% lower than those of plants grown with nitrate. However, plant growth was entirely normal in all cases and mean dry weights of plants were generally highest when both nitrogen sources were present. Organic nitrogen contents were highest in ammonium-grown plants.

Perhaps the need for an adequate cation accumulation (particularly potassium and calcium) explains the occurrence of *P. media* and *P. major* at relatively high pH values in the field.

Interspecific tolerance of nitrogen source and aluminium

I. H. RORISON AND R. E. SPENCER *Unit of Comparative Plant Ecology (NERC), Department of Botany, The University, Sheffield S10 2TN*

The growth of plants in response to nitrogen source and to aluminium is well documented. Amounts of available nitrate and of ammonium nitrogen vary according to soil reaction and plants occurring on acidic and on calcareous soils differ in their response to the two sources of nitrogen. However, although single factors may have a dominant influence in a habitat it is unlikely that they will operate in isolation. Thus, acidic soils are characterized chemically not only by high concentrations of ammonium but also of aluminium and manganese and by very low concentrations of available nutrients such as calcium, potassium and phosphorus. Neutral to calcareous soils, although lacking toxic concentrations of polyvalent cations (aluminium, manganese, iron), do have relatively higher concentrations of nitrate, calcium and potassium. An experiment was designed to measure the early seedling growth of three grasses whose response to two of the major pH-linked chemical factors involved (nitrogen source and aluminium), taken both individually and in combination, might be expected to differ. All three species showed one or more of four basic responses (Rorison 1980) in accordance with their ecological distribution. *Deschampsia flexuosa* was completely insensitive to ammonium toxicity and to aluminium in combination with ammonium. It was susceptible to aluminium in the presence of nitrate but at considerably higher concentrations of aluminium than the other two species. *Holcus lanatus* was less tolerant of the acidity factors than *D. flexuosa* but more tolerant than *Bromus erectus*. It was susceptible to aluminium toxicity in the presence of nitrate to a degree which varied according to the concentration of both aluminium and nitrogen. There was also evidence of the relaxation of ammonium toxicity on the addition of aluminium. *Bromus erectus* was highly susceptible to aluminium toxicity in the presence of nitrate while the overwhelmingly toxic effect of ammonium alone masked any combined effect of ammonium and aluminium that there might have been.

It was concluded that, in acidic conditions, both ammonium and aluminium can inhibit growth. Aluminium is most effectively toxic in the presence of nitrate. When ammonium is the predominant nitrogen source it appears to reduce the activity of aluminium. In acidic soils, ammonium is not only the predominant nitrogen source and inhibitory on its own but there is relatively little nitrate to provide a further nitrogen source or to enhance aluminium toxicity.

Finally, two questions were posed: Do we need to revise our ideas as to the critical inorganic toxins in soils? Is ammonium rather than aluminium the prime candidate?

REFERENCES

Rorison I.H. (1980) The effects of soil acidity on nutrient availability and plant response. In *The Effect of Acid Precipitation on Terrestrial Ecosystems* (ed. T. C. Hutchinson), pp. 283–304. Plenum Publishing Corporation, New York.

Nitrogen source and intraspecific tolerance of populations of *Holcus lanatus* L. to aluminium during seedling growth

S. P. MCGRATH *Rothamsted Experimental Station, Harpenden, Herts*

Holcus lanatus occurs in many habitats over a range of soil pH values. Chemical analysis of soil samples showed that the form of nitrogen and the amount of extractable aluminium varies greatly across a range of sites where *H. lanatus* grows. Ammonium was the predominant form of nitrogen only in soils of pH <4.5—because nitrification is inhibited at low pH (Cornfield 1952; Weber & Gainey 1962). Low levels of ammonium-N were detected in soils of higher pH, while nitrate-N was present in small quantities at low pH but accumulated with increasing pH. Soils of pH <4.5 contained large amounts of extractable aluminium, but aluminium solubility falls off sharply with increasing pH. Variation in these two pH-linked chemical factors may affect the growth responses of widely distributed *Holcus* populations.

To demonstrate differences in physiological adaptations, *Holcus* genotypes were grown on an acid and a calcareous soil. When grown on acid soil, root yields of a genotype from calcareous soil were about one half those of plants from acid soil; shoot yields were not reduced as much. When these genotypes were grown on calcareous soil yield of both roots and shoots of acid soil genotypes was depressed approximately equally. Inhibition of root growth on acid soil could be due to susceptibility to ammonium and/or aluminium in the calcicole genotype. On calcareous soils, whole plant growth of the calcifuge genotypes was depressed by lime-induced chlorosis.

The effects of nitrogen source and aluminium on two calcifuge and two calcicole genotypes were examined in more detail by growing plants in nutrient solutions with nitrogen supplied as ammonium, nitrate or ammonium plus nitrate with and without added aluminium.

None of the genotypes showed tolerance of exclusive ammonium nutrition. However, all except the most calcicole genotype tolerated half nitrogen supplied as ammonium. All genotypes grew well in nitrate, but when aluminium was added with nitrate, root growth was reduced in all the genotypes except that from the most acid soil. Only the genotype from calcareous soil showed aluminium toxicity in ammonium plus nitrate. When plants were grown with ammonium, aluminium toxicity was not evident; here growth reduction due to ammonium alone could be masking any effects of aluminium. It is possible, therefore, that where ammonium predominates, ammonium toxicity could be a more potent acid soil factor than aluminium.

Whether or not aluminium toxicity was observed in this species depended on the genotype tested—one acid soil genotype was fully tolerant, but other genotypes may be tolerant if a proportion of the nitrogen supply is ammonium, or completely intolerant (strict calcicole type).

One of the adaptations of *H. lanatus* genotypes to acid soils appears to be tolerance of a proportion of nitrogen supply as ammonium, and the ability to tolerate aluminium under these conditions. However, sensitivity to complete nutrition may exclude this species from the most extremely acid soils. Tolerance to ammonium-N is the only one of the hypothetical responses of species to nitrogen source and aluminium given by Rorison (1980) that these genotypes of *H. lanatus* do not display.

Whether or not aluminium tolerance (and possibly tolerance to other metals) is detected will depend on the form of nitrogen supplied in culture solutions. The use of nitrate alone in test solutions may have led to overestimation of the importance of aluminium as a factor reducing growth on acid soils.

REFERENCES

Cornfield A.H. (1952) The mineralisation of the nitrogen of soils during incubation: influence of pH, total nitrogen, and organic carbon contents. *Journal of the Science of Food and Agriculture*, **3**, 343–9.

Rorison I.H. (1980) The effects of soil acidity on nutrient availability and plant response. In *The Effect of Acid Precipitation On Terrestrial Ecosystems* (eds T. C. Hutchinson & M. Havas), pp. 283–304. Plenum Publishing Corporation, New York.

Weber D.F. & Gainey P.L. (1962) Relative sensitivity of nitrifying organisms to hydrogen ions in soils and solutions. *Soil Science*, **94**, 138–45.

Influence of nitrogen on spatio-temporal interactions between a collembola species and a soil fungus.

M. A. LEONARD *Dept. of Biological Sciences, Wolfson Ecology Laboratory, University of Exeter, Exeter, Devon*

The roles of microorganisms in the decomposition of litter in both terrestrial and marine environments have been studied in detail but as yet comparatively little work has been carried out on the interactions between the components of the soil community.

It has previously been shown that the growth rate and the nitrogen content of soil fungi in cultures are proportional to the nitrogen content of the medium. Similarly, the fecundity of collembola is positively related to the nitrogen content of their fungal food (Booth & Anderson 1979). The grazing impact of the collembola is, however, strongly inhibitory to both fungal growth and sporulation (Hanlon & Anderson 1979).

This situation provides a model system in which study and quantification of the influence of the nitrogen resource on the interactions between the fungus and the grazer may be made. Glass beads of approximately 150 μm size were used to construct a simple, two-dimensional matrix, their use allowing the accurate control of nitrogen and other nutrient levels as well as the recovery of mycelium. All experiments were carried out at ambient temperature.

The simple system produced a single grazing interface at which the level of nitrogen was defined. The results using this system have shown that at high levels of nitrogen (20–200 mg l^{-1}) the weight of fungus recovered from the grazed systems was significantly lower than that from ungrazed controls. Recoveries dropped to below 10% after 90 days (83 grazing). At the lowest level tested (2 mg l^{-1}), however, the weights recovered were fairly constant over the period studied. Similarly the numbers of collembola recovered were constant, whilst in the higher nitrogen treatments the populations, in response to the higher nitrogen supply to their fungal food, exhibited increased reproduction and subsequently declined to extinction due to the overgrazing of their food resource. The equilibrium observed at the lowest level has not yet been fully tested, or its stability over long periods of time determined.

The glass beads have been replaced with ceramic column packing to produce a matrix analogous to the soil. This system contained pore spaces, some small enough to act as refuges for the fungus from grazing. An experiment using 20 mg l^{-1} nitrogen has shown that the population increase was smaller than that observed in the simple system and was maintained by an increase in the recoverable fungus from the grazed system. The addition of spatial heterogeneity to the interaction led to enhanced growth of the fungus by a factor of 74% over control levels, compared to the decrease of 90% over a period of 90 days. The importance of spatial heterogeneity in the interaction between the two components is therefore very important in maintaining coexistence and will be subject to further study.

REFERENCES

Booth R.G. & Anderson J.M. (1979) The influence of fungal food quality on the growth and fecundity of *Folsomia candida* (Collembola: Isotomidae). *Oecologia (Berlin)*, **38**, 317–23.

Hanlon R.D.G. & Anderson J.M. (1979) The effects of collembola grazing on microbial activity in decomposing leaf litter. *Oecologia (Berlin)*, **38**, 93–9.

Planktonic nitrogen cycling in coastal waters

P. H. BURKILL, N. J. P. OWENS AND R. F. C. MANTOURA *Natural Environment Research Council, Institute for Marine Environmental Research, Prospect Place, The Hoe, Plymouth, PL1 3DH, Devon*

Some preliminary results of a multi-disciplinary study on the biology, chemistry and physics of nutrient cycling in Carmarthen Bay (51° 31′ N, 04° 30′ W) were presented as a series of displays and a short film that describes the major seasonal sources, sinks and fluxes in the cycling of nitrogen within a grid measuring 12.8 × 16 km.

The pool of total nitrogen varied seasonally, with spring levels of 29 mmol N m^{-3} falling to 12 mmol N m^{-3} in summer reflecting a net export of nitrogen from the water column during this period. The chemical speciation of this pool also varied; whereas in spring, nitrate predominated, by summer this form had been assimilated by the growing phytoplankton, leaving dissolved organic nitrogen as the major constituent. The concentrations of some nutrients such as nitrate and ammonia showed anomalies when grid concentrations were compared with bulk Bristol Channel water of the same salinity, thus indicating bay-specific effects on nitrogen cycling activity.

Phytoplankton nitrogen requirements were studied using both direct ^{15}N and indirect ^{14}C uptake experiments together with simulation models constructed from these physiological measurements and suitable measurements of incident photosynthetically available radiation (PAR) and the attenuation of PAR in the water column. During a period of nitrogen depletion, the summer chlorophyll *a* distribution varied within the grid between 1.6 and 3.2 mg chl-*a* m^{-3}, while primary production and nitrogen demand (averaged throughout the water column) varied between 22 and 39 mmol C m^{-2} day^{-1} and 36 and 300 μmmol N m^{-3} day^{-1}, respectively.

Zooplankton excretion experiments carried out on shipboard in summer showed ammonia to be the main metabolite, with excretion rates varying between 86 and 663 μmol NH_3-N g dry $wt.^{-1}$ day^{-1}, highly dependent on the size of the animal. Summer zooplankton standing stock, determined using an acoustically telemetered dual net sampling system, varied between 15 and 155 mg dry wt. m^{-3}. Concurrently, the remineralization *in situ* of nitrogen as ammonia by zooplankton was calculated to vary between 6 and 71 μmol NH_3-N m^{-3} day^{-1} with the smallest size-class (100–280 μm in length) contributing >75% of the ammonia released by the zooplankton community.

Microbial nitrogen transformations were studied and revealed high activities, particularly in summer when nitrification varied between 0.2 and 2.8 mmol N m^{-3} day^{-1}. Experiments on the uptake of ^{14}C-glucose, as an index of hereotrophic potential, showed V_{max} rates up to 0.3 mmol C m^{-3} day^{-1} in early summer.

Turnover times of ammonia in the water column have been estimated assuming that the cycling process is in steady-state. In summer, minimal turnover times, assuming that phytoplankton take up ammonia as a sole nitrogen source, varied between one and 20 days. Corresponding maximum turnover times, due to remineralization processes by zooplankton and bacteria, varied between seven and 90 days and one and 16 days, respectively. These data suggest the importance of the smallest heterotrophic components, bacteria and microzooplankton, in the remineralization of nitrogen in the water column.

AUTHOR INDEX

Figures in italics refer to pages where full references appear

SUBJECT INDEX

Related names appear in parentheses